SPRINGER HANDBOOK OF AUDITORY RESEARCH

Series Editors: Richard R. Fay and Arthur N. Popper

SPRINGER HANDBOOK OF AUDITORY RESEARCH

Volume 1: The Mammalian Auditory Pathway: Neuroanatomy
Edited by Douglas B. Webster, Arthur N. Popper, and Richard R. Fay

Volume 2: The Mammalian Auditory Pathway: Neurophysiology
Edited by Arthur N. Popper and Richard R. Fay

Volume 3: Human Psychophysics
Edited by William Yost, Arthur N. Popper, and Richard R. Fay

Volume 4: Comparative Hearing: Mammals
Edited by Richard R. Fay and Arthur N. Popper

Volume 5: Hearing by Bats
Edited by Arthur N. Popper and Richard R. Fay

Volume 6: Auditory Computation
Edited by Harold L. Hawkins, Teresa A. McMullen, Arthur N. Popper, and Richard R. Fay

Volume 7: Clinical Aspects of Hearing
Edited by Thomas R. Van De Water, Arthur N. Popper, and Richard R. Fay

Volume 8: The Cochlea
Edited by Peter Dallos, Arthur N. Popper, and Richard R. Fay

Volume 9: Development of the Auditory System
Edited by Edwin W Rubel, Arthur N. Popper, and Richard R. Fay

Volume 10: Comparative Hearing: Insects
Edited by Ronald Hoy, Arthur N. Popper, and Richard R. Fay

Volume 11: Comparative Hearing: Fish and Amphibians
Edited by Richard R. Fay and Arthur N. Popper

Volume 12: Hearing by Whales and Dolphins
Edited by Whitlow W.L. *Au, Arthur N. Popper, and Richard R. Fay*

Volume 13: Comparative Hearing: Birds and Reptiles
Edited by Robert Dooling, Arthur N. Popper, and Richard R. Fay

Volume 14: Genetics and Auditory Disorders
Edited by Bronya J.B. Keats, Arthur N. Popper, and Richard R. Fay

Volume 15: Integrative Functions in the Mammalian Auditory Pathway
Edited by Donata Oertel, Richard R. Fay, and Arthur N. Popper

Volume 16: Acoustic Communication
Edited by Andrea Simmons, Arthur N. Popper, and Richard R. Fay

Volume 17: Compression: From Cochlea to Cochlear Implants
Edited by Sid P. Bacon, Richard R. Fay, and Arthur N. Popper

Volume 18: Speech Processing in the Auditory System
Edited by Steven Greenberg, William Ainsworth, Arthur N. Popper, and Richard R. Fay

Volume 19: The Vestibular System
Edited by Stephen M. Highstein, Richard R. Fay, and Arthur N. Popper

Volume 20: Cochlear Implants: Auditory Prostheses and Electric Hearing
Edited by Fan-Gang Zeng, Arthur N. Popper, and Richard R. Fay

Volume 21: Electroreception
Edited by Theodore H. Bullock, Carl D. Hopkins, Arthur N. Popper, and Richard R. Fay

Continued after index

Theodore H. Bullock
Carl D. Hopkins
Arthur N. Popper
Richard R. Fay
Editors

Electroreception

With 118 illustrations and two color illustrations

Springer

Theodore H. Bullock
Department of Neurosciences
School of Medicine
University of California, San Diego
La Jolla, CA 92093-0240, USA
tbullock@ucsd.edu

Carl D. Hopkins
Department of Neurobiology & Behavior
Cornell University
Ithaca, NY 14583, USA
cdh8@cornell.edu

Arthur N. Popper
Department of Biology
University of Maryland
College Park, MD 20742, USA
apopper@umd.edu

Richard R. Fay
Parmly Hearing Institute and Department of Psychology
Loyola University of Chicago
Chicago, IL 60626, USA
rfay@wpo.it.luc.edu

Cover illustration: Gymnotiform fishes from South America utilize electroreception for passive sensing of prey, for active sensing objects detected as distortions in their own electric fields, and for sensing electric communication signals generated from their electric organs. A few of the 27 known genera of gymnotiforms are illustrated: *Electrophorus*, *Gymnotus*, *Microsternarchus*, *Brachyhypopomus*, *Hypopomus*, *Racenisia*, *Hypopygus*, *Steatogenys*, *Rhamphichthys*, and *Gymnorhamphichthys* (see J.S. Albert and W.G.R. Crampton, p. 364, for key).

Library of Congress Cataloging-in-Publication Data
Electroreception / Theodore H. Bullock (editor) . . . [et al.]
p. cm.
Includes bibliographical references and index.
ISBN 0-387-23192-7
1. Electroreceptors. I. Bullock, Theodore Holmes.
QP447.5.E44 2005
573.8'7—dc22 2004057843

ISBN 10: 0-387-23192-7 Printed on acid-free paper
ISBN 13: 978-0387-23192-1

Printed in the United States of America. (EB)

9 8 7 6 5 4 3 2 1

springeronline.com

Volume Dedication

This volume is dedicated to the memories of two true pioneers in the study of electroreception, Walter Heiligenberg and Thomas Szabo. The contributions that Walter and Tom made to our understanding of electroreception are truly monumental, and their discoveries, and those of the students and others they influenced, permeate this volume.

Thomas Szabo (d. 1994) was the director of the Laboratory of Sensory Physiology at the CNRS in Paris. Along with many co-workers, Thomas was a pioneer in electroreception, especially in its peripheral and central histological basis. Thomas not only did wonderful work in the laboratory, but he also did extensive field work both in Africa and South America. Most importantly, perhaps, he trained a long list of younger workers.

Walter Heiligenberg (d. 1994) was a student of Konrad Lorenz and Hans-Jochem Autrum. Walter began a career in behavioral physiology with insects and teleosts, switched to electroreception, and led a large group at the Scripps Institution of Oceanography who worked out the cells, pathways, and physiology of the jamming avoidance response—probably the best-understood piece of vertebrate elective behavior.

Chapters Dedication

Each author in this volume dedicates his or her chapter to Theodore Holmes Bullock, a pioneer in the discovery of electroreception and a true champion for understanding the diversity of organisms that possess this wonderful sense. Many of us have worked in Ted Bullock's laboratory in La Jolla or have collaborated with him from afar. All of us are inspired by conversations with Ted and by his writing, his lectures, his letters, and his e-mails. He continues to excite, to instruct, and to urge us to rethink old ideas and replace them with new. Many of the topics and discoveries reported in our chapters were in some measure inspired or influenced by Ted's papers, lectures, remarks, or comments.

James S. Albert
Joseph Bastian
Curtis C. Bell
David Bodznick
Angel Ariel Caputi
Bruce A. Carlson
Sheryl Coombs
William G.R. Crampton
Richard R. Fay
Michael H. Hofmann
Carl D. Hopkins
Jørgen Mørup Jørgensen
Masashi Kawasaki
Omar Macadar
Leonard Maler
John C. Montgomery
Mark E. Nelson
R. Glenn Northcutt
Arthur N. Popper
Lon A. Wilkens
Harold H. Zakon
Günther K.H. Zupanc

Series Preface

The *Springer Handbook of Auditory Research* presents a series of comprehensive and synthetic reviews of the fundamental topics in modern auditory research. The volumes are aimed at all individuals with interests in hearing research including advanced graduate students, postdoctoral researchers, and clinical investigators. The volumes are intended to introduce new investigators to important aspects of hearing science and to help established investigators to better understand the fundamental theories and data in fields of hearing that they may not normally follow closely.

Each volume presents a particular topic comprehensively, and each serves as a synthetic overview and guide to the literature. As such, the chapters present neither exhaustive data reviews nor original research that has not yet appeared in peer-reviewed journals. The volumes focus on topics that have developed a solid data and conceptual foundation rather than on those for which a literature is only beginning to develop. New research areas will be covered on a timely basis in the series as they begin to mature.

Each volume in the series consists of a few substantial chapters on a particular topic. In some cases, the topics will be ones of traditional interest for which there is a substantial body of data and theory, such as auditory neuroanatomy (Vol. 1) and neurophysiology (Vol. 2). Other volumes in the series deal with topics that have begun to mature more recently, such as development, plasticity, and computational models of neural processing. In many cases, the series editors are joined by a co-editor having special expertise in the topic of the volume.

Richard R. Fay, Chicago, Illinois
Arthur N. Popper, College Park, Maryland

Volume Preface

This volume represents a slightly different approach for books in the *Springer Handbook of Auditory Research*—it is not about hearing. At the same time, this volume is about a major sensory system that has evolved multiple times in the history of the vertebrates and shares many similarities in detection and processing with the auditory system. Thus, the series editors concluded that investigators in the hearing sciences would value learning about the electrosensory system, and so they invited two of the world's leaders in that field, Professor Theodore H. Bullock and Professor Carl D. Hopkins, to collaborate on this volume. Indeed, it is anticipated that future volumes in the SHAR series might cover other topics that, although not directly on the topic of hearing, could provide unique insights into sensory systems that could benefit those of us in the hearing sciences.

This volume, like our recent volume on The Vestibular System (SHAR Vol. 19, 2004), is also unlike most other SHAR volumes. Rather than considering a small area within the hearing sciences, it takes a broader view and provides an overview that encompasses a field. Thus, this volume not only includes chapters on physiology, signal processing, receptors, and related topics but also gives the reader a broader historic, behavioral, and taxonomic overview of the field. In effect, someone reading this whole volume will understand not only how electroception works but also its evolution and how animals use electroreception in their daily lives.

The volume starts with a brief historic overview by Bullock and Hopkins (Chapter 1) that gives a personal understanding as to the earliest discoveries in this field. Chapter 2 by Zupanc and Bullock continues this historic perspective but also introduces the reader to the diverse species that produce and detect electric currents. Jørgensen (Chapter 3) provides an exciting overview of the receptors involved in electroreception, while Bell and Maler (Chapter 4) extend the system into the brain and explain the central anatomy and physiology of electroreception as well as potential parallels to the auditory system. In Chapter 5, Northcutt considers the ontogeny of the electric sense and provides a context within which one can view the evolution of electroreception in the vertebrates.

Electrosensory systems can be "divided" into low-frequency and high-frequency types, and these different systems are considered in the next several chapters. In Chapter 6, Bodznick and Montgomery describe the physiology of low-frequency systems, while Kawasaki discusses the physiology of high-frequency systems in Chapter 7. After a discussion of plasticity in the electrosensory system by Bastian and Zakon (Chapter 8), several subsequent chapters consider the behaviors of fishes with different types of electrosensory systems. In Chapter 9, Wilkens and Hofmann discuss the behavior of fishes with low-frequency systems, while in Chapter 10, Hopkins takes a parallel course with the behavior of fishes that use high-frequency systems. Finally, Nelson (Chapter 11) analyzes target detection and provides models to help understand how electroreceptive fishes are able to analyze the information they are sensing.

In the remaining chapters, authors consider several additional and important topics that help in understanding electrosensory systems. It has long been held that the electrosensory and lateral line senses are phylogentically related, and this, as well as functional similarities and differences, are considered in Chapter 12 by Coombs and Montgomery. In Chapter 13, Albert and Crampton describe how molecular techniques are used to explore the systematic relationships among and between electrosensory fishes. Finally, in Chapter 14, Caputi, Carlson, and Macadar explore how the organs that emit electric signals are controlled by the central nervous system.

Theodore H. Bullock, La Jolla, California
Carl D. Hopkins, Ithaca, New York
Arthur N. Popper, College Park, Maryland
Richard R. Fay, Chicago, Illinois

Contents

Contributors

JAMES S. ALBERT
Department of Biology, University of Louisiana at Lafayette, Lafayette, LA 70504, USA

JOSEPH BASTIAN
Department of Zoology, University of Oklahoma, Norman, OK 73019, USA

CURTIS C. BELL
Neurological Sciences Institute, Oregon Health Sciences University, Beaverton, OR 97006, USA

DAVID BODZNICK
Department of Biology, Wesleyan University, Middletown, CT 06459-0170, USA

THEODORE H. BULLOCK
Department of Neurosciences, School of Medicine, University of California, San Diego, La Jolla, CA 92093-0240, USA

ANGEL ARIEL CAPUTI
Department of Comparative Neurophysiology, Instituto de Investigaciones Biológicas Clemente Estable, CP 11600 Montevideo, Uruguay

BRUCE A. CARLSON
Department of Biology, University of Virginia, Charlottesville, VA 22904, USA

SHERYL COOMBS
Department of Biological Sciences, Bowling Green State University, Bowling Green, OH 43402, USA

WILLIAM G.R. CRAMPTON
University of Florida, Gainesville, FL 32611-7800, USA

MICHAEL H. HOFMANN
Institute of Zoology, University of Bonn, 53115 Bonn, Germany

CARL D. HOPKINS
Department of Neurobiology & Behavior, Cornell University, Ithaca, NY 14853, USA

JØRGEN MØRUP JØRGENSEN
Department of Zoophysiology, University of Aarhus, DK 8000 Aarhus C, Denmark

MASASHI KAWASAKI
Department of Biology, University of Virginia, Charlottesville, VA 22904, USA

OMAR MACADAR
Department of Neurophysiology, Instituto de Investigaciones Biológicas Clemente Estable, CP 11600 Montevideo, Uruguay

LEONARD MALER
Department of Cell and Molecular Medicine, University of Ottawa, Ottawa, ON K1H 8M5, Canada

JOHN C. MONTGOMERY
Leigh Marine Laboratory and School of Biological Sciences, University of Auckland, Auckland 1020, New Zealand

MARK E. NELSON
Beckman Institute, University of Illinois, Urbana, IL 61801, USA

R. GLENN NORTHCUTT
Neurobiology Unit, Scripps Institution of Oceanography, and Department of Neurosciences, School of Medicine, University of California, San Diego, La Jolla, CA 92093-0240, USA

LON A. WILKENS
Center for Neurodynamics, Department of Biology, University of Missouri–St. Louis, St. Louis, MO 63121, USA

HAROLD H. ZAKON
Section of Neurobiology, University of Texas, Austin, TX 78712, USA

GÜNTHER K.H. ZUPANC
School of Engineering and Science, International University Bremen, D-28725 Bremen, Germany

1

Explaining Electroreception

THEODORE H. BULLOCK AND CARL D. HOPKINS

What is electroreception? It may call to mind stories about ambiguous conference titles that invite speculation by guests in the hotel elevator, or a column in the paper about microchips implanted under the skin, or some new-fangled weapons control. It has a more interesting history than any of these, however. In the early 1950s there was no such term or phenomenon to be named. Then Harry Grundfest in New York, collaborating with an aquarist with unusual tropical fish, became curious about the many species of tropical New World fish related to the familiar electric eel, which can discharge pulses of hundreds of volts. He found that relatively tiny cousins—smaller than a pencil—are discharging small fractions of a volt all the time. The old German literature had shown they have electric organs, but very small ones, called pseudoelectric organs. What can they do with such feeble, sustained discharges? Some species are going at several hundred pulses per second, night and day. So far, there was no reason to suspect a new sense modality. No functions had been discovered and the research being done was effector centered. Then Hans Lissmann at Cambridge (on the Cam) thought up a key experiment. By now it was known that another large order of freshwater fishes in Africa (mormyriforms) is all electric, meaning that all species tested have electric organs and they discharge pulses all the time. Hans kept a few of the large, eel-like *Gymnarchus* long before they became readily available and patiently taught them that certain objects mean food. He found (Lissman and Machin 1958) that the fish can discriminate between a porous porcelain container with aquarium water and one in which 20% of the liquid is distilled water—and that discrimination could not have been possible by chemical, mechanical, or visible clues but must have been the result of electrical conductivity of the fish's maintained 300-Hz sinusoidal discharges! Here then was strong evidence for an object-detection function requiring fantastic sensitivity to small differences in the local intensity of the fish's field in the surrounding water. Hence a sensory system was necessary, specialized for high sensitivity—electroreceptors! Möhres (1957) in Tübingen soon showed evidence of a social communicating function as well, using another mormyriform fish, *Gnathonemus*, that changes the repetition rate of its brief pulses drastically to stimuli of ethological interest such as the proximity of a conspecific fish.

We now had, not simply a function for the puny electric organs, but two classes of functions: object detection and social communication, but no details about either one. The kinds of questions raised by this level of discovery were: What is the code for one's own electric organ discharge (EOD) and for the neighbor's? How does a fish tell one from the other? What is the repertoire of signals that any individual of the species would understand? Where and how, in the brain, are these signals generated after the preceding signal had been read and interpreted? At this point in the narrative we are into the 1960s and well into the phenomenon of electroreception in a few favorable species of the New World, tropical gymnotiforms.

Descriptive natural history was still important and determined which species and behaviors to study. We had learned that among the species there were "wave fish" and "pulse fish"—species whose EOD, when converted from electricity into sound, is a quasisinusoidal, steady tone, in the range from about 200 to 2000 Hz, the intervals between single EODs quite uniform and about equal to the duration of the discharge versus those species with much less frequent EODs, long and irregular intervals, making sputtering sounds usually between 1 and 100 Hz, when the electric signal is sent to a loudspeaker. We had learned that each individual of the wave species has its own preferred EOD frequency, maintained over hours and days, but under brain control and labile to special stimuli, whereas pulse species are not obviously individual; are very low in repetition rate at rest; and easily, widely, speeded up by many stimuli, among them EODs of other individuals, manifesting the communicative function.

We had learned that sharks, rays, and catfish show electroreception of slowly changing electric currents, less than 20 Hz, even in species that lack EODs and electric organs—opening the general question of which other fishes or taxa might have this sensory modality. This was answered some decades later: quite a number of nonteleost taxa have it, including lampreys, ratfish, lungfish, sturgeons, and paddle fish, even some salamanders. Reciprocally, some teleosts with electric organs and presumably functional EODs, such as the stargazers (Uranoscopids), seem to lack the sense modality. We are not through with natural history, either in the phylogenetic distribution or in the functional uses of electroreception.

For years we were pessimistic about identifying the sense cells or sensory organs. Skin senses, such as touch, tickle, pain, cold, and heat, have been worked out only slowly and haltingly in the most frequently studied forms—laboratory mammals. Electrophysiological recordings showed the afferent nerve fibers must be in certain branches of the lateral line nerve, most numerous on the head. Suspected were the ampullae of Lorenzini, characteristic of elasmobranchs and a few other taxa and with a checkered history of functional assignment (Bullock 1974). Kalmijn (1974) provided the crucial evidence that in fact these ampullae are quite specialized electroreceptors, and although they can respond to cold and to some mechanical events, these responses are adventitious and unusual in nature. Somewhat similar organs occur in catfish and other species but on substantial grounds are not considered homologous. I (T.H.B.) well remember the day when, with my co-worker, Shiko Chichibu, we encountered for the first time, in a gymnotiform fish at the Goeldi Museum in Belem, an afferent unit that was sensitive to electrical events but fired at a frequency unrelated to the frequency of the stimulus, which had to be low (below 30 Hz) and could be simply the DC field around a freshly killed fish—similar to the class of non-Lorenzini receptors in siluriforms (catfish).

Another whole line of sense organs apparently evolved in the teleosts with electric organs and EODs, called tuberous organs. They typically respond to fast or brief electrical events, up to components of several kiloHerz, whereas the ampullary organs, of both the elasmobranch type and the siluriform type, respond to low-frequency components, best below 5 Hz. The tuberous receptors are specialized into subcategories, usually two in each species, differing in the dynamics of response and hence in their functional roles.

This is the "bottom line" of a long series of researches, with surprises and repeated "Ah, so's." The variety of sense organs bespeaks a corresponding complexity in central connections and maplike representations, funneling toward a midbrain of higher neurons with rather complex input requirements for firing—relatively high-level "recognition units" as well as circuits that reduce the sensitivity to signals coming from the animal's own movements. Each of these topics and ancillary ones has had a dramatic history of discoveries and is developed in the chapters that follow in this volume. They are all incompletely understood and are ready for another advance in understanding, which usually means a bit more adequate description at the next lower level—cellular, membrane, or molecular and sometimes at a higher level, such as circuit, assemblage, topographic, or algorithmic, leading to behavioral, ethological, and ecological levels.

In addition to these developments in the chain of integrative levels of organization and function, there are striking contributions to the "whence" class of questions—the ontogeny within each taxon and the evolution among taxa. Each has benefited from dramatic steps such as learning how to keep and to breed some species, how to raise young fish of each group, and how to supply modern techniques for DNA analysis.

Many of the findings explaining electroreception impinge on broad questions in general biology, for which these fish provide especially favorable exemplars. One chapter deals, not with electroreceptors, but with the electric organs and the EODs as the signals for sensory analysis, their fields, and their control, synchrony, and modulation. Convergent and parallel evolution are illustrated as well as species diversity and adaptive specialization. The field is still young, so that instead of a wide choice of authors, the writers of each chapter tend to be world authorities pioneering on their respective fronts.

References

Bullock TH (1974) An essay on the discovery of sensory receptors and the assignment of their functions together with an introduction to electroreceptors. In: Fessard A (ed), Handbook of Sensory Physiology, Vol. III/3: Electroreceptors and Other Specialized Receptors in Lower Vertebrates. Berlin: Springer-Verlag, pp. 1–12.

Kalmijn AJ (1974) The detection of electric fields from inanimate and animate sources other than electric organs. In: Fessard A (ed), Handbook of Sensory Physiology, Vol. III/3: Electroreceptors and Other Specialized Receptors in Lower Vertebrates. Berlin: Springer-Verlag, pp. 147–200.

Lissmann HW, Machin KE (1958) The mechanism of object location in *Gymnarchus niloticus* and similar fish. J Exp Biol 35:451–486.

Möhres FP (1957) Elektrische Organe im Dienste der Revierabgrenzung. Naturwissenschaften 44:431–432.

2

From Electrogenesis to Electroreception: An Overview

GÜNTHER K.H. ZUPANC AND THEODORE H. BULLOCK

1. Introduction

Electric fish have fascinated scientists and non-scientists alike for thousands of years, but only in recent times, with the advent of sophisticated neurobiological tools, have we been able to appreciate the anatomical structure and the physiological mechanisms of electric systems. Still, analysis of these systems has led to more than a dissection of their individual components. Rather, a consensus has been attained among researchers in the field in their attempt to arrive at a synthesis by integrating the results obtained through various approaches, thereby achieving an unprecedented degree of understanding of the function of electric systems. A major step toward this aim was the publication of *Electroreception*, edited by Theodore H. Bullock and Walter Heiligenberg, in 1986 (Bullock and Heiligenberg 1986). Like *Electroreception*, the present book aims to draw a larger picture by illuminating the many facets of electric systems. Access to these subareas, covered in detail in the chapters that follow, will be facilitated by familiarity with the language and the major concepts used by researchers in the area of electrogenesis and electroreception. It is the goal of this overview to provide such an introductory guide.

After taking a look at the history of research on electric fish, this chapter summarizes the approaches used to classify and characterize electric organ discharges. This is followed by an overview of the distribution of electroreceptive and electric animals, and a discussion of the current opinions on the phylogenetic development of electric systems. Then, the main features of two of the major components of electric systems—electroreceptors and electric organs—are described. However, both electrosensory and electromotor systems are not static but exhibit a remarkable degree of plasticity, which bears important consequences for the function of these systems. Therefore, some key discoveries related to the dynamics of these systems are reviewed in another section of this chapter. Then, as one of the most fascinating aspects of electric systems, four important functions are discussed—passive electrolocation, active electrolocation, species recognition, and intraspecific communication. The power of electric fish as excellent model systems for the study of the neural basis of behavior is exemplified by introducing the reader to one of the best-studied vertebrate behavioral patterns, the jamming avoidance response. Yet, the enormous potential of electric fish is not limited to neuroethological studies, but includes other areas of biological research as well. Two of these areas are highlighted at the end of this chapter: the study of cellular mechanisms of neuronal plasticity and investigations of the molecular structure of the cholinergic synapse.

2. Electric Fish: A Brief History

Electric fish have been known to humans since ancient times (for a review, see Moller 1995). A number of Greek and Roman writers described the numbing power of the electric shock of several strongly electric fishes, such as the Mediterranean electric torpedo ray (*Torpedo torpedo*) and the African electric catfish (*Malapterurus electricus*). Electrical stimulation using these species was employed as an early form of electrotherapy for the treatment of pain and a variety of diseases, including headache and gout. However, the physical nature and the biological origin of the numbing power remained obscure for another 2000 years. A "frigorific principle" (causing pain similar to the one observed after exposure of parts of the body to extreme cold), a "magnetizing effect," the effect of a special poison, and a mechanical effect were postulated to explain the nature of the shocks produced by strongly electric fish.

In the eighteenth century, evidence was obtained against the then dominating hypothesis of a mechanical nature and in favor of an electrical nature of the shocks generated by the fish. In 1769, the American physician Edward Bancroft demonstrated that, when he elicited discharges by touching an electric eel with a rod, a second person with his hands in water several meters away from the fish received a shock. Thus, no direct contact of a person with an electric fish is necessary to receive a shock. Instead, the discharges of the fish can be transmitted solely by water as a conducting medium—a result that strongly argues against a mechanical nature of the shocks produced by the fish.

Probably the first detailed analysis of the biophysical properties of the electric organ discharges (EODs) of electric fish was performed by John Walsh (1726–1795), a fellow of the Royal Society and a member of the English Parliament (Piccolino and Bresadola 2002). At about the same time, in 1773, John Hunter published a detailed anatomical description of the electric organ of *Torpedo*.

In the nineteenth century, a number of scientists demonstrated that the individual cells (electrocytes) composing the electric organ are, in many electric fishes, embryonically derived from muscles. At approximately the same time, the weakly electric discharges of electric rays, several species of the genus *Mormyrus*, and other mormyrids were discovered. Obviously, the discharges of these fishes are too weak to be employed for defense or stunning prey. Therefore, many scholars of that time believed that the discharges of weakly electric fish lack any function. Such speculations appeared to receive support from the failed attempts to detect discharges from many mormyrids and gymnotids, despite the fact that in fishes of these taxonomic families organs were identified that resembled the electric organs of their electric relatives. Hence the term "pseudoelectric organ" was coined to indicate what was thought to represent an incomplete stage of phylogenetic development compared to that of strongly electric organs.

An important step toward a more detailed analysis of electric signals was made at the end of the nineteenth century and the beginning of the twentieth century when the first electric organ discharges of electric fish were recorded. The foundation of modern research on electric systems was laid by Hans Lissmann (1909–1995) of the University of Cambridge in a series of seminal papers published in the 1950s. In the first study, he showed that the African mormyriform fish *Gymnarchus niloticus* produces weakly electric discharges—an ability that had already been suggested more than 100 years earlier, based on the discovery in this species of differentiated tissue resembling the electric organ of fish from which discharges could be recorded (Lissmann 1951). Then, in two following investigations published in 1958 (Lissmann 1958; Lissmann and Machin 1958), Lissmann presented a model that convincingly proposed a function of the weakly electric discharges. According to this model, which was based on a detailed analysis of the biophysical properties of the electric field of *Gymnarchus niloticus*, weakly electric fish locate objects differing in conductivity from that of the surrounding water by analysis of the self-generated dipole-like electric field.

At about the same time, Harry Grundfest of Columbia University in New York discovered that, contrary to the expectation, "pseudoelectric" fishes (i.e., species possessing "pseudoelectric" organs) produce electric discharges. Although these species generate their discharges more or less continuously, night and day, their electric activity usually remains undetected because of the low voltage. Therefore, to monitor the discharges of these weakly electric fish, their signals have to be amplified with high gain. Over the following years, Grundfest concentrated on investigating the structure and function of the electric organs that produce the weakly electric discharges. This work was continued primarily by his collaborator Michael Bennett who, by the beginning of the 1970s, put forward a detailed model of how electric organs generate discharges (for review, see Bennett 1971).

Around 1960, work on electroreceptors began. These studies were inspired by Lissmann's model that postulated a specific and highly sensitive "electric sense," mediated by electroreceptors distributed over the fish's body surface. Such a new sensory modality had already been suggested by experiments performed by John Walsh, which were reported by Alessandro Volta in 1782 (Piccolino and Bresadola 2002). Walsh found that, when the recording electrodes were short-circuited outside the water tank, an electric eel would become agitated, approach the electrodes, and produce electric discharges. As Walsh correctly concluded, this behavioral response requires that the fish is able to detect conducting objects through a specialized electric sense. Such a notion was, however, rejected by many prominent scholars, including the influential German neurophysiologist Johannes Müller and his student Emil Du Bois-Reymond, because they believed that electricity stimulates sense organs and nerves in an unspecific way—as other noxious stimuli do.

The hypothesis of a specific electric sense was revitalized, and confirmed, only in the twentieth century. In 1961, Theodore Bullock and colleagues (Bullock et al. 1961) of the University of California at Los Angeles, and Alfred Fessard and Thomas Szabo (Fessard and Szabo 1961) of the C.N.R.S. in Paris, published results of physiological experiments that demonstrated the existence of a class of specialized electroreceptors in the lateral line system of gymnotiform and mormyriform fish. At about the same time, Richard Murray (1960) of the University of Birmingham provided experimental evidence that the ampullae of Lorenzini of elasmobranchs, first described by Lorenzini in 1678 in *Torpedo*, respond with high sensitivity to electrical stimuli. Final establishment of electroreception as a sensory modality on its own rights was greatly accelerated not only by many subsequent physiological investigations confirming and extending these earlier findings but also by behavioral studies demonstrating that biologically significant signals act as adequate stimuli for the electroreceptors. As illuminated throughout this book, such signals include the fish's own EODs, discharges produced by conspecifics, as well as other electric cues originating from both biotic and abiotic sources.

3. Classification and Characterization of Electric Organ Discharges

Based on the voltage of the emitted discharge, electric fish are commonly classified as weakly electric or strongly electric. Weakly electric fish produce EODs from several hundred millivolts to a few volts. Strongly electric fish generate discharges of up to several hundred volts as recorded from the skin of fish out of the short-circuiting water.

A second classification scheme is based on the discharge pattern. Pulse-type species produce brief electric pulses, followed by relatively long and often highly variable intervals of silence. As an example, a brief EOD sequence and the waveform of an individual pulse of the elephant nose (*Gnathonemus petersii*) are shown in Figures 2.1A and 2.1B. Wave-type species, on the other hand, discharge in a quasi-sinusoidal manner, with the duration of the electric pulses comparable to the duration of the interpulse interval. As examples of such wavelike discharges, the EODs of the two gymnotiform species, the brown ghost, *Apteronotus leptorhynchus* (Fig. 2.2A) and the knifefish, *Eigenmannia* sp. (Fig. 2.2B), are shown in Figures 2C, C' and 2 D, D'.

The EODs of electric fish can be monitored by recording electrodes placed near the head and tail, or dorsum and ventrum, respectively. In their simplest form, such electrodes consist of stainless steel rods. The discharges are amplified—particularly in weakly electric fish, in which the intensity of the signal is

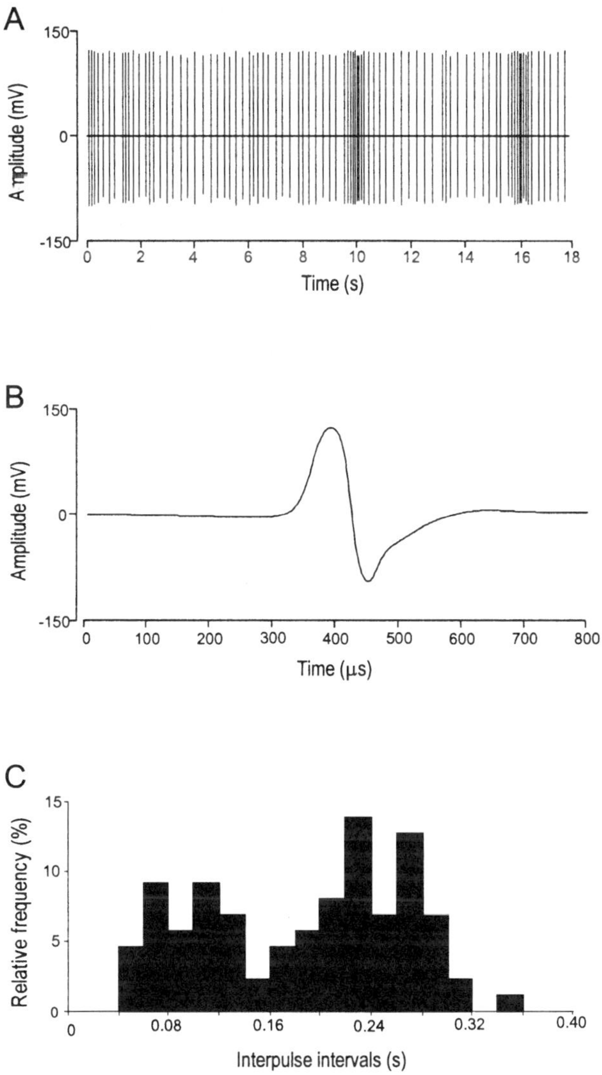

FIGURE 2.1. Electric organ discharge of the elephant nose (*Gnathonemus petersii*). The presentation at a compressed time scale (**A**), showing the individual discharge pulses as *vertical lines*, reveals the rather irregular discharge rhythm. The expanded time scale (**B**) resolves the details of the waveform of an individual pulse. The frequency distribution histogram (**C**) is based on analysis of the interpulse intervals of the discharge sequence shown in (**A**). This histogram suggests two major classes of interpulse intervals—short intervals between 0.04 and 0.16 s and long intervals between 0.16 and 0.36 s. (Courtesy of G.K.H. Zupanc and J.R. Banks.)

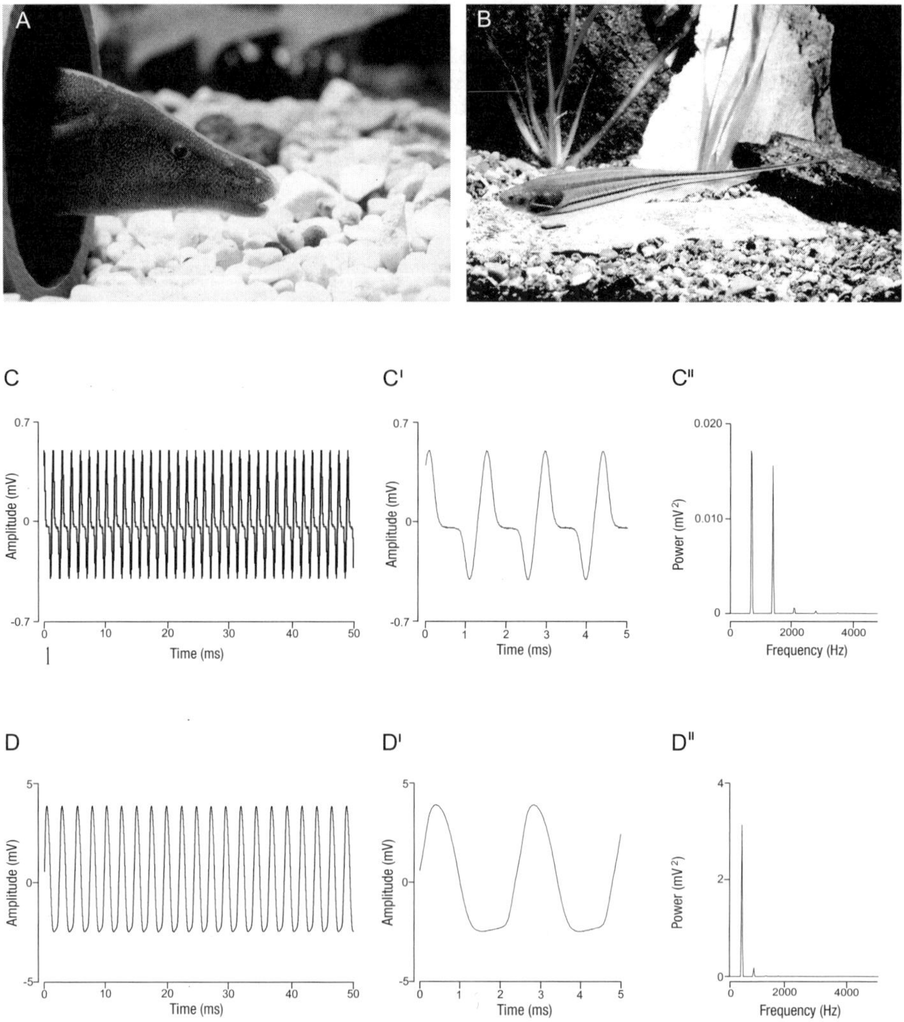

FIGURE 2.2. Two gymnotiform fish, *Apteronotus leptorhynchus* (**A**) and *Eigenmannia* sp. (**B**), and their EODs (**C**, **C'**, **C'' and D**, **D'**, **D''**, respectively). Although each of the two species produces wavelike EODs, their discharges differ in frequency, waveform, and spectral composition, as evident from the oscilloscope traces at compressed time scales (**C**, **D**), at expanded time scales (**C'**, **D'**), and from the Fourier spectra (**C''**, **D''**). (Courtesy of G.K.H. Zupanc and J.R. Banks.)

too low to be detected without proper amplification—and displayed on an oscilloscope. By convention, the positive electrode is positioned at the head or dorsum, whereas the negative electrode is placed at the tail or ventrum. As a characteristic feature of the species examined, the polarity of the initial phase of the EOD is described as head-positive or head-negative (or dorsum-positive or dorsum-negative).

Another important feature of the EOD is its rhythm. In wave-type species, the rhythm can be characterized by the EOD frequency, which is determined by the number of discharges produced within 1s. Measurement of this parameter is often supplemented by Fourier analysis (Fig. 2C", D"). The resulting frequency spectra provide information about the harmonic composition of the discharge pulse.

In pulse-type species, the intervals between individual pulses are often quite variable, thus resulting in a nonperiodic pattern of discharge (Fig. 2.1A). To avoid confusion with the parameter "EOD frequency" in wave-type species, in this chapter "pulse repetition rate" is used as a designator to characterize the discharge rhythm. Since this latter parameter reflects only the average number of pulses produced per second, but does not give any indication of the variability of the interpulse intervals, often the frequency distribution of the interpulse interval lengths is analyzed and plotted in form of interpulse interval histograms (Fig. 2.1C). Such histograms have revealed that the mode of discharge pattern (for instance, a highly rhythmic discharge pattern versus a more random discharge pattern) changes with the behavioral context, and may be used by the fish to encode information important for communication (see Section 9.4).

4. Phylogeny of Electroreceptors and Electric Organs

Electroreceptors exhibit a widespread phylogenetic distribution among fishes, representing at least four independent inventions in the course of evolution. They occur in most non-teleost fishes (especially Agnatha, Elasmobranchii, Holocephali, Chondrostei, Polypteri, and Dipnoi) and in four orders of teleosts—the siluriforms (catfishes), the gymnotiforms (knifefishes), the mormyriforms (elephant nose fishes), and in one subfamily (Xenomystinae) of the osteoglossiforms (for reviews, see Bodznick and Boord 1986; Northcutt 1986; Zakon 1986). In addition, electroreceptors have been found in the larval stages of aquatic apodan and urodele amphibians (for review, see Fritzsch and Münz 1986) and on the bill of *Platypus* (Scheich et al. 1986), a primitive Australian mammal. This taxonomic distribution suggests that electroreception is an ancestral vertebrate feature, common to all of the original vertebrates, but was lost in most vertebrate groups in the course of further evolution.

In contrast to electroreceptors, electric organs are not considered an ancestral vertebrate feature. They are believed to have evolved later in a number of taxa, possibly to complement the already existent electroreceptors. Several lines of evidence make it likely that they have evolved at least six times independently

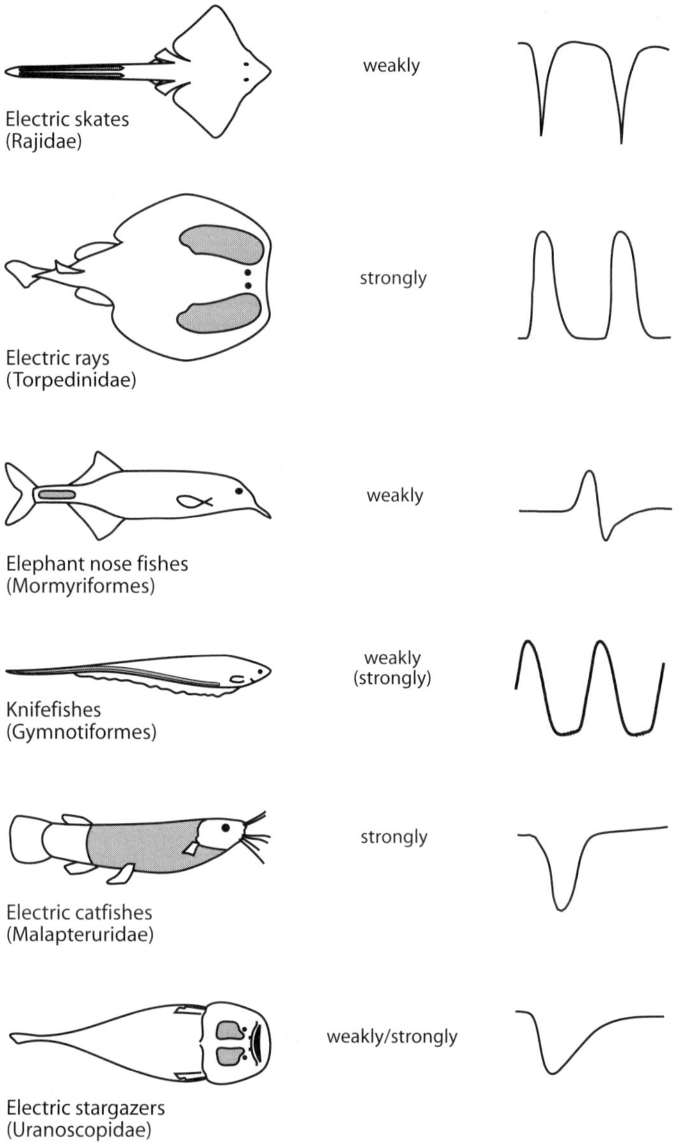

FIGURE 2.3. Representatives of six groups of electric fish. In addition to a line drawing of a fish and an indication of the location of the electric organ (*gray tint*), an oscilloscope trace of one or two characteristic pulses of the adult EOD is shown. These traces are arranged such that the positive polarity at the head is up. The classification of the intensity of the discharge is based on the amplitude of the EOD. Fish emitting EODs of less than a few volts are classified as weakly electric, whereas those that discharge at higher voltages are designated as strongly electric. Electric eels (Electrophoridae) within

among fishes (Bennett 1971; Bass 1986)—twice among chondrichthians, namely in torpedinoids (about 38 species) and rajoids (more than 200 species); as well as four times among actinopterygians (ray-finned fishes), namely in all of the estimated 200 species of mormyriforms, all of the estimated 130 species of gymnotiforms, several (including the well-known strongly electric African catfishes, Malapteruridae) of the 34 families of catfishes, and 1 (stargazers, Uranoscopidae) of the approximately 148 families of perciforms (Fig. 2.3). All of the electric fish except the members of the family Uranoscopidae are also electroreceptive. This contrasts with the large number of species that are electroreceptive, but lack the capability to produce electric discharges.

Among catfishes, the strongly electric *Malapterurus* has been known for a long time. In recent years, several other catfishes emitting weakly electric discharges have been identified within the families of the Mochocidae (including *Synodontis* and others), Clariidae, Bagridae, Siluridae, and Plotosidae (Hagedorn et al. 1990; Baron et al. 1994a,b, 1996; V.D. Baron, personal communication). The discovery of the electric discharges of these fishes only in recent years can be attributed to two facts: First, these species discharge not continuously, as most electric fish do, but very sporadically, primarily during agonistic encounters or in response to mechanical stimulation. Second, in many (although not all) of these fishes the voltage of their discharges is significantly lower than the voltage produced by the weakly electric representatives among the gymnotiforms and mormyriforms. Nevertheless, the electric activity of the weakly electric catfishes does not just reflect unspecialized electric potentials, such as respiratory potentials or electromyograms, but also shows sufficient synchrony to be considered as specialized for electric discharge. In terms of their sporadic appearance, the discharges of the weakly electric catfishes exhibit a certain degree of resemblance with those of strongly electric fish, rather than the EODs of the more or less incessantly discharging weakly electric species among the Gymnotiformes and Mormyriformes.

The identification of weakly electric species among the catfishes leaves the exciting possibility that there are more (probably predominantly sporadically) discharging weakly electric fish to be discovered in taxonomic groups traditionally viewed as electrically "silent." Indeed, Baron and Pavlov (2003) recently described specialized electrogenerating activity in two species of *Polypterus*. This genus has been known for some time to possess ampullary electroreceptors

---

FIGURE 2.3. *Continued*
the order Gymnotiformes can produce both weakly and strongly electric discharges, but all other species of this order generate weakly electric discharges, hence the designation "weakly (strongly)." The EOD of stargazers is intermediate between weakly electric and strongly electric, and is hence referred to as "weakly/strongly." Several lines of evidence suggest that electric organs have evolved independently six times in the groups represented here: twice among the Chondrichthyes (the electric skates and electric rays), once within the Osteoglossomorpha (the elephant nose fishes), and three times among the Euteleostei (the knifefishes, the electric catfishes, and the electric stargazers). (Modified after Szabo 1970 and Bass 1986.)

(Roth 1973; Jørgensen 1982; for review, see Northcutt 1986). The discovery of electric discharges in *Polypterus* is of special interest because this genus belongs to the oldest living branch of the Osteichthyes. This can be used as an indication that the capability for both electroreception and electrogenesis may have been present, at least in some species, already very early during the phylogenetic development of fishes.

It is possible that the sporadic discharges of the catfishes and the two species of *Polypterus* represent a primitive stage in the evolution of electric organs, resulting from the rather poor synchronization of the activity of individual cells. Similar stages may have existed in the ancestors of the continuously discharging electric fish. Electric discharges have developed by synchronization of muscle or nerve cell activity, and there is good evidence that this development took place independently in a number of fish groups. Interestingly, lampreys show well-synchronized muscle action potentials associated with ventilation movements (Kleerekoper and Sibakin 1956, 1957). These synchronized potentials are larger than those typically recorded during opercular movements, and may thus reflect such an early step in the development of electric organs.

5. Electroreceptors

The perception of electric signals is mediated by specific sensory cells called electroreceptors (for review, see Zakon 1986; see also Jørgensen, Chapter 3). Together with support cells, the electroreceptor cells form encapsulated end organs known as the electroreceptor organs. These receptor organs are located in invaginations of the fish's epidermis, but communicate with the body's surface through canals filled with mucous substrate or loosely packed epithelial cells. Typically, electroreceptor organs are distributed over most of the fish's body surface, with especially high densities in the head region.

Two anatomically distinct classes of electroreceptor organs can be distinguished: ampullary and tuberous (Fig. 2.4). Ampullary receptors are found in most non-teleost fishes (except in Myxiniformes and Holostei), in four orders of Teleostei, and in the aquatic forms of apodan and urodelian amphibians. Tuberous receptors have been identified only in two teleostean orders, the Gymnotiformes and the Mormyriformes (for reviews, see Bullock 1982; Zakon 1986).

Ampullary receptors are maximally sensitive to DC and low-frequency AC electric fields of both biological and nonbiological sources. One function, particularly well examined in elasmobranchs, is the detection of prey animals. This is possible because, as numerous experiments have shown, elasmobranchs can detect the bioelectric field that surrounds any living animal, even if this animal is buried in sand. As a second function, it has been proposed that certain electroreceptive species, such as sharks, can detect the earth's magnetic field using their ampullary receptor organs, which are known as the ampullae of Lorenzini in these fishes. When the fish, surrounded by sea water—which acts as a highly

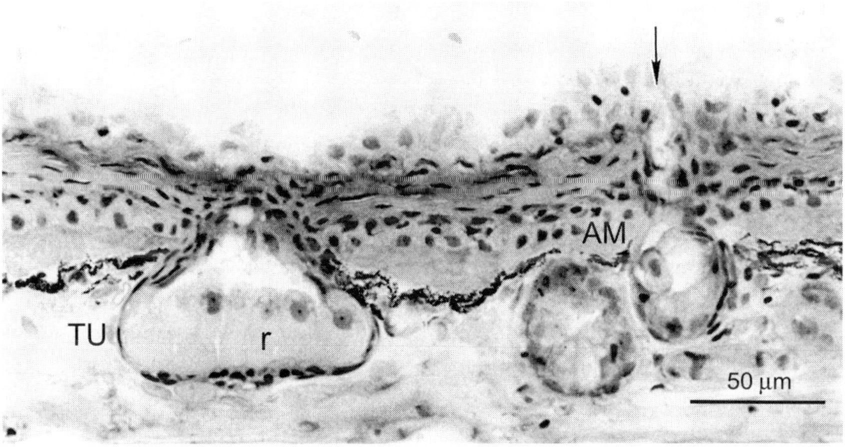

FIGURE 2.4. Two types of electroreceptor organs in the skin of the knifefish *Eigenmannia* sp. In the left half of the photomicrograph, a tuberous (TU) electroreceptor organ, composed of individual receptor cells (one is indicated by r), is shown. In the right half of the photomicrograph, an ampullary electroreceptor organ (AM) is visible. This latter receptor organ type is characterized by a long, large-diameter canal (*arrow*). (Courtesy of G.K.H. Zupanc and F. Kirschbaum.)

conductive medium—move in any direction (except parallel) to the field lines defining the earth's magnetic field, current flow is induced. This physical phenomenon of electromagnetic induction is observed whenever a closed conductive circuit is moved within a magnetic field. It has been hypothesized that the highly sensitive electroreceptors within the ampullae of Lorenzini can detect the tiny alterations in voltage associated with the induced current. As a third function, ampullary receptors have been shown to be involved in the detection of low-frequency components associated with electric signals used for electrocommunication in wave-type gymnotiform fish (Heiligenberg et al. 1991; Metzner and Heiligenberg 1991).

Tuberous receptors are tuned to frequencies within or near the EOD frequency of the respective species, and sometimes exhibit even fine-tuning to the "private" frequency bands of individual fish within the species-specific range of frequencies. Evidence for such a fine-tuning has been obtained in several wave-type gymnotiform species in which mature males and females discharge at different frequencies.

Based on their physiological response to electric stimuli, tuberous receptors can be further divided into two subcategories. The first subcategory, called the "rapid-timing" class, comprises receptors that follow each EOD cycle (in wave-type species) or pulse (in pulse-type species) with a single spike, thus mirroring the timing of the EOD. The second subcategory, called the "amplitude-modulated" class, consists of receptors whose probability of firing depends on the EOD amplitude. These latter receptors thus encode information about the amplitude of the electric stimulus.

In wave-type gymnotiforms, these two subtypes of the tuberous receptors are commonly known as T-type receptors (that encode the firing of the positive transition of the EOD) and P-type receptors (that encode the instantaneous EOD amplitude). In fishes of the family Mormyridae among the Mormyriformes, the corresponding receptor subtypes are called knollenorgans and mormyromasts. Knollenorgans fire a single spike time locked to the EOD pulse. In mormyromasts, the number of spikes generated is a function of stimulus intensity. As shown in detail below, both the rapid-timing electroreceptors and the amplitude-modulated receptors play important roles in active electrolocation and in electrocommunication.

6. Electric Organs

The electric organs of the vast majority of electric fish are derived from various types of muscles, and hence are termed myogenic (see also Macadar, Caputi, and Carlson, Chapter 14). The only exception are members of the family Apteronotidae within the order Gymnotiformes, where the electric organs are formed by modified spinal motor axons; the latter organs are therefore referred to as neurogenic. The cells comprising the electric organs are commonly called electrocytes. In many species, they are stacked behind each other along the anterio–posterior body axis, thus forming a column of cells (Fig. 2.5). De-

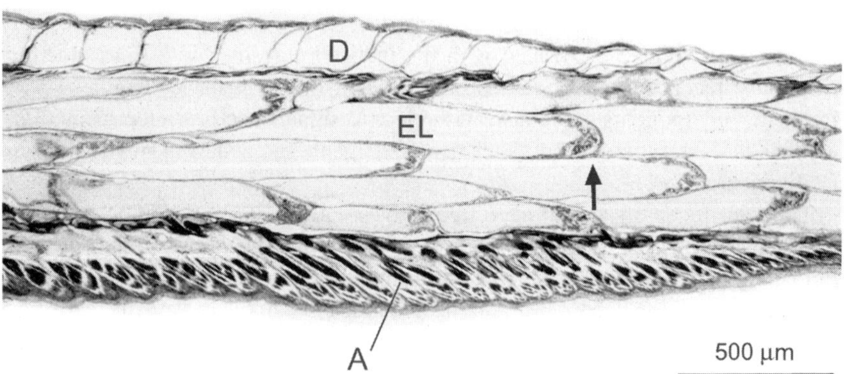

FIGURE 2.5. Photomicrograph of a parasagittal section through the tail of the knifefish, *Eigenmannia* sp. Both muscle fibers of the anal fin (A) and dermis (D) with scales are visible. The electrocytes (EL) are distinguished by their long, slender shape. They are arranged in series along the longitudinal axis of the body. At the caudal end, where the membrane exhibits pronounced evaginations (*arrow*), the electrocytes are innervated by axons of spinal motoneurons. (Courtesy of G.K.H. Zupanc and F. Kirschbaum.)

pending on the species, between a few columns and several hundreds of such columns, enclosed within tubes of connective tissue, run parallel to each other—thus resembling rolls of coins wrapped in paper, all facing the same way (for reviews, see Bennett 1971; Bass 1986).

The synchronous activity of the electrocytes defines the output of the electric organ—the EOD. The number of electrocytes arranged in series determines the voltage; the number of electrocytes arranged in parallel determine the current of the EOD. Typically, individual electrocytes generate potentials on the order of 100 mV. In the electric eel (*Electrophorus electricus*), more than 100 columns, with 5000 to 10,000 electrocytes each, are responsible for the generation of the EOD. Thus, when these electrocytes are activated synchronously, their potentials summate to generate an EOD of several hundreds of volts.

Some electric fish have more than one electric organ. *Narcine*, a member of the strongly electric marine torpedinoids, possesses a main organ that produces pulses of 25 to 35 V, and a smaller accessory organ that generates weak EODs of 0.1 to 1.0 V. *Electrophorus* has three electric organs called Main, Hunter's, and Sachs organs (Fig. 2.6). The Main organ and the anterior part of the Sachs organ are responsible for the high-voltage EODs, whereas the Sachs organ and the posterior part of the Hunter's organ are active during low-voltage emission.

In most electric fishes, the electric organ is innervated by either cranial or spinal motoneurons, collectively referred to as electromotoneurons. An exception is *Astroscopus,* in which the electric organ is composed of extraocular eye musculature and innervated by a specialized portion of the oculomotor complex of the midbrain.

The pattern by which the electromotoneurons innervate the individual electrocytes varies greatly among species. Usually, the axons of the electromotoneurons terminate on one of two distinguishable faces of the electrocytes. In some cases, the innervation site exhibits a distinct evagination of the cell membrane; this morphological specialization is referred to as a stalk.

Additional complications occur in mormyriforms. In one group of fishes within this order, the stalk is the result of a direct outpocketing of the posterior face of the electrocytes. This pattern is subsumed under the term nonpenetrating stalk morphology. In the second group, the stalk emerges from one face, usually the posterior side, but penetrates the electrocyte to actually emerge opposite to the side of origin, so that the electrocyte is typically innervated anteriorly. The latter geometrical design of the electrocytes is referred to as penetrating stalk morphology.

Reconstruction of the evolution of the electric organs of mormoyriform fishes based on phylogenetic analysis of mitochondrial and nuclear DNA suggests that the stalkless electrocytes represent the primitive condition, whereas the penetrating stalked electrocytes evolved once early in the history of modern mormyrids (Alves-Gomes and Hopkins 1997; Sullivan et al. 2000; see also Orti and Sullivan, Chapter 13). This step was presumably followed by multiple independent reversals to the nonpenetrating stalked electrocytes in certain clades.

The geometry of the electrocytes and the complexity of the stalk system

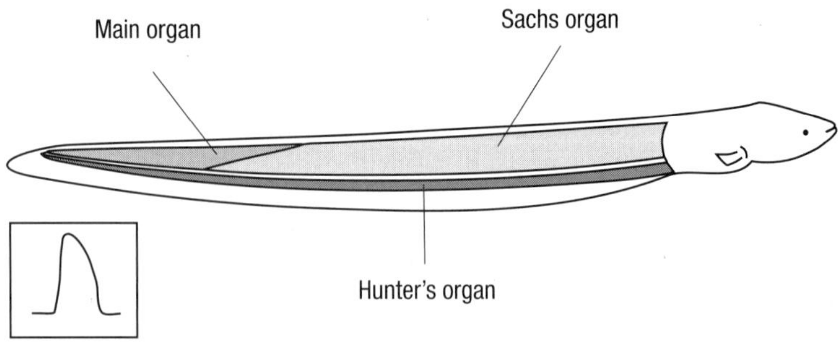

FIGURE 2.6. The South American electric eel (*Electrophorus electricus*) with its three electric organs. Whereas the Main organ and the anterior part of the Hunter's organ produce strong electric discharges, the Sachs organ and the posterior part of the Hunter's organ generate only weak discharges. A discharge pulse is shown in the *inset*. (Modified after Szabo 1970 and Bass 1986.)

largely determine the waveform of the EOD. This makes it possible to predict the EOD waveform based on the activity of an individual electrocyte. On the other hand, the rhythm of the EOD is controlled by central structures of the electromotor pathway. In the two best-studied groups of electric fish, the gymnotiforms and mormyriforms, this structure is a midline nucleus in the reticular formation—the pacemaker or command nucleus, which fires in a one-to-one fashion with the electric organs (for review, see Dye and Meyer 1986). These command spikes of the pacemaker nucleus are conveyed to the electric organ via a medullary relay nucleus and electromotor neurons in the spinal cord. In many species, the pacemaker nucleus and the medullary relay nucleus have merged to a single structure, which, although different from the "pure" pacemaker nucleus, is also referred to as the pacemaker nucleus. The latter structure is comprised of both pacemaker cells and relay cells as its two principal cellular components.

7. Development of Electric Organs

Besides the electric organ of adult electric fish, in both mormyriforms and gymnotiforms additional electric organs have been identified in larval stages (see also Northcutt, Chapter 5). The existence of these larval organs is restricted to a rather short period of early life and precedes that of the adult electric organ. One species in which the development of the larval and adult electric organ has been particularly well examined is *Pollimyrus isidori* (Denizot et al. 1978, 1982) (Fig. 2.7A–D). This was made possible following the successful breeding of this mormyriform, as well as other electric fish, in captivity (Kirschbaum 1975, 1979).

In *Pollimyrus isidori*, the larval electric organ first appears in 8-day-old fish,

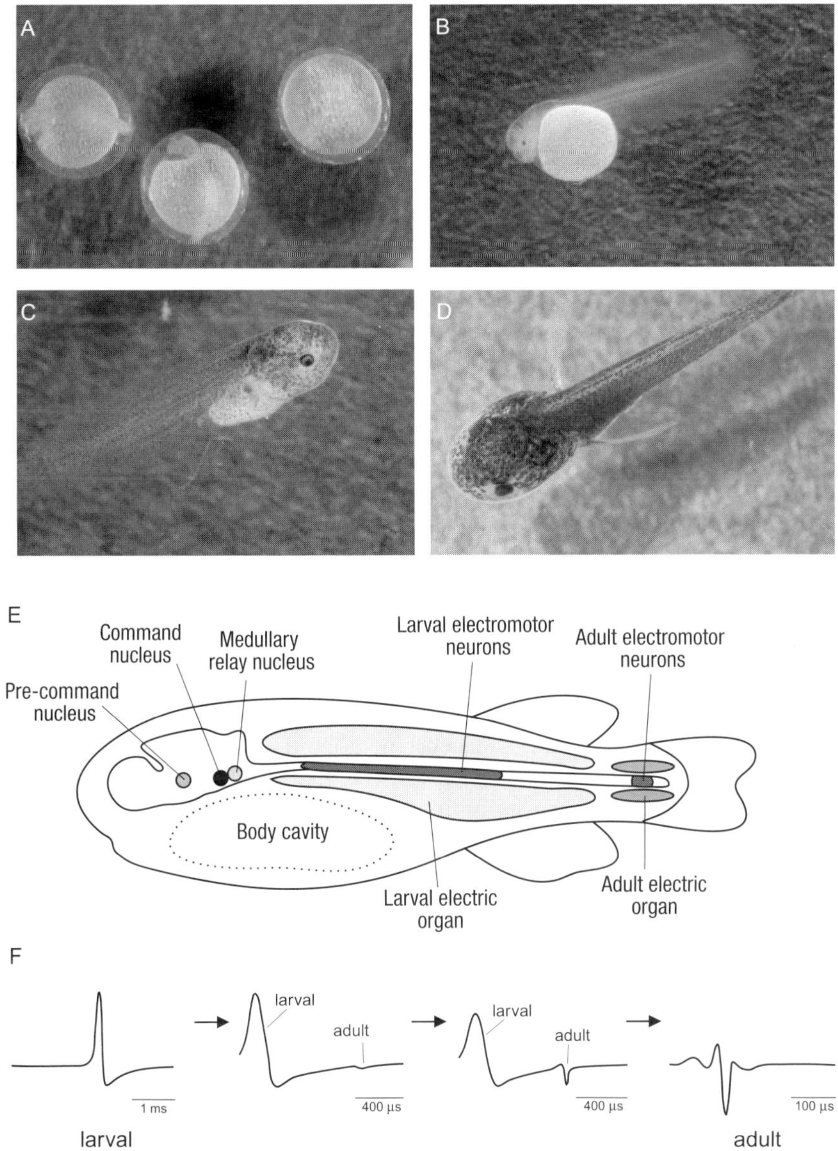

FIGURE 2.7. Development of the mormyriform fish *Pollimyrus isidori*. (**A**) Egg stage, 45 hours after fertilization. (**B**) Early larval stage, 90 hours after fertilization. (**C**). Late larval stage, 8 days after fertilization and shortly before free swimming. At this stage, the fish start to generate electric organ discharges by their larval electric organ. (**D**) Free-swimming stage. The juvenile shown is 15 days old. (**E**) Schematic diagram of the larval and adult electric organ. In the larval stage, the larval electric organ accounts for up to 50% of the total musculature of the fish. During further development, this larval organ is replaced by the adult electric organ, which is much smaller and restricted to a region near the tail of the fish. (**F**) Development of the electric organ discharge from the larval, head-positive form (*left*) to the adult, head-negative form (*right*). Two intermediate developmental stages reveal the emerging adult discharge, which follows the larval peak by several hundred microseconds. (**A–D**, G.K.H. Zupanc; **E**, modified after Denizot et al. 1978; **F**, modified after Denizot et al. 1978, 1982; Hopkins 1981.)

and begins to degenerate as the adult electric organ gradually differentiates (Fig. 2.7E). The EOD generated by the larval organ is characterized by a rather long duration and low amplitude—1 ms and approximately 10 mV, respectively. This contrasts with the adult EOD which is, with 100 μs and 1 V, both shorter and more powerful. As Figure 2.7F shows, the larval EOD is also distinguished by its head-positive polarity, which is opposite the head-negative polarity of the adult fish.

In 15-mm juveniles, which are approximately 40 to 50 days old, the adult EOD begins to develop. It follows the larval EOD with a lag of approximately 1 ms. The larval and adult EODs coexist for about 40 days, during which the larval EOD gradually diminishes, while the adult EOD becomes more and more dominant. After this period of coexistence, only the adult EOD persists.

The larval and adult EODs are generated by two different electric organs. Whereas the larval organ extends in the form of its four longitudinal tubes from the caudal end of the skull to the rostral beginning of the caudal peduncle, the adult organ is much more compact, being restricted to the region of the caudal peduncle. Each of the two organs is innervated by its own set of spinal electromotoneurons.

The transient existence of a larval electric organ is paralleled by the transient existence of two types of larval tuberous electroreceptor organs (Bensouilah et al. 2002). The tuberous electroreceptors appear with the start of the emissions of the larval EOD at day 8, and degenerate at the time when the larval electric organ degenerates. Thus, *Pollimyrus isidori*, and probably many other mormyrids as well, appear to possess a complete larval electric system, comprising both a larval electric organ and larval electroreceptor organs.

It has been speculated that the larval EOD may serve as a communication signal to help the parental fish identify the offspring in species with broodcare. In addition, the larvae may use their functional larval electric system to electrolocate exogenous food sources.

8. Dynamics of the Electrosensory and Electromotor System

In wave-type gymnotiform fish, the EOD is extremely stable in both frequency and waveform under constant environmental conditions. This is reflected by a very low coefficient of variation [= (standard deviation/mean) · 100], which, in the case of the EOD frequency of *Eigenmannia* sp., is approximately 0.3% over 10 min (Bullock 1969) and 0.04% over 1000 EOD cycles (Bullock et al. 1975). As such, the EOD frequency is one of the most robust phenomena known in biology.

Despite this enormous regularity, the EOD can be modulated by a variety of factors (see also Bastian and Zakon, Chapter 8). An immediate effect is exerted by ambient water temperature (Coates et al. 1954; Enger and Szabo 1968; Boudinot 1970; Feng 1976; Zupanc et al. 2003). In *Eigenmannia* sp., the EOD frequency increases (decreases) by approximately 10 to 20 Hz with an increase (decrease) in temperature of 1°C (Fig. 2.8A). In *Apteronotus leptorhynchus*, this effect is even more pronounced, with a 1°C temperature change causing an alteration in EOD frequency of roughly 30 Hz (Fig. 2.8B). This effect also depends on the rapidity of the temperature change. The temperature-dependent alterations in EOD frequency are likely to be mediated by a direct action of ambient temperature on the activity of the pacemaker nucleus, as suggested by experiments in which the temperatures of the head and trunk were changed independently. Then, the EOD frequency varied with the temperature of the head, but not with the temperature of the trunk (Coates et al. 1954; Enger and Szabo 1968).

Long-term changes in the EOD have been found in a number of fishes and are commonly associated with sexual maturation, particularly in those species that exhibit a sexually dimorphic discharge. Such a sexual dimorphism was first discovered by Carl Hopkins in *Sternopygus macrurus* (Hopkins 1972). In this species, sexually mature males discharge on average at 67 Hz, whereas sexually mature females discharge on average at 120 Hz. The mean EOD frequency of sexually immature fish is somewhat intermediate, namely at 93 Hz. The development of this sexual dimorphism is controlled by steroid hormones. Treatment of *Sternopygus* with the androgens testosterone or 5α-dihydrotestosterone leads

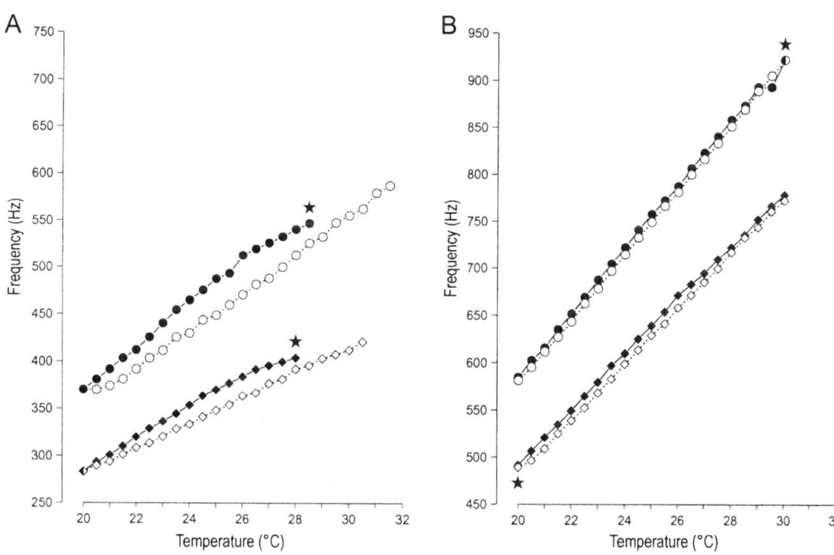

FIGURE 2.8. Effect of water temperature on EOD frequency of *Eigenmannia* sp. (**A**) and *Apteronotus leptorhynchus* (**B**). For each of the two species, the results for two different individuals are shown. Temperature and frequency at the start of the experiment are indicated by *asterisks*. The *closed symbols* and *solid lines* indicate frequency values obtained when the ambient temperature is lowered. The *open symbols* and *dotted lines* indicate frequency values measured when raising the water temperature. (Courtesy of G.K.H. Zupanc and G. Engler.)

to a pronounced decrease in EOD frequency, whereas administration of estradiol causes a slight increase in EOD frequency (Meyer 1983). This effect appears to be produced at the level of the pacemaker nucleus. In *Apteronotus leptorhynchus*, in which sexually mature males discharge at higher frequencies than sexually mature females, the effect of steroid hormones is reversed: implantation of estrogens causes a gradual decrease in EOD frequency, whereas treatment with 5α-dihydrotestosterone results in a slight increase (Fig. 2.9).

The hormonal effects on EOD frequency are paralleled by alterations in receptor tuning. In *Sternopygus*, in which the tuberous receptors are closely tuned to the EOD frequency of the individual fish, hormonal treatment leads, in the course of days, to alterations in receptor tuning such that the best frequency of the receptors tracks—with a certain delay—the EOD frequency (Meyer and Zakon 1982; Zakon and Meyer 1983). This effect is independent of the presence of an exogenous electric field (Keller et al. 1986) and the pacemaker nucleus as a potential source of internal reference (Ferrari and Zakon 1989). Instead, the

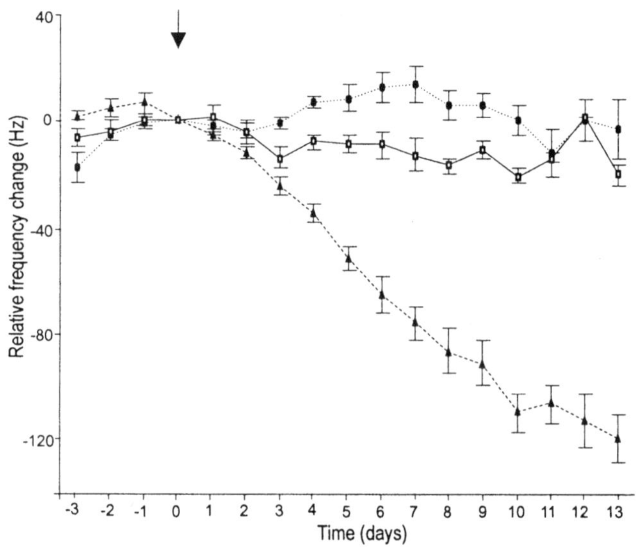

FIGURE 2.9. Effect of steroid hormones on the EOD frequency of *Apteronotus leptorhynchus*. Fish received implants of 5α-dihydrotestosterone (13 individuals, *circles* and *dotted lines*) or 17β-estroadiol (12 individuals, *triangles* and *dashed lines*), or were sham-treated (12 individuals, *squares* and *closed lines*). In each of the three groups, the EOD frequency was measured on four consecutive days before the operation (days −3 through 0) and on 13 days following the surgery (days 1 through 13). The fish received the implants immediately following the EOD frequency measured on day 0 (indicated by *arrow*). The frequency measured on each day was then normalized relative to the frequency determined on day 0 and averaged over the number of fish examined. Also indicated is the standard error of the mean. The diagram shows a pronounced decrease in EOD frequency in the estradiol-treated fish, and a slight frequency increase in the 5α-dihydrotestosterone-treated fish, compared to the control animals. (Courtesy of G.K.H. Zupanc.)

androgens appear to affect electroreceptor tuning directly through the electroreceptors, probably by altering receptor cell membrane conductances.

Besides the discharge rate, hormones can also influence the waveform of the EOD. Such an effect was first discovered by Andrew Bass and Carl Hopkins in the mormyrid fish *Brienomyrus brachyistius* (long biphasic) (Bass and Hopkins 1983). (The descriptor "long biphasic" is used to distinguish this species from other, morphologically indistinguishable species producing a different EOD.) In mature males, the average EOD duration is 2.3 ms, compared to 0.9 ms in mature females and immature specimens. Treatment of females with the androgen 17α-methyltestosterone causes a twofold, reversible increase in EOD duration over a 2- to 3-week period. Since in some vertebrates testosterone can be converted to 17β-estroadiol or another androgen, 5α-dihydrotestosterone, the androgen specificity of the effect on the EOD can be tested by administration of the latter two metabolites. Such experiments have demonstrated that only fish treated with 5α-dihydrotestosterone show a marked change in EOD duration, whereas the effect of estradiol is rather minimal. Thus, the hormone effect appears to be androgen specific.

Similar findings were made in a pulse-type gymnotiform fish, *Hypopomus occidentalis* (Hagedorn and Carr 1985). In breeding populations in the natural habitat, males are distinguished from females and immature males by a number of characteristics, including the larger size and the thicker tail in males compared to females (Fig. 2.10A, B). Both males and mature females produce biphasic

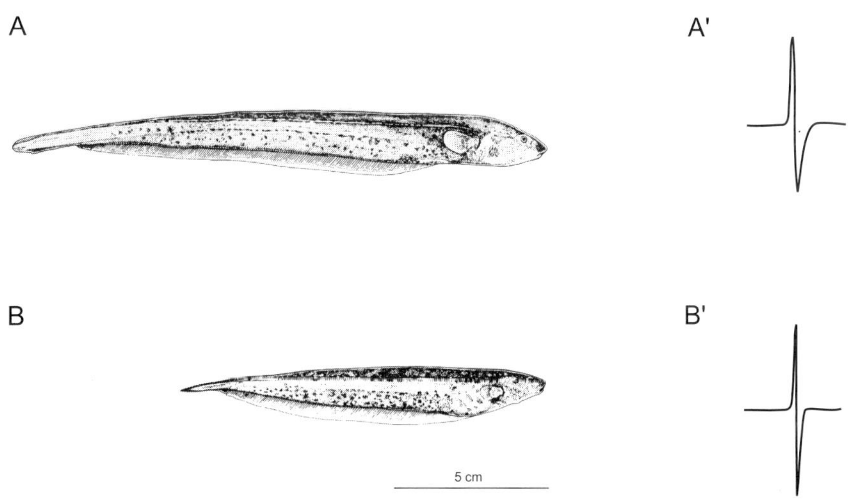

FIGURE 2.10. Sexual dimorphisms in *Hypopomus occidentalis*. Males are larger and have a thickened tail (**A**). Their EOD pulse is asymmetrical, which is caused by the longer duration of the second, head-negative phase, compared to the first, head-positive phase (**A'**). In contrast, females are smaller and have a slender tail (**B**). Their EOD pulse is of shorter duration and exhibits a symmetrical form (**B'**). (Modified after Hagedorn and Carr 1985.)

EOD pulses, the first phase of which is head positive and the second phase head negative. However, while in mature females and immature males the duration of the first and second phases are equal, thus resulting in a symmetrical pulse, in mature males the EOD pulse is asymmetrical, which is caused by a longer duration of the second phase compared to the first phase (Fig. 2.10A', B').

Treatment of females with 5α-dihydrotestosterone leads to the development of broader tails and asymmetric EOD pulses, thus mimicking the sexually dimorphic characteristics of mature males. Histological examination has shown that the enlargement of the tail is caused by an increase in the size of individual electrocytes compared to those of saline-treated (control) fish.

Intracellular recordings have demonstrated that the differences in EOD waveform between 5α-dihydrotestosterone-treated females and untreated females can be accounted for by changes in the responses of individual electrocytes. In *Hypopomus*, the caudal face of the electrocytes is innervated by the electromotoneurons. When the fish generate an EOD, the caudal face is first depolarized, producing the head-positive phase of the diphasic pulse. This is followed by depolarization of the rostral face, which causes the head-negative phase. Differential recordings across the caudal face of the electrocytes have yielded responses following this pattern in both hormone- and saline-treated fish. However, such recordings have also revealed longer responses from the rostral face of 5α-dihydrotestosterone-treated fish compared to those of control fish. This suggests that androgens act at the level of individual electrocytes, inducing the sexually dimorphic changes of the electric organ, which, in turn, result in the sexually dimorphic electric signal.

9. Functions of Electric Systems

Besides the prey-hunting function of strongly electric fish, electric systems are involved in one or more of the following behavioral functions: passive electrolocation, active electrolocation, species recognition, and intraspecific communication. In the following, we provide an overview of these four functions.

9.1 Passive Electrolocation

Passive electrolocation is exhibited by animals that themselves do not possess electric organs, but are able to perceive electric fields emanating from physical or biological sources through their electroreceptors (see also Bodznick and Montgomery, Chapter 6; Wilkens, Chapter 9). The animals use their electroreception ability for prey location and, possibly, for detection of orientation cues.

We owe most of our knowledge about these functional roles to the landmark studies of Ad Kalmijn (for review, see Kalmijn 1974). Research in this area was stimulated by the observation that sharks and rays can make well-aimed responses toward live prey, such as flatfish, buried in sand and invisible to the predator. Attack is typically initiated at a few tens of centimeters and followed by uncovering and devouring of the prey.

As further analysis has shown, such prey animals—like many other aquatic organisms—are the source of weak, low-frequency AC or modulated DC electric fields. Typical field intensities measured at a distance of 10 cm are on the order of approximately 0.2 μ V/cm. The exact source of these steady bioelectric fields is unclear, but in marine fishes relatively strong electric signals emerge from the head region, especially from the mouth cavity and from the opercular slits.

To test the hypothesis that the elasmobranchs' prey attack response is triggered by bioelectric fields from the prey animal, but not by any other cues, a variety of approaches have been employed. One of these approaches is illustrated in Figure 2.11. Smooth dogfish (*Mustelus canis*) were attracted to the research area by liquefied herring released from a tube in the center of a polyvinyl plate. Prey fields were simulated by passing direct current through a pair of electrodes to the right or left of the odor source. A second pair of electrodes was fitted to the other side. To these latter electrodes, which served as a control, no current was applied. In almost all trials, the dogfish attacked the electrodes between which current was passed. Rarely, they bit the control electrodes or the odor source (Kalmijn 1982).

The distance at which the dogfish initiate attack allows investigators to estimate the stimulus strength perceived by the dogfish. Such estimates have suggested that dogfish can detect electric fields with intensities as low as 5 nV/cm. This sensitivity is high enough to trigger attack at distances several tens of centimeters from the prey animal.

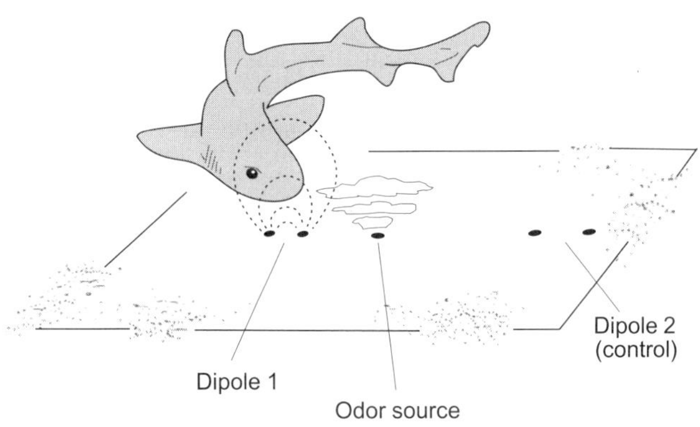

FIGURE 2.11. Feeding attack of smooth dogfish (*Mustelus canis*) triggered by mimicking the bioelectric field of prey animals. The shark is attracted by liquefied herring released from a tube in the center of a polyvinyl plate set into the bottom of the test area. The electric field of prey is simulated by passing direct current through one of two pairs of electrodes located to the right or left of the odor source. The "silent" electrode pair serves as a control. In the vast majority of the experiments, the dogfish attacked the electrodes through which current was applied. (Modified after Kalmijn 1982.)

The enormous sensitivity of the ampullae of Lorenzini, which mediate the perception of electric signals in elasmobranchs, may also enable these fishes to detect electric fields from inanimate sources, such as fields induced by the drift of ocean currents or by the swimming of the fish through the Earth's magnetic field. In both instances of electromagnetic induction, the voltage gradients range between 0.05 and 0.5 μV/cm, which are thus well above detection threshold. Elasmobranchs may extract information from the direction and the polarity of these uniform electric fields for the purpose of orientation. Such a notion is supported by conditioning experiments in which stingrays have demonstrated their ability to orient relative to uniform electric fields that are similar to those produced by ocean currents (Kalmijn 1982).

9.2 Active Electrolocation

As mentioned above (see Section 2), Hans Lissmann convincingly demonstrated in the 1950s that the weakly electric fish *Gymnarchus niloticus* can detect objects differing in conductivity from that of the surrounding water by analyzing distortions of the self-generated three-dimensional electric field (Lissmann and Machin 1958). A detailed analysis of this phenomenon, which is commonly referred to as "active electrolocation," has shown that objects with impedances (ohmic resistances and/or capacitive reactances) differing from the impedance of the surrounding water cast "electric shadows" on the body surface of the fish (see also Nelson, Chapter 11). This shadow, or "electric image," consists of an area in which the density of the current lines defining the fish's electric field has changed in a systematic manner. Objects with impedance values lower than that of the surrounding water cause more current flowing through them and thus "attract" the electric current lines, whereas objects with higher impedance values reduce the electric current flow and thus "repel" the electric current lines (Fig. 2.12). These changes in current flow result in the first case in more current flowing through the electroreceptor organs in that part of the skin located opposite to the object. In the second case, the opposite effect occurs—the current flow through the corresponding electroreceptors is reduced compared to the current flow in the absence of an object.

The electric image of an object depends on its electrical properties, size, and shape and on its distance from the fish. Investigations by Gerhard von der Emde and co-workers, using conditioning approaches (von der Emde 1990; von der Emde and Ringer 1992; von der Emde et al. 1998), have shown that the fish not only can detect the presence of objects with impedances different from that of the water, but also discriminate different object distances and analyze independently their capacitive and resistive properties. The fish might use the latter ability to discriminate between living and inanimate objects in the environment, because living objects are distinguished by a considerable capacitive component, while inanimate objects possess mainly resistive properties (Heiligenberg 1973; von der Emde 1990). Although the effective range of active electrolocation is limited to a few centimeters around the fish, it appears to have developed in all

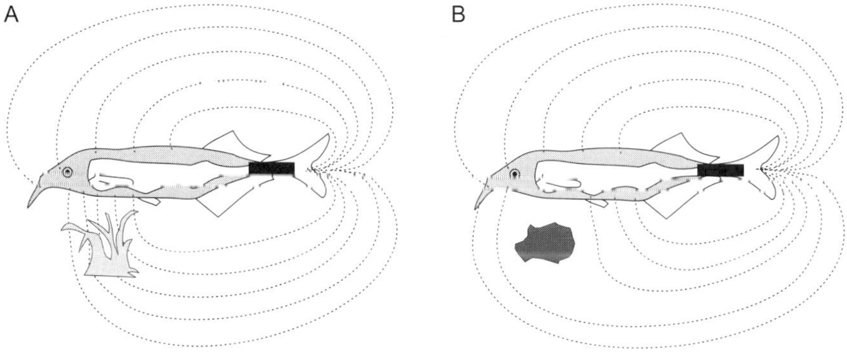

FIGURE 2.12. Distortions of the electric field of *Gnathonemus petersii* caused by natural objects with impedances differing from that of the surrounding water. (**A**) Water plants as good conductors "attract" the electric field lines, and thus result in more current flowing through the electroreceptor organs in that part of the skin located opposite to the object. (**B**) Stones as isolators repel the electric field lines and thus result in a reduced current flow in the corresponding electroreceptors compared to the current flow in the absence of an object. In addition to the field lines, the location of the electric organ in the caudal peduncle (*black bar*) and the electroreceptive body surface of the fish (*gray shade*) are shown. (Modified after von der Emde 1999.)

weakly electric fish examined thus far as a powerful behavioral mechanism to control distance to objects and body posture relative to substrate (Meyer et al. 1976).

9.3 Species Recognition

A number of field studies have revealed an enormous diversity in the EODs of different sympatric species (Hopkins and Heiligenberg 1978; Hopkins 1980, 1981; Kramer et al. 1981; Sullivan et al. 2002; see also Hopkins, Chapter 10). This diversity arises not only from the differences between wave-type and pulse-type species, but also from several invariant features of the signals produced within each of the two groups. In wave-type species, such invariant properties are the fundamental frequency and the waveform. In pulse-type species, a specific-specific classification is possible based on waveform and peak spectral frequency of the pulse, as well as on pulse repetition rate.

Especially in species-rich areas, unambiguous identification of the species is sometimes not possible using just one parameter. Instead, a combination of two or three of the invariant features is necessary. Kramer and associates (Kramer et al. 1981) found in a small part of the Solimões near Manaus 43 gymnotiform species. Of these, 11 were pulse-type fishes and 32 belonged to the group of wave-type species. Of the 32 wave-type species, 28 occupied a frequency range between 300 and 1800 Hz, which yields a hypothetical frequency band of 0.09

octaves for each species, if the individual species were evenly spread out over the entire frequency range. However, the frequency range occupied by *Eigenmannia* alone spans 1.2 octaves. Thus, additional parameters, such as EOD waveform, are necessary to improve species identification.

In pulse-type species, unambiguous identification is often possible using just waveform. This has even allowed investigators to distinguish species by EODs that are morphologically virtually indistinguishable.

It is tempting to speculate that by adopting divergent and species-specific EODs, electric fish are able to recognize members of their own species, and that selection on EOD waveforms as mate recognition signals may have been involved in the radiation of sympatric weakly electric species. As an important step toward establishing species identity based on the electric signals emitted, weakly electric fish have increased their sensory sensitivity to cues of the EOD of their own species, while filtering out unwanted EODs. Such species-specific filtering mechanisms have been demonstrated in a number of fishes. In *Hypopomus*, a pulse-type species, Hopkins and Heiligenberg (1978) found that two out of five electroreceptor types respond maximally to spectral frequencies characteristic of the peak power of the species-specific EOD, thus acting as EOD filters. In wave-type fishes, such as *Sternopygus macrurus*, *Eigenmannia virescens*, and *Apteronotus albifrons*, tuberous receptors are commonly tuned to frequencies in the range of their own species (Hopkins 1976).

The species-recognition hypothesis is corroborated further by the results of behavioral experiments. Studies employing conditioning paradigms have shown that *Eigenmannia* sp. can discriminate electric signals differing in frequency (Kramer and Kaunzinger 1991) or waveform (Kramer and Zupanc 1986; Kramer and Otto 1991; Kramer 1999), thus demonstrating that this wave-type species is capable of using these cues to distinguish between different EODs. In addition, it has been shown that *Sternopygus macrurus* males produce electric courtship signals preferentially to electric sine waves in the frequency range of females of their own species, but less to sinusoidal signals outside this range (Hopkins 1972). This suggests that *Sternopygus* males can distinguish between their own species and other sympatric species, as well as between males and females, on the basis of frequency alone.

9.4 Intraspecific Communication

A fourth function of the electric system examined in great details involves the generation and perception of electric signals for the purpose of communication with conspecifics (see also Hopkins, Chapter 10). As the mode of communication differs between pulse-type and wave-type species, we first take a look at mormyrids, all of which discharge in a pulse-type manner. Then, in the second part of this section, some principles of intraspecific communication in wave-type species among gymnotiforms are discussed.

9.4.1 Pulse-Type Discharges

The EOD signals of mormyrids consist of very brief, stereotyped pulses, separated by a sequence of rather long and variable interpulse intervals. A sample of such a signal sequence, recorded from a *Gnathonemus petersii*, is shown in Figure 2.1A at a more compressed time scale. Figure 2.1B reveals the waveform of an individual pulse generated by this fish at an expanded time scale. As the oscilloscope trace (Fig. 2.1A) and the corresponding interpulse interval histogram (Fig. 2.1C) demonstrate, this sequence can be divided into two major classes of signals: pulses separated by relatively short intervals and pulses followed by somewhat longer periods of "silence." A large number of studies have shown that both signal parameters—the waveform of the individual pulse and the sequence of interpulse intervals—play important roles in intraspecific communication.

The functional significance of the pulse waveform for sex recognition was first examined by Hopkins and Bass in a series of experiments using *Brienomyrus brachyistius* (Hopkins 1981; Hopkins and Bass 1981). During the breeding season, males of this species occupy territories in shallow streams and produce courtship EODs consisting of a burst of EOD pulses at a pulse repetition rate of 100 to 150 Hz. This behavior, called "rasp," can be used as an indicator of the male's response to the presumptive presence of a female when playing electric signals into the water. Indeed, EODs recorded from females are more effective in eliciting rasps than discharges from males or other electric fish living in the same habitat.

The relative significance of the EOD waveform versus the sequence of the interpulse interval could be elucidated by generating artificial signal sequences based on digitized EOD pulses. Through this approach, the EOD waveform and the sequence of interpulse intervals could be varied independently of each other. These experiments have demonstrated that males use the EOD waveform to recognize a female EOD signal. Even a completely "scrambled" interpulse interval sequence is as effective in evoking rasps as the natural interpulse interval sequence, as long as it is based on female pulses. This role of the EOD waveform in sex recognition may be widespread, and possibly even universal, as sex differences in EOD waveform have been found in a large number of species.

Whereas the EOD waveform is highly stereotyped and undergoes alterations only in the course of weeks and months (see Section 8), the sequence of interpulse intervals is highly variable and subject to modulations, particularly in the context of social interactions between two or more fish. This can readily be demonstrated by a simple experiment (Fig. 2.13A). Two fish are kept isolated in separate tanks. Their EODs are recorded by electrodes placed closely to each fish and, on closure of an electric switch, transmitted into the other tank. Through this approach, cues other than electric ones can be excluded to play a role during stimulation. Moreover, recording of the EOD of one fish through a second pair of electrodes allows the investigator to monitor the discharges not

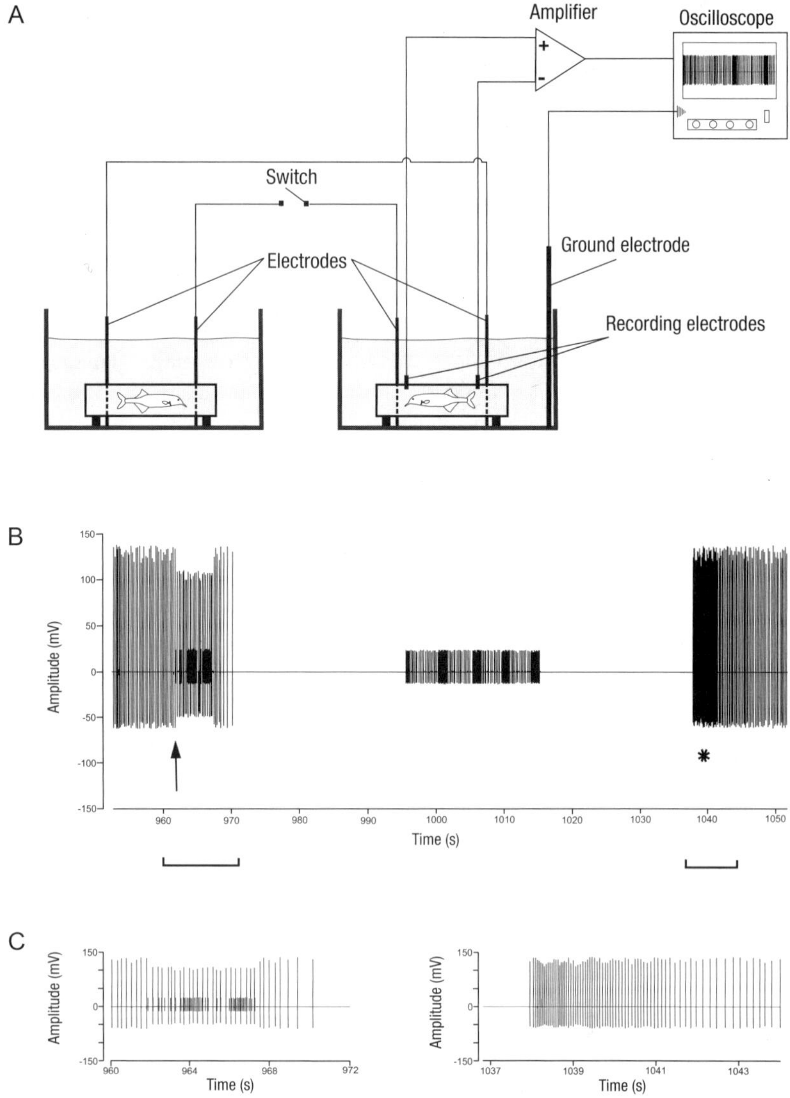

FIGURE 2.13. (**A**) Experimental setup to demonstrate electrical interactions between two elephant nose fish (*Gnathonemus petersii*). A single fish is kept in each tank. The two aquaria are electrically connected via wires attached to electrodes, which are placed near the fish. This connection can be controlled by an electrical switch. In addition, the electric organ discharges of one of the two fish are monitored through recording electrodes. The signals of the second fish are visible as pulses of significantly lower amplitude. These discharges are transmitted into the tank of the first fish as long as the electrical circuit connecting the two tanks is closed. (**B**) Modulation of the discharge

only of the individual in the tank but also of the fish in the neighboring tank, because its discharges are transmitted via the electric circuit into the aquarium. The EODs of this latter fish can easily be identified by their smaller amplitude compared to the signals produced by the first fish.

The result of such an experiment is shown in Figure 2.13B and—in an expanded time plot—in Figure 2.13C. Within a few seconds after transmission of the neighbor's signals into the tank, the fish started to modulate the discharge pattern. First, it reduced its discharges to almost half the pulse repetition rate compared to the signal sequence generated before stimulation. Second, this brief initial phase was followed by a complete cessation of the EOD—a pattern lasting for more than 1 min. The neighboring fish discharged only during part of this phase, and its EOD pulses are clearly distinguished by their lower amplitude. Third, the return to the production of EODs was characterized by an initial generation of a burst of pulses, followed by a slowing down of the discharge rhythm. A similar discharge pattern is frequently observed during aggressive encounters (Fig. 2.14A–E), suggesting that the fish presumably interpreted the signals of the neighbor as those of an intruder trying to enter its territory.

In general, mormyrids can modulate the sequence of their pulse intervals in a variety of ways. Four basic patterns have been described (for review, see Hopkins 1986).

1. Complete cessations of the electric discharges often accompany aggressive encounters.
2. Modulations of the pulse repetition rate, typically occurring in the form of transient accelerations of the discharge rate, appear to play important roles in the context of agonistic behavior and courtship.
3. Long-term shifts in the baseline EOD repetition rate or in the regularity of the discharge pattern have been observed as responses on detection of a conspecific or an intruder, and in behavioral situations of intense aggression. They include a switching of the discharge rhythm from a rather random interval sequence to a train of highly regular intervals.
4. The final pattern of EOD modulations involves a duetting between two fish such that the discharge pattern of one individual influences the pattern of a nearby fish.

FIGURE 2.13. *Continued*
rhythm of *Gnathonemus petersii* by EODs of a conspecific, as demonstrated by using the setup shown in (**A**). A few seconds after the electrical circuit is closed (indicated by *arrow*), and thus stimulating the first fish with the discharges of the second, the first fish slows down its discharge rhythm and, another few seconds later, even completely ceases discharging. During this period of silence, the neighboring fish generates a salvo of discharge pulses, to which the first fish responds with a delay of some 20 s (*). This response of the first fish is distinguished by the production of discharges at an extremely high pulse repetition rate within the first few seconds. Two parts of the recording of (**B**), as defined by the *brackets* below the time scale, are shown at expanded times scales in (**C**) (Courtesy of G.K.H. Zupanc and D. Meissner.)

FIGURE 2.14. The production by *Gnathonemus petersii* of specific EOD patterns in various behavioral situations. At rest, the fish likes to stay in caves or, in this case, in a clay pipe (**A**). In such a situation, a mean pulse repetition rate of approximately 8 Hz, with the pulses produced at rather irregular intervals, is typical. If a second fish enters the territory, both fish stay initially at a short distance from each other (**B**). This aggressive behavior is often accompanied by a cessation of the EOD. Then, the fish frequently adopt a head-to-tail stance alongside each other, with their "chin" appendage (sometimes incorrectly called a "nose") rigidly projecting forward (**C**). If the intruder fails to move off, the territory holder attacks and rams the intruder (**D**). These latter two stages are characterized both by an increase in the EOD firing rate and by a regularization of the discharge pattern, during which an interpulse interval of 8 ms alternates with an interpulse interval of 16 ms. This is often followed by longer discharge salvos with interpulse intervals of 8 msec or 16 ms. Finally, the intruder is beaten by the territory-holder. Now, the defeated fish has turned light brown, and each of the two fish has curled in its chin appendage (**E**). The fleeing fish decreases its mean discharge rate to approximately 8 pulses per second, but in contrast to a fish at rest, the sequence of the interpulse intervals is rather regular. The behavioral significance of the different discharge patterns can be explored further by playing recorded EOD sequences of artificial pulse trains to an isolated fish via an "electric fish dummy" consisting of electric dipole electrodes (**F**). (Courtesy of G.K.H. Zupanc.)

A well-examined pattern of the latter category is the echo response, which is characterized by a short (12 to 14 ms in *Gnathonemus petersii*) latency of pulse generation in response to the EOD pulse of a conspecific. A second response type in this category, termed preferred latency avoidance, involves an avoidance of a discharge within a certain latency following the pulse production of a nearby fish. In *Pollimyrus isidori*, preferred latency avoidance appears to be a typical response of sexually mature females, whereas males tend to produce an echo response (Lücker and Kramer 1981).

The production of specific EOD displayed in different situations of social interaction suggests a role in intraspecific communication. One approach to test this hypothesis has been to play back recorded EOD sequences to isolated individuals via electric dipoles, or to stimulate the fish with artificially generated sequences of interpulse intervals and observe the behavioral response (Fig. 2.14F). Such experiments have, among other results, demonstrated clear differences in the response of *Gnathonemus petersii* to signal trains recorded from resting and attacking fish (Kramer 1979). In contrast to the resting discharge, the attack discharge, distinguished by a higher mean pulse repetition rate and by short sequences of EOD bursts, elicits a larger number of attacks on the dipole electrodes.

9.4.2 Wave-Type Discharges

Similar to the EOD of pulse-type species, the discharges of wave-type species are highly stable, even over hours and days. Long-term changes occur especially during periods of sexual maturation and often result in sexually dimorphic signals. In the two best-examined wave-type species, *Eigenmannia* sp. and *Apteronotus leptorhynchus*, such a dimorphism exists in EOD frequency. In *Eigenmannia* sp., sexually mature females discharge at higher frequencies than males, whereas in *Apteronotus leptorhynchus* the males produce EODs of higher frequencies. Moreover, in *Eigenmannia* sp., females generate a quasi-sinusoidal signals, while the EOD waveform of males is more distorted, with the positive half-wave being markedly shorter than the negative half-wave. As mentioned earlier, conditioning experiments have demonstrated that *Eigenmannia* sp. is capable of distinguishing signals differing in EOD frequency and/or waveform (see Section 9.3). This includes the ability to discriminate male and female EODs, thus suggesting that the fish may identify the sex of a conspecific based on sexually dimorphic characteristics of the signal.

Despite the relative stability of the EOD, short-term alterations occur in the form of transient frequency and/or amplitude modulations and are commonly—but not always—observed in the context of social interactions. In *Apteronotus leptorhynchus*, two forms of such EOD modulations have been described and analyzed in detail (for review, see Zupanc 2002). One form are "chirps," complex frequency and amplitude modulations, lasting between some 10 and a few hundred milliseconds. The term "chirp," coined by Bullock (1969), has been adopted to describe the quality of the modulation, as perceived by the human observer when transforming the electric signal into an audible one. Based on a number of biophysical properties, chirps can be divided into four distinct categories, as shown in Figures 2.15A–D. The second form of EOD modulation is referred to as "gradual frequency rise" (Fig. 2.16E). Signals of this type consist of a rise in EOD frequency, which is less steep than the rises associated with chirp-like modulations. They last between some hundreds of milliseconds and more than 1 min, followed by a gradual decline back to baseline frequency.

In *Apteronotus*, analysis of the behavioral function of chirping is eased by the availability of a "chirp assay". This assay consists of stimulation of an isolated fish with a sinusoidal electric signal in the frequency range of the fish's own EOD (Larimer and MacDonald 1968; Dye 1987; Maler and Ellis 1987; Zupanc and Maler 1993). Such a stimulation regimen, believed to mimic the EOD, and thus the presence, of a neighboring fish, evokes chirps in males of *Apteronotus* reliably and at high rate. By contrast, females do not respond to external electrical stimuli by the production of chirping behavior.

Stimulation of males of *Apteronotus leptorhynchus* with an electric sine wave evokes almost exclusively chirps of type 2 (Engler and Zupanc 2001). Under optimal conditions, the fish can produce more than 50 chirps per second during such external electric stimulation (Zupanc and Maler 1993). By comparison, in the absence of detectable electric signals, typically only one chirp is generated every 10 min during the daytime. The composition of the different chirp types produced under such spontaneous conditions varies among different fish, but is highly robust in a given individual. Most fish kept under nonbreeding conditions generate preferentially type 1 chirps, but there are also a few individuals that produce predominantly type 2 chirps or a significant number of type 3 chirps (Engler et al. 2000).

It has been hypothesized that the "strong" type 1 chirps function as advertisement signals, marking the presence of a fish in its territory, and decreasing the tendency of a neighboring conspecific to challenge the fish through physical attack (Engler and Zupanc 2001). As soon as a neighboring fish is detected, this chirp type is replaced by the production of type 2 chirps, which appear to function as aggressive signals. Using the chirp assay, the stimulation threshold above which the fish starts chirping has been determined. The field intensities of these threshold values vary between 1 μV/cm and 1 mV/cm, depending on the individual (Dunlap et al. 1998; Engler and Zupanc 2001). These values are equivalent to field intensities of the discharges of a 6-cm-long fish measured at distances of approximately 2 to 20 cm, or those of a larger fish measured at somewhat greater distances (Knudsen 1975). This shows that a potential rival has to get relatively close to evoke chirping.

In *Eigenmannia* sp., the function of chirping has been studied particularly in the context of reproduction. While this seasonally breeding species almost never chirps outside the breeding season, the rate of chirping dramatically increases in the course of sexual maturation. During the night of spawning, the male chirps almost incessantly. Further, when isolated gravid females of *Eigenmannia virescens* are stimulated with recorded male chirps, they will spawn in the vi-

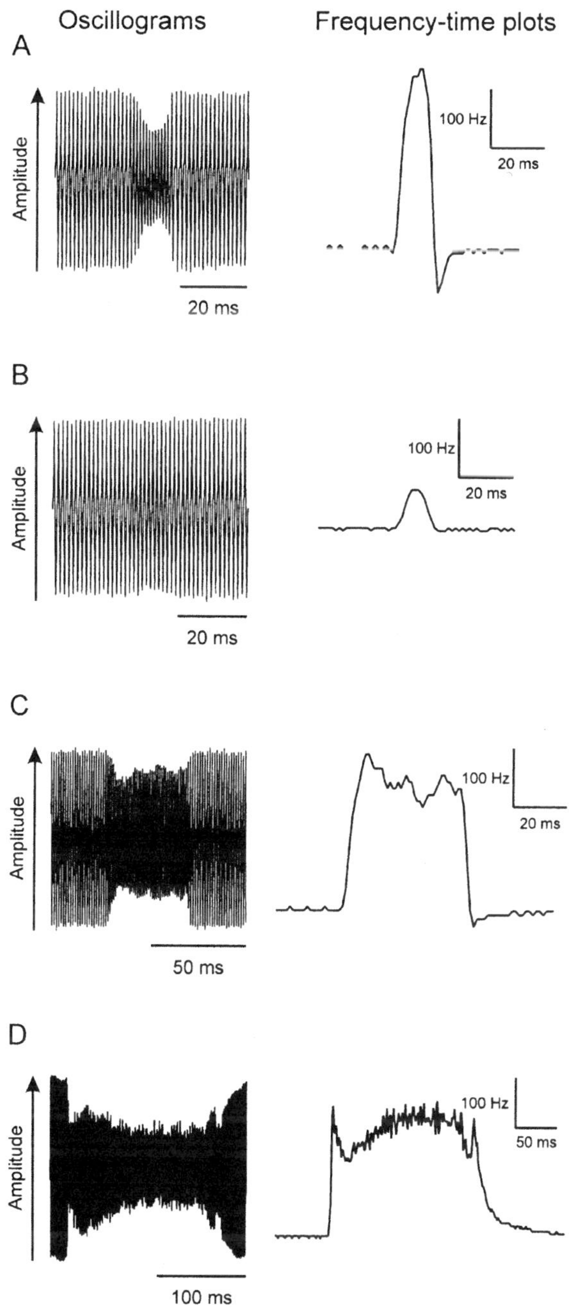

FIGURE 2.15. Chirp types produced by *Apteronotus leptorhynchus*. (**A**) Type 1 chirp. (**B**) Type 2 chirp. (**C**) Type 3 chirp. (**D**) Type 4 chirp. On the left-hand side oscillograms are shown, on the right-hand side frequency–time plots. The four chirp types differ in terms of duration, degree of amplitude decrease, amount of frequency increase, and/or presence/absence of an undershooting of the EOD frequency at the end of the modulatory phase. (Modified after Zupanc 2002.)

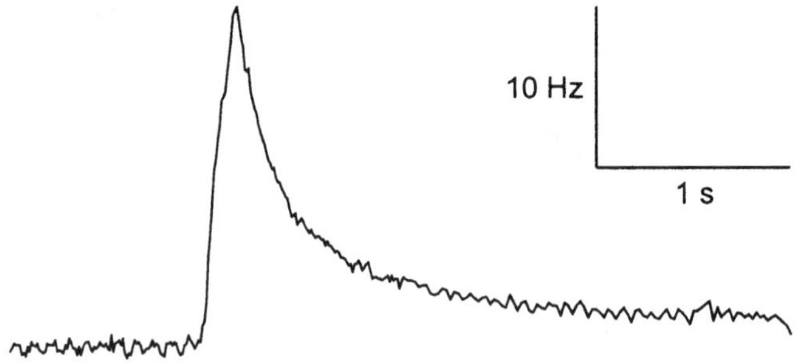

FIGURE 2.16. Frequency–time plot of a gradual frequency rise. Modulations of this type may last for many seconds. They are characterized by a relatively fast increase in discharge frequency, followed by a slow return to baseline frequency. (Modified after Zupanc 2002.)

cinity of the stimulation electrodes (Hagedorn and Heiligenberg 1985). This suggests that chirp-like EOD modulations may function as courtship displays, triggering reproductive behaviors particularly in females. In addition, the incessant production of chirps during the night of spawning may also exert a tonic effect on the motivation of the receiver, including long-lasting alterations in the fish's endocrine status.

10. Mechanisms of Neural Control of Behavior: The Jamming Avoidance Response

The certainly most intensively studied behavior of electric fish is the jamming avoidance response (JAR) of high-frequency wave-type species (for reviews, see Heiligenberg 1991; Metzner 1999). Studies pioneered especially by the laboratory of Walter Heiligenberg have established the JAR as one of the most adequately known examples of behavior among vertebrates in terms of the underlying neuronal types and connectivity (see also Bell and Maler, Chapter 4; Kawasaki, Chapter 7).

The JAR, discovered by Akira Watanabe and Kimihisa Takeda (Watanabe and Takeda 1963), involves a shift of the EOD frequency of a fish "A" away from an interfering signal of a neighboring fish, called "B," that discharges at a similar frequency (Fig. 2.17). In *Eigenmannia* sp.—the genus most widely used for such studies—this response leads to the following two behavioral patterns: If the frequency of fish "B" is higher than the frequency of fish "A," and thus the frequency difference (= frequency of fish "B" minus frequency of fish "A") is positive, then fish "A" lowers its discharge frequency. If, on the other hand, the frequency of fish "B" is lower than the frequency of fish "A," and thus the frequency difference is negative, then fish "A" raises its discharge frequency. Such frequency shifts are evoked as long as the absolute difference between the

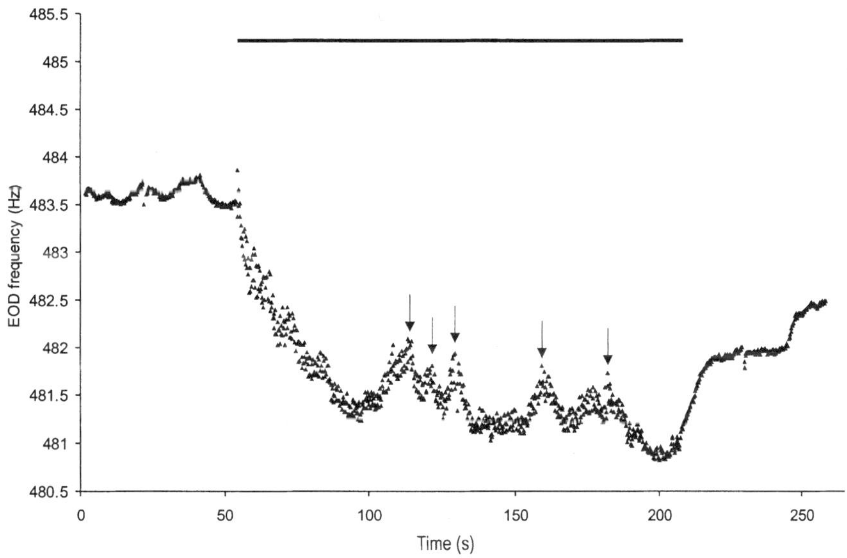

FIGURE 2.17. Jamming avoidance response of the knifefish *Eigenmannia* sp. In the experiment, a fish was initially stimulated with an extrinsic sinusoidal signal 1.6 Hz above its baseline frequency. Immediately after the beginning of the stimulation (indicated by the *black bar*), the fish shifted its EOD frequency away from the stimulus frequency. On reaching an EOD frequency approximately 3.5 Hz below the stimulus frequency, the fish made several attempts to return to baseline frequency by raising its EOD frequency (indicated by *arrows*). However, as soon as the difference between the fish frequency and the stimulus frequency became smaller than approximately 3 Hz, the fish lowered its frequency again. On termination of the stimulation, the fish started to return to its baseline EOD frequency. Each data point represents the mean EOD frequency of 100 instantaneous, cycle-by-cycle, frequency measurements. (Courtesy of G.K.H. Zupanc and D. Meissner.)

two signals is smaller than approximately 20 Hz. Within this range, the most effective frequencies are in the range of a few Hertz (approximately 3 to 4 Hz) above and below the fish's EOD frequency.

The JAR is thought to serve as a behavioral mechanism to maintain "private" frequency bands within a group of fish. Such undisturbed frequency domains appear to be necessary for the fish's electrolocation capabilities to function properly. Indeed, experiments by Walter Heiligenberg in the 1970s showed that electrolocation of objects is impaired if interfering signals are present (Heiligenberg 1973). As a second function, the JAR appears to enhance waveform discrimination, as first suggested by Bernd Kramer (Kramer 1999). To test this hypothesis, in *Eigenmannia* individual fish were trained to discriminate electric signals differing in waveform only. Whenever the frequency of the stimulus signal was identical to the EOD frequency, the fish performed a JAR. On the other hand, whenever the frequency difference between the fish's discharge and the test signal was sufficiently large, no JAR was observed in such conditioning experiments.

Behavioral experiments have shown that *Eigenmannia*'s decision whether to lower or to raise its EOD frequency in response to a jamming signal is based solely on afferent information, specifically the beat pattern caused by the mixing of the fish's EOD and the signal of the neighboring fish. Somewhat surprisingly, the fish does not make use of the frequency of the pacemaker nucleus as an internal reference. Behavioral experiments have also revealed that the fish needs to be exposed to a mixture of its own signal and the interfering signal simultaneously in more than just one patch of its body surface—where the electroreceptors are located—in order to execute a JAR. The information about the correct direction of the frequency shift is extracted from a comparison of the amplitude and phase of the mixed signal in areas of the body surface that are differentially contaminated by the interfering signal. In experiments, this requires that the fish's signal and the interfering signal are presented through a so-called separate geometry arrangement of the stimulus electrodes. This can be shown by silencing the fish's electric organ with curare and substituting the EOD by a sinusoidal signal delivered through electrodes placed into the mouth and at the tail. Then, a JAR can be elicited as long as the neighbor's EOD mimic is applied through a geometrically separate pair of electrodes. If, on the other hand, under identical geometry conditions the two signals are first added electronically, and the mixed signal is then presented through a pair of electrodes in the fish's mouth and at the tail, the fish fails to produce a JAR.

Detailed analysis has shown that the mixing of the fish's EOD and the discharge of a neighboring fish causes modulation of both amplitude and phase of the mixed signal. The pattern of this amplitude and phase modulation contains information about the sign of the frequency difference between the signal of the fish and that of the neighbor, and thus determines whether the fish raises or lowers its EOD frequency. However, since no pure representation of the fish's signal is available anywhere on its body surface (typically, in any part of the body the fish's EOD is contaminated by the interfering signal), the fish has to sample amplitude and phase information from many points of the body surface and perform pairwise comparisons to obtain information about the sign of the frequency difference. Thus, using a distributed system of pairwise amplitude/phase comparisons, the fish can determine whether it has to raise or to lower its EOD frequency in order to shift away from the frequency of the interfering signal in the correct direction.

The electrosensory information relevant for the execution of the JAR is largely, and possibly exclusively, carried by the two subtypes of the tuberous receptors, namely the P-type and the T-type receptors. P-type receptors encode the amplitude, whereas T-type receptors encode the phase of the electric signal. These two types of information are relayed via separate pathways to the electrosensory lateral line lobe. In this hindbrain structure, amplitude and phase information are processed separately—referred to as parallel processing.

The next station along the pathway mediating the processing of electrosensory information is the torus semicircularis. A distinct feature of this midbrain structure is the convergence of amplitude and phase information. Based on this convergence, the sign of the frequency difference between the fish's EOD and the interfering signal is computed, although the encoding of this message is still ambiguous in the sense that the sign of the frequency difference depends on the position of the neighboring fish relative to the fish's own position.

Unambiguous encoding of the sign of this frequency difference is achieved in the final station of the electrosensory processing pathway, the nucleus electrosensorius in the diencephalon. A dorsally located part of this nucleus contains neurons that respond specifically in jamming situations leading to an increase in EOD frequency. This neuronal cluster projects to the dorsomedial part of the central posterior/prepacemaker nucleus in the dorsal thalamus. Axons originating from the latter structure terminate on pacemaker cells within the pacemaker nucleus. A second, more ventrally situated part of the nucleus electrosensorius contains neurons that respond such that the fish lowers its EOD during the JAR. Neurons of this portion send axons to the sublemniscal prepacemaker nucleus in the mesencephalon, which, in turn, innervates relay cells within the pacemaker nucleus.

Supporting experimental evidence for the differential roles of the central posterior/prepacemaker nucleus and the sublemniscal prepacemaker nucleus has been obtained through lesioning studies: Lesions of the dorsomedial portion of the central posterior/prepacemaker nucleus selectively abolish increases of the EOD frequency, whereas lesions of the sublemniscal prepacemaker nucleus result in abolition of decreases of the EOD frequency. Since lesioning of the latter structure also reduces the resting frequency of the EOD, it is thought that neurons in the sublemniscal prepacemaker nucleus are tonically active. As input to the sublemniscal prepacemaker nucleus is mediated by the inhibitory transmitter γ-aminobutyric acid (GABA), activation of the ventral portion of the nucleus electrosensorius leads to a reduced level of activity of the sublemniscal prepacemaker nucleus, which in turn results in a lower level of excitatory drive to relay cells and thus to a decrease of the EOD frequency. On the other hand, stimulation of the dorsal portion of the nucleus electrosensorius causes an increase in excitation of both the dorsomedial part of the central posterior/prepacemaker nucleus and the pacemaker cells in the pacemaker nucleus. As a result, the EOD frequency increases.

11. Beyond Neuroethology: Electric Fish, the Study of Neuronal Plasticity, and the Molecular Characterization of Cholinergic Transmission

Progress in experimental biology is crucially dependent on the selection of favorable model systems. Electric fish have proven to be just such an excellent choice not only in ethology and neuroethology, but also in two other areas of biology: the study of neuronal plasticity and the molecular analysis of cholinergic transmission.

11.1 Neuronal Plasticity

The study of neuronal plasticity, especially in relation to behavioral plasticity, has played a central role in many neuroethological investigations using weakly electric fish. A major portion of this research has focused on the development of sex differences in the EODs through the action of steroid hormones (see Section 8). Other investigations have particularly addressed how seasonal alterations in electric behavior are accommodated by structural reorganization and biochemical modification of the underlying neural networks (for review, see Zupanc and Maler 1997).

In addition to such neuroethological work, weakly electric fish have also proven to be excellent model systems to study cellular mechanisms governing the generation of new neurons in the adult central nervous system and the regeneration of central and peripheral structures. The superiority of weakly electric fish over other fishes for such research is based on the enormous body of information available on the anatomy and physiology of central structures. Originally, this wealth of data was established to understand how electric discharges are controlled and how electrosensory information is processed at central levels. Based on these results, the first comprehensive neuroanatomical atlas of the brain of a fish species, *Apteronotus leptorhynchus*, was created (Maler et al. 1991). This atlas, together with the favorable size of the brain, enabled investigators for the first time to quantitatively map proliferation zones in the entire adult brain of a vertebrate species (Zupanc and Horschke 1995). The detailed knowledge about brain structures has also formed the basis to follow the fate of the newborn cells, and to place the results in a meaningful functional context. Furthermore, such studies have led to the identification of a number of key cellular mechanisms that provide fish with their unsurpassed potential to regenerate brain tissue after injuries (for reviews, see Zupanc 1999, 2001; Zupanc and Clint 2003).

The second system of weakly electric fish extensively exploited for regeneration studies has been the tail of gymnotiforms (for reviews, see Waxman and Anderson 1986; Zakon and Unguez 1999). These studies have capitalized on the fish's enormous ability to regenerate the spinal cord and the electric organ after amputation of the tail. This regenerative capability encompasses not only the regrowth of lesioned axons but also the generation and differentiation of whole neurons, and the formation of new electrocytes. This latter ability, observed in both central and peripheral structures, distinguishes fish fundamentally from mammalian species. Identification of the factors that cause this difference between fish and mammals is not only of immense interest to basic researchers, but also bears a tremendous potential to develop novel therapeutic approaches aimed at a replacement of neurons lost to injury or degenerative disease by newly generated ones.

11.2 Cholinergic Synapse

Research in the area of synaptic transmission employing electric fish was initiated after the pharmacologist Wilhelm Feldberg, the electrophysiologist Alfred Fessard, and the biochemist David Nachmansohn demonstrated in 1939 at the Station Biologique d'Arcachon in France that the transmitter released by the electromotor fibers innervating the electric organ of the electric ray, *Torpedo marmorata*, is the same as the one mediating transmission at the neuromuscular junction—acetylcholine (Feldberg et al. 1939/40; Feldberg and Fessard 1942). This turned out to be a landmark discovery, since the myogenic electric organ offers the enormous advantage to be several hundred times richer in cholinergic synapses than muscle.

Capitalizing on the electric organ of strongly electric fish as a particularly rich source of cholinergic synapses, Nachmansohn—who, after his emigration to the United States in 1939, switched to the electric eel as a model system—succeeded in the isolation and molecular characterization of the enzyme that hydrolyzes acetylcholine, acetylcholinesterase (for a historical account, see Whittaker 1992). Subsequently, in the 1970s and 1980s, the electric organ of *Torpedo marmorata* and *Torpedo californica* provided the basis for the isolation, purification, and complete sequencing of the acetylcholine receptor by several groups (for a historical account, see Changeaux 1993). This was a milestone in the history of neurobiology, since the acetylcholine receptor was the first neurotransmitter receptor to be isolated, thus paving the path for the molecular characterization of the structurally similar acetylcholine receptors in other species, including humans, and of other, related neurotransmitter receptors, such as the receptors for GABA and glycine.

12. Perspectives

Throughout the history of neuroethology, research on electrogenesis and electroreception has set new standards for the exploration of neural mechanisms of natural behavior. However, the significance of electroreceptive animals in general, and electric fish in particular, goes far beyond that of opening new vistas just in neuroethology. Electric fish played a crucial role in the discovery of electric phenomena in animals during the infancy of physiology. In the last half a century, the success in the molecular characterization of the processes occurring during transmission at the chemical synapse would not have been possible without the use of electric fish. It does not take much to predict that research based on electric systems will continue to have a major impact on the development of neurobiology in general, and neuroethology in particular, also in the future.

Acknowledgments. We thank Daniela Meissner for her assistance in the preparation of the figures, and Marianne M. Zupanc for critically reading through the manuscript.

References

Alves-Gomes J, Hopkins CD (1997) Molecular insights into the phylogeny of mormyriform fishes and the evolution of their electric organs. Brain Behav Evol 49:324–350.

Baron VD, Pavlov DS (2003) Discovery of specialized electrogenerating activity in two species of *Polypterus* (Polypteriformes, Osteichthyes). J Ichthyol 43:S259–S261.

Baron VD, Morshnev KS, Olshansky VM, Orlov AA (1994a) Electric organ discharges of two species of African catfish (*Synodontis*) during social behaviour. Anim Behav 48:1472–1475.

Baron VD, Orlov AA, Golubtsov AS (1994b) African *Clarias* catfish elicits long-lasting weak electric pulses. Experientia 50:644–647.

Baron VD, Orlov AA, Golubtsov AS (1996) Detection of electric organ discharges in African catfish *Auchenoglandis occidentalis* (Siluriformes: Bagridae). Dokl Biol Sci 349:377–379.

Bass AH (1986) Electric organs revisited: evolution of a vertebrate communication and orientation organ. In: Bullock TH, Heiligenberg W (eds), Electroreception. New York: John Wiley & Sons, pp. 13–70.

Bass AH, Hopkins CD (1983) Hormonal control of sexual differentiation: changes in electric organ discharge waveform. Science 220:971–974.

Bennett MVL (1971) Electric organs. In: Hoar WS, Randall DJ (eds), Fish Physiology, Vol. 5: Sensory Systems and Electric Organs. New York: Academic Press, pp. 347–491.

Bensouilah M, Schugardt C, Roesler R, Kirschbaum F, Denizot J-P (2002) Larval electroreceptors in the epidermis of mormyrid fish: I. Tuberous organs of type A and B. J Comp Neurol 447:309–322.

Bodznick D, Boord RL (1986) Electroreception in chondrichthyes: central anatomy and physiology. In: Bullock TH, Heiligenberg W (eds), Electroreception. New York: John Wiley & Sons, pp. 225–256.

Boudinot M (1970) The effect of decreasing and increasing temperature on the frequency of the electric organ discharge in *Eigenmannia* sp. Comp Biochem Physiol 37:601–603.

Bullock TH (1969) Species differences in effect of electroreceptor input on electric organ pacemakers and other aspects of behavior in electric fish. Brain Behav Evol 2:85–118.

Bullock TH (1982) Electroreception. Annu Rev Neurosci 5:121–170.

Bullock TH, Heiligenberg W (eds) (1986) Electroreception. New York: John Wiley & Sons.

Bullock TH, Hagiwara S, Kusano K, Negishi K (1961) Evidence for a category of electroreceptors in the lateral line of gymnotid fishes. Science 134:1426–1427.

Bullock TH, Behrend K, Heiligenberg W (1975) Comparison of the jamming avoidance responses in gymnotoid and gymnarchid electric fish: a case of convergent evolution of behavior and its sensory basis. J Comp Physiol A 103:97–121.

Changeux J-P (1993) Chemical signaling in the brain. Sci Am November: 58–62.

Coates CW, Altamirano M, Grundfest H (1954) Activity in electrogenic organs of knifefishes. Science 120:845–846.

Denizot J-P, Kirschbaum F, Westby GWM, Tsuji S (1978) The larval electric organ of the weakly electric fish *Pollimyrus (Marcusenius) isidori* (Mormyridae, Teleostei). J Neurocytol 7:165–181.

Denizot J-P, Kirschbaum F, Westby GWM, Tsuji S (1982) On the development of the adult electric organ in the mormyrid fish *Pollimyrus isidori* (with special focus on the innervation). J Neurocytol 11:913–934.

Dunlap KD, Thomas P, Zakon HH (1998) Diversity of sexual dimorphism in electrocommunication signals and its androgen regulation in a genus of electric fish, *Apteronotus*. J Comp Physiol A 183:77–86.

Dye J (1987) Dynamics and stimulus-dependence of pacemaker control during behavioral modulations in the weakly electric fish, *Apteronotus*. J Comp Physiol A 161: 175–185.

Dye JC, Meyer JH (1986) Central control of the electric organ discharge in weakly electric fish. In: Bullock TH, Heiligenberg W (eds), Electroreception. New York: John Wiley & Sons, pp. 71–102.

Enger PS, Szabo T (1968) Effect of temperature on the discharge rates of the electric organ of some gymnotids. Comp Biochem Physiol 27:625–627.

Engler G, Zupanc GKH (2001) Differential production of chirping behavior evoked by electrical stimulation of the weakly electric fish, *Apteronotus leptorhynchus*. J Comp Physiol A 187:747–756.

Engler G, Fogarty CM, Banks JR, Zupanc GKH (2000) Spontaneous modulations of the electric organ discharge in the weakly electric fish, *Apteronotus leptorhynchus*: a biophysical and behavioral analysis. J Comp Physiol A 186:645–660.

Feldberg W, Fessard A (1942) The cholinergic nature of the nerves to the electric organ of the torpedo (*Torpedo marmorata*). J Physiol 101:200–216.

Feldberg W, Fessard A, Nachmansohn D (1939/40) The cholinergic nature of the nervous supply to the electrical organ of the torpedo (*Torpedo marmorata*). J Physiol 97:3P–5P.

Feng AS (1976) The effect of temperature on a social behavior of weakly electric fish *Eigenmannia virescens*. Comp Biochem Physiol A 55:99–102.

Ferrari MB, Zakon HH (1989) The medullary pacemaker nucleus is unnecessary for electroreceptor tuning plasticity in *Sternopygus*. J Neurosci 9:1354–1361.

Fessard A, Szabo T (1961) Mise en évidence d'un récepteur sensible à l'électricité dans la peau des Mormyres. CR Acad Sci (Paris) 253:1859–1860.

Fritzsch B, Münz H (1986) Electroreception in amphibians. In: Bullock TH, Heiligenberg W (eds), Electroreception. New York: John Wiley & Sons, pp. 483–496.

Hagedorn M, Carr C (1985) Single electrocytes produce a sexually dimorphic signal in South American electric fish, *Hypopomus occidentalis* (Gymnotiformes, Hypopomidae). J Comp Physiol A 156:511–523.

Hagedorn M, Heiligenberg W (1985) Court and spark: electric signals in the courtship and mating of gymnotoid fish. Anim Behav 33:254–265.

Hagedorn M, Womble M, Finger TE (1990) Synodontid catfish: a new group of weakly electric fish. Behavior and anatomy. Brain Behav Evol 35:268–277.

Heiligenberg W (1973) Electrolocation of objects in the electric fish *Eigenmannia* (Rhamphichthyidae, Gymnotoidei). J Comp Physiol 87:137–164.

Heiligenberg W (1991) Neural Nets in Electric Fish. Cambridge, MA: MIT Press.

Heiligenberg W, Keller CH, Metzner W, Kawasaki M (1991) Structure and function of neurons in the complex of the nucleus electrosensorius of the gymnotiform fish *Eigenmannia*: detection and processing of electric signals in social communication. J Comp Physiol A 169:151–164.

Hopkins CD (1972) Sex differences in electric signaling in an electric fish. Science 176: 1035–1037.

Hopkins CD (1976) Stimulus filtering and electroreception: tuberous electroreceptors in three species of gymnotoid fish. J Comp Physiol A 111:171–207.

Hopkins CD (1980) Evolution of electric communication channels of mormyrids. Behav Ecol Sociobiol 7:1–13.

Hopkins CD (1981) On the diversity of electric signals in a community of mormyrid electric fish in West Africa. Am Zool 21:211–222.

Hopkins CD (1986) Behavior of Mormyridae. In: Bullock TH, Heiligenberg W (eds), Electroreception. New York: John Wiley & Sons, pp. 527–576.

Hopkins CD, Bass AH (1981) Temporal coding of species recognition signals in an electric fish. Science 212:85–87.

Hopkins CD, Heiligenberg WF (1978) Evolutionary designs for electric signals and electoreceptors in gymnotoid fishes of Surinam. Behav Ecol Sociobiol 3:113–134.

Jørgensen JM (1982) Fine structure of the ampullary organs of the bichir *Polypterus senegalus* Cuvier, 1829 (Pisces: Brachiopterygii) with some notes on the phylogenetic development of electroreceptors. Acta Zool (Stockholm) 63:211–217.

Kalmijn AJ (1974) The detection of electric fields from inanimate and animate sources other than electric organs. In: Fessard A (ed), Handbook of Sensory Physiology, Vol. III/3: Electroreceptors and Other Specialized Receptors in Lower Vertebrates. Berlin: Springer-Verlag, pp. 147–200.

Kalmijn AJ (1982) Electric and magnetic field detection in elsmobranch fishes. Science 218:916–918.

Keller CH, Zakon HH, Sanchez DY (1986) Evidence for a direct effect of androgens upon electroreceptor tuning. J Comp Physiol A 158:301–310.

Kirschbaum F (1975) Environmental factors control the periodical reproduction of tropical electric fish. Experientia 31:1159–1160.

Kirschbaum F (1979) Reproduction of the weakly electric fish *Eigenmannia virescens* (Rhamphichtyidae, Teleostei) in captivity: I. Control of gonadal recrudescence and regression by environmental factors. Behav Ecol Sociobiol 4:331–355.

Kleerekoper H, Sibakin K (1956) An investigation of the electrical "spike" potentials produced by the sea lamprey (*Petromyzon marinus*) in the water surrounding the head region. I. J Fish Res Board Can 13:375–383.

Kleerekoper H, Sibakin K (1957) An investigation of the electrical "spike" potentials produced by the sea lamprey (*Petromyzon marinus*) in the water surrounding the head region. II. J Fish Res Board Can 14:145–151.

Knudsen EI (1975) Spatial aspects of the electric fields generated by weakly electric fish. J Comp Physiol 99:103–118.

Kramer B (1979) Electric and motor responses of the weakly electric fish, *Gnathonemus petersii* (Mormyridae), to play-back of social signals. Behav Ecol Sociobiol 6:67–79.

Kramer B (1999) Waveform discrimination, phase sensitivity and jamming avoidance in a wave-type electric fish. J Exp Biol 202:1387–1398.

Kramer B, Kaunzinger I (1991) Electrosensory frequency and intensity discrimination in the wave-type electric fish *Eigenmannia*. J Exp Biol 161:43–59.

Kramer B, Otto B (1991) Waveform discrimination in the electric fish *Eigenmannia*: sensitivity for the phase differences between the spectral components of a stimulus wave. J. Exp. Biol. 159:1–22.

Kramer B, Zupanc GKH (1986) Conditioned discrimination of electric waves differing only in form and harmonic content in the electric fish, *Eigenmannia*. Naturwissenschaften 73:679–680.

Kramer B, Kirschbaum F, Markl H (1981) Species specificity of electric organ discharges in a sympatric group of gymnotoid fish from Manaus (Amazonas). In: Szabó T, Czéh G (eds), Advances in Physiological Science, Vol. 31. Sensory Physiology of Aquatic Lower Vertebrates. Budapest: Pergamon Press/Akadémiai Kiadó, pp. 195–219.

Larimer JL, MacDonald JA (1968) Sensory feedback from electroreceptors to electromotor pacemaker centers in gymnotids. Am J Physiol 214:1253–1261.

Lissmann HW (1951) Continuous electrical signals from the tail of a fish, *Gymnarchus niloticus* Cuv. Nature 167:201–202.

Lissmann HW (1958) On the function and evolution of electric organs in fish. J Exp Biol 35:156–191.

Lissmann HW, Machin KE (1958) The mechanism of object location in *Gymnarchus niloticus* and similar fish. J Exp Biol 35:451–486.

Lücker H, Kramer B (1981) Development of a sex difference in the preferred latency response in the weakly electric fish, *Pollimyrus isidori* (Cuvier et Valenciennes) (Mormyridae, Teleostei). Behav Ecol Sociobiol 9:103–109.

Maler L, Ellis WG (1987) Inter-male aggressive signals in weakly electric fish are modulated by monoamines. Behav Brain Res 25:75–81.

Maler L, Sas E, Johnston S, Ellis W (1991) An atlas of the brain of the electric fish *Apteronotus leptorhynchus*. J Chem Neuroanat 4:1–38.

Metzner W (1999) Neural circuitry for communication and jamming avoidance in gymnotiform electric fish. J Exp Biol 202:1365–1375.

Metzner W, Heiligenberg W (1991) The coding of signals in the electric communication of the gymnotiform fish *Eigenmannia*: from electroreceptors to neurons in the torus semicircularis of the midbrain. J Comp Physiol A 169:135–150.

Meyer DL, Heiligenberg W, Bullock TH (1976) The ventral substrate response: a new postural control mechanism in fishes. J Comp Physiol. A 109:59–68.

Meyer JH (1983) Steroid influences upon the discharge frequencies of a weakly electric fish. J Comp Physiol A 153:29–37.

Meyer JH, Zakon HH (1982) Androgens alter the tuning of electroreceptors. Science 217:635–637.

Moller P (1995) Electric Fishes: History and Behavior. London: Chapman & Hall.

Murray RW (1960) Electrical sensitivity of the ampullae of Lorenzini. Nature 187:957.

Northcutt RG (1986) Electroreception in nonteleost bony fishes. In: Bullock TH, Heiligenberg W (eds), Electroreception. New York: John Wiley & Sons, pp. 257–285.

Piccolino M, Bresadola M (2002) Drawing a spark from darkness: John Walsh and electric fish. Trends Neurosci 25:51–57.

Roth A (1973) Electroreceptors in Brachiopterygii and Dipnoi. Naturwissenschaften 60: 106.

Scheich H, Langner G, Tidemann C, Coles RB, Guppy A (1986) Electroreception and electrolocation in platypus. Nature 319:401–402.

Sullivan JP, Lavoué S, Hopkins CD (2000) Molecular systematics of the African electric fishes (Mormyroidea: Teleostei) and a model for the evolution of their electric organs. J Exp Biol 203:665–683.

Sullivan JP, Lavoué S, Hopkins CD (2002) Discovery and phylogenetic analysis of a riverine species flock of African electric fishes (Mormyridae: Teleostei). Evolution 56:597–616.

Szabo T (1970) Elektrische Organe und Elektrorezeption bei Fischen. In: Rheinisch-Westfälische Akademie der Wissenschaften, Vorträge, N 205. Opladen: Westdeutscher Verlag, pp. 7–40.

von der Emde G (1990) Discrimination of objects through electrolocation in the weakly electric fish, *Gnathonemus petersii*. J Comp Physiol A 167:413–421.

von der Emde G (1999) Active electrolocation of objects in weakly electric fish. J Exp Biol 202:1205–1215.

von der Emde G, Ringer T (1992) Electrolocation of capacitive objects in four species of pulse-type weakly electric fish: I. Discrimination performance. Ethology 91:326–338.

von der Emde G, Schwarz S, Gomez L, Budelli R, Grant K (1998) Electric fish measure distance in the dark. Nature 395:890–894.

Watanabe A, Takeda K (1963) The change of discharge frequency by a.c. stimulus in a weak electric fish. J Exp Biol 40:57–66.

Waxman SG, Anderson MJ (1986) Regeneration of central nervous system structures: *Apteronotus* spinal cord as a model system. In: Bullock TH, Heiligenberg W (eds), Electroreception. New York: John Wiley & Sons, pp. 183–208.

Whittaker VP (1992) The Cholinergic Neuron and Its Target: The Electromotor Innervation of the Electric Ray *Torpedo* as a Model. Boston: Birkhäuser.

Zakon HH (1986) The electroreceptive periphery. In: Bullock TH, Heiligenberg W (eds), Electroreception. New York: John Wiley & Sons, pp. 103–156.

Zakon HH, Meyer JH (1983) Plasticity of electroreceptor tuning in the weakly electric fish, *Sternopygus dariensis*. J Comp Physiol A 153:477–487.

Zakon HH, Unguez GA (1999) Development and regeneration of the electric organ. J Exp Biol 202:1427–1434.

Zupanc GKH (1999) Neurogenesis, cell death and regeneration in the adult gymnotiform brain. J Exp Biol 202:1435–1446.

Zupanc GKH (2001) Adult neurogenesis and neuronal regeneration in the central nervous system of teleost fish. Brain Behav Evol 58:250–275.

Zupanc GKH (2002) From oscillators to modulators: behavioral and neural control of modulations of the electric organ discharge in the gymnotiform fish, *Apteronotus leptorhynchus*. J Physiol (Paris) 96:459–472.

Zupanc GKH, Clint SC (2003) Potential role of radial glia in adult neurogenesis of teleost fish. Glia 43:77–86.

Zupanc GKH, Horschke I (1995) Proliferation zones in the brain of adult gymnotiform fish: a quantitative mapping study. J Comp Neurol 353:213–233.

Zupanc GKH, Maler L (1993) Evoked chirping in the weakly electric fish *Apteronotus leptorhynchus*: a quantitative biophysical analysis. Can J Zool 71:2301–2310.

Zupanc GKH, Maler L (1997) Neuronal control of behavioral plasticity: the prepacemaker nucleus of weakly electric gymnotiform fish. J Comp Physiol A 180:99–111.

Zupanc GKH, Banks JR, Engler G, Beason RC (2003) Temperature dependence of the electric organ discharge in weakly electric fish. In: Ploger BJ, Yasukawa K (eds), Exploring Animal Behavior in Laboratory and Field: An Hypothesis-Testing Approach to the Development, Causation, Function, and Evolution of Animal Behavior. Amsterdam: Academic Press, pp. 85–94.

3

Morphology of Electroreceptive Sensory Organs

JØRGEN MØRUP JØRGENSEN

1. Introduction

It is now clear that special sensory organs that are sensitive to weak electric fields arose very early in the evolution of vertebrates. With the exception of hagfishes, all other anamniotic craniate classes, including lampreys, cartilaginous fishes, and a number of bony fishes as well as urodele and caecilian amphibians possess electroreceptors. Electroreceptive sense organs have also been identified among mammals.

We know that reception of weak electric fields plays an important role for many aquatic vertebrates. Such fields are used during feeding activity including prey detection and registration of key elements in the environment, as well as in social communication (Bullock 1982; Bullock and Heiligenberg 1986). It is therefore enigmatic why many fishes apparently have lost this valuable sense during evolution.

The structure, or morphology, of electroreceptive sense organs provides a necessary foundation for understanding their function. This chapter briefly describes the different types of electroreceptive organs found in various vertebrates.

2. Superficial Electroreceptive Organs in Lampreys

Adult lampreys have epidermal sensory organs called end buds on the head and trunk. These end buds contain sensory cells that are directly exposed to the surrounding water and have no duct. Each sensory organ consists of 3 to 30 slender sensory cells that are separated from one another by supporting cells (Fig. 3.1). The apical surface of each sensory cell is equipped with a hair bundle consisting of 80 to 90 microvilli. Presynaptic synaptic bodies lie opposite to afferent nerve boutons (Whitear and Lane 1981).

The ammocete larvae of lampreys and newly metamorphosized lampreys do not possess end buds, although they appear to have a sense for weak electric

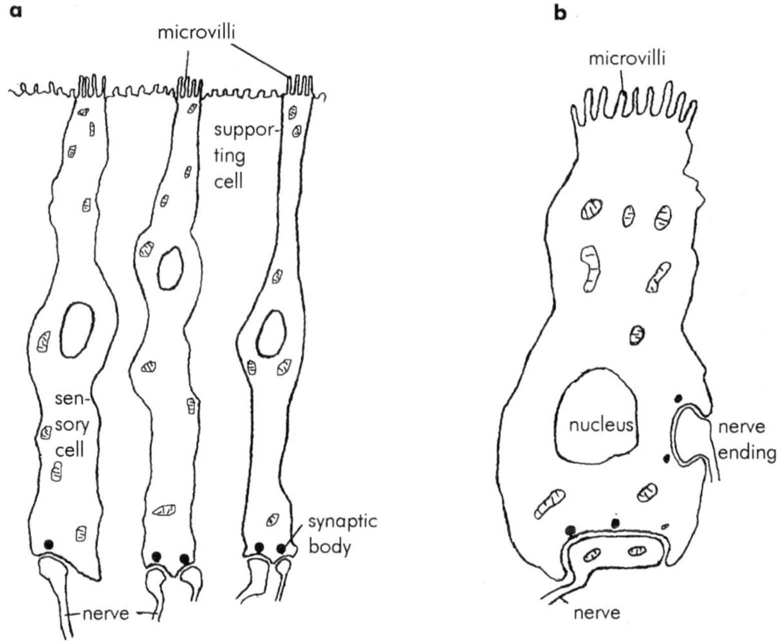

FIGURE 3.1. Outlines of (**a**) sensory cells in the lamprey end bud and (**b**) multivillous cells from the lamprey ammocete larva. Not drawn to scale.

fields (Ronan and Bodznick 1986). Instead of end buds, they have a number of scattered multivillous sensory cells (Whitear and Lane 1983), which have many of the characteristics of the sensory cells belonging to the octavolateralis system (Fig. 3.1). Such multivillous cells are found in small groups in the gill vent papillae and on the tail (Steven 1951).

Solid indication of an electroreceptive sensory system in adult lampreys is the presence of a dorsal octavolateralis nucleus in the medulla oblongata and direct recordings from single fibers (Bodznick and Northcutt 1981; Bodznick and Preston 1983), although there are no behavioral data as yet to support the physiology and anatomy. Ammocetes probably use their multivillous cells for electroreception, but definitive proof has not yet been obtained.

3. Ampullary Organs

3.1 General Structure

Electroreceptive ampullary organs are characterized by a tubelike structure, with a pore in the skin, and the sensory epithelium lying at the blind end of the tube.

The canal leading from the sensory epithelium to the environment varies considerably in length. It is filled with a mucous jelly with a very low electrical resistance. The wall of the ampullary duct is composed of flattened cells that are connected to each other by tight junctions and desmosomes, creating a layer with high electrical resistance (Waltman 1966; Fields et al. 1993; see Bodznick and Montgomery, Chapter 6).

3.2 Non-Teleost Ampullae: The Ampullae of Lorenzini

The ampullae of Lorenzini are defined here as ampullary sense organs that project to a dorsal octavolateral nucleus in the medulla oblongata (Northcutt 1986) and that are excited by cathodal stimuli (see Bodznick and Montgomery, Chapter 6). With this definition, the organs of Lorenzini include the electroreceptive organs in non-teleostean fishes and the ampullary organs in amphibians.

The morphology of the sensory cells in the ampullae of Lorenzini is characterized by a narrow apical area exposed to the ampullary lumen. Typically, a short cilium extends from a basal body. In many ampullae the cilium is surrounded by a few microvilli. Basally, synaptic sheets form contact with a single or a few afferent nerve endings.

3.2.1 Ampullary Organs in Cartilaginous Fishes (Chondrichthyes)

Cartilaginous fishes include sharks, skates, and rays, as well as chimeras. The ampullae in the marine species have generally long canals that lead to sensory ampullae situated deep in the dermis. However, the canal length as well as shape may vary considerably within the same individual (Andres and von Düring 1988). The pores are found in the head region of all species and in the pectoral fins of rays and skates. The sensory epithelium covers the interior of the ampullary end, which often forms several blind ending sacs called diverticulae.

The sensory ampullae are located in clusters, and the total number of ampullary organs varies considerably in different species (Bodznick and Boord 1986). Some species may have only 148 (*Heterodontus francisci*) or 162 ampullae (*Torpedo marmorata*), whereas other taxa have more than 2000. The maximum number has been found in a hammerhead shark (*Sphyrna lewini*) according to the compilation of Bodznick and Boord (1986).

The sensory ampullae have different shapes, as shown in Figure 3.2. A new classification of the different types, partly following Dotterweich (1932) and Andres and von Düring (1988), could be as follows. The ampulla may be *simple* with one enlarged chamber and, in some cases, a few diverticulae. This type is found in electric rays belonging to the genus *Torpedo*. In the shark *Hexanchus* a number of simple ampullae form a group, which together may be termed *assembled ampullae*.

More widespread is a *lobular* type, with 2 to 80 diverticulae arranged side by side, all of which open into a central ampulla. This type is found in most

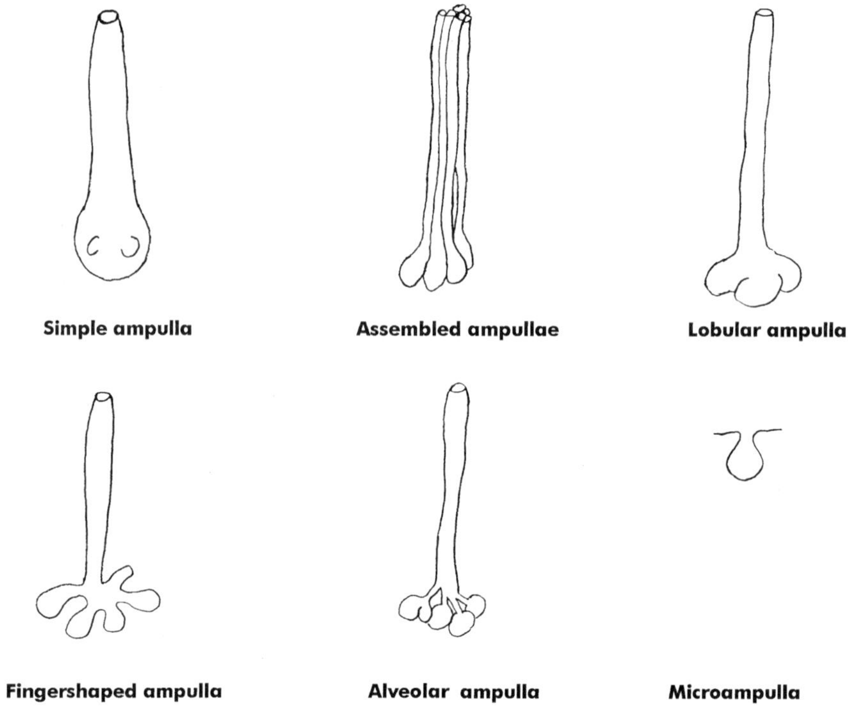

FIGURE 3.2. Different types of ampullae of Lorenzini known from various cartilaginous fishes. Not drawn to scale.

sharks, rays, and skates. A section of such an ampulla shows the chambers (Fig. 3.3). In chimeras the chambers may be slender pockets forming a *finger-shaped ampulla.*

Some sharks, such as the piked dogfish (*Squalus acanthias*), have chambers that are connected to the central canal by ducts; this type may be termed *alveolar*. In permanently freshwater elasmobranchs, such as *Potamotrygon*, short ducts end in simple *microampullae*.

In a detailed study of 40 species of rays belonging to the genus *Raja,* Raschi (1986) found that the majority of ampullary organs are located on the ventral side of the animal. Raschi also found that in fish-eating species the ampullary pores are distributed over a large part of the body compared with those species that feed primarily on invertebrates. With regard to the size of each ampulla and the number of diverticulae, Raschi (1986) found that deep-dwelling species have larger ampullae with more chambers than shallow-water species.

3.2.2 Ampullary Organs in Bichirs and Reedfish (Polypteriniformes)

Extant fishes belonging to this superorder are a small group of African fishes of the genera *Polypterus* and *Calamoichthys*. The ampullary organs in these

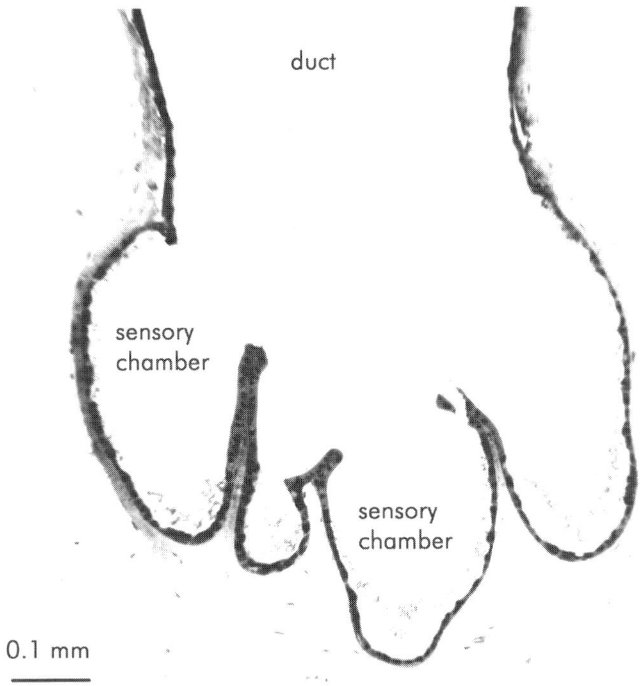

FIGURE 3.3. Micrograph of a section from an ampulla from the spotted dogfish (*Scyliorhinus canicula*).

species are distributed on the head, with the highest density on the snout region in front of the eyes (Northcutt 1986). The ampullary organs occur as simple tubes in the skin, except where an ampullary organ seems to divide, forming two organs. The sensory epithelium consists of electroreceptor cells separated by supporting cells. The apical part of the sensory cell is characterized by having a single cilium surrounded by 8 to 10 microvilli. Basally in the receptor cell, synaptic sheets or rounded synaptic bodies are seen opposite to the afferent nerve endings (Roth and Tscharntke 1976; Jørgensen 1982).

3.2.3 Ampullary Organs in Sturgeons and Paddlefishes (Acipenseriformes)

The ampullary organs of sturgeons have been examined in the genera *Scaphirhynchus* and *Acipenser* by Weisel (1978), Teeter et al. (1980) and Jørgensen (1980). Similar organs are found on the rostrum of the paddlefish *Polyodon spathula* (Jørgensen et al. 1972). In all species examined, the ampullary organs occur on the head, including on the rostrum, and in some, on the opercula. The distribution in *Scaphirynchus* was mapped by Northcutt (1986). The ampullary organs form groups of 5 to 30 single organs (Figs. 3.4 and 3.5). Occasionally, two to five ampullae may share a common pore. The pear-shaped sensory cells

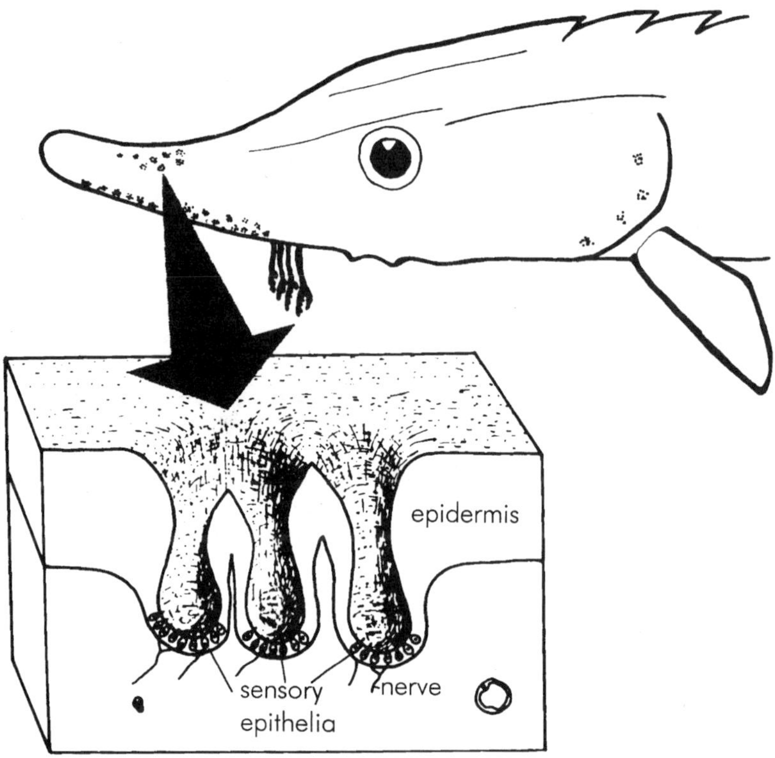

FIGURE 3.4. The head of a sturgeon (*Acipenser ruthenus*) with ampullary organs on the snout and gill cover. (From Jørgensen 1980.)

are apically equipped with a cilium, and a single nerve fiber ramifies below the sensory epithelium of each ampulla and forms several afferent boutons in contact with the sensory cells. These boutons have synaptic sheets opposite the afferent boutons.

Recent studies of electroreception in the paddlefish have shown convincingly that this species uses its sensory ampullae in passive detection and feeding of plankton (Gurgens et al. 2000; Wilkens et al. 2001; also see Wilkens, Chapter 9).

3.2.4 Ampullary Organs in Lungfishes (Dipnoi)

The extant lungfishes have ampullary organs embedded in a thick epidermis. Their distribution is unusual among other non-teleost bony fishes, since they are found not only on the head, but also on the trunk and tail fin (Northcutt 1986). The large pear-shaped sensory cells are equipped with an apical cilium, which may be surrounded by a few microvilli (Fig. 3.6). Basally in the cells synaptic sheets surrounded by synaptic vesicles are found opposite to the afferent nerve

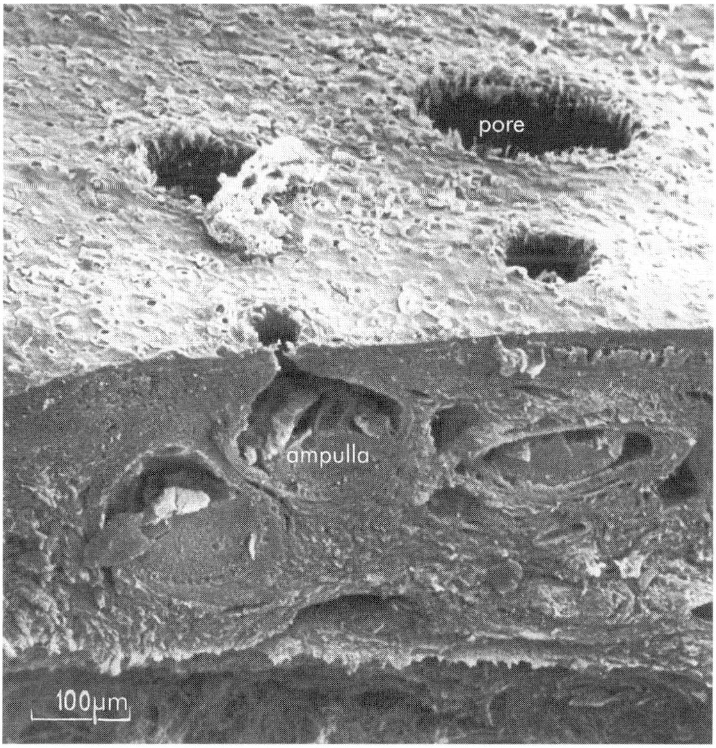

FIGURE 3.5. Scanning electron micrograph of a part of the rostrum of the paddlefish (*Polyodon spathula*). The pores open into jelly-containing ampullae, which are lined at the bottom of with a sensory epithelium. (From Jørgensen et al. 1972).

endings (Jørgensen 1984). Electrophysiologically, electroreception has been demonstrated for the South American lungfish *Lepidosiren*.

3.2.5 Ampullary Organ in the Coelacanth (Coelacanthiformes)

In the saltwaterliving coelacanth (Fig. 3.7) the ampullary organ is morphologically very different from that of other fishes. The sensory cells are situated in the rostral organ (Fig. 3.7), which is an irregular chamber in the ethmoid region. It has three pairs of openings to the surrounding water. As in all other ampullary organs, the tubes leading to the pores are filled with a jellylike substance. In a part of the chamber are found deep crypts, which are lined by a sensory epithelium consisting of sensory and supporting cells. Although no well-fixed tissue has been examined, it seems clear that the sensory cells are very much like the ampullary sensory cells found in other non-teleost fishes (Jørgensen 1991).

Additional evidence for electroreception is that the rostral organs are inner-

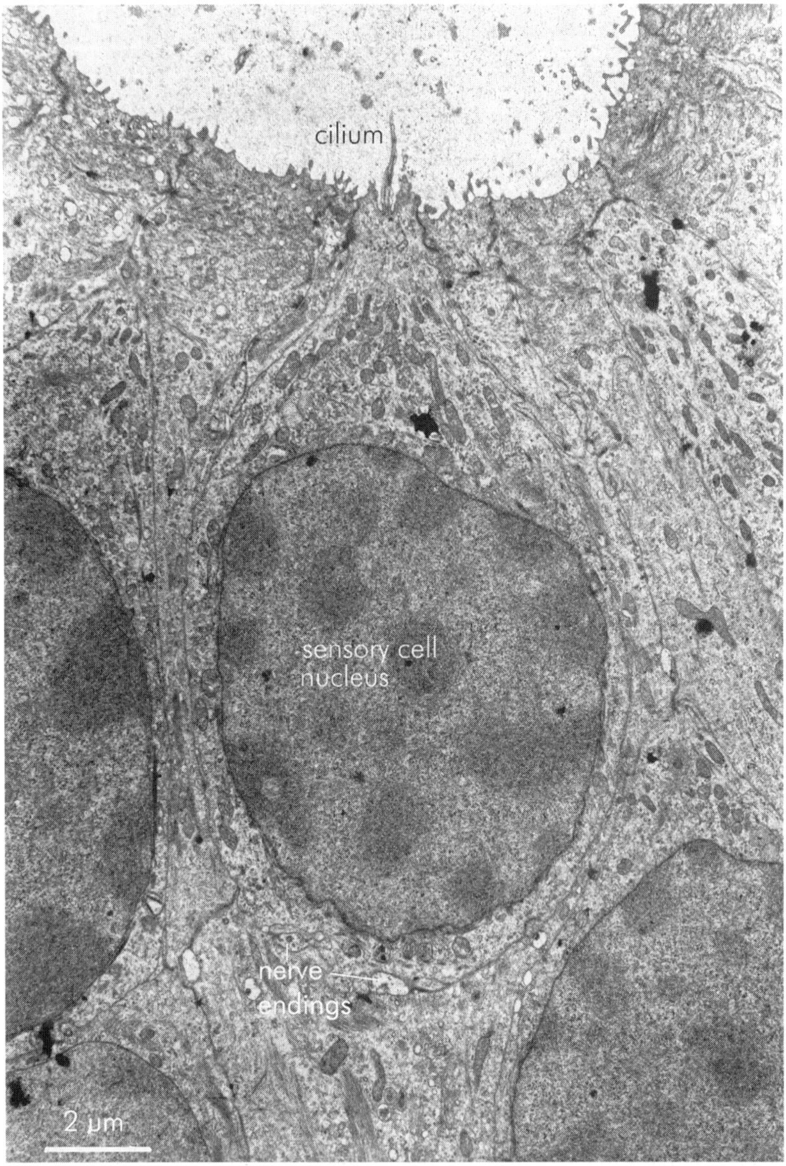

FIGURE 3.6. An ampullary sensory cell from an African lungfish (*Protopterus annectens*) as seen in the transmission electron microscope. A cilium surrounded by a few microvilli is seen apically. Small nerve endings are seen at the basal end of the sensory cell. (From Jørgensen 1984.)

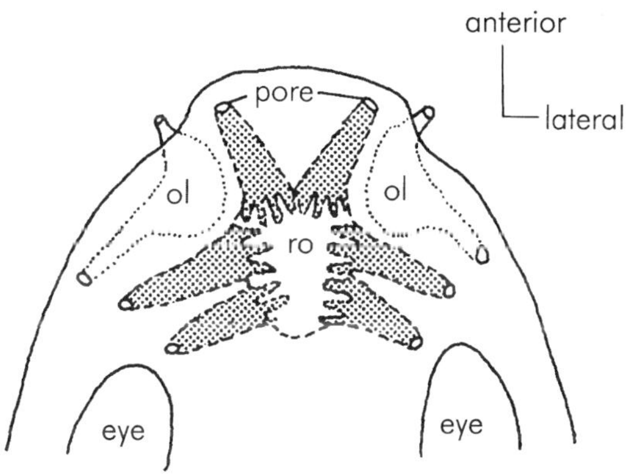

FIGURE 3.7. Drawing of a horizontal section through the anterior part of the head of the coelacanth. The rostral organ (ro) is located medially to the olfactory organs (ol). (From Jørgensen 1991).

vated by branches of the superficial ophthalmic ramus of the anterodorsal lateral line nerve, which projects to a distinct dorsal nucleus in the medulla oblongata (Northcutt 1980; Northcutt and Bemis 1993).

3.2.6 Ampullary Organs in Caecilian and Urodele Amphibians

A number of species belonging to caecilian and urodele amphibians have ampullae that are located on the head (Northcutt 1992). Each ampulla contains 10 to 30 pear-shaped sensory cells, all of which are innervated by a single afferent nerve fiber.

In larval *Ichthyophis* (Caecilian) each sensory cell has an apical single cilium surrounded by a few microvilli (Wahnschaffe et al. 1985), while the urodeles show a more varied pattern. In the axolotl (*Ambystoma mexicanum*) the sensory cells have approximately 200 microvilli around an eccentric cilium (Northcutt et al. 1994). The European cave salamander (*Proteus anguinus*) has some sensory cells with an apical cilium, while others possess only a basal body (Istenic and Bulog 1984). Electroreceptive sensory cells of other urodeles belonging to the genera *Triturus* and *Salamandra* have no cilia or basal bodies, but possess numerous long microvilli (Fritzsch and Wahnschaffe 1983).

Basally in the electroreceptive cells, very large synaptic spheres up to 3 μm in diameter are found opposite to afferent nerve endings (Fritsch and Wahnschaffe 1983). Electrophysiological evidence of electroreception and the role of electroreception during feeding has been demonstrated in the axolotl (*Ambystoma mexicanum*) (Himstedt et al. 1982; Münz et al. 1982). The nerve fibers from the amphibian ampullary organs project to a dorsal nucleus in the medulla (Fritzsch and Münz 1986).

3.3 Ampullary Organs in Teleosts

Ampullary organs are not found very widely in teleosts (Bullock et al. 1982). However, they have been demonstrated in species belonging to the orders Osteoglossiformes (family Mormyridae, with many electroreceptive species; family Gymnarchidae, with one species; family Notopteridae, with two species with electroreceptive organs—*Xenomystus* and *Papyrocranus*—and other genera without electroreception), Siluriformes (many species in all studied families), and Gymnotiformes (many species belonging to all families).

With few exceptions, these electroreceptive teleosts are all freshwater fishes, and a main characteristic of their ampullary organs is that the ampullary duct is rather short (Fig. 3.8a). In contrast, in the few electroreceptive saltwater teleosts, for example, *Plotosus*, which belongs to the order Siluriformes, the ampullary ducts are long, as in cartilaginous fishes (Bennett and Obara 1986).

Nerve fibers from the teleost electroreceptors project to a lateral lobe in the

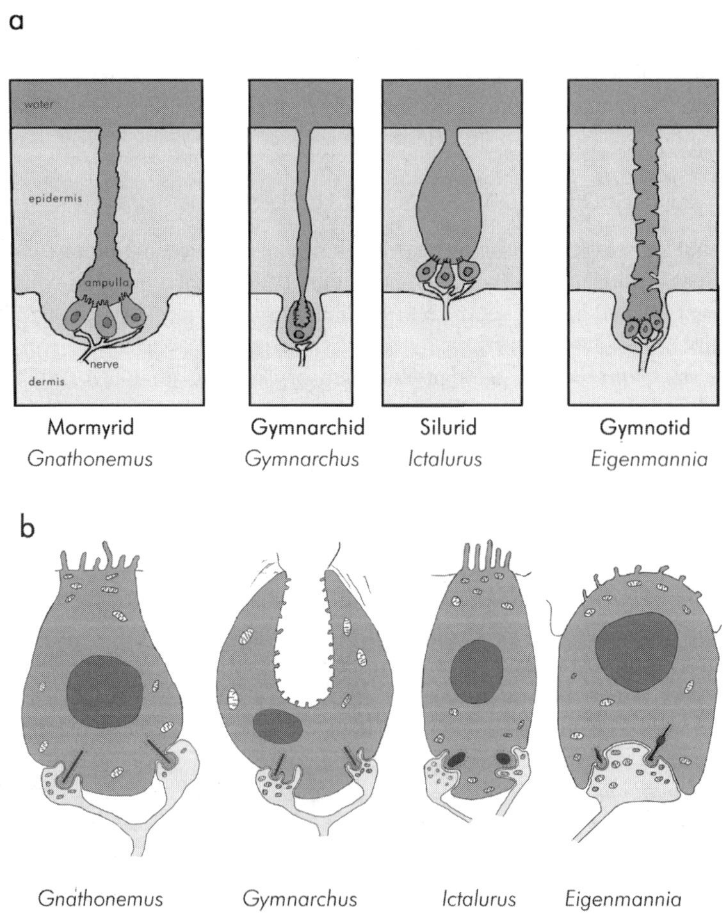

FIGURE 3.8. (**a**) Teleost ampullary organs. (**b**) Sensory cells from representatives of four different families. Not drawn to scale. (See color insert.)

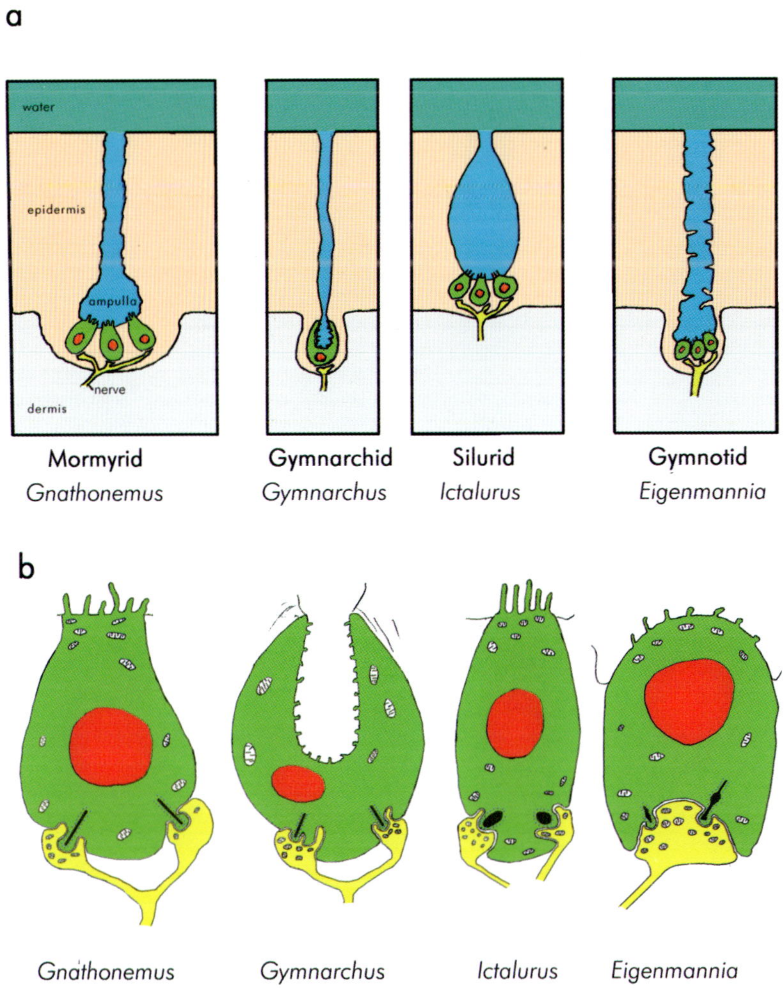

FIGURE 3.8. (**a**) Teleost ampullary organs. (**b**) Sensory cells from representatives of four different families. Not drawn to scale.

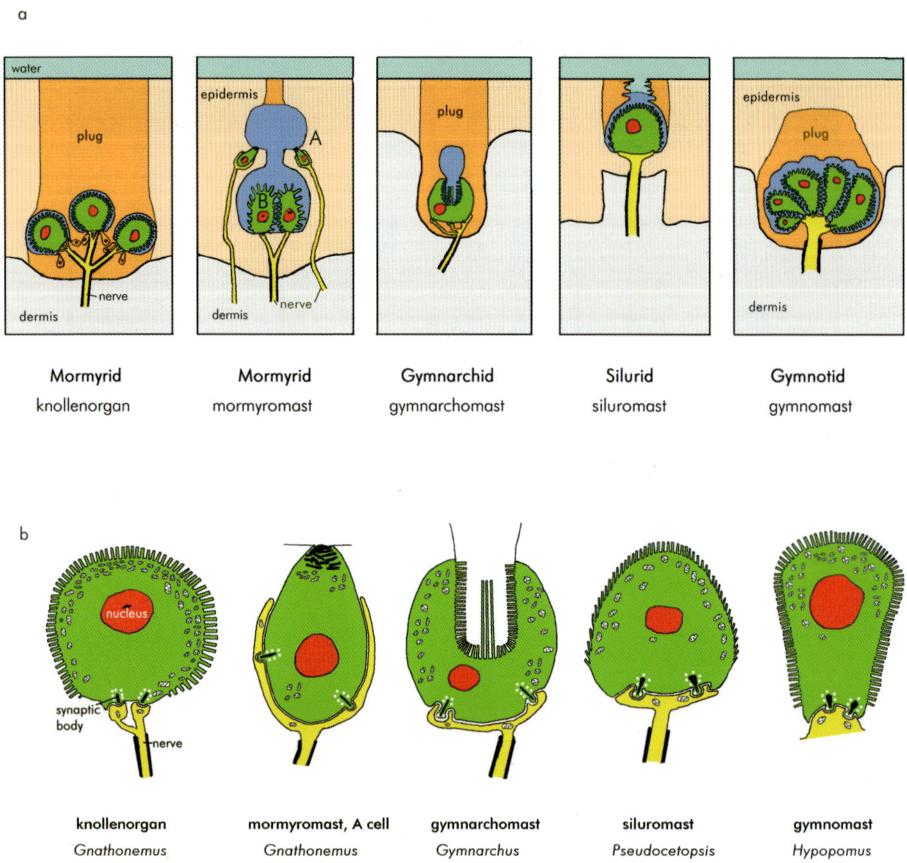

FIGURE 3.12. (**a**) Teleost tuberous organs. (**b**) Five different tuberous sensory organs from teleost fishes.

medullary octavolateralis area, which is different from the dorsal nucleus in the non-telostean fishes (see Bell and Maler, Chapter 4). Another difference between teleosts and non-teleosts is that whereas the former are excited primarily by anodal fields, the non-teleosts are cathode sensitive. This naturally leads to the hypothesis that the electroreceptive organs of these two main groups have an independent evolutionary history. Further, the line leading to osteoglossiform fishes is so far from the gymnotiform and siluriform line that Bullock et al. (1982) suggested that these teleost lines have evolved ampullary organs independently.

3.3.1 Ampullary Organs in Mormyrid Teleosts

Mormyrid ampullary organs were examined with light microscopy by Harder (1968) and at the electron microscopic level by Szamier and Bennett (1974). They are distributed all over the head as well as dorsally and ventrally along the body. The ampullary ducts in two species of *Gnathonemus* have a length of about 150 μm or less and a diameter of 20 μm. Each organ contains from three to eight egg- or pear-shaped sensory cells (Fig. 3.8a), separated by supporting cells. Each sensory cell has a number apical microvilli (Fig. 3.8b), which reach for a short distance into the ampullary canal. The nucleus is found in the center of the sensory cell. In addition to mitochondria and vesicles in all parts of the cytoplasm, the sensory cell has about 20 to 30 synaptic sheets opposite to the afferent nerve boutons, which contact the basal or lateral parts of the cell. One single nerve fiber divides to contact all of the sensory cells in a single ampullary organ.

3.3.2 Ampullary Organs in the Teleost *Gymnarchus*

Gymnarchid ampullary organs were described by Szabo (1965) and in detail by Mullinger (1969). They are distributed all over the head and body of the fish. The ampullary lumen is 10 to 20 μm in diameter and only one sensory cell is generally found in the ampulla (Fig. 3.8a). The shape of the sensory cell is unique (Fig. 3.8b) in that the ampullary lumen extends almost half way down into the apical part of the cell. The surface of the cell has a sparse number of short microvilli that are about 1 μm in length and 0.1 μm in diameter. Synaptic sheets surrounded by numerous vesicles are found opposite to several synaptic boutons, all of which are connected to a single nerve fiber.

3.3.3 Ampullary Organs in Silurid Teleost Fishes

Silurid ampullary organs are most numerous in the head skin, but are also found along the body side. In *Plotosus* they are primarily found at the base of the fin rays. Mullinger (1964) described in detail the ampullae of the brown bullhead catfish *Ictalurus (Amiurus) nebulosus*; Friedrich-Freksa (1930) and Szabo (1974) described the ampullae of *Plotosus*. Jørgensen (1992) examined the ampullae

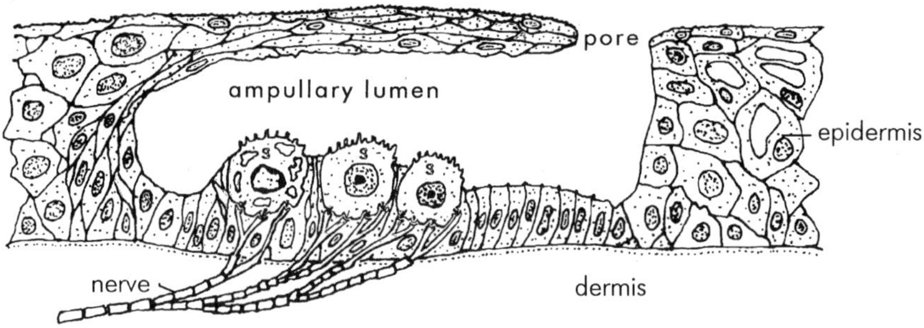

FIGURE 3.9. An ampullary organ from the transparent catfish (*Kryptopterus bicirrhus*).

of the transparent catfish (*Kryptopterus bicirrhus*), Gelinek (1978) the ampullae of the South American paddlefish *Sorubim lima*, Srivastava and Seal (1981) ampullary organs in several Indian catfishes—*Clarias, Heteropneustes, Rita*, and *Mystus*—and Andres et al. (1988) the ampullae of *Pseudocetopsis* sp.

In the freshwater silurids, the ampullae are situated in the epidermis (Figs. 3.8a and 3.9) only a few reach deep into the dermis. In *Pseudocetopsis* usually only one or two sensory cells are found in the bottom of the ampullae, while the electroreceptive ampullae of *Ictalurus (Amiurus)* contain 5 to 15 sensory cells, and *Kryptopterus* has 10 to 30 sensory cells. These cells in all of the species mentioned are equipped with a sparse number of short microvilli (Fig. 3.10b). The marine species, for example, *Plotosus*, have long ducts, reaching about 3 cm deep into the dermis, and the sensory cells have numerous long microvilli.

3.3.4 Ampullary Organs in Gymnotid Teleost Fishes

Gymnotid ampullary organs have been described from a number of species, by Lissmann and Mullinger (1968), Szamier and Wachtel (1969, 1970), and Szabo (1974). Generally the ampullary organs are found scattered on the head and in bands along the side of the body. Between two and eight pear-shaped sensory cells protrude slightly into the ampullary lumen (Fig. 3.8), exposing about 15% to 20% of the sensory cell surface to the lumen. Irregularly distributed microvilli that are less than 1 μm in length and with a core of microfilaments are seen on the surface. Numerous mitochondria are found around the central or luminally placed nucleus. Basally, 10 to 15 synaptic bodies are found opposite a single large nerve bouton, which measures about 5 μm in diameter. These dark-staining presynaptic bodies or rods have a peculiar structure consisting of a roundish body with a rodlike structure pointing centrally into the cell and a foot process that reaches into the synaptic lobe of the sensory cell which projects into the bouton. These synaptic bodies are surrounded by numerous clear vesicles with a diameter of 200 to 300 Å.

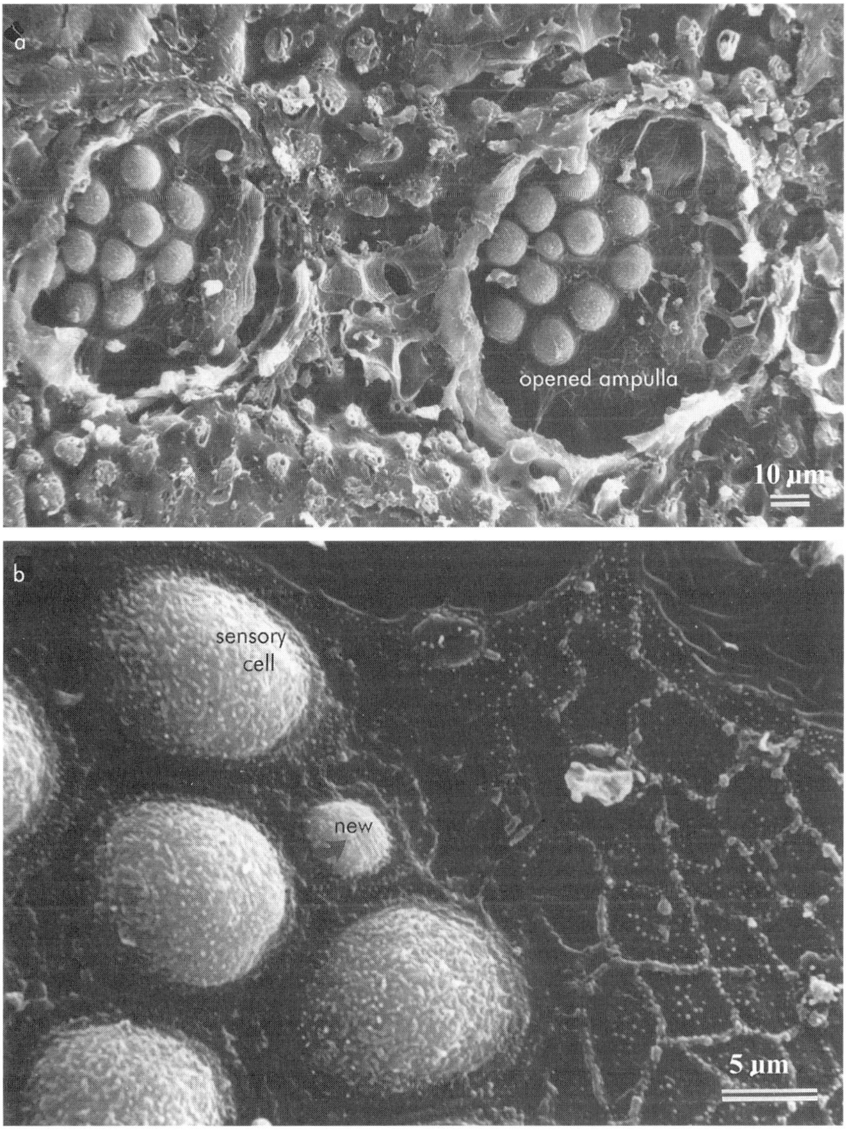

FIGURE 3.10. The ampullary organs of the transparent catfish, seen in the scanning electron microscope. (**a**) Two organs are shown, which have been opened to reveal the sensory cells. (**b**) A few sensory cells with numerous surface microvilli. A newborn cell is seen in the middle of the picture (*new*).

3.3.5 Ampullary Organs in Notopterid Fishes

Fish belonging to the order Osteosteglossiformes also contain the family Notopteridae, with many species in Asia and a few in Africa. Only the African species are known to possess electroreceptive organs. The ampullary organs in the African knifefish *Xenomystys nigri* (Fig. 3.11a) were examined by Jørgensen and Bullock (1987). They are distributed all over the head and along the body but are particularly numerous slightly dorsal to the horizontal septum. Each organ has from four to eight pear-shaped sensory cells that apically have 40 to

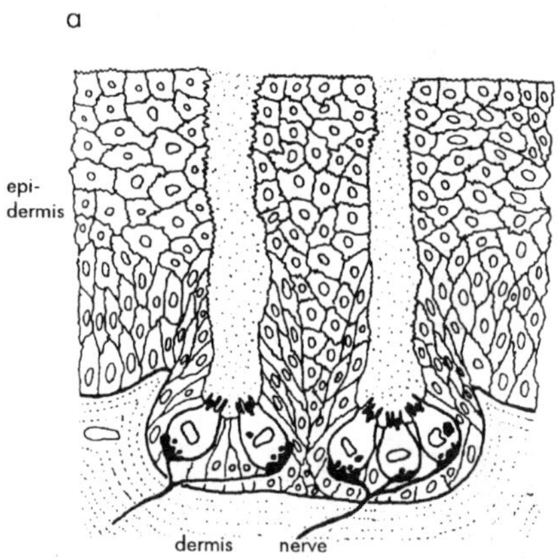

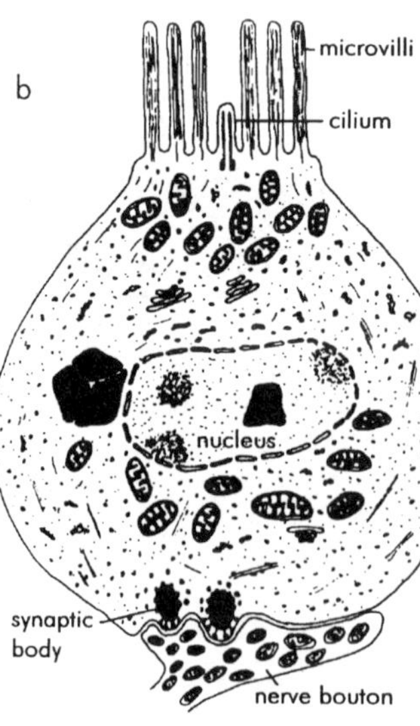

FIGURE 3.11. (**a**) Drawing of two ampullary organs from the African knifefish (*Xenomystus nigri*). (**b**) An isolated sensory cell. (From Jørgensen and Bullock 1987.)

60 microvilli and a short cilium, measuring up to 1 μm (Fig. 3.11b). So far this is the only teleost species found to have an apical cilium on its electroreceptive cells. The cilium arises from a basal body and it is found mostly between the microvilli. These microvilli are about 2.5 μm long and have a diameter of 0.2 to 0.3 μm. Mitochondria are seen primarily in the apical part of the cell and below the centrally located nucleus. One or a few round synaptic bodies surrounded by clear vesicles are found opposite to the single nerve bouton.

4. Tuberous Electroreceptive Organs

Tuberous organs have so far been demonstrated only in teleost fishes. They differ from ampullary organs in that they do not have a duct from the surface of the skin to the sensory organ. Instead the sense organ is covered by a plug of loosely packed epidermal cells. The sensory cells open into an intraepidermal cavity. The sensory cells are different from the ampullary sensory cells, both in structure and function (see Hopkins, Chapter 10).

4.1 Tuberous Organs in Mormyrids

Two distinct types of tuberous organs are found in mormyrids, the so-called knollenorgan and the mormyromast (Szabo 1974). The fine structure of the knollenorgan was described by Derbin and Szabo (1968). Each knollenorgan is composed of three to four sensory cells and a number of supporting cells. These sensory cells lie in a narrow cavity, or sensory chamber (Fig. 3.12a), that is limited externally by flattened epidermal cells. Below the sensory cells are found a number of supporting cells. Some pear-shaped supporting cells, situated below the sensory cells, are filled with rough endoplasmic reticulum, indicating that they are involved in production of the jelly that fills the sensory chamber. The sensory cells are large, up to 40 μm in diameter, with an apical or central nucleus. Most of the surface is covered by a large number of microvilli up to 3 μm in length and 0.1 to 0.15 μm in diameter (Fig. 3.12b). A large number of mitochondria are found below the microvilli. Basally, the sensory cell sits on a platform in which the synapses are found. All sensory cells in an organ are innervated by a single nerve fiber that divides to form several boutons on the individual sensory cells.

The mormyromast has been described by Szabo and Wersäll (1970) and Bell et al. (1989). This type of sense organ is a complicated structure involving at least two different sensory cells with separate innervation. The mormyromast has two chambers that are connected by a narrow canal (Fig. 3.12a). The shallower chamber contains a type A sensory cell (Fig. 3.12b), while the deeper chamber has a type B cell, which looks like the sensory cell of the knollenorgan. Six to 13 type A cells lie in a circle in the outer (superficial) chamber and these are separated by supporting cells. The apical part of this pear-shaped A cell is

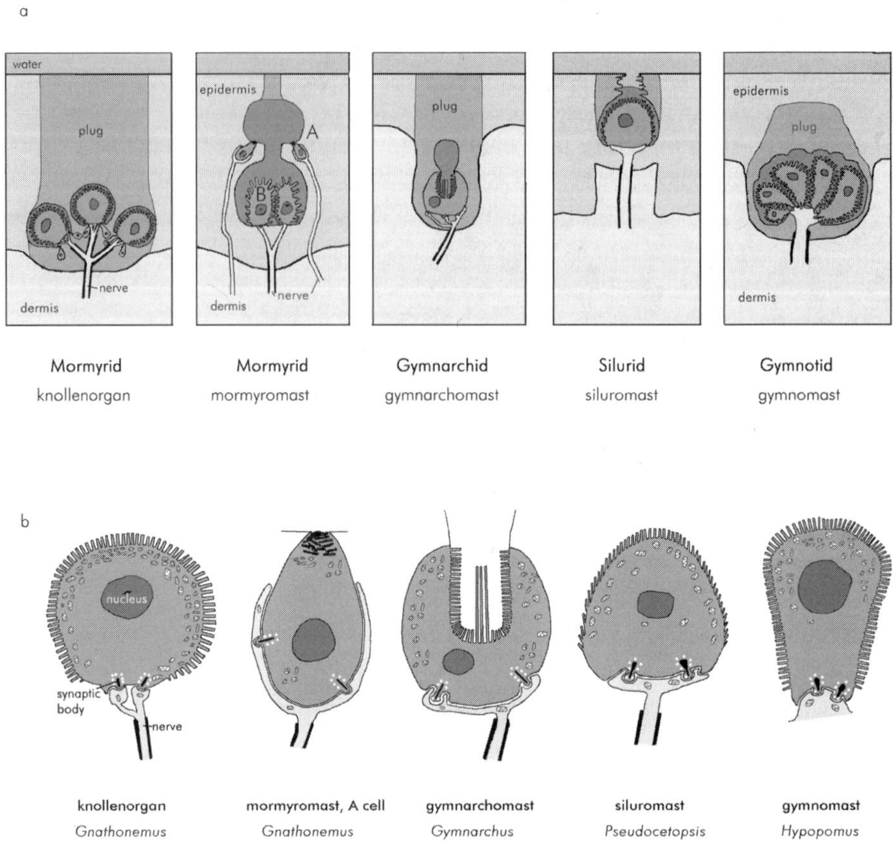

FIGURE 3.12. (**a**) Teleost tuberous organs. (**b**) Five different tuberous sensory cells from teleost fishes. (See color insert.)

characterized by numerous membrane foldings (Fig. 3.12b). The A cell is partially surrounded by fingerlike projections of the afferent nerve which stretches up along the lateral part of the cell. Here and there synaptic sheets indicate the position of synapses. The type B sensory cell is situated on a platform, with the majority of the cell surface exposed to the deep chamber. Numerous long and slender microvilli stretch out into the chamber jelly. The B cell forms basally contact with nerve endings, which originate from a single fiber.

4.2 Tuberous Organ in Gymnarchus

Gymnarchus has a peculiar sensory cell in the tuberous organ, which has been called a gymnarchomast (Mullinger 1969; Szabo 1974). The single sensory cell opens into a jelly-filled chamber. The cell has an apical depression filled with numerous microvilli. The microvilli originating from the bottom of the depression form a long projection into the apical cavity. The cell nucleus is located basally. A single nerve fiber splits to from several nerve endings at the basolateral part of the sensory cell. In some gymnarchomasts more sensory cells are seen located together, and organs with more than one sensory cell have been termed gymnarchomast type II (Szabo 1974). However, it is uncertain whether this type of organ is a further developmental stage of the gymnarchomast type I.

4.3 Silurid Tuberous Organ

In silurids, tuberous organs have been described only from one species, *Pseudocetopsis* (Andres et al. 1988). This organ, the siluromast, is located rather superficially in the epidermis (Fig. 3.12a). It contains a jelly-filled cavity with a single spherical sensory cell that has a diameter of about 25 μm and apically numerous microvilli that increase in height toward the cell apex (Fig. 3.12b). The nucleus has a central position. A single afferent nerve flattens out at the base of the sensory cell.

4.4 Gymnotid Tuberous Organ

In Gymnotids the tuberous organ, the gymnomast, is described in a number of species (Szabo 1965; Wachtel and Szamier 1966; Lissmann and Mullinger 1968; Szamier and Wachtel 1970; Bennett et al. 1989). A cavity contains 25 to 35 elongated sensory cells (Fig. 3.12a). The exposed lateral and apical surfaces of these cells are enlarged by varying numbers of microvilli that stretch out into the gymnomast chamber (Fig. 3.12b). In some species a single nerve fiber splits to innervate all sensory cells with several nerve boutons; in other species such as *Hypopomus* the nerve ending is one single large knob that forms synapses with all the sensory cells in a single gymnomast.

5. Monotreme Electroreceptive Organs

The monotreme mammals include the platypus (*Ornithorhynchus anatinus*) and two species of spiny anteaters (*Tachyglossus aculeata* and *Zaglossus bruijnii*). Electroreception has been demonstrated in all three species (Scheich et al. 1986; Manger et al. 1997; Pettigrew et al. 1998). In all cases the bill or snout contains specialized mucous glands with electroreceptive nerve endings (Andres and von Düring 1988). The platypus has about 40,000 electroreceptive mucous glands whereas *Zaglossus* may have about 3000 and *Tachyglossus* only about 100 (Manger et al. 1997).

In platypus the electroreceptive organs are distributed in narrow stripes on the entire bill, while the spiny anteaters have their organs on the outermost part of the snout. The sense organs always consist of a large coiled mucous gland (Fig. 3.13) where naked nerve endings opens into a capsule of epidermal cells, which forms around the middle portion of the mucous gland.

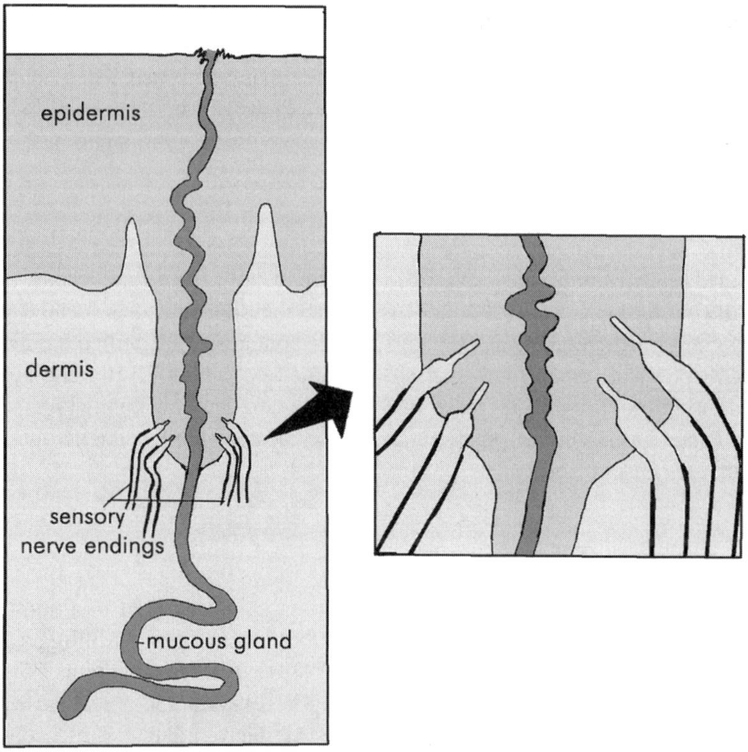

FIGURE 3.13. The electroreceptive mucous gland from the bill of the platypus (*Ornithorhynchus anatinus*). Naked sensory nerve endings are located in the middle portion of the gland.

6. Summary

The morphology of electroreceptive organs in lampreys and their larvae, different fishes, aquatic amphibians, and monotreme mammals is described. The sense organs vary from superficial end buds in lampreys, to ampullary organs and tuberous organs in many fishes and amphibians, to specialized mucous glands in the monotremes. The sensory cells are quite different. Some have a bundle of apical microvilli, as in lampreys and many teleost fishes. Cartilaginous and non-teleost fishes have a single apical cilium that in some species is surrounded by a few microvilli. Caecilian and urodele amphibians may have both a cilium and microvilli. Only the monotremes lack specialized sensory cells, possessing instead naked nerve endings connected to mucous glands. In all electroreceptive organs examined, only afferent nerve endings have been identified.

References

Andres KH, von Düring M (1988) Comparative anatomy of vertebrate electroreceptors. Prog Brain Res 74:113–131.

Andres KH, von Düring M, Petrasch E (1988) The fine structure of ampullary and tuberous electroreceptors in the South American blind catfish *Pseudocetopsis* spec. Anat Embryol 177:523–535.

Bell CC, Zakon H, Finger TE (1989) Mormyromast electroreceptor organs and their afferent fibers in mormyrid fish. J Comp Neurol 286:391–407.

Bennett MVL, Obara S (1986) Ionic mechanisms and pharmacology of electroreceptors. In: Bullock TH, Heiligenberg W (eds), Electroreception. New York: John Wiley & Sons, pp. 157–181.

Bennett MVL, Sandri C, Akert K (1989) Fine structure of the tuberous electroreceptor of the high-frequency electric fish, *Sternarchus albifrons* (gymnotiformes). J Neurocytol 18:265–283.

Bodznick D, Boord R (1986) Electroreception in Chondrichthyes. Central anatomy and physiology. In: Bullock TH, Heiligenberg W (eds), Electroreception. New York: John Wiley & Sons, pp. 225–256.

Bodznick D, Northcutt, RG (1981) Electroreception in lampreys: evidence that the earliest vertebrates were electroreceptive. Science 212:465–467.

Bodznick D, Preston DG (1983) Physiological characterization of electroreceptors in the lampreys *Ichthyomyzon unicuspis* and *Petromyzon marinus*. J Comp Physiol 152:209–217.

Bullock TH (1982) Electroreception. Annu Rev Neurosci 5:121–170.

Bullock TH, Heiligenberg W (1986) Electroreception. New York: John Wiley & Sons,

Bullock TH, Northcutt RG, Bodznick DA (1982) Evolution of electroreception. Trends Neurosci 5:50–53.

Derbin C, Szabo T (1968) Ultrastructure of an electroreceptor (knollenorgan) in the Mormyrid fish *Gnathonemus petersii*. J Ultrastruct Res 22:469–484.

Dotterweich H (1932) Bau und Funktion der Lorenzini'schen Ampullen. Zool Jb Abt Allg Zool 50:347–418.

Fields RD, Bullock TH, Lange GD (1993) Ampullary sense organs, peripheral, central and behavioral electroreception in chimeras (*Hydrolagus*, Holocephali, Chondrichthyes). Brain Behav Evol 41:269–289.

Friedrich-Freksa H (1930) Lorenzinischen Ampullen bei dem Siluroiden *Plotosus anguillaris* Bloch. Zool Anz 87:49–66.

Fritzsch B, Münz H (1986) Electroreception in amphibians. In: Bullock TH, Heiligenberg W (eds), Electroreception. New York: Wiley & Sons, pp. 483–496.

Fritzsch B, Wahnschaffe U (1983) The electroreceptive ampullary organs of urodeles. Cell Tissue Res 229:483–503.

Gelinek E (1978) On the ampullary organs of the South-American paddle-fish *Sorubim lima* (Siluroidea, Pimelodidae). Cell Tissue Res 190:357–369.

Gurgens C, Russell DF, Wilkens LS (2000) Electrosensory avoidance of metal obstacles by the paddlefish. J Fish Biol 57:277–290.

Harder W (1968) Zum Aufbau der epidermalen Sinnesorgane der Mormyridae (Mormyriformes, Teleostei). Z Zellforsch 89:212–224.

Himstedt W, Kopp J, Schmidt W (1982) Electroreception guides feeding behaviour in amphibians. Naturwissenschaften 69:552–553.

Istenic L, Bulog B (1984) Some evidence for the ampullary organs in the European cave salamander *Proteus anguinus* (Urodela, Amphibia). Cell Tissue Res 235:393–402.
Jørgensen JM (1980) The morphology of the Lorenzinian ampullae of the sturgeon *Acipenser ruthenus* (Pisces: Chondrostei). Acta Zool (Stockh) 61:87–92.
Jørgensen JM (1982) Fine structure of the ampullary organ of the bichir *Polypterus senegalus* Cuvier, 1929 (Pisces: Brachiopterygii) with some notes on the phylogenetic development of electroreceptors. Acta Zool (Stockh) 63:211–217.
Jørgensen JM (1984) On the morphology of the electroreceptors of the two lungfish: *Neoceratodus forsteri* Krefft and *Protopterus annectens* Owen. Vidensk Meddr Dansk Naturh Foren 145:77–85.
Jørgensen JM (1991) Ciliated sensory cells in the rostral organ of the coelacanth *Latimeria chalumnae* (Smith 1939). Acta Zool (Stockh) 72:121–124.
Jørgensen JM (1992) The electrosensory cells of the ampullary organ of the transparent catfish (*Kryptopterus bicirrhus*). Acta Zool (Stockh) 73:79–83.
Jørgensen JM, Bullock TH (1987) Organization of the ampullary organs of the African knife fish *Xenomystus nigri* (Teleostei: Notopteriade). J Neurocytol 16:311–315.
Jørgensen JM, Flock Å, Wersäll J (1972) The Lorenzinian ampullae of *Polyodon spathula*. Z Zellforsch 130:362–377.
Lissmann HW, Mullinger AM (1968) Organization of ampullary electric receptors in Gymnotidae (Pisces). Proc R Soc Lond B 169:345–378.
Manger, PR, Collins R, Pettigrew JD (1997) Histological observations on presumed electroreceptors and mechanoreceptors in the beak skin of the long-beaked echidna, *Zaglossus bruijnii*. Proc R Soc Lond B 264:165–172.
Mullinger AM (1964) The fine structure of ampullary electric receptors in *Amiurus*. Proc R Soc Lond B 160:345–359.
Mullinger, AM (1969) The organization of ampullary sense organs in the electric fish, *Gymnarchus niloticus*. Tissue Cell 1:31–52.
Münz H, Claas B, Fritzsch B (1982) Electrophysiological evidence of electroreception in the axolotl *Siredon mexicanum*. Neurosci Lett 28:107–111.
Northcutt RG (1980) Anatomical evidence of electroreception in the coelacanth (*Latimeria chalumnae*). Zbl Vet Med C Anat Histol Embryol 9:289–295.
Northcutt RG (1986) Electroreception in nonteleost bony fishes. In: Bullock TH, Heiligenberg W (eds), Electroreception. New York: John Wiley & Sons, pp. 257–285.
Northcutt RG (1992) Distribution and innervation of lateral line organs in the axolotl. J Comp Neurol 325:95–123.
Northcutt RG, Bemis WE (1993) Cranial nerves of the coelacanth, *Latimeria chalumnae* [Osteichthyes: Sarcopterygii: Actinistia], and comparisons with other craniata. Brain Behav Evol 42 (Suppl 1):1–76.
Northcutt RG, Catania KC, Criley BB (1994) Development of lateral line organs in the axolotl. J Comp Neurol 340:480–514.
Pettigrew JD, Manger PR, Fine SLB (1998) The sensory world of the platypus. Philos Trans R Soc Lond B 353:1199–1210.
Raschi W (1986) A morphological analysis of the ampullae of Lorenzini in selected skates (Pisces, Rajoidei). J Morphol 189:225–247.
Ronan MC, Bodznick D (1986) End buds: non-ampullary electroreceptors in adult lampreys. Comp Physiol A 158:9–15.
Roth A, Tscharntke H (1976) Ultrastructure of the ampullary electroreceptors in lungfish and brachiopterygii. Cell Tissue Res 173:95–108.

Scheich H, Langner G, Tidemann C, Coles RB, Guppy A (1986) Electroreception and electrolocation in platypus. Nature 319:401–402.

Srivastava CBL, Seal M (1981) Electroreceptors in Indian catfish teleosts. Adv Physiol Sci 31:1–11.

Steven DM (1951) Sensory cells and pigment distribution in the tail of the ammocoete. Q J Micr Sci 92:233–247.

Szabo T (1965) Sense organs of the lateral line system in some electric fish of the Gymnotidae, Mormyridae and Gymnarchidae. J Morphol 117:229–250.

Szabo T (1974) Anatomy of the specialized lateral line organs of electroreception. In: Fessard A (ed), Handbook of Sensory Physiology III/3. Berlin: Springer-Verlag, pp. 13–58.

Szabo T, Wersäll J (1970) Ultrastructure of an electroreceptor (Mormyromast) in a Mormyrid fish, *Gnathonemus petersii*. II. J Ultrastruct Res 30:473–490.

Szamier RB, Bennett MVL (1974) Special cutaneous receptor organs of fish. VII. Ampullary organs of Mormyrids. J Morphol 143:365–384.

Szamier RB, Wachtel AW (1969) Special cutaneous receptor organs of fish. III. The ampullary organs of *Eigenmannia*. J Morphol 128:261–290.

Szamier RB, Wachtel AW (1970) Special cutaneous receptor organs of fish. VI. Ampullary and tuberous organs of *Hypopomus*. J Ultrastruct Res 30:450–471.

Teeter JH, Szamier RB, Bennett MVL (1980) Ampullary electroreceptors in the sturgeon *Scaphirhynchus platorynchus* (Rafinesque). J Comp Physiol 138:213–223.

Wachtel AW, Szamier RB (1966) Special cutaneous receptor organs of fish: the tuberous organs of *Eigenmannia*. J Morphol 119:51–80.

Wahnschaffe U, Fritzsch B, Himstedt W (1985) The fine structure of the lateral-line organs of larval *Ichthyophis* (Amphibia: Gymnophiona). J Morphol 186:369–377.

Waltman B (1966) Electrical properties and fine structure of the ampullary canals of Lorenzini. Acta Physiol Scand 66 (Suppl 264):1–59.

Weisel GF (1978) The integument and caudal filament of the shovelnose sturgeon, *Scaphirhynchus platorynchus*. Am Midland Nat 100:178–189.

Whitear M, Lane EB (1981) Bar synapses in the end buds of lamprey skin. Cell Tissue Res 216:445–448.

Whitear M, Lane EB (1983) Multivillous cells: epidermal sensory cells of unknown function in lamprey skin. J Zool 201:259–272.

Wilkens LA, Wettring B, Wagner E, Wojtenek W, Russell D (2001) Prey Detection in selective plankton feeding by the paddlefish: Is the electric sense sufficient? J Exp Biol 204:1381–1389.

4

Central Neuroanatomy of Electrosensory Systems in Fish

Curtis C. Bell and Leonard Maler

1. Introduction

This chapter is an introduction to the central nervous system (CNS) anatomy of electrosensory systems in fish. The neuroanatomy of these systems has been extensively studied in a variety of taxa and all of this information cannot be included in a short chapter such as this one. More complete descriptions of the neuroanatomy of different taxa may be found in Bullock and Heiligenberg (1986), and references to more recent neuroanatomical work are included in this chapter.

The purpose of this chapter is to provide the basic knowledge of anatomy needed for understanding the physiological and behavioral studies of fish electrosensory systems. Such studies have taken advantage of the attractive features that electrosensory systems offer for gaining insight into various fundamental issues in neuroscience, including the preservation and analysis of temporal information, the integration of time and intensity information, the descending control of sensory processing, the storage of sensory patterns, and the control of motor activity. This chapter emphasizes the anatomical features that have attracted investigators to the study of electrosensory systems, paying particular attention to features that are common to the different taxa.

The subjects covered by this chapter reflect the proportions of scientific effort that have been devoted to the different electrosensory systems. Thus, the CNS anatomy of elasmobranch, mormyriform, and gymnotiform fishes receives more attention because more effort has been devoted these taxa. Similarly, the initial stages of electrosensory information processing receive more attention than higher levels because so much more work has been done on the initial stages.

2. First-Order Electrosensory Structures and the Central Termination of Afferent Fibers from Electroreceptors

Primary afferent fibers from electroreceptors, from mechanical lateral line receptors, and from eighth nerve end organs terminate in cerebellum-like structures of the medulla in many different vertebrates (Montgomery et al. 1995; Bell 2002). Such structures include the dorsal octavolateral nuclei that receive input from electroreceptors in non-neopterygian fish and in some aquatic amphibians, the medial octavolateral nuclei that receive input from mechanical lateral line and from eighth nerve end organs and are found in almost all fish and aquatic amphibians (Coombs and Montgomery, Chapter 12), the electrosensory lateral line lobes (ELLs) that receive input from electroreceptors in electroreceptive teleosts, and the dorsal cochlear nuclei that receive input from the cochlea in mammals.

The different cerebellum-like first order structures share a number of common features. They all have a deep layer, a layer of principal cells, and a molecular layer. Primary afferents terminate in the deep layer. The principal cells have both basilar and apical dendrites (Fig. 4.1). The basilar dendrites descend into the deeper layers and receive input from primary afferent fibers, either directly or indirectly via interneurons. The apical dendrites of principal cells ascend into the molecular layer, where they are contacted by parallel fibers that arise from a mass of granule cells. These granule cells receive input from a rich

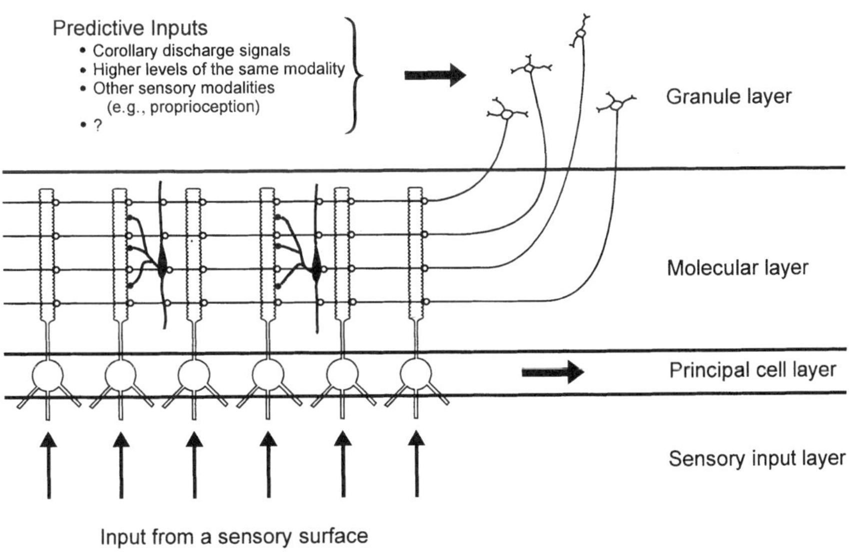

FIGURE 4.1. Schematic drawing showing major features of cerebellum-like sensory structures. Inhibitory stellate cells of the molecular layer are shown in *black*. (From Bell et al. 1997a).

variety of sources that include corollary discharge signals associated with motor commands, descending input from higher levels of the same modality, and input from other sensory modalities (Montgomery et al. 1995; Bell 2002). These various signals can be used to predict changes in input from the periphery to the deeper layers. Evidence from studies of mormyrid, gymnotid, and elasmobranch fish indicates that these cerebellum-like sensory structures can generate predictions about sensory patterns based on associations between predictive signals conveyed by parallel fibers and particular patterns of sensory input to the deeper layers. The predictions are subtracted from the actual input, allowing unexpected or novel input to stand out. (Bell et al. 1997a; Bastian and Zakon, Chapter 8).

Fish that have electroreceptors but that lack electric organs have only a single type of electroreceptor, the ampullary type (Jørgensen, Chapter 3), and the first-order electrosensory structures of such fish have only a single region representing the electroreceptive periphery. In most cases the representation is somatotopically organized. Some fish with electric organs as well as electroreceptors—electric rays and electric catfish—also have only a single type of electroreceptor and a single map of the electroreceptive periphery. However, other fish with electric organs, namely fish of the mormyriform and gymnotiform orders, have three or more types of electroreceptors and multiple maps of the electroreceptive periphery in the medulla. Commissural connections between the maps on the two sides of the brain are present in most of the different systems, and have been shown to be both point-to-point and inhibitory in several different taxa.

The following sections first describe the first-order structures of non-neopterygian fish and then describe those of neopterygian fish (advanced bony fish—holosteans and teleosts).

2.1 The Dorsal Octavolateral Nucleus (DON) of Non-neopterygian Fish and the Termination of Afferent Fibers from Electroreceptors

Thirty years ago the electrosense was believed to be an exotic sense found in only a few groups of fish. Since then, however, the electrosense has been shown to be a fundamental vertebrate sense that is present in all non-neopterygian fish, some teleost taxa, and two of the major groups of aquatic amphibians (Bullock et al. 1983; Zupanc and Bullock, Chapter 2). The non-neopterygian fish include the Petromyzontidae (lampreys), the Chondrichthyes (elasmobranchs and holocephalans), the Chondrostei (sturgeons and paddle fish), the Cladistia (bichir), the Dipneusti (lungfish), and the Crossopterygii (coelacanth) (Bodznick and Boord 1986; Northcutt 1986; Ronan 1986; Hofmann et al. 2002). The myxinoids (hagfish), a sister group of the vertebrates, lack electroreception (Ronan 1986).

Electroreceptors in non-neopterygian fish are of the ampullary type that responds to low-frequency electrical signals (Jørgensen, Chapter 3). The receptors are quite similar both morphologically and functionally among non-neopterygian fish. In all of these fish, the primary afferent fibers from electroreceptors enter the medulla via a dorsal branch of the anterior lateral line nerve to terminate within the dorsal octavolateral nucleus (DON; Fig. 4.2A). The DON is found in all non-neopterygian fish as well as in aquatic amphibia (Montgomery et al. 1995).

The termination of electroreceptor afferents in the DON is somatotopically organized in some, but not all, non-neopterygian fish. In the skate, afferent fibers from caudal clusters of electroreceptors terminate dorsally in DON and those from more rostral clusters terminate ventrally. A similar pattern of ter-

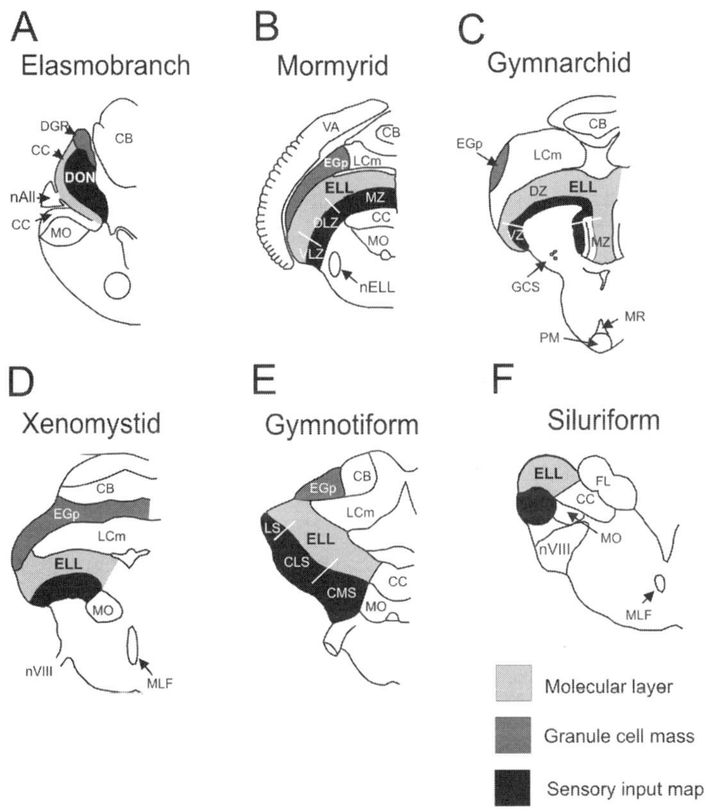

FIGURE 4.2. First-order electrosensory structures in different fish. The regions of central termination for primary afferent fibers from the periphery are shown in black. The granule cell masses that give rise to the parallel fibers of the molecular layers of these structures are shown in *dark gray*. The molecular layers of these structures are shown in *light gray*. (Adapted from Bell et al. 1997a).

mination is present in sharks and holocephalans, the other taxa within the Chondrichthyes (Bodznick and Boord 1986). Afferent fibers from electroreceptors do not terminate somatotopically in the DON of lampreys (Ronan 1986), and the issue of somatotopy has not yet been investigated in the non-neopterygian bony fish (Chondrostei, Cladistia, Dipneusti, and Crossopterygii).

The DON has only two main regions, a deeper cell rich region and an overlying molecular region or layer. The cell-rich region in elasmobranchs has large multipolar neurons with smooth basilar dendrites near the somas and spine-covered apical dendrites extending up into the molecular layer (Paul and Roberts 1977; Bodznick and Boord 1986). In the skate, the large multipolar neurons are efferent neurons with axons that project to the mesencephalon (Bodznick and Boord 1986) and are referred to as ascending efferent neurons. Large multipolar neurons in other non-neopterygian fish are probably efferent neurons also (Ronan 1986; Northcutt 1986).

The molecular layer of the DON, also known as the cerebellar crest, contains parallel fibers that arise from an external granule cell mass, the dorsal granular ridge or eminentia granularis (Fig. 4.2A). The parallel fibers course in a rostro–caudal direction, synapsing on the apical dendrites of large multipolar interneurons and on the dendrites of molecular layer stellate cells.

Relatively little is known about cellular connectivity in the DON. Primary afferent fibers in the skate make monosynaptic contact with ascending efferent neurons (Bodznick and Boord 1986). Immunocytochemistry reveals inhibitory interneurons in the cell-rich layers that probably mediate the inhibitory receptive fields of ascending efferent neurons (Duman and Bodznick 1996). Immunocytochemistry in the skate also indicates inhibitory commissural cells in the cell-rich layers that have axons projecting to the contralateral DON as well as inhibitory stellate cells of the molecular layer. Commissural connections have also been demonstrated in the lamprey (Ronan 1986).

2.2 The ELLs and the Termination of Afferent Fibers from Electroreceptors in Mormyriform and Xenomystine Fish

The order Mormyriformes and the order Notopteriformes are sister groups within the teleost cohort Osteoglossomorpha. The order Mormyrifomes are all electric fish and includes two families, the Mormyridae and the Gymnarchidae. All of the many species within the Mormyridae have electric organ discharges (EODs) of the pulse type, whereas the single species within the Gymnarchidae, *Gymnarchus niloticus*, has an EOD of the wave type (Zupanc and Bullock, Chapter 2). The order Notopteriformes includes one family of electroreceptive fish the Xenomystinae (African knifefish). These fish do not have electric organs and have only the ampullary type of electroreceptors. (See Braford 1986 for a discussion of the evolution of electroreception in Mormyriformes and Notopteriformes.)

2.2.1 The ELL of Mormyrid Fish

Fish of the family Mormyridae have three types of electroreceptors: mormyromast; knollenorgan; and ampullary (Jørgensen, Chapter 3). Mormyromast electroreceptors and their afferents are responsible for active electrolocation, that is, for sensing the transcutaneous current evoked by the fish's own EOD and the modulation of this current by external objects. The mormyromast electroreceptor has two distinct types of separately innervated electroreceptors, types A and B (Szabo 1974; Bell et al. 1989; Kawasaki, Chapter 7). The A and B types of afferents respond differently to waveform distortions caused by capacitive objects (von der Emde and Bleckman 1992), and this may account for the fish's ability to distinguish capacitive and resistive objects (von der Emde 1990). Knollenorgan electroreceptors are responsible for electrocommunication, that is, for sensing the high-frequency EODs of other fish. Ampullary electroreceptors are responsible for passive electrolocation of low-frequency external voltages, as in non-neopterygian fish.

The primary afferent fibers from electroreceptors in mormyrid fish enter the medulla through both the anterior and posterior lateral line nerves. Each of four different types of electroreceptor afferents (mormyromast A, mormyromast B, knollenorgan, and ampullary) terminates somatotopically within a distinct region of the mormyrid ELL, with rostral body being represented rostrally and caudal body caudally in ELLs (Bell and Szabo 1986).

The mormyrid ELL includes a cortex and a nucleus. The cortex has the shape of an inverted cup and is divided into three zones by longitudinal breaks in the cellular layers: a ventrolateral zone (VLZ) where ampullary afferents terminate (Bell 1982), a dorsolateral zone (DLZ) where B type mormyromast afferents terminate; and a medial zone (MZ) where A type mormyromast afferents terminate (Fig. 4.2B; Bell 1990). The ventral body is represented laterally in the MZ but dorsally in the VLZ and DLZ. Knollenorgan afferents terminate in the nucleus of ELL (nELL; Bell and Grant 1989). The somatotopy within each of the cortical maps is finely grained, whereas the somatotopy within the nucleus is quite rough.

The ELL cortex of mormyrids and gymnotids, unlike the DON of non-neopterygian fish, has several distinct lamina and a variety of well-differentiated cell types. The layers of the mormyrid ELL are as follows: molecular, ganglionic, plexiform, granular, intermediate, and deep fiber (Fig. 4.3). Parallel fibers of the molecular layer arise from an external granule cell mass, the eminentia granularis posterior (Figs. 4.2B and 4.3), and course in the transverse plane. The mormyrid ELL has as many as 20 different types of cells, and six different types of afferents from other central structures (Grant et al. 1996; Meek et al. 1996; Han et al. 1999; Mohr et al. 2003).

Primary afferent fibers terminate on granular cells in all three zones of the ELL cortex. The synapses are of a mixed, electrical–chemical, type (Bell et al. 1989). The granular cells of the mormyrid (and gymnotid) ELL do not form

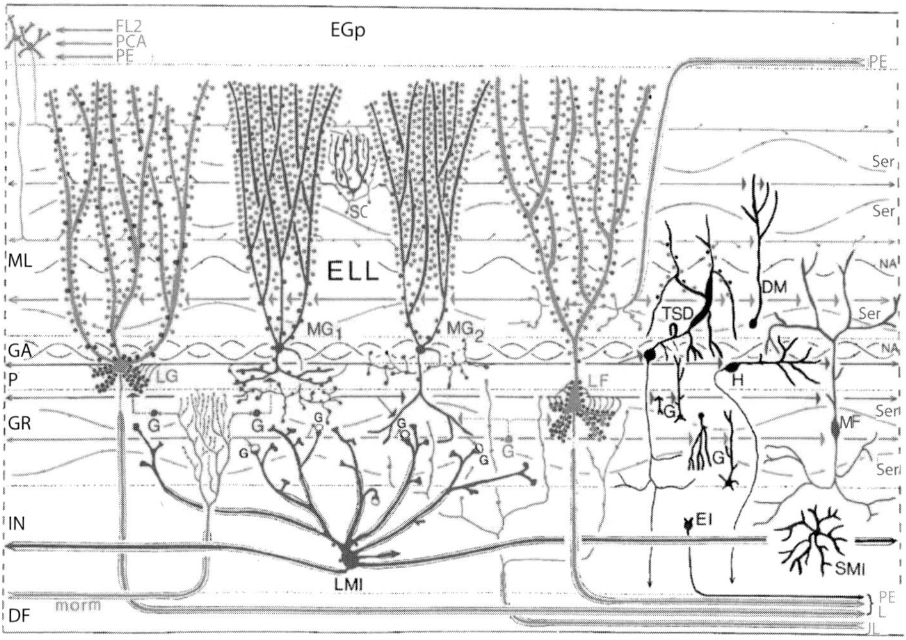

FIGURE 4.3. Histological structure of the mormyrid ELL. Known inhibitory cells and connections are shown in black. Known excitatory cells and connections are shown in gray. (From Meek et al. 1999.)

parallel fibers and should not be confused with the granule cells of the eminentia granularis (see later) or of the cerebellum, which do give rise to parallel fibers. Axons of ELL granular cells in the mormyrid ascend vertically to terminate in the plexiform, ganglionic, and deep molecular layers of ELL (Figs. 4.3 and 4.4). The granular cells relay the information from electroreceptors to higher order cells that are immediately external to them, thus preserving the spatial information conveyed by electroreceptor afferents.

Immunocytochemistry shows that many of the granular cells are γ-aminobutyric-ergic (GABAergic) (Bell, Yang, and Meek, unpublished observations) and the hypothesis is that all of them release inhibitory transmitter at their terminals (Mohr et al. 2003). In addition, unusual contacts are present between ELL granular cells and ELL cells that may function as excitatory electrical synapses (Meek, personal communication). These hypothesized inhibitory and excitatory actions of ELL granular cells are diagrammed in Figure 4.4 as separate inhibitory and excitatory cells.

The medium ganglionic cell (MG cell) is the most numerous higher order cell of ELL and is probably of central importance for ELL function (Figs. 4.3 and 4.4; Meek et al. 1996). MG cells may be described as "Purkinje-like" because: (1) they are GABAergic inhibitory cells; (2) they have spine covered apical

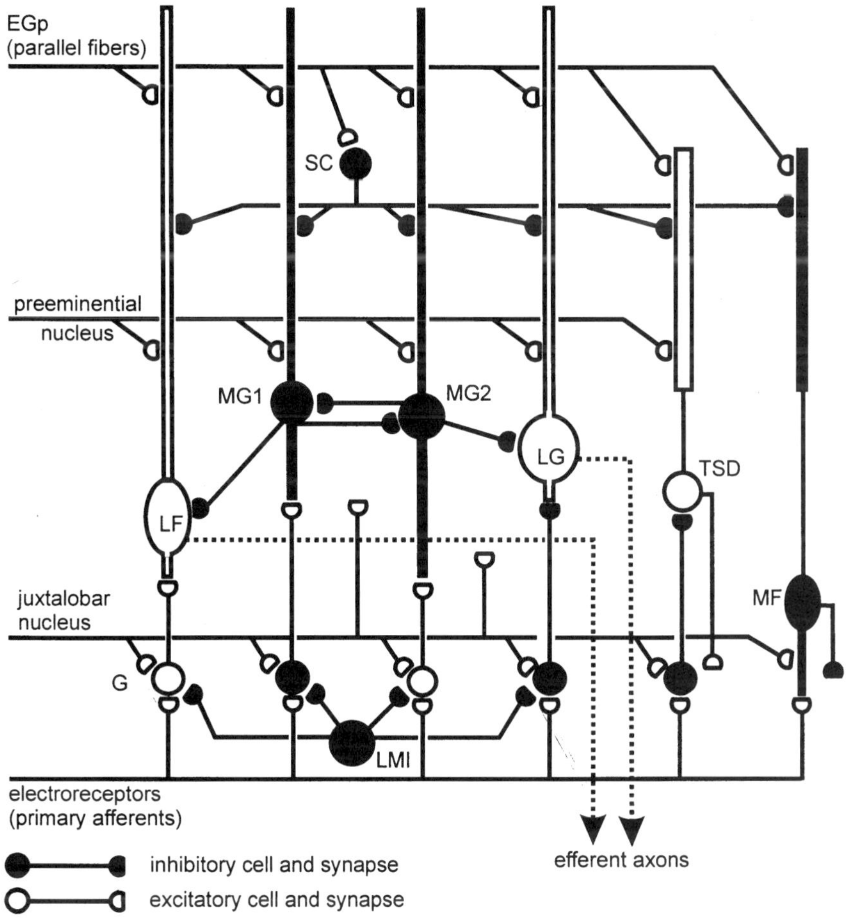

FIGURE 4.4. Diagram of cells and connectivity in mormyrid ELL. The four inputs to ELL are shown at the left: primary afferents from electroreceptors; fibers from the juxtalobar nucleus; fibers from the preeminential nucleus; and parallel fibers from EGp. The efferent axons from the structure are shown at right as *dotted lines*. (Adapted from Mohr et al. 2003.)

dendrites contacted by parallel fibers; and (3) their axons terminate locally on efferent cells within a few hundred microns of their cell bodies. The latter property is a general feature of the actinopterygian (ray-finned fishes—Chondrostei, Holostei, Teleostei) cerebellum in which Purkinje cell axons do not leave the cortex but rather terminate locally on efferent cells that convey information away from the cortex and are the equivalent of cerebellar nucleus cells in mammals (Meek 1998).

The cell bodies of MG cells are in the ganglionic layer and their apical dendrites extend throughout the molecular layer (Meek et al. 1996). MG cells are of two morphologically distinct types, MG1 and MG2. MG1 cells have basilar dendrites that are confined to the plexiform layer and axonal arbors that descend into the granular layer, whereas MG2 cells have basilar dendrites that descend into the granular layer and axonal arbors that are confined to the plexiform layer (Meek et al. 1996; Han et al. 1999). Recent physiological experiments indicate that MG1 cells are inhibited by electrosensory stimuli in the center of their receptive field (I cells), whereas MG2 cells are excited by such stimuli (E cells; Mohr et al. 2003).

ELL efferent cells, which convey electrosensory information from ELL to higher levels of the electrosensory system, are also of two main types: large ganglionic cells with cell bodies in the ganglionic layer and large fusiform cells with cell bodies at the boundary between the granular and plexiform layers or in the granular layer itself (Figs. 4.3 and 4.4; Grant et al. 1996; Meek et al. 1999). Both cell types have apical dendrites in the molecular layer as well as basilar dendrites. The basilar dendrites of large ganglion cells are in the plexiform layer, whereas those of large fusiform cells are in the granular layer. Large ganglionic cells are inhibited by electrosensory stimuli (I cells) whereas large fusiform cells are excited by such stimuli (E cells; Bell et al. 1997b).

MG cells synapse on other MG cells and on efferent cells (Meek et al. 1996). The axonal arbors of MG cells suggest that MG1 cells synapse on large fusiform cells whereas MG2 cells synapse on large ganglion cells. This hypothesis, that I type MG cells inhibit E type efferent cells and vice versa, is included in the circuit diagram of Figure 4.4. This hypothesis has not yet been confirmed but makes functional sense since inhibition of E or I type efferent cells by MG cells of the same type would cancel the sensory responses of efferent cells.

Other important interneurons of the mormyrid ELL besides granular cells and MG cells include the following: (1) Large multipolar interneurons (LMI in Figs. 4.3 and 4.4). These large GABAergic neurons of the intermediate layer have myelinated dendrites that form large inhibitory terminals on granular cells (Meek et al. 2001). They appear to mediate a rapidly acting lateral inhibition at the initial stages of sensory processing in ELL that could enhance the small differences in EOD-evoked afferent latency that are believed to convey the critical information about the electrical images of objects (Han et al. 2000). (2) Medium fusiform cells (MF in Figs. 4.3 and 4.4). These GABAergic neurons were referred to as "small fusiform cells" in a previous publication (Han et al. 1999). They have somas and basilar dendrites in the granular layer and apical dendrites in the deep molecular layer. Their axons terminate widely in the granular layer (Mohr et al. 2003). (3) Thick smooth dendrite cells (TSD in Figs. 4.3 and 4.4). These cells of the ganglionic and plexiform layers have a rich dendritic plexus in the deep molecular layer and recurrent dendrites that can reach the granular layer. The axons terminate widely in the granular layer (Han et al. 1999; Mohr et al. 2003). (4) Stellate cells (SC in Figs. 4.3 and 4.4). These GABAergic cells of the molecular layer are like the molecular layer interneurons of the cerebellum and of other cerebellum-like structures. (5) Interzonal cells. Reciprocal interzonal connections link somatotopically corresponding points in the two mormyromast zones of ELL (Bell et al. 1981; Mohr et al. 2003).

The nELL has no connections with the cortex of the ELL and is a far simpler structure than the cortex (Fig. 4.2B). Primary afferent fibers from knollenorgan electroreceptors terminate on the round, adendritic cells of this nucleus with large synaptic terminals of a mixed chemical–electrical morphology (Szabo et al. 1983). All cells of the nELL send their axons to the anterior exterolateral nucleus of the mesencephalon via the lateral lemniscus (see below). The large electrical synapses made by knollenorgan afferents, the adendritic cell bodies, and the long thin initial segment are all structural features that enhance the faithful transmission of temporal information (Carr 1986).

2.2.2 The ELL of Gymnarchid Fish

Gymnarchus niloticus has ampullary electroreceptors and two types of tuberous electroreceptor—the S type, which signals the phase of the wavelike EOD of these fish, and the O type, which signals the amplitude of the EOD (see Jørgensen, Chapter 3; Kawasaki, Chapter 7). Afferent fibers from all three types of electroreceptors enter the brain via both anterior and posterior lateral line nerves and terminate in ELLs. Fibers from the anterior body terminate anteriorly and those from the posterior body terminate posteriorly.

The ELL of *Gymnarchus* consists of bilateral lobes in the roof of the medulla (Fig. 4.2C). Each lobe is divided into three zones: a dorsal zone (DZ), a medial zone (MZ), and a ventral zone (Bass and Hopkins 1982). The interior of each lobe contains a cluster of adendritic giant neurons that are about 40 mm in diameter (GCS in Fig. 4.2C). Primary afferent fibers of the S type terminate in the MZ, and most of these S type fibers give off collaterals that terminate on the giant cells (Kawasaki and Guo 1996). The termination regions for the O and ampullary types of afferent fibers are not known, but physiological results suggest that the MZ receives input from these two types of afferent fibers, as well as from the S type.

The MZ of the *Gymnarchus* ELL is unusual in that it has two large cell layers, an inner cell layer and an outer cell layer. The outer cell layer is continuous with the single large cell layers of the dorsal and ventral zones. Intracellular labeling shows that cells in both the inner and outer cell layers of the medial zone have apical dendrites that extend up into the molecular layer (Kawasaki and Guo 1998). Cells in both layers also have basilar dendrites.

The round, adendritic giant cells in the interior of the ELL are similar in morphology to cells of the nELL in mormyrids, and like them are involved in the transmission of precise temporal information. The thick axons of individual giant cells project bilaterally to the MZ of the cortex, where they branch widely in the inner cell layer. The input from giant cells interacts in the inner cell layer with direct input from S type electroreceptor afferents to compute differences in phase between different skin regions (Kawasaki and Guo 1996; Kawasaki, Chapter 7).

2.2.3 The ELL of Xenomystine Fish

The Xenomystinae include two species, *Xenomystus nigri* and *Papycranus afer*. Both species are electroreceptive. Afferent fibers from electroreceptors in *Xenomystus* enter the brain via both anterior and posterior lateral line nerves, as in mormyrids and *Gymnarchus,* and terminate in the ELL (Braford 1986). Fibers from the anterior body terminate medially in the ELL, and fibers from the posterior body terminate laterally. The ELL of xenomystids is a bilateral structure and has three layers: a deep layer where primary afferent fibers terminate, a layer of large cells, and a molecular layer (Fig. 4.2D).

2.3 The ELLs and the Termination of Afferent Fibers from Electroreceptors in Gymnotiform and Siluriform Fish

The gymnotiforms and siluriforms are closely related sister orders of teleost fish. The gymnotiforms are all electric fish with at least three different types of electroreceptors. The siluriforms include one strongly electric species (*Malapterurus ellctricus*; Bennett 1971) and a few weakly electric species (Hagedorn et al. 1990; Baron et al. 1994), but most siluriforms do not have electric organs. Siluriforms have only the ampullary electroreceptors.

2.3.1 The ELL of Gymnotiform Fish

Gymnotiform fish are divided into wave and pulse species based on their pattern of EOD (Zupanc and Bullock, Chapter 2; Macadar et al., Chapter 14). The wave species emit a continuous highly regular EOD, whereas the pulse species emit pulselike EODs with interpulse intervals exceeding pulse duration. Almost all of the morphological and physiological work on the CNS of gymnotiform fish has been done with wave species. Thus, the descriptions of the gymnotiform nervous system in this chapter refer to wave fish, unless there is an explicit reference to pulse fish.

Gymnotiform fish have two classes of electroreceptors: ampullary and tuberous. Ampullary receptors are, as in Mormyridae, responsible for passive detection of low-frequency external electric fields (Jørgensen, Chapter 3). Tuberous receptors of pulse species code for the amplitude of the EOD mainly by the number of spikes they produce per EOD (pulse coders; Jørgensen, Chapter 3), although there may also be tuberous receptors that use a latency code. Tuberous receptors in wave fish include phase coders (T units) and "probability" coders (P units). The T unit discharge is precisely synchronized with the EOD and increases in EOD amplitude result in shortened response latency (or alternatively, in a decrease in the phase of the response with respect to the EOD cycle). P units discharge in an irregular probabilistic manner in response to the unperturbed EOD. Increases or decreases in EOD amplitude (amplitude modulations [AMs]) lead to corresponding changes in discharge probablility.

Primary electroreceptor afferents enter the medulla only from the anterior lateral line nerve and terminate in the deep layers of the ELL (Carr and Maler 1986). The gymnotiform ELL is a bilateral structure protruding laterally from the dorsal medulla and capped by the molecular layer of the caudal lobe of the cerebellum and the eminentia granularis posterior (EGp; Fig. 4.2E). Each ELL is divided into four segments: medial (MS), centromedial (CMS), centrolateral (CLS), and lateral (LS) segments. Each segment is similarly differentiated into several distinct laminae: a deep fiber layer containing primary afferent fibers; a deep neuropil where the electroreceptor afferents terminate on spherical cells and on the basal dendrites of pyramidal cells and interneurons; a granular layer containing mainly two classes of interneurons; a plexiform layer containing the efferent axons of pyramidal cells; a pyramidal cell layer containing the projection neurons of the ELL; the stratum fibrosum, a compact layer of myelinated fibers conveying direct feedback to the ELL; and a molecular layer where direct and indirect feedback pathways terminate on the apical dendrites of pyramidal cells (Fig. 4.5).

Ampullary receptors terminate topographically within the MS, whereas tuberous receptors (T and P units) trifurcate as they enter the ELL and terminate as three topographic maps within the CMS, CLS, and LS (Carr et al. 1982; Heiligenberg and Dye 1982). T units terminate mainly on spherical cells (Maler 1979; Fig. 4.5). The spherical cells are round and adendritic or paucidendritic like the cells of the nucleus of the mormyrid ELL and the giant cells of the gymnarchid ELL. All three cell types are engaged in the faithful transmission of temporal information. The number, morphology, and biochemistry of cells in the T-unit pathway vary with the EOD frequency of the genus, with the number of T units and spherical cells decreasing in higher frequency species (Losier and Matsubara 1990a,b).

There are two major categories of ELL projection neurons associated with P-unit input: basilar and non-basilar pyramidal cells (Fig. 4.5). Basilar pyramids have basal dendrites in the deep neuropil layer that receive direct P-unit input (Maler 1979). These cells are excited by increases in EOD amplitude (E cells; Saunders and Bastian 1984). Non-basilar pyramidal cells lack a basal dendrite and are inhibited by P-unit input disynaptically. These cell are inhibited by increases in EOD amplitude (I cells; Saunders and Bastian 1984). Both basilar and non-basilar pyramidal cells extend dendrites into the molecular layer. Bastian and co-workers (Bastian and Zakon, Chapter 8; Bastian and Courtwright 1991) have described subtypes of both basilar and non-basilar pyramidal cells, which differ in terms of their location (superficial versus deep), morphology, physiology, and biochemistry (Zupanc et al. 1992; Berman et al. 1995).

Three classes of interneurons receive strong P-unit input: ovoid cells and type 1 and type 2 granular interneurons (Fig. 4.5). Ovoid cells are GABAergic cells which receive mixed gap junctional and chemical synaptic input from P units and are not influenced by feedback input. These cells have widespread projections in the ipsilateral and contralateral ELLs that terminate on the basal trunk of basilar pyramidal cells (Bastian et al. 1993; Maler and Mugnaini 1994). Although both the in vivo (Bastian et al. 1993) and in vitro (Berman and Maler

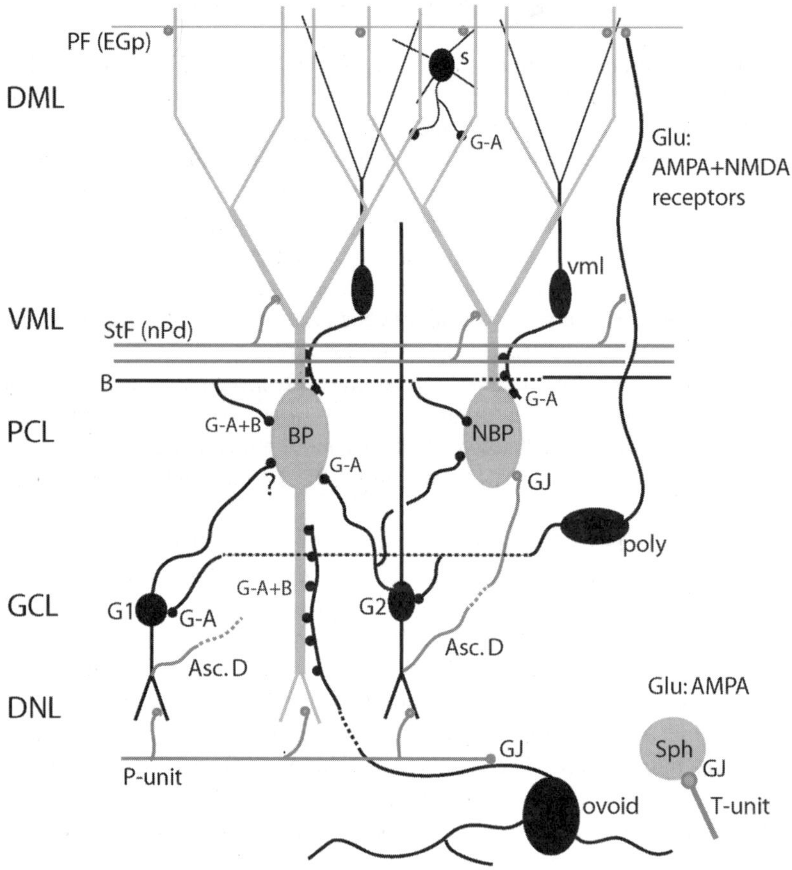

FIGURE 4.5 Diagram of cells and connectivity in gymnotid ELL. The drawing shows major elements of ELL circuitry within one tuberous segment of the gymnotiform ELL. The deep basilar pyramidal cells are not included in this diagram for lack of space (they are found within the granular layer and therefore do not receive inhibitory input from G1 and G2 interneurons). Inhibitory elements are shown in *black* and excitatory ones in *gray*. Basilar pyramidal cells (BP) have a basal dendrite that ramifies in the deep neuropil where it receives glutamatergic input from P units. Both basilar and non-basilar (NBP) cells have dendrites that extend into the molecular layer (VML+DML), where they receive glutamatergic input from the dorsal preeminential nucleus (nPd; direct feedback via StF) and EGp (indirect feedback via parallel fibers). GABAergic bipolar cells from the preeminential nucleus (B via StF) inhibit through both GABA-A and GABA-B receptors. Molecular layer interneurons (stellate and vml cells) also inhibit both types of pyramidal cells. Granular (G1 and G2) and ovoid interneurons receive input primarily from P units. Both BP and NBP cells receive inhibitory input from G1 (transmitter unknown) and G2 (GABAergic, GABA-A receptors) interneurons. The GABAergic ovoid cell projects (bilaterally) to the basal trunk of distant (*dashed line*) basilar pyramidal cells (GABA-A and B receptors). Ascending dendrites of granular interneurons make gap junction contacts on the somata and somatic dendrites (*not shown*) of distant (*dashed line*) non-basilar pyramidal cells. Polymorphic cells (*poly*) receive primarily excitatory feedback input via their apical dendrites in the molecular layer. They project (bilaterally) on distant (*dashed line*) granular interneurons. Spherical cells receive input only from phase coders (*T units*).

1999) physiology of ovoid cells and their synaptic contacts have been studied in detail, there is still no clear understanding of the role of this interneuron.

Both type 1 and 2 granular interneurons receive direct P-unit input to their basal dendrites (Fig. 4.5). Type 2 granular interneurons also have apical dendrites that receive feedback input in the molecular layer (Maler 1979). Both types of granular cells project to pyramidal cells and electron microscopic analysis has demonstrated that their synaptic contacts are suggestive of inhibitory synapses (Maler et al. 1981). More recently Maler and Mugnaini (1994) have demonstrated that type 2 granular cells are GABAergic while type 1 cells are not. Berman and Maler have confirmed that type 2 cells are inhibitory and utilize GABA-A receptors. These authors suggested that type 1 cells are also inhibitory, but did not establish the transmitter. It has been proposed that the type 1 interneurons are responsible for center-surround inhibition of pyramidal cells (Berman and Maler 1999; Bastian et al. 2002).

The ELL of pulse gymnotiforms is roughly similar to that of wave gymnotiforms (Réthelyi and Szabo 1973b) and has the same four segments—medial, centromedial, centrolateral, and lateral—each with a somatotopically organized map of the electroreceptive periphery. Spherical cells are present and primary afferent fibers contact these spherical cells with mixed chemical–electrical synapses (Castello et al. 1998).

2.3.2 The ELL of Siluriform Fish

The afferent fibers from electroreceptors in Siluriforms enter the medulla via both the anterior and posterior lateral line and terminate in the deeper layers of the ELL (Fig. 4.2F). The afferents form a single map with anterior body mapped medially and posterior body mapped laterally The layering of ELLs in siluriforms is similar to that in gymnotiforms, although simpler. There are two types of pyramidal cells with morphological and physiological properties similar to basilar and non-basilar pyramidal cells (McCreery 1977; Finger 1986).

3. Ascending and Descending Connections of First-Order Electrosensory Structures

Ascending electrosensory pathways to the mesencephalon are similar in all electroreceptive fish. These pathways are also similar to the pathways that convey mechanical lateral line and auditory information in fish (Coombs and Montgomery, Chapter 12). The efferent axons of large cells in the cerebellum-like first-order structures of electrosensory systems enter the lateral lemniscus where they ascend to terminate in the mesencephalon. The projections are predominantly contralateral but a clear ipsilateral component is present in all fish. The bilateral projections arise from the branching of individual efferent axons in mormyrids (Bell et al. 1981), *Gymnarchus* (Kawasaki and Guo 1998), and gymnotids (Carr and Maler 1986), and this may be true in other fish also.

In most cases, the ascending axons give off collaterals to nuclei that in turn project back to the first-order structures, providing a short feedback loop. Such feedback projections are both direct, to the first-order structure itself, and indirect, via the eminentia granularis. These nuclei that project back to the first-order structures receive not only the collaterals of ascending lemniscal fibers, but also input from still higher centers, including the mesencephalic nuclei where efferent axons from the first-order structure terminate.

The term "descending connection" is used here to refer to all of the input from other central structures to the first-order structures, that is, for all of the input except for the primary afferent input from the periphery. Such central input is prominent in the first-order structures of fish electrosensory systems, indicating the presence of strong central modulation of sensory processing at this first stage.

3.1 Connections of the DON in Non-neopterygian Fish

Fibers from the DON terminate in the optic tectum and in the lateral mesencephalon in non-neopterygian fish (Boord and Northcutt 1982; Ronan 1986; Schmidt and Bodznick 1987; Hofmann et al. 2002). The region of the mesencephalon where the fibers terminate is known as the lateral mesencephalic nucleus in elasmobranchs and the torus semicircularis ("torus") in lampreys and non-neopterygian bony fish. The direct projection from first-order electrosensory structures to the tectum does not occur in electroreceptive teleost fish, and is thus a clear difference between electrosensory pathways in teleost and non-teleost fish.

Ascending and descending connections of the DON have been examined in most detail in the skate (Bodznick and Boord 1986: Fig. 4.6). In the skate, efferent axons from the DON project not only to the torus and optic tectum but also to a "nucleus B," the nucleus of the lateral lemniscus, and possibly to the paralemniscal nucleus. Nucleus B (Smeets 1982) is an elongated nucleus in the medulla that receives a strong projection from the ipsilateral DON and projects back to the ipsilateral DON, providing a major feedback loop. Other inputs to nucleus B are not known.

The paralemniscal nucleus is of particular interest from a comparative perspective. Efferent axons from the DON that ascend to the mesencephalon may give off collaterals to this nucleus. The paralemniscal nucleus projects back to both the DON and the eminentia granularis, and it receives input from the lateral mesencephalic nucleus where efferent axons from the DON terminate. Thus, the connections of this nucleus are very similar to those of the preeminential nucleus, a major feedback structure within the electrosensory system of electroreceptive teleosts (see Section 3.2). A small group of cells in lampreys, which are located near the ascending axons from the DON and which project back to the DON, appear to have a similar connectivity (Ronan 1986).

The eminentia granularis of the skate relays information from a rich variety of sources to the parallel fibers in the molecular layer of the DON. Physiological

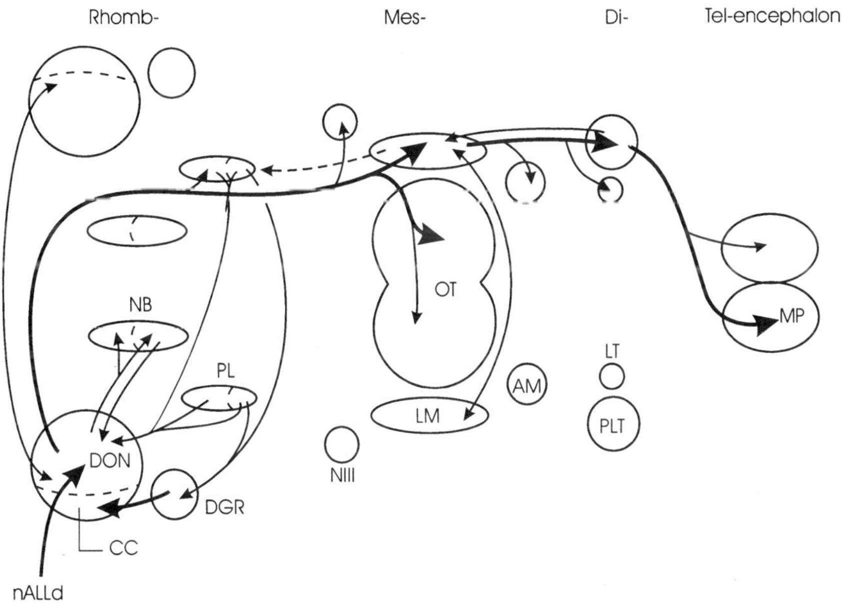

FIGURE 4.6. Ascending and descending connections of the elasmobranch electrosensory system. Corresponding structures on both sides of the brain are shown at four levels of the neuraxis: the rhombencephalon, mesencephalon, diencephalon, and telencephalon. (From Bodznick and Boord 1986.)

recordings from the eminentia granularis show that the eminentia granularis receives electrosensory input, proprioceptive input, and corollary discharge input associated with motor commands that drive respiration (Hjelmstad et al. 1996; Chapter 6, Bodznick and Montgomery). The structures that have been demonstrated anatomically as projecting to the eminentia granularis include the paralemniscal nucleus, the spinal cord, and two brain stem nuclei, nucleus F and nucleus K (Bodznick and Boord 1986). The paralemniscal nucleus could be a source of electrosensory input and the spinal cord could be a source of proprioceptive input, but nothing is known about the type of information conveyed by nucleus F and nucleus K.

3.2 Ascending and Descending Connections of First-Order Structures in Mormyrid Fish

The connections of the cortex of the mormyrid ELL are quite distinct from those of the the nucleus of the ELL and are discussed first. Ascending axons from the cortex of the ELL project bilaterally to the lateral nucleus of the torus semicircularis ("torus") in the mesencephalon (Fig. 4.7). The ascending axons give off collaterals to the dorsal preeminential nucleus and the medial ventral nucleus

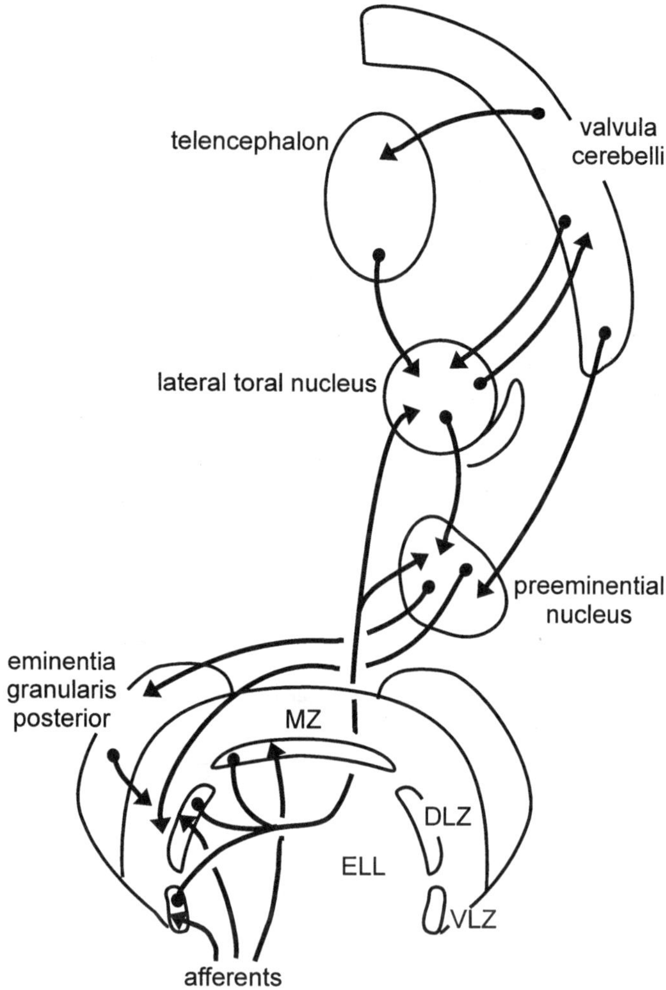

FIGURE 4.7. Major ascending and descending connections of the mormyrid electrosensory system. (Adapted from Bell and Szabo 1986.)

of the torus. The dorsal preeminential nucleus (referred to in subsequent paragraphs as simply "the preeminential nucleus") is a center for feedback from higher order electrosensory structures to the ELL in electroreceptive teleosts. A ventral preeminential nucleus plays a similar role in the auditory and lateral line system (Bell 1981; Montgomery et al. 1995: Coombs and Montgomery, Chapter 12). The projections to the preeminential and lateral toral nuclei are somatotopically organized. Somatotopically corresponding points within all three zones of ELL cortex project to a single point in both the preeminential and lateral toral nuclei, suggesting that information from ampullary receptors and from the two types of sensory cells in mormyromast receptors is combined in these higher-level nuclei (Bell et al. 1981).

The major descending connections to the cortex of the ELL are from the eminentia granularis posterior (EGp), the preeminential nucleus, and the medial juxtalobar nucleus. Neuromodulatory inputs include serotoninergic (Grant et al. 1989) and noradrenergic (Meek et al. 1993) fibers (Fig. 4.3).

Granule cells of EGp relay information from many different sources to the parallel fibers of ELL, as described earlier in elasmobranchs. These sources of input to EGp in the mormyrid include: the spinal cord; the preeminential nucleus; brain stem proprioceptive nuclei; mechanical lateral line receptors; cells in the intermediate layer of the ELL; and two nuclei, the paratrigeminal command associated nucleus and the lateral juxtalobar nucleus, that transmit electric organ corollary discharge (EOCD) signals associated with the motor command that drives the EOD (Bell et al. 1983); see Section 5). EOCD, proprioceptive, lateral line, and electrosensory signals in EGp have also been demonstrated physiologically (Bell et al. 1992).

The information from all of the different inputs to EGp is combined and transformed by the intrinsic circuitry of EGp into the patterns of parallel fiber activity. The intrinsic circuitry of EGp includes unipolar brush cells and GABAergic Golgi cells (unpublished findings of Bell, Meek, and Wang).

The preeminential nucleus of mormyrids receives input from the ELL, from the lateral toral nucleus, and the valvula cerebelli (Fig. 4.7), and projects back both directly and indirectly to the ELL. The major direct projection is a topographically organized projection to the deep molecular layer of the ELL (Bell et al. 1981). A second, less prominent, direct projection to the ELL arises from large GABAergic cells at the medial edge of the preeminential nucleus and enters the ELL from below (not shown in Fig. 4.7). The mode of termination of this second direct projection is not known. The preeminential nucleus also has a major indirect projection to the ELL via a strong connection to EGp. These three projections from the preeminential nucleus to the ELL, two direct and one indirect, arise from three discrete populations of cells within the nucleus (Mohr and Bell, unpublished observations).

The juxtalobar nucleus is located at the anterior ventral margin of the ELL. The cells in the medial half of this nucleus send axons to the mormyromast regions of the ELL, where they terminate predominantly in the granular layer. These fibers convey an electric organ corollary discharge (EOCD) signal (Bell and von der Emde 1995; see Section 5).

Axons from the nELL terminate in the anterior exterolateral nucleus of the torus, after giving off collaterals to the medial ventral nucleus of the torus (Bell et al. 1981; Haugedé-Carré 1980). The medial ventral nucleus also receives input from the ELL cortex, and this nucleus is therefore one of the few sites where information from mormyromast, ampullary, and knollenorgan receptors can converge.

The descending input to the nELL is entirely inhibitory and arises from GABAergic cells in a sublemniscal nucleus located at the level of the preemi-

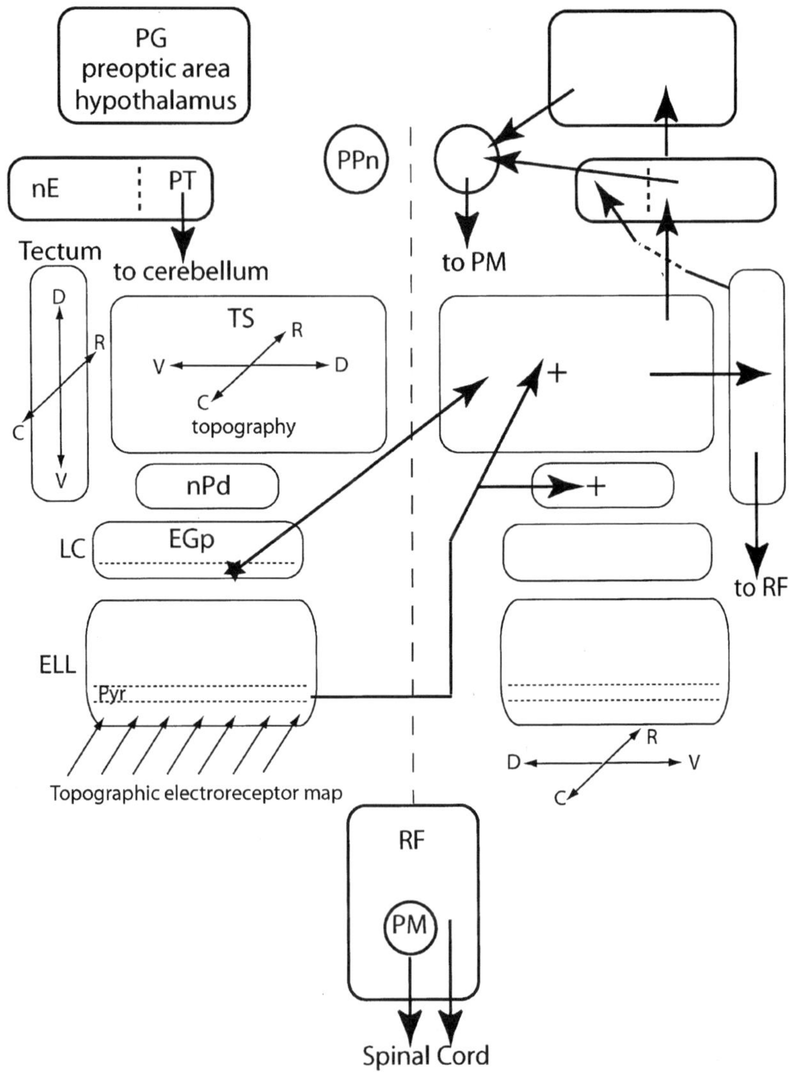

FIGURE 4.8. Major ascending and motor control pathways of the gymnotiform electrosensory system. Contralateral pathways are omitted as is the small sublemniscal prepacemaker nucleus. Within ELL the mapping of dorsal body (D) medially and ventral body (V) laterally that is shown in the figure is true for the medial and centrolateral segments of ELL, but not for the centromedial or lateral segments where the mapping is with dorsal body lateral and ventral body medial. The torus semicircularis (TS) receives electrosensory input from ELL and a minor input from eurydendroid cells (star) of the caudal lobe of the cerebellum (LC) which is associated with the eminentia granularis posterior (EGp). The TS projects topographically to the optic tectum which in turn

nential nucleus (Bell et al. 1981; Mugnaini and Maler 1987b). The descending input conveys an EOCD signal that elicits a powerful and precisely timed inhibition (Bell and Grant 1989). The inhibition completely blocks the entry into the CNS of the reafferent response of knollenorgan afferent fibers to the fish's own EOD (see Kawasaki, Chapter 7).

3.3 Ascending and Descending Connections of First-Order Structures in Gymnarchid and Xenomystine Fish

The central electrosensory pathways have not been extensively studied in these two groups of fish, but some of the major connections are known and are similar to those in mormyrids. The efferent cells from the ELL project bilaterally to a lateral region of the mesencephalic torus in both groups. The ascending axons give off collaterals to the preeminential nucleus in *Gymnarchus* (Kawasaki and Guo 1998) but do not appear to do so in *Xenomystus* (Braford 1986). The preeminential nucleus receives input from the torus in both groups and has been shown to project back to ELL in *Xenomystus*.

3.4 Ascending and Descending Connections of First-Order Structures in Gymnotiform and Siluriform Fish

3.4.1 Connections of the ELL in Gymnotiform Fish

The efferent axons of both spherical cells (T-unit input) and pyramidal cells (P-unit input) project from the gymnotiform ELL to the torus semicircularis ("torus") of the mesencephalon via the lateral lemniscus (Fig. 4.8). The axons of pyramidal cells but not spherical cells give off collaterals to the preeminential nucleus as they ascend. The preeminential projection preserves the overall topography of the ELL, so that lateral ELL projects dorsally and medial ELL projects ventrally. Thus, the segregation of ELL into four segments is preserved in the gymnotid preeminential nucleus (Maler et al. 1982). This is in contrast to what occurs in the mormyrid where the three zones of the ELL converge onto a single map in the preeminential nucleus.

The four ELL maps converge in the torus, in contrast to the segregation of these maps in the projection to the preeminential nucleus. Rostral body maps rostrally and caudal body maps caudally, with the dorsoventral axis of the body

---

FIGURE 4.8. *Continued*
projects to various brain stem regions. These include the reticular formation (which projects to the spinal cord) and the pretectum (PT, which projects to the corpus cerebelli). The TS also projects (non-topographically) to the nucleus electrosensorius (nE). The nE projects to the prepacemaker nucleus (PPn) as well as various hypothalamic nuclei. The PPn also receives input from hypothalamic, preglomerular (PG) and reticular formation (RF) cell groups. The PPn is the main source of input to the medullary pacemaker nucleus (PM). +, Excitatory connections; −, inhibitory connections.

mapping mediolaterally in the torus (Carr and Maler 1986). The gymnotiform torus is a complex structure with eight major layers (12 layers altogether) and about 50 cell types (Carr and Maler 1981, 1985). T-units project exclusively to layer 6, pyramidal cells representing P-unit input (CMS, CLS, and LS segments of the ELL) project mainly to laminae 5 and 7 (lesser projections to 3, 8C, 8D, and 9). The medial segment of ELL (ampullary region) projects mainly to laminae 3 and 7 (Rose and Call 1992).

The preeminential nucleus in gymnotiforms, as in mormyrids, receives input from both the ELL and the torus (Fig. 4.9). Two types of cells in the preeminential nucleus project back directly to ELL: GABAergic, bipolar cells in the medial part of the nucleus (labeled B in Fig. 4.9) and glutamatergic stellate cells in the central nucleus. Both cell types send thin myelinated axons via the stratum fibrosum to the ELL. The bipolar axons terminate in a spatially diffuse manner within the ELL pyramidal cell layer (Maler and Mugnaini 1994).

Stellate cells project back to the ELL in a reciprocal topographic manner and terminate on the proximal spines of pyramidal cell apical dendrites as well as on molecular layer interneurons (Maler 1979, 1981; Fig. 4.5). The projection is excitatory (Berman et al. 1997) and has been hypothesized to mediate an attention "searchlight" (Bratton and Bastian 1990), similar to that proposed for cortico–thalamic feedback projections (Crick 1984). Several cell types within the preeminential nucleus (not including bipolar or stellate cells) affect ELL indirectly via a projection to the EGp where they terminate in a diffuse manner (Sas and Maler 1987).

EGp granule cells bifurcate and send parallel fibers to the dorsal molecular layer of ELL where they terminate on spines of pyramidal cell apical dendrites (Maler 1979). The indirect pathway from the preeminential nucleus to ELL via EGp is involved in gain control of pyramidal cells (Bastian 1986). The EGp of gymnotiforms, like that of mormyrids, receives proprioceptive information from brain stem centers (Sas and Maler 1987; Bastian 1995).

Relatively little is known about the electrosensory pathways in pulse gymnotiforms. It is known, however, that spherical cells from the ELL of pulse fish project to a separate nucleus known as the magnocellular mesencephalic nucleus (Réthelyi and Szabo 1973a; Castello et al. 1998). Each EOD of the pulse fish evokes a brief and extremely short latency response in the magnocellular nucleus that appears to mark the occurrence of the fish's own EOD.

3.4.2 Ascending and Descending Connections of the First-Order Structure in Siluriform Fish

The ascending and descending connections of ELLs in siluriform fish are similar to those in gymnotiform fish. The ELL projects to both the preeminential nucleus and to the torus. The torus projects back to the preeminential nucleus and the preeminential nucleus projects back to the ELL both directly and indirectly via EGp (Finger 1986).

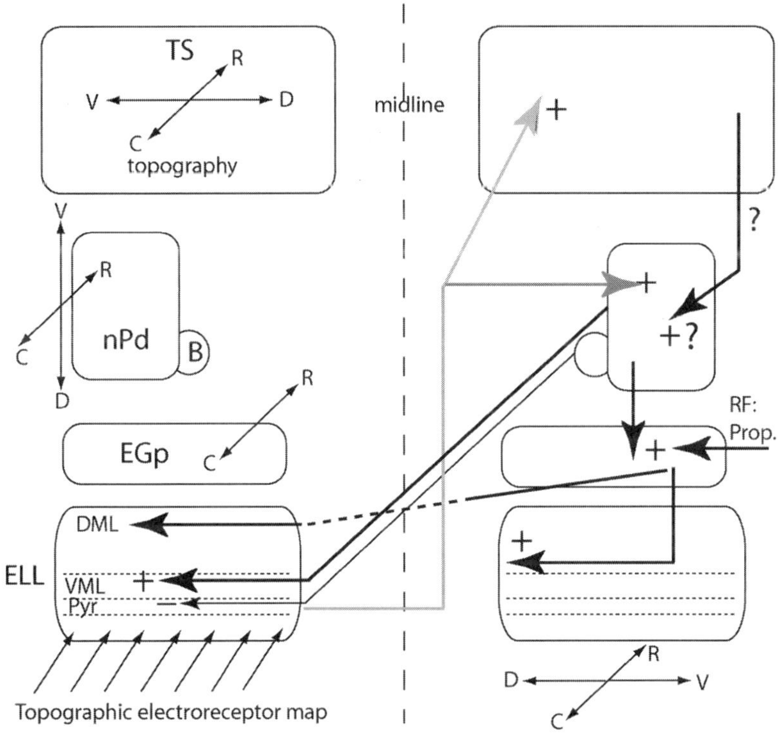

FIGURE 4.9. Major descending pathways of the gymnotid electrosensory system. The diagram shows the feedback pathways associated with one tuberous segment of the gymnotiform ELL. The dorsal–ventral topography alternates in mirror image fashion across the ELL and nPd maps; the orientation shown (in ELL and nPd) would be seen in the medial and centrolateral segments and would be reversed in the centromedial and lateral segments. Smaller ipsilateral projections (both from and to ELL) are also found but are not included in this diagram. The ELL projects topographically to nPd and TS (*gray line*). TS has massive feedback projections to nPd but their topography and transmitters have not been characterized (thus: ?). A subset of nPd neurons project back topographically to the ventral molecular layer of ELL. In addition, GABAergic bipolar cells of nPd project back to the ELL pyramidal cell layer (*thin black line, small arrow*). Different populations of nPd cells project to EGp; only a rostrocaudal topography has been described for this pathway. The EGp also receives brain stem (including proprioceptive) input. The EGp projects (parallel fibers) nontopographically across the ELL dorsal molecular layer. This projection is bilateral: ipsilateral fibers terminate within the ventral DML, while contralateral fibers terminate in the dorsal DML.

4. Higher-Order Electrosensory Pathways and Structures

The lemniscal pathways that convey electrosensory information to the mesencephalon are similar not only to the auditory–lateral line pathways of fish but also to the auditory pathways of terrestrial vertebrates (Montgomery et al. 1995). The torus semicircularis of fish, where the axons from most first-order electrosensory structures terminate, is possibly homologous to the inferior colliculus of birds and mammals. The connections at a still higher order, however, particularly those that convey information from the mesencephalon to the telencephalon through the diencephalon, are only poorly understood in fish and appear to be different in different species of fish. Thus, the similarity of these higher order connections, beyond the mesencephalon, to those in terrestrial vertebrates is unclear.

As stated previously, the DON of non-neopterygian fish projects not only to the lateral mesencephalon or torus but also to the optic tectum, whereas the ELL of teleosts does not project to the optic tectum. However, the torus of teleosts receives input from the ELL in mormyrids and gymnotiforms and does project to the optic tectum. The optic tectum is not only a visual processing center but also a center for motor control, particularly for orienting movements (Springer et al. 1977). The direct connections to the tectum in non-neopterygian fish and the indirect connections in teleost fish provide some of the necessary circuitry for electrosensory modulation of ordinary motor behavior.

The electrosensory system also modulates EOD behavior in electric teleosts. Such modulation includes the jamming avoidance response (JAR) in gymnotid wave fish (Heiligenberg 1991) and *Gymnarchus* (Kawasaki 1993), the novelty response in gymnotid pulse fish (Caputi et al. 2002) and mormyrid fish (Szabo and Fessard 1965; Post and von der Emde 1999), and the echo response in mormyrid fish (Russell et al. 1974). The central pathways responsible for the JAR in gymnotid fish have been well explored but the central pathways responsible for novelty responses in gymnotids and mormyrids and for the echo response in mormyrids remain obscure.

4.1 Higher-Order Electrosensory Structures in Non-neopterygian Fish

The only non-neopterygian fish in which the electrosensory pathways beyond the mesencephalon have been studied anatomically is the skate (Bodznick and Boord 1986). As described previously, the major projection of the DON in skates is to the lateral mesencephalic nucleus (Fig. 4.6). This projection is somatotopically organized, with the rostral body being represented rostrally and the caudal body being represented caudally in the mesencephalon. The lateral mesencephalic nucleus has a descending connection to the paralemniscal nucleus, and this nucleus appears to project back to the DON and the eminentia granularis, as described previously.

The major ascending connection from the lateral mesencephalic nucleus is to the lateral posterior nucleus of the diencephalon (Fig. 4.6). This connection is reciprocal, the diencephalic nucleus projecting back to the mesencephalic nucleus. The diencephalic nucleus also has an ascending projection to the medial pallium of the telencephalon. Physiological findings are in accord with these anatomical results in showing that responses to electrosensory stimuli can be recorded throughout much of the medial pallium (Bodznick and Northcutt 1984).

4.2 Higher-Order Electrosensory Pathways and Structures in Mormyrid Fish

Almost nothing is known about electrosensory structures and pathways beyond the mesencephalon in *Gymnarchus* or *Xenomystus*, but some work has been done on the higher-order electrosensory pathways of mormyrid fish. One of the striking features of the central electrosensory pathways in mormyrid fish is the nearly complete separation of the knollenorgan and mormyromast–ampullary pathways. The mormyromast–ampullary pathways that originate in ELL cortex are described first followed by the knollenorgan pathways that originate in nELL. The description of these two distinct electrosensory systems is followed by a brief description of the extraordinary cerebellum of mormyrid fish.

4.2.1 Mormyromast–Ampullary Pathways

The major destination of efferent axons from the ELL cortex is the lateral nucleus of the torus semicircularis in the mesencephalon (Fig. 4.7). The projection is somatotopically organized with the anterior ELL (representing the anterior body) projecting to the anterior lateral nucleus and posterior body (representing the posterior body) projecting to the posterior lateral nucleus. The lateral toral nucleus in turn projects to the preeminential nucleus, the valvula cerebelli, the optic tectum, the lateral toral nucleus on the contralateral side, the caudal preglomerular nucleus, and the dorsal preglomerular nucleus. The caudal and dorsal preglomerular nuclei, as described by Wullimann and Northcutt (Wullimann and Northcutt 1990), were previously known as the posteroventral thalamic and dorsal anterior pretectal nuclei, respectively (Stendell 1914; Finger et al. 1981). The projections of the lateral nucleus to the preeminential nucleus, the valvula, the optic tectum, and the contralateral lateral toral nucleus are topotopic, that is, point-to-point, preserving of spatial relationships among adjacent regions.

The valvula cerebelli is a part of the cerebellum that is found only in actinopterygian fish (Meek 1998) and is especially developed in mormyriform fish, where it covers most of the dorsal surface of the brain (see later). The projection from the lateral nucleus to the valvula cerebelli is massive. The fibers terminate in the granule layer of the valvula as mossy fibers. Physiological recordings identified three electrosensory regions of the mormyrid valvula, a dorsomedial "ampullary" region, a ventrolateral "mormyromast" region, and a lateral "knollenorgan" region (Russell and Bell 1978). The lateral toral nucleus projects to regions of the valvula that include the "ampullary" and "mormyromast" regions but also extends somewhat beyond the borders of these physiologically identified regions. Although the "ampullary" and "mormyromast" regions of the valvula had physiological differences suggestive of activation by the two different types of electroreceptors, such electroreceptor specificity could not be shown definitively, and there is as yet no indication of separate ampullary and mormyromast regions in the lateral nucleus.

The caudal and dorsal preglomerular nuclei receive input from the lateral nucleus and project as mossy fibers to the same regions of the valvula as those that receive direct input from the lateral nucleus. Thus, there are both direct and indirect connections between the lateral nucleus and the valvula.

Physiological recordings show that electrosensory information reaches the telencephalon (Prechtl et al. 1998), but the anatomical pathway by which it does so is not yet certain. The mormyrid valvula has an unusual direct projection to the telencephalon that might convey electrosensory information to the forebrain (Wullimann and Rooney 1990). A second possible pathway involves the medial ventral nucleus of the torus, which receives input from the ELL. The medial ventral nucleus of the torus projects to the ventral preglomerular nucleus that in turn projects to the telencephalon (von der Emde and Prechtl 1999). Finally, further pathway tracing experiments with more complete labeling of connections may establish a connection between the lateral toral nucleus and the telencephalon through one of the preglomerular nuclei since such a pathway has been established for mechanical lateral line and auditory information (von der Emde and Prechtl 1999). (See Braford 1995; and Braford and McCormick 1992 for a discussion of ascending sensory connections to the telencephalon in fish.)

The lateral toral nucleus receives input from the ELL cortex, from the valvula cerebelli, from the caudal lobe of the cerebellum, from large cells along the medial edge of the preeminential nucleus, from the medial funicular nucleus of the caudal brain stem that receives somatosensory input from the spinal cord, and from the area dorsalis pars centralis of the telencephalon (Finger et al. 1981). The regions of the valvula that project to the lateral toral nucleus are the same as those that receive input from this nucleus. These regions also project to the preeminential nucleus.

Two features of the central pathways associated with the mormyromast-ampullary component of the mormyrid electrosensory system stand out: the descending control of sensory processing and recurrent feedback loops. Most of the descending connections referred to in the preceding paragraphs are massive and are conveyed by large fiber tracts. Thus, higher levels of the system, such as the valvula cerebelli and telencephalon, are only a few synapses away from the first stage of processing in the ELL and must exert strong effects on this first stage. These two features, descending control of sensory processing and recurrent feedback loops, are also present in the other electrosensory systems described here.

4.2.2 Knollenorgan Pathways

Primary afferent fibers from knollenorgan receptors terminate only in the NELL, and fibers from the NELL terminate mainly in the anterior exterolateral nucleus (ELa) of the torus semicircularis ELa. The knollenorgan pathway as a whole appears to be entirely concerned with the temporal analysis of the EODs of other fish (Friedman and Hopkins 1998; Xu-Friedman and Hopkins 1999). There is no somatotopy in the projection from the nELL to ELa and individual axons from the nELL terminate widely in the ELa. The ELa is remarkably simple in its histology and connectivity. It receives input only from the nELL and projects only to the exterolateral posterior nucleus (ELp), another toral nucleus that is immediately posterior to the ELa. The ELa has only two types of neurons, large cells and small cells, both of which are adendritic and both of which receive synaptic input from the nELL via mixed electrical–chemical synapses. The axons from the nELL terminate first on the large cells and then on the small cells. The large cells are GABAergic (Mugnaini and Maler 1987a) and branch widely in the nucleus, terminating in clusters of inhibitory terminals on the somas of small cells. The small neurons project to ELp.

The circuitry of the ELa appears to be well suited for analyzing the small differences in EOD duration that signal sexual and species differences in mormyrid fish (Hopkins and Bass 1981; Friedman and Hopkins 1998; Kawasaki, Chapter 7). The ELa appears to be a site where the coding of temporal information by the exact timing of action potentials is transformed into the coding of such information by a place or frequency code. In this respect, the ELa is like the medial zone of the gymnarchid ELL (Kawasaki and Guo 1996), layer 6 of the gymnotid torus (Carr and Maler 1986), and nucleus laminaris of the barn owl (Carr 1986).

The posterior exterolateral toral nucleus (ELp) projects to the medial ventral nucleus of the torus; to the isthmic granule nucleus located lateral to the preeminential nucleus; and to the subpreeminential nucleus, a cell group just ventral to the preeminential nucleus (HaugedJ-CarrJ 1979; Finger et al. 1981). The connections of the subpreeminential nucleus are not known. The isthmic granule nucleus projects as mossy fibers to a lateral region of the valvula that has been shown to respond physiologically to input from knollenorgan electroreceptors (Russell and Bell 1978) and to have a role in communication (Szabo and Moller 1984).

4.2.3 Mormyrid Cerebellum

The cerebellum of actinopterygian fish (chondrosteans, holosteans, and teleosts) has three main parts: a caudal or vestibulolateral lobe associated with vestibular, lateral line, and electrosensory systems; a centrally located corpus cerebelli; and an anterior valvula cerebelli (Meek 1998). The valvula cerebelli is unique to actinopterygian fish, other vertebrates having only a corpus cerebelli and a vestibulolateral lobe. All three parts of the cerebellum are large in mormyriform fish. The valvula cerebelli and corpus cerebelli of mormyriform fish are unusual not only for their large size but also for the extraordinary regularity of their histological structure (Nieuwenhuys and Nicholson 1969; Meek and Nieuwenhuys 1991).

The mormyrid valvula is of truly extraordinary extent, covering most of the dorsal surface of the brain. At least half the surface area of the valvula is devoted to the processing of electrosensory and auditory-lateral line information. The reciprocal connections between the lateral toral nucleus and the valvula were described in preceding paragraphs, as were the connections to the preeminential nucleus. Reciprocal connections also exist between another region of the valvula and the mediodorsal toral nucleus, a nucleus that receives auditory and lateral line information from the brain stem (Haugedé-Carré 1980; Bell 1981). The fact that large regions of the cerebellum of mormyrid fish are clearly involved in the control of sensory processing, via reciprocal connections with mesencephalic sensory structures, argues that the cerebellum is not only involved in motor control, but may also modulate sensory and other types of nonmotor information processing (Paulin 1993; Kelly and Strick 2003).

The caudal lobe of the mormyrid cerebellum has two parts, an anterior caudal lobe associated with the eighth nerve and lateral line systems, and a posterior caudal lobe associated with the electrosensory system. The parallel fibers of the posterior caudal lobe arise from a dorsal portion of the EGp, and this part of EGp receives input from the same sources as the granule cells that give rise to the parallel fibers of the ELL. Intracellular recording from Purkinje cells of the caudal lobe shows that these cells respond to electrosensory, proprioceptive, and EOCD inputs that arrive at the cells via the parallel fibers (Campbell and Bell, unpublished observations). The same recordings show that the climbing fibers to these cells also respond to electrosensory and EOCD signals.

Efferent cells from the posterior caudal lobe in mormyrids ascend to terminate mainly in the preeminential nucleus and the lateral toral nucleus (Bell et al. 1981). Thus, the connections of the posterior caudal lobe and the ELL are very similar. Both receive parallel fiber input from the EGp and both project to the preeminential and lateral toral nuclei. This caudal lobe circuit, with connections that are so similar to those of the ELL, is also present in gymnotids (Carr and Maler 1981). The caudal lobe projections to sensory nuclei illustrate, as do the valvula connections in mormyrids, that the cerebellum is not exclusively a motor structure.

4.3 Higher-Order Electrosensory Pathways and Structures in Gymnotiform Fish

The torus of gymnotiform fish, like the lateral toral nucleus of mormyrids, receives input not only from ELL but also from the caudal lobe of the cerebellum, the tectum, a somatosensory nucleus of the brain stem, and the telencephalon (Carr and Maler 1981; Fig. 4.9).

Carr and Maler (1985) have provided a thorough description of the morphology of toral neurons. Some cells have small dendritic arbors, whereas others have widely spreading dendritic arbors. Presumably cells in the first class would have relatively restricted receptive fields while those in the second class would integrate P-unit input over extensive spatial regions. Moreover, some cells have dendritic arbors confined to a single lamina while others have dendritic arbors in two or more laminae. The intrinsic circuitry of the torus is still essentially unknown.

Detailed analysis of the jamming avoidance response (JAR) of *Eigenmannia* by Heiligenberg and colleagues (Heiligenberg 1991) has demonstrated that this behavior requires a combination of two types of spatial comparison: comparison of the phase (timing) of the EOD at different body locations and comparison of the amplitude of the EOD at different body locations. The T units provide the temporal information on EOD timing required for the first computation. Carr and colleagues (Carr et al. 1986) have analyzed the toral circuitry in layer 6 required for this computation. There are two cell types in layer 6: large adendritic cells (giant cells) and small cells with thin dendrites. Spherical cell axons terminate topographically onto the somata of the giant cells and onto the dendrites of the small cells; gap junction contacts are made in both cases. The giant cells send thick axons that cover most of the ipsilateral and contralateral layer 6, terminating with gap junction contacts onto the somata of the small cells. Each small cell therefore receives input from spherical cells (on its dendrites) representing the phase or timing of the EOD within a small patch of skin. The same granule cell also receives input (on its soma) from giant cells representing the EOD phase from a different patch of skin. All possible spatial phase comparisons are made in this way by the small cells of layer 6. The small layer 6 cells then project to adjacent toral laminae where their phase comparison information is presumably combined with input related to local EOD amplitude (P-unit input relayed by ELL pyramidal cells).

The torus has three targets associated with different functions: the preeminential nucleus (electrosensory feedback; Fig. 4.9), the optic tectum (object localization and motor control; Fig. 4.8), and the nucleus electrosensorius (electrocommunication; Fig. 4.8) (Carr et al. 1981). The projection to preeminentialis is exclusively ipsilateral and massive, as in mormyrids. Thus, again as in mormyrids, the results of higher level processing are fed back to the ELL, where they may be presumed to have strong effects on initial processing.

The tectum of gymnotiform fish is highly developed and there is general conservation of cell types previously identified in other teleosts (Sas and Maler 1986a). Retinal projections to the tectum are sparse (Sas and Maler 1986b). Tectal circuitry has been co-opted by massive input from the torus that is in register with the retinotopic map (Sas and Maler 1986a). The tectum of *Apteronotus leptorhynchus* projects to brain stem regions (Heiligenberg and Rose 1987) that in turn project to the spinal cord (Behrend and Donicht 1990), as in other teleosts. The tectum therefore provides a pathway for electrosensory control of motor output.

The torus also projects to the nucleus electrosensorius (nE), a complex cluster of subnuclei on the lateral aspect of the pretectum (diencephalon), and to the adjacent pretectal area (Keller et al. 1990; Fig. 4.8). The toral projection to the nE is bilateral and diffuse. Spatial topography does not appear to be well preserved in the nE. Detailed anatomical and physiological studies of the nE have shown that it consists of clusters of functionally defined cells with specific connectivity; these clusters are all related to electrocommunication (Keller and Heiligenberg 1989; Heiligenberg 1991). The nE projects to the hypothalamus (Keller et al. 1990) and the prepacemaker nucleus (see later).

Interestingly, the pretectal region medial to nE, that receives toral and tectal input, projects to the caudal cerebellum (Keller et al. 1990; Fig. 4.8). Earlier studies had already revealed that cells within this region of the cerebellum had large complex electrosensory receptive fields and responded to movement of objects with such receptive fields (Bastian 1975).

5. Motor Pathways Related to the EOD

Most of the recent work on the central motor pathways associated with the generation of EODs has been done in mormyriform and gymnotiform electric fish, and this section is concerned only with these two groups of fish. Very little work has been published in recent years on the central motor pathways related to the EOD in other groups of electric fish. A review by Bennett (1971) describes early work on the weakly electric rays, the strongly electric torpedo, the strongly electric stargazer (Astroscopus), and the strongly electric catfish (Malapterurus).

In all mormyriform and gymnotiform fish, the EOD is initiated in a midline medullary nucleus known as the "command" nucleus in mormyrids and as the "pacemaker" in gymnotids and *Gymnarchus*. The term "command" is used in mormyrids because the marked irregularity of inter-EOD intervals in mormyrid fish makes the term "pacemaker" inappropriate. In all of these fish, the nucleus that initiates the EOD excites a relay nucleus, either directly (in gymnotids and mormyrids) or indirectly (in *Gymnarchus*). The relay nucleus is also located on the midline in the medulla. The axons of the relay nucleus descend to the spinal cord to activate electromotoneurons. In both mormyriform and gymnotiform fish, the activities of command and pacemaker nuclei are modulated by descending input from mesencephalic and diencephalic nuclei.

This section reviews only a few aspects of the EOD motor systems in mormyriform and gymnotiform electric fish. Other chapters in this book should be consulted for more complete descriptions (Macadar et al., Chapter 14; Kawasaki, Chapter 7).

5.1 EOD-Related Structures in Mormyrid Fish

The command nucleus is located just beneath the relay nucleus in mormyrids (Szabo 1957b; Bell et al. 1983). The axons of the command nucleus branch, one branch ascends vertically to excite the relay nucleus, the other branch travels laterally and caudally to excite the bulbar command associated nucleus (BCA; Fig. 4.10). Axons from the BCA ascend. After giving off a collateral to the relay nucleus, the axons from the BCA continue to ascend. They terminate in the paratrigeminal command associated nucleus (PCA) and the mesencephalic command associated nucleus (MCA).

The projection from the BCA to the PCA and MCA is the beginning of the mormyrid EOCD pathway (Fig. 4.10; Bell et al. 1983). The EOCD signals in this pathway prepare central electrosensory structures for the (re)afferent sensory input evoked by the EOD (Bell 1986). Axons from the PCA project to the EGp, providing one source of the EOCD driven parallel fiber activity that is believed to underlie the generation of EOCD-driven, memory-like expectations in the ELL (Bell et al. 1997b; Bastian and Zakon, Chapter 8). Axons from the MCA descend to terminate in juxtalemniscal and sublemniscal regions near the preeminential nucleus. Along the way the axons of MCA give off collaterals that terminate in a dorsal part of the ventral posterior toral nucleus (Carlson 2002;

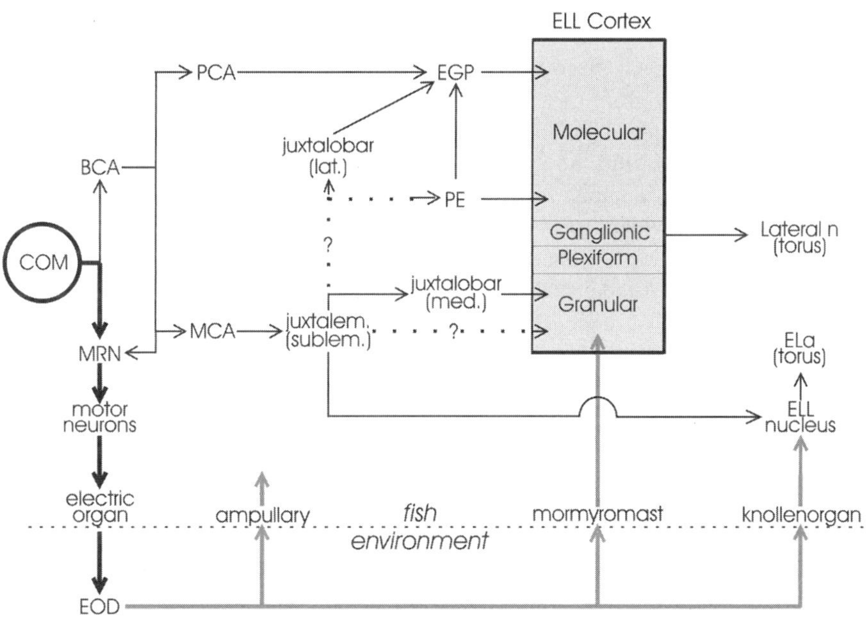

FIGURE 4.10. EOD motor command and corollary discharge paths in mormyrid fish. The motor pathway from the command nucleus to the EOD is shown as a thick black line, and the reafferent pathway from electroreceptors to termination of primary afferent fibers in ELL is shown as a thick gray line. The corollary discharge pathway and electrosensory projections to the torus are shown as thin black lines. Precommand nuclei in the mesencephalon and diencephalon are not shown. (Adapted from Bell et al. 1995.)

Macadar, Caputi, and Carlson, Chapter 11). These toral cells project to precommand nuclei of the mesencephalon and appear to mediate an EOCD-driven inhibition of the precommand neurons following each EOD command. The ventral posterior toral cells are therefore responsible, at least in part, for the interval between EODs.

Inhibitory neurons of the sublemniscal region that receive input from MCA project to the nELL, where they are responsible for the EOCD-driven inhibition of nELL that was described previously (Bell et al. 1981). Excitatory neurons of the juxtalemniscal region project to the juxtalobar nucleus, a small, compact nucleus at the anterior ventral limit of ELL (Bell and von der Emde 1995). Neurons in the lateral part of the juxtalobar nucleus project to EGp, and are another of the inputs responsible for EOCD driven activity in parallel fibers (Bell et al. 1981). Neurons in the medial part of the juxtalobar nucleus project to the cellular layers of the mormyromast regions of ELL, where they selectively enhance EOD-evoked reafferent input from mormyromast receptors and probably also provide a timing pulse for the decoding of reafferent latency as a measure of stimulus intensity (Hall et al. 1995).

The timing of the EOD motor command is preserved with remarkable fidelity through the different stages of the EOCD pathway. A comparison of the timing of the discharge of electromotoneurons, as recorded with an electrode near the tail of the fish, with the timing of spikes in the juxtalobar nucleus showed only 50 μs of variability (Bell and von der Emde 1995). The electromotoneurons are two synapses away from the command nucleus and the juxtalobar cells are five synapses away from the command nucleus in the opposite direction, for a total of seven synapses between the two measured sites. Thus, timing information is preserved with an accuracy of better than 50 μs across five to seven synaptic levels. In contrast to some of the other neural systems in which temporal information is well preserved, preliminary findings concerning the EOCD pathway suggests that chemical synapses as well electrical synapses are present (Bell and von der Emde 1995), that the axons joining different nuclei within the pathway are of only moderate thickness, and that cells in some of the nuclei have extended dendritic arbors (Bell, unpublished observations).

The EOCD pathways in mormyrid fish serve a variety of functions, as described in the preceding paragraphs and in more detail elsewhere (Bell 1986; Hall et al. 1995). To summarize:

1. EOCD effects on precommand nuclei of the EOD motor system modulate inter EOD intervals.
2. The EOCD inhibits the first central stage of the knollenorgan pathway, blocking the potentially disruptive reafferent response of these receptors.
3. The EOCD enhances EOD-evoked reafferent input from mormyromast electroreceptors in ELL.
4. A precisely timed EOCD signal to ELL probably serves to decode reafferent latency as a measure of stimulus intensity.
5. EOCD-driven activity in parallel fibers of the ampullary and mormyromast

zones of ELL can elicit memory-like expectations of sensory input that serve to remove predictable components of the sensory responses.

EOCD signals are also recorded in the valvula and other electrosensory structures beyond ELL (Russell and Bell 1978; Bell et al. 1995), but the functions of these higher-level EOCD inputs are not known.

5.2 EOD-Related Structures in Gymnarchus

The command nucleus that initiates the EOD in *Gymnarchus* is located just beneath the relay nucleus, as in mormyrids. Curiously, the command nucleus in *Gymnarchus* does not project directly to the relay nucleus but rather projects to a lateral relay nucleus that may be homologous to the bulbar command associated nucleus of mormyrid fish (Szabo 1957a; Kawasaki 1994). The lateral relay nucleus projects to the relay nucleus that in turn projects to the electromotoneurons of the spinal cord. All of these connections are exclusive, neither the command nucleus, nor the lateral relay nucleus, nor the relay nucleus project to any other central structures. Thus, the anatomy suggests that *Gymnarchus* lacks the corollary discharge pathway associated with the EOD motor command that is so prominent in mormyrid fish (Kawasaki, Chapter 7). Physiological and behavioral studies also indicate an absence of EOCD effects in *Gymnarchus* (Kawasaki 1996).

5.3 EOD-Related Structures in Gymnotiform Fish

The EOD of gymnotiform fish is initiated in a medullary pacemaker nucleus that excites the relay neurons. Relay neurons send their axons down the spinal cord to excite electromotoneurons that drive the electric organ to discharge (see Macadar et al., Chapter 14).

The EOD of high-frequency species normally maintains a constant frequency and amplitude. During social interactions the EOD frequency and amplitude may, however, change. One well known EOD modulation is the jamming avoidance response (JAR); as its name suggest, the JAR prevents the fish's electrolocation system from being jammed by the EOD of neighbors. The JAR has been thoroughly reviewed recently (Heiligenberg 1991; Metzner 1999; Kawasaki, Chapter 7) and is not discussed further here. High-frequency gymnotiforms (especially *A. leptorhynchus*) also modulate their EOD for purposes of electrocommunication. Various types of transient (chirps) and gradual (slow rises) EOD frequency modulations have been described in the context of agonistic, courtship, and mating behaviors (Hagedorn and Heiligenberg 1985; Dulka et al. 1995; Dunlap and Larkins-Ford 2003).

EOD frequency modulations are controlled by a diencephalic prepacemaker nucleus (PPn) and a rhombencephalic sublemniscal prepacemaker nucleus (SPPn) (for a recent review see Metzner, 1999). The SPPn is responsible for the EOD decelerations of the JAR in *Eigenmannia*, but for more complex communication signals (not the JAR) in apteronotids. The PPn is a complex nucleus associated with the central posterior diencephalic nucleus (CP; this nucleus is also found in the diencephalons of non-electroreceptive teleosts) consisting of two subdivisions: PPnG and PPnC (Kawasaki et al. 1988; Zupanc and Maler 1997). Stimulation of PPnG causes gradual (G) increases in EOD frequency (JAR and slow rises) while stimulation of the PPnC causes transient EOD accelerations known as chirps (C). The PPn/CP complex proper receives input from nE, thus closing the loop between electrosensory input and electromotor output (Keller et al. 1990). In addition, PPn/CP also receives extensive input from various non-electrosensory regions including the hypothalamus, and stimulation of these regions can modulate chirping (Zupanc and Maler 1997). The non-electrosensory input to PPn is associated with monoaminergic and peptidergic pathways emanating from the brain stem, basal forebrain, and hypothalamus (Wong 1997a,b; Zupanc and Maler 1997; Correa and Zupanc 2002) which are believed to underlie the ability of hormones, and "motivational" state, to modulate the propensity of an animal to produce chirps or slow rises (Maler and Ellis 1987; Dulka et al. 1995; Bastian et al. 2001; Dunlap and Larkins-Ford 2003).

6. Summary and Conclusions

Electrosensory systems are attractive sites in which to examine some general issues of information processing in the nervous system, including *the preservation and analysis of temporal information*; *the integration of time and intensity information*; *the descending control of sensory processing*; *the storage of sensory patterns*; and *the control of motor activity*. This chapter has provided the anatomical knowledge that is needed for understanding how electrosensory systems can help clarify these general issues.

The neural systems that are specialized for the *preservation of temporal information* include: the knollenorgan (communication) and EOCD pathways of mormyrid fish; the phase comparison system of *Gymnarchus*; and the spherical cell or phase comparison pathways of gymnotid fish. Some of these systems are characterized by large electrical synapses, round adendritic cells, and thick axons but others, such as the EOCD pathway of mormyrid fish, do not have these features. Circuitry for *analysis of temporal information* has been identified in layer 6 of the gymnotid torus (Carr and Maler 1986), the medial zone of the ELL of *Gymnarchus* (Kawasaki and Guo 1996), the ELa of mormyrid fish (Xu-Friedman and Hopkins 1999), and may be present in the ELL of mormyrid fish (Hall et al. 1995).

The JAR is one of the best-studied examples of the *integration of time and intensity information* in sensory-motor systems and the circuitry responsible for this behavior has been described in both gymnotiforms (Heiligenberg 1991) and *Gymnarchus* (Kawasaki and Guo 1996).

The *descending control of sensory processing* is anatomically very prominent in electrosensory systems, as it is in most vertebrate sensory systems. Moreover, the functions of this descending input are perhaps better understood in parts of the electrosensory systems than in most other systems. Examples from electrosensory systems include: the different effects of EOCD input on sensory processing in the mormyrid ELL (Bell 1986); the generation of predictions about sensory patterns by descending parallel fiber input to cerebellum-like first order structures (Bell 1982; Bodznick 1993; Bastian 1995; Bell et al. 1997a); and the varied roles of descending preeminential input to the ELL, as studied in gymnotid fish (Chacron et al. 2003; Doiron et al. 2003).

The anatomy of cerebellum-like first-order structures in the different electrosensory systems is very similar and the anatomical projections to the granule cells of these structures indicate the types of predictive signals that the parallel fibers convey. These first-order structures and their connections appear to mediate the *storage of sensory patterns* through an associative process between parallel fiber signals and patterns of sensory input.

Finally, this chapter has described some of the anatomical pathways involved in the generation, control, and distribution of EOD motor commands. The simplicity of EOD behavior, the ability to evoke such behaviors while recording from central cells, and the many results that have already been obtained indicate that electrosensory systems are useful sites in which to examine *the control of motor activity*.

Abbreviations

AM	amplitude modulation
AM	anterior mesencephalic nucleus
AMPA	amino-3-hydroxy-5-methylisoxazolepropionic acid glutamate receptors
Asc-D	ascending dendrite of granular interneurons
B	bipolar cell of preeminential nucleus
BCA	bulbar command associated nucleus
BP	basilar pyramidal cell
C	caudal (body)
CB	cerebellum
CC	cerebellar crest
CLS	centrolateral segment
CMS	centromedial segment
COM	EOD motor command nucleus
CP	central posterior diencephalic nucleus
D	dorsal (body)
DF	deep fiber layer
DGR	dorsal granular ridge
Di	diencephalon

DLZ	dorsolateral zone
DM	deep molecular layer interneuron
DML	dorsal molecular layer
DNL	deep neuropil layer
DON	dorsal octavolateral nucleus
EGp	eminentia granularis posterior
EI	efferent neuron of the intermediate layer
ELa	anterior exterolateral nucleus of the torus
ELL	electrosensory lateral line lobe
Elp	posterior exterolateral nucleus of the torus
EOCD	electric organ corollary discharge
EOD	electric organ discharge
EPSP	excitatory postsynaptic potential
FL	facial lobe
G	granular cell
G1	type 1 granular interneuron
G2	type 2 granular interneuron
GA	ganglionic layer
G-A	GABAergic synapse with GABA-A receptors
G-A+B	GABAergic synapse with both GABA-A and GABA-B receptors
GABA	γ-aminobutyric acid
GCL	granular cell layer
GCS	giant cells
GJ	gap junction synapse
Glu	glutamatergic synapse
GR	granular layer
H	horizontal cell
IN	intermediate layer
IPSP	inhibitory postsynaptic potential
JAR	jamming avoidance response
JL	juxtalobar nucleus
juxtalem	juxtalemniscal region
L	lateral toral nucleus
LC	caudal lobe of the cerebellum
LCm	molecular layer of caudal lobe of cerebellum
LF	large fusiform cell
LG	large ganglion cell
LM	lateral mesencephalic nucleus
LMI	large multipolar interneuron
LS	lateral segment
LT	lateral tuberal nucleus
MCA	mesencephalic command nucleus
Mes	mesencephalon
MG1	medium ganglion cell type 1
MG2	medium ganglion cell type 2

ML	molecular layer
MLF	medial longitudinal fasciculus
MO	medial octavolateral nucleus
MP	medial pallium
MRN	medullary relay nucleus
MS	medial segment
MZ	medial zone
NA	noradrenergic fibers
nALLd	dorsal root of anterior lateral line nerve
NB	nucleus B
NBP	non-basilar pyramidal cell
nE	nucleus electrosensorius
nELL	nucleus of ELL
Nlll	nucleus of lateral lemniscus
NMDA	*N*-methyl-D-aspartate glutamate receptors
nPd	dorsal preeminential nucleus
nVIII	eighth nerve
OT	optic tectum
ovoid	ovoid cell
P	plexiform layer
PCA	paratrigeminal command associated nucleus
PCL	pyramidal cell layer
PE	preeminential nucleus
pf	parallel fibers
PL	paralemniscal nucleus
PLT	posterior lateral thalamic nucleus
PM	pacemaker nucleus
poly	polymorphic cell
PPn	prepacemaker nucleus
PPnC	chirp-inducing part of prepacemaker nucleus
PPnG	gradual rise inducing part of prepacemaker nucleus
Prop	proprioceptive afferents
PT	pretectum
P-unit	probability coding afferent
Pyr	pyramidal cells
R	rostral (body)
RF	reticular formation
Rhomb	rhombencephalon
S	stellate interneuron
SC	stellate cell
Ser	serotoninergic fibers
SF	small fusiform cell
SMI	small multipolar neuron of the intermediate layer
Sph	spherical cell
SPPn	sublemniscal prepacemaker nucleus

StF	stratum fibrosum
Sublem	sublemniscal nucleus
TS	torus semicircularis
T unit	time (or phase) coding afferent
V	ventral (body)
VA	valvula cerebelli
VML	ventral molecular layer
vml	neuron of ventral molecular layer.

References

Baron VD, Orlov AA, Golubtsov AS (1994) African *Clarias* catfish elicits long-lasting weak electric pulses. Experientia 50:644–647.

Bass AH, Hopkins CH (1982) Comparative aspects of brain organization of an African "wave" electric fish, *Gymnarchus niloticus*. J Morphol 174:313–334.

Bastian J (1975) Receptive fields of cerebellar cells receiving exteroceptive input in a gymnotid fish. J Neurophysiol 38:285–300.

Bastian J (1986) Gain control in the electrosensory system mediated by descending inputs to the electrosensory lateral line lobe. J Neurosci 6:553–562.

Bastian J (1995) Pyramidal-cell plasticity in weakly electric fish: a mechanism for attenuating responses to reafferent electrosensory inputs. J Comp Physiol A 176:63–78.

Bastian J, Courtright J (1991) Morphological correlates of pyramidal cell adaptation rate in the electrosensory lateral line lobe of weakly electric fish. J Comp Physiol A 168: 393–407.

Bastian J, Courtright J, Crawford J (1993) Commissural neurons of the electrosensory lateral line lobe of *Apteronotus leptorhynchus*: morphological and physiological characteristics. J Comp Physiol A 173:257–274.

Bastian J, Schniederjan S, Nguyenkim J (2001) Arginine vasotocin modulates a sexually dimorphic communication behavior in the weakly electric fish *Apteronotus leptorhynchus*. J Exp Biol 204:1909–1923.

Bastian J, Chacron MJ, Maler L (2002) Receptive field organization determines pyramidal cell stimulus-encoding capability and spatial stimulus selectivity. J Neurosci 22:4577–4590.

Behrend K, Donicht M (1990) Descending connections from the brainstem to the spinal cord in the electric fish *Eigenmannia*. Brain Behav Evol 35:227–239.

Bell CC (1981) Some central connections of medullary octavolateral centers in a mormyrid fish. In: Fay RR, Popper AN, Tavolga WN (eds), Hearing and Sound Communication in Fishes. New York: Springer-Verlag, pp. 383–392.

Bell CC (1982) Properties of a modifiable efference copy in an electric fish. J Neurophysiol 47:1043–1056.

Bell CC (1986) Electroreception in mormyrid fish: central physiology. In: Bullock TH, Heiligenberg W (eds), Electroreception. New York: John Wiley & Sons, pp. 423–452.

Bell CC (1990) Mormyromast electroreceptor organs and their afferents in mormyrid electric fish: III. Physiological differences between two morphological types of fibers. J Neurophysiol 63:319–332.

Bell CC (2002) Evolution of cerebellum-like structures. Brain Behav Evol 59:312–326.

Bell CC, Grant K (1989) Corollary discharge inhibition and preservation of temporal information in a sensory nucleus of mormyrid electric fish. J Neurosci 9:1029–1044.

Bell CC, Szabo T (1986) Electroreception in mormyrid fish: central anatomy. In: Bullock TH, Heiligenberg W (eds), Electroreception. New York: John Wiley & Sons, pp. 375–421.

Bell CC, von der Emde G (1995) Electric organ corollary discharge pathways in mormyrid fish: II. The medial juxtalobar nucleus. J Comp Physiol A 177:463–479.

Bell CC, Finger TE, Russell CJ (1981) Central connections of the posterior lateral line lobe in mormyrid fish. Exp Brain Res 42:9–22.

Bell CC, Libouban S, Szabo T (1983) Pathways of the electric organ discharge command and its corollary discharges in mormyrid fish. J Comp Neurol 216:327–338.

Bell CC, Zakon H, Finger TE (1989) Mormyromast electroreceptor organs and their afferent fibers in mormyrid fish: I. Morphology. J Comp Neurol 286:391–407.

Bell CC, Grant K, Serrier J (1992) Sensory processing and corollary discharge effects in the mormyromast regions of the mormyrid electrosensory lobe: I. Field potentials and cellular activity in associated structures. J Neurophysiol 68:843–858.

Bell C, Dunn K, Hall C, Caputi A (1995) Electric organ corollary discharge pathways in mormyrid fish: I. The mesencephalic command associated nucleus. J Comp Physiol A 177:449–462.

Bell C, Bodznick D, Montgomery J, Bastian J (1997a) The generation and subtraction of sensory expectations within cerebellum-like structures. Brain Behav Evol 50:17–31.

Bell CC, Caputi A, Grant K (1997b) Physiology and plasticity of morphologically identified cells in the mormyrid electrosensory lobe. J Neurosci 17:6409–6422.

Bennett MVL (1971) Electric organs. In: Hoar WS, Randall DJ (eds), Fish Physiology. New York: Academic Press, pp. 347–491.

Berman NJ, Maler L (1999) Neural architecture of the electrosensory lateral line lobe: adaptations for coincidence detection, a sensory searchlight and frequency-dependent adaptive filtering. J Exp Biol 202:1243–1253.

Berman NJ, Hincke MT, Maler L (1995) Inositol 1,4,5-trisphosphate receptor localization in the brain of a weakly electric fish (*Apteronotus leptorhynchus*) with emphasis on the electrosensory system. J Comp Neurol 361:512–524.

Berman NJ, Plant J, Turner R, Maler L (1997) Excitatory amino acid transmission at a feedback pathway in the electrosensory system. J Neurophysiol 78:1869–1881.

Bodznick D (1993) The specificity of an adaptive filter that suppresses unwanted reafference in electrosensory neurons of the skate medulla. Biol Bull 185:312–314.

Bodznick D, Boord RL (1986) Electroreception in Chondrichthyes: central Anatomy and Physiology. In: Bullock TH, Heiligenberg W (eds), Electroreception. New York: John Wiley & Sons, pp. 225–256.

Bodznick D, Northcutt RG (1984) An electrosensory area in the telencephalon of the little skate, *Raja erinacea*. Brain Res 298:117–124.

Boord RL, Northcutt RG (1982) Ascending lateral line pathways to the midbrain of the clearnose skate, *Raja eglanteria*. J Comp Neurol 207:274–282.

Braford MR (1986) African knife fishes: the Xenomystines. In: Bullock TH, Heiligenberg W (eds), Electroreception. New York: John Wiley & Sons, pp. 453–464.

Braford MR (1995) Comparative aspects of forebrain organization in the ray-finned fishes: touchstone or not? Brain, Behav Evol 46:259–274.

Braford MR, McCormick CA (1992) Brain organization in teleost fish: lessons from the electrosense. J Comp Physiol A 173:704–708.

Bratton B, Bastian J (1990) Descending control of electroreception. II. Properties of nucleus praeeminentialis neurons projecting directly to the electrosensory lateral line lobe. J Neurosci 10:1241–1253.

Bullock TH, Heiligenberg W (1986) Electroreception. New York: John Wiley & Sons.

Bullock TH, Bodznick DA, Northcutt RG (1983) The phylogenetic distribution of electroreception: evidence for convergent evolution of a primitive vertebrate sense modality. Brain Res Rev 6:25–46.

Caputi AA, Aguilera PA, Castello ME (2002) Probability and amplitude of novelty responses as a function of the change in contrast of the reafferent image in *G. carapo*. J Exp Biol 206:999–1010.

Carlson BA (2002) Neuroanatomy of the mormyrid electromotor control system. J Comp Neurol 454:440–455.

Carr CE (1986) Time coding in electric fish and barn owls. Brain Behav Evol 28:122–133.

Carr CE, Maler L (1981) Laminar organization of the afferent and efferent systems of the torus semicircularis of gymnotiform fish: morphological substrates for parallel processing in the electrosensory system. J Comp Neurol 203:649–670.

Carr CE, Maler L (1985) A Golgi study of the cell types of the dorsal torus semicircularis of the electric fish *Eigenmannia*: functional and morphological diversity in the midbrain. J Comp Neurol 235:207–240.

Carr CE, Maler L (1986) Electroreception in gymnotiform fish: central anatomy and physiology. In: Bullock TH, Heiligenberg W (eds), Electroreception. New York: John Wiley & Sons, pp. 319–374.

Carr CE, Maler L, Sas E (1982) Peripheral organization and central projections of the electrosensory organs in gymnotiform fish. J Comp Neurol 211:139–153.

Carr CE, Maler L, Taylor B (1986) A time-comparison circuit in the electric fish midbrain. II. Functional morphology. J Neurosci 6:1372–1383.

Castello ME, Caputi A, Trujillo-Cenoz O (1998) Structural and functional aspects of the fast electrosensory pathway in the electrosensory lateral line lobe of the pulse fish *Gymnotus carapo*. J Comp Neurol 401:549–563.

Chacron MJ, Doiron B, Maler L, Longtin A, Bastian J (2003) Non-classical receptive field mediates switch in a sensory neuron's frequency tuning. Nature 423:77–81.

Correa SA, Zupanc GK (2002) Connections between the central posterior/prepacemaker nucleus and hypothalamic areas in the weakly electric fish *Apteronotus leptorhynchus*: evidence for an indirect, but not a direct, link. J Comp Neurol 442:348–364.

Crick F (1984) Function of the thalamic reticular complex: the searchlight hypothesis. Proc Natl Acad Sci USA 81:4586–5490.

Doiron B, Chacron MJ, Maler L, Longtin A, Bastian J (2003) Inhibitory feedback required for network oscillatory responses to communication but not prey stimuli. Nature 421:539–543.

Dulka JG, Maler L, Ellis W (1995) Androgen-induced changes in electrocommunicatory behavior are correlated with changes in substance P-like immunoreactivity in the brain of the electric fish *Apteronotus leptorhynchus*. J Neurosci 15:1879–1890.

Duman CH, Bodznick D (1996) A role for GABAergic inhibition in electrosensory processing and common mode rejection in the dorsal nucleus of the little skate, *Raja erinacea*. J Comp Physiol A 179:797–807.

Dunlap KD, Larkins-Ford J (2003) Diversity in the structure of electrocommunication signals within a genus of electric fish. J Comp Physiol A 189:153–161.

Finger TE (1986) Electroreception in catfish: behavior, Anatomy, and electrophysiology.

In: Bullock TH, Heiligenberg W (eds), Electroreception. New York: John Wiley & Sons, pp. 287–318.

Finger TE, Bell CC, Russell CJ (1981) Electrosensory pathways to the valvula cerebelli in mormyrid fish. Exp Brain Res 42:23–33.

Friedman MA, Hopkins CD (1998) Neural substrates for species recognition in the time-coding electrosensory pathway of mormyrid electric fish. J Neurosci 18:1171–1185.

Grant K, Clausse S, Libouban S, Szabo T (1989) Serotinergic neurons in the mormyrid brain and their projection to the preelectromotor and primary electrosensory centers: immunohistochemical study. J Comp Neurol 281:114–128.

Grant K, Meek J, Sugawara Y, Veron M, Denizot JP, Hafmans J, Serrier J, Szabo T (1996) Projection neurons of the mormyrid electrosensory lateral line lobe: morphology, immunocytochemistry and synaptology. J Comp Neurol 375:18–42.

Hagedorn M, Heiligenberg W (1985) Court and spark: electric signals in the courtship and mating behavior of gymnotid fish. Anim Behav 33:254–265.

Hagedorn M, Womble M, Finger TE (1990) Synodontid catfish: a new group of weakly electric fish. Behavior and anatomy. Brain Behav Evol 35 :268–277.

Hall JC, Bell C, Zelick R (1995) Behavioral evidence of a latency code for stimulus intensity in mormyrid electric fish. J Comp Physiol A 177:29–39.

Han VZ, Bell CC, Grant G, Sugawara Y (1999) In vitro studies of the mormyrid electrosensory lobe: I. Morphology of cells and circuits. J Comp Neurol 404:359–374.

Han VZ, Grant K, Bell CC (2000) Rapid activation of GABAergic interneurons and possible calcium independent GABA release in the mormyrid electrosensory lobe. J Neurophysiol 83:1592–1604.

Haugedé-Carré F (1979) The mesencephalic exterolateral posterior nucleus of the mormyrid fish *Bryenomyrus niger*: efferent connections studied by the HRP method. Brain Res 178:79–84.

Haugedé-Carré F (1980) Contribution a l'Jtude des connexions du torus semicircularis et du cervelet chez certains mormyrides. Ph. D. Thesis. L'Univ. Pierre et Marie Curie, Paris.

Heiligenberg W (1991) Neural Nets in Electric Fish. Cambridge, MA: MIT Press.

Heiligenberg W, Dye J (1982) Labeling of electrosensory afferents in a gymnotid fish by intracellular injection of HRP: the mystery of multiple maps. J Comp Physiol 148: 287–296.

Heiligenberg W, Rose GJ (1987) The optic tectum of the gymnotiform electric fish, *Eigenmannia*: labeling of physiologically identified cells. Neuroscience 22:331–340.

Hjelmstad G, Parks G, Bodznick D (1996) Motor corollary discharge activity and sensory responses related to ventilation in the skate vestibulolateral cerebellum: implications for electrosensory processing. J Exp Biol 199:673–681.

Hofmann MH, Wojtenek W, Wilkens LA (2002) Central organization of the electrosensory system in the paddle fish (*Polyodon spathula*). J Comp Neurol 446:25–36.

Hopkins CD, Bass AH (1981) Temporal coding of species recognition signals in an electric fish. Science 212:85–87.

Kawasaki M (1993) Independently evolved jamming avoidance responses employ identical computational algorithms: a behavioral study of the African electric fish, *Gymnarchus niloticus*. J Comp Physiol A 173:9–22.

Kawasaki M (1994) The African wave-type electric fish, *Gymnarchus niloticus*, lacks corollary discharge mechanisms for electrosensory gating. J Comp Physiol A 174: 133–144.

Kawasaki M (1996) Comparative analysis of the jamming avoidance response in African and South American wave-type electric fishes. Biol Bull 191:103–108.

Kawasaki M, Guo YX (1996) Neuronal circuitry for comparison of timing in the electrosensory lateral line lobe of the African wave-type electric fish *Gymnarchus niloticus*. J Neurosci 16:380–391.

Kawasaki M, Guo YX (1998) Parallel projection of amplitude and phase information from the hindbrain to the midbrain of the African electric fish *Gymnarchus niloticus*. J Neurosci 18:7599–7611.

Kawasaki M, Maler L, Rose GJ, Heiligenberg W (1988) Anatomical and functional organization of the prepacemaker nucleus in gymnotiform electric fish: the accommodation of two behaviors in one nucleus. J Comp Neurol 276:113–131.

Keller CH, Heiligenberg W (1989) From distributed sensory processing to discrete motor representations in the diencephalon of the electric fish, *Eigenmannia*. J Comp Physiol A 164:565–576.

Keller CH, Maler L, Heiligenberg W (1990) Structural and functional organization of a diencephalic sensory-motor interface in the gymnotiform fish, *Eigenmannia*. J Comp Neurol 293:347–376.

Kelly RM, Strick PL (2003) Cerebellar loops with motor cortex and prefrontal cortex of a nonhuman primate. J Neurosci 23:8432–8444.

Losier BJ, Matsubara JA (1990a) Light and electron microscopical studies on the spherical neurons in the electrosensory lateral line lobe of the gymnotiform fish, *Sternopygus*. J Comp Neurol 298:237–249.

Losier BJ, Matsubara JA (1990b) Comparison of calbindin D 28K and cytochrome c oxidase in electrosensory nuclei of high- and low-frequency weakly electric fish (Gymnotiformes). Cell Tissue Res 260:29–39.

Maler L (1979) The posterior lateral line lobe of certain gymnotiform fish. Quantitative light microscopy. J Comp Neurol 183:323–363.

Maler L, Ellis WG (1987) Inter-male aggressive signals in weakly electric fish are modulated by monoamines. Behav Brain Res 25:75–81.

Maler L, Mugnaini E (1994) Correlating gamma-aminobutyric acidergic circuits and sensory function in the electrosensory lateral line lobe of a gymnotiform fish. J Comp Neurol 345:224–252.

Maler L, Sas EK, Rogers J (1981) The cytology of the posterior lateral line lobe of high frequency weakly electric fish (*Gymnotoidei*): differentiation and synaptic specificity in a simple cortex. J Comp Neurol 195:87–139.

McCreery DB (1977) Two types of electroreceptive lateral lemniscal neurons of the lateral line lobe of the catfish *Ictalurus nebulosus*: connections from the lateral line nerve and steady-state frequency response characteristics. J Comp Physiol A 113: 317–339.

Meek J (1998) Holosteans and teleosts. In: Nieuwenhuys R, Ten Donkelaar HJ, Nicholson C (eds), The Central Nervous System of Vertebrates. Berlin: Springer-Verlag, pp. 759–937.

Meek J, Nieuwenhuys R (1991) Palisade pattern of mormyrid Purkinje cells: a correlated light and electron microscopic study. J Comp Neurol 306:156–192.

Meek J, Joosten HWJ, Hafmans TGM (1993) Distribution of noradrenaline-immunoreactivity in the brain of the mormyrid teleost *Gnathonemus petersii*. J Comp Neurol 328:145–160.

Meek J, Grant K, Sugawara S, Hafmans TGM, Veron M, Denizot JP (1996) Interneurons of the ganglionic layer in the mormyrid electrosensory lateral line lobe: morphology, immunocytochemistry, and synaptology. J Comp Neurol 375:43–65.

Meek J, Grant, K., and Bell, C. (1999) Structural organization of the mormyrid electrosensory lateral line lobe. J Exp Biol 202:1291–1300.

Meek J, Hafmans TGM, Han VZ, Bell CC, Grant K (2001) Myelinated dendrites in the mormyrid electrosensory lobe. J Comp Neurol 431:255–275.

Metzner W (1999) Neural circuitry for communication and jamming avoidance in gymnotiform electric fish. J Exp Biol 202:1365–1375.

Mohr C, Roberts PD, Bell CC (2003) Cells of the mormyromast region of the mormyrid electrosensory lobe: I. Responses to the electric organ corollary discharge and to electrosensory stimuli. J Neurophysiol 90:1193–1210.

Montgomery JC, Coombs S, Conley RA, Bodznick D (1995) Hindbrain sensory processing in lateral line, electrosensory, and auditory systems: a comparative overview of anatomical and functional similarities. Audit Neurosci 1:207–231.

Mugnaini E, Maler L (1987a) Cytology and immunohistochemistry of the nucleus exterolateralis anterior of the mormyrid brain: possible role of GABAergic synapses in temporal analysis. Anat Embryol 176:313–336.

Mugnaini E, Maler L (1987b) Cytology and immunohistochemistry of the nucleus of the lateral line lobe in the electric fish *Gnathonemus petersii* (Mormyridae) brain. Evidence suggesting that GABAergic synapses mediate an inhibitory corollary discharge. Synapse 1:32–56.

Nieuwenhuys R, Nicholson C (1969) Aspects of the histology of mormyrid fishes. In: Llinas R (ed) Neurobiology of Cerebellar Evolution and Development. American Medical Association, Chicago, pp. 135–169.

Northcutt RG (1986) Electroreception in Nonteleost Bony Fishes. In: Bullock TH, Heiligenberg W (eds), Electroreception. New York: John Wiley & Sons, pp. 257–286.

Paul DH, Roberts BL (1977) Studies on a primitive cerebellar cortex I. The anatomy of the lateral-line lobes of the dogfish, *Scyliorhinus canicula*. Proc R Soc Lond B 195: 453–466.

Paulin MG (1993) The role of the cerebellum in motor control and perception. Brain Behav Evol 41:39–50.

Post N, von der Emde G (1999) The "novelty response" in an electric fish: response properties and habituation. Physiol Behav 68:115–128.

Prechtl JC, von der Emde G, Wolfart J, Karamursel S, Akoev GN, Andrianov YN, Bullock TH (1998) Sensory processing in the pallium of a mormyrid fish. J Neurosci 18: 7381–7393.

Réthelyi M, Szabo T (1973a) A particular nucleus in the mesencephalon of a weakly electric fish, *Gymnotus carapo* (Gymnotidae) I. Light microscopic structure. Exp Brain Res 17:229–241.

Réthelyi M, Szabo T (1973b) Neurohistological analysis of the lateral lobe in an electric fish, *Gymnotus carapo* (Gymnotidae). Exp Brain Res 18:323–339.

Ronan MC (1986) Electroreception in Cyclostomes. In: Bullock TH, Heiligenberg W (eds), Electroreception. New York: John Wiley & Sons, pp. 209–224.

Rose GJ, Call SJ (1992) Differential distribution of ampullary and tuberous processing in the torus semicircularis of *Eigenmannia*. J Comp Physiol A 170:253–261.

Russell CJ, Bell CC (1978) Neuronal responses to electrosensory input in the mormyrid valvula cerebelli. J Neurophysiol 41:1495–1510.

Russell CJ, Myers JP, Bell CC (1974) The echo response in *Gnathonemus petersii.* J Comp Physiol 92:181–200.

Sas E, Maler L (1986a) The optic tectum of gymnotiform teleosts *Eigenmannia viriscens* and *Apteronotus leptorhynchus*: a Golgi study. Neuroscience 18:215–246.

Sas E, Maler L (1986b) Retinofugal projections in a weakly electric gymnotid fish (*Apteronotus leptorhynchus*). Neuroscience 18:247–259.

Sas E, Maler L (1987) The organization of afferent input to the caudal lobe of the cerebellum of the gymnotid fish *Apteronotus leptorhynchus.* Anat Embryol 177:55–79.

Saunders J, Bastian J (1984) The physiology and morphology of two classes of electrosensory neurons in the weakly electric fish *Apteronotus leptorhynchus.* J Comp Physiol A 154:199–209.

Schmidt AW, Bodznick D (1987) Afferent and efferent connections of the vestibulolateral cerebellum of the little skate, *Raja erinacea.* Brain Behav Evol 30:282–302.

Smeets WJAJ (1982) The afferent connections of the tectum mesencephali in two chondrichthyans, the shark, *Scyliorhinus canicula*, and the ray, *Raja clavata.* J Comp Neurol 205:139–152.

Springer AD, Easter SS, Agranoff BW (1977) The role of the optic tectum in various visually mediated behaviors of goldfish. Brain Res 128:393–404.

Stendell W (1914) Die Faseranatomie des Mormyridengehirns. Abh Senckenb Naturforsch Ges 36:3–40.

Szabo T (1957a) Organisation particuliere de la commande nerveuse centrale de la décharge chez un poisson électrique, *Gymnarchus niloticus.* Compt Rend Acad Sci Paris 248:3488–3489.

Szabo T (1957b) Un noyau particulier dans la formation reticuleJ bulbaire de certains poissons Jlectriques (Mormyrides). Arch Anat Microsc Morphol Exp 46:81–92.

Szabo T (1974) Anatomy of the specialized lateral line organs of electroreception. In: Fessard A (ed), Handbook of Sensory Physiology. New York: Springer-Verlag, pp. 13–58.

Szabo T, Fessard A (1965) Le fonctionnement des electrorecepteurs etudies chez les Mormyres. J Physiol (Paris) 57:343–360.

Szabo T, Moller P (1984) Neuroethological basis for electrocommunication. In: Bolis L, Keynes RD, Maddrell SHP (eds), Comparative Physiology of Sensory Systems. Cambridge, UK: Cambridge University Press, Cambridge, pp. 455–474.

Szabo T, Ravaille M, Libouban S, Enger DS (1983) The mormyrid rhombencephalon: I. Light and EM investigations on the structure and connections of the lateral line lobe nucleus with HRP labeling. Brain Res 266:1–19.

von der Emde G (1990) Discrimination of objects through electrolocation in the weakly electric fish, *Gnathonemus petersii.* J Comp Physiol A 167:413–421.

von der Emde G, Bleckman H (1992) Extreme phase sensitivity of afferents which innervate mormyromast electroreceptors. Naturwissenschaften 79:131–133.

von der Emde G, Prechtl JC (1999) Anatomical connections of auditory and lateral line areas of the dorsal telencephalon (Dm) in the osteoglossomorph teleost, *Gnathonemus petersii.* Brain Res 818:355–367.

Wong CJH (1997a) Connections of the basal forebrain of the weakly electric fish, *Eigenmannia viriscens.* J Comp Neurol 389:49–64.

Wong CJH (1997b) Afferent and efferent connections of the diencephalic prepacemaker nucleus in the weakly electric fish, *Eigenmannia viriscens*: interactions between the electromotor system and the neuroendocrine axis. J Comp Neurol 383:18–41.

Wullimann MF, Northcutt RG (1990) Visual and electrosensory circuits of the diencephalon in mormyrids: an evolutionary perspective. J Comp Neurol 297:537–552.
Wullimann MF, Rooney DJ (1990) A direct cerebello-telencephalic projection in an electrosensory mormyrid fish. Brain Res 520:354–357.
Xu-Friedman MA, Hopkins CD (1999) Central mechanisms of temporal analysis in the knollenorgan pathway of mormyrid electric fish. J Exp Biol 202:1311–1318.
Zupanc G, Maler L (1997) Neuronal control of behavioral plasticity: the prepacemaker nucleus of weakly electric fish. J Comp Physiol A 180:99–111.
Zupanc GKH, Airey JA, Maler L, Sutko JL, Ellisman MH (1992) Immunohistochemical localization of ryanodine binding protein in the central nervous system of gymnotiform fish. J Comp Neurol 325:135–151.

5

Ontogeny of Electroreceptors and Their Neural Circuitry

R. GLENN NORTHCUTT

1. Introduction

Ontogenetic data are particularly efficacious in our attempts to understand how complex structures come into being and how these structures have changed over their phylogenetic history. This is particularly true for the lateral line system. In many adult fishes and amphibians this system consists of two or more receptor classes with distinct morphologies and functions (Zupanc and Bullock, Chapter 2; Jørgensen, Chapter 3). These receptors are distributed over the body in complex patterns (Fritzsch and Münz 1986; Northcutt 1986, 1989; Zakon 1986; Coombs et al. 1988) and innervated by a large number of cranial nerves that are closely associated secondarily with cranial nerves of totally different functions (Northcutt 1989; Song and Northcutt 1991; Northcutt and Bemis 1993; Piotrowski and Northcutt 1996). Finally, two parallel series of closely associated neural centers and their interconnections process information from lateral line electroreceptors or mechanoreceptors (see Bullock and Heiligenberg 1986; Coombs et al. 1989).

Ontogenetic studies have provided the key to understanding how receptor classes, as well as their distribution and innervation, are organized. The lateral line sensory receptors, as well as the cranial nerves that innervate these receptors, arise from a series of localized thickenings termed dorsolateral placodes, which form within the inner layer of the cephalic ectoderm following neurulation (Stone 1922; Metcalfe 1985; Northcutt et al. 1994; Northcutt and Brandle 1995; Schlosser and Northcutt 2000; Schlosser 2002a; Northcutt 2003; Gibbs and Northcutt 2004). Not only is a dorsolateral placode the basic ontogenetic unit responsible for the formation of the lateral line system, but in addition the unique pattern of elongation or migration that each placode undergoes over a portion of the body is responsible for the complex distribution of lateral line receptors into lines and fields (reviewed by Northcutt 1989).

Although the ontogeny and phylogeny of the mechanoreceptive neuromasts appear to be fairly straightforward across fishes and amphibians, the phylogeny of electroreceptors is more complex (McCormick 1982; Bullock et al. 1983;

Northcutt 1986, 1997; New 1997; Zupanc and Bullock, Chapter 2). Primitively, the lateral line system of vertebrates (lampreys and jawed fishes) consisted of electroreceptive ampullary organs and mechanoreceptive neuromasts. The ampullary organs appear to have been lost with the origin of neopterygian bony fishes (gars, bowfins, and teleosts) and reevolved at least twice among different groups of teleosts (osteoglossomorphs and ostariophysians). For this reason, the development of primitive electroreceptive ampullary organs is described first, followed by a description of the development of the derived electroreceptors (teleost ampullary and tuberous receptors). After this, the information on the genetic basis for the induction and morphogenesis of these receptors is summarized. Finally, the very scant literature on the development of the lateral line neural centers and their connections is touched on. Unfortunately, the role of ontogenetic studies in understanding the phylogeny of the lateral line system (Finger 1986; Northcutt 1989, 1997; Webb 1989a; Schlosser 2002b) is beyond the scope of this review.

2. Embryonic Origins of Electroreceptors

The embryonic origin of primitive (nonteleost) ampullary organs is fairly well established, based on descriptive (Fritzsch and Bolz 1986; Northcutt et al. 1994) and experimental (Northcutt et al., 1995) studies in salamanders. Although experimental studies are lacking for other vertebrates that possess primitive ampullary organs, there is considerable descriptive information on the development of these organs in elasmobranchs (Ruud 1920; Disler 1977; Northcutt et al. 1997), sturgeons (Disler 1971; Nikolskaya 1983; Gibbs and Northcutt 2004), and paddlefish (Nikolskaya 1983; Bemis and Northcutt 2004). The development of these organs has not been described in lampreys, bichirs, coelacanths, lungfishes, and apodan amphibians. For this reason, the development of primitive ampullary organs in axolotls will serve as a general model for how these receptors develop, and this description will be supplemented with pertinent information about the development of primitive ampullary organs from other taxa.

Amazingly, there are few descriptive studies on the development of teleost electroreceptors (Kirschbaum and Denizot 1975; Srivastava and Seal 1981; Roth 1986, 1993, 1994; Vischer 1989a,b, 1995; Vischer et al. 1989; Bensouilah et al. 2002; Northcutt 2003) and no experimental studies. Given the limited data, it is not surprising that there is no agreement regarding their embryonic origin. Two hypotheses have been proposed: (1) Teleost electroreceptors are induced from general ectoderm by the lateral line nerves that innervate these receptors (Roth 1986, 1993; Vischer et al. 1989; Bever and Borgens 1991a,b; Vischer 1995). (2) Teleost electroreceptors arise from lateral line (dorsolateral) placodes (Northcutt 2003), as they do in other fishes and amphibians. Details of each of these hypotheses are considered in turn.

2.1 Development of Primitive Electroreceptors

An out-group analysis of the development of the lateral line systems of elasmobranchs, bony fishes, and amphibians (Northcutt 1989, 1997) suggests that the lateral line system of the earliest jawed fishes arose from six pairs of lateral line placodes (Fig. 5.1). These placodes form a dorsolateral series that are divided into preotic and postotic groups by the octaval placode. The octaval placode is usually termed the otic placode but will be called the octaval placode here to distinguish it from an otic lateral line placode that gives rise to a line of neuromasts that passes around the lateral edge of the inner ear. Following invagination, the octaval placode forms the hair cell receptors of the membranous labyrinth of the inner ear and the sensory ganglia of the octaval nerve which innervates these receptors.

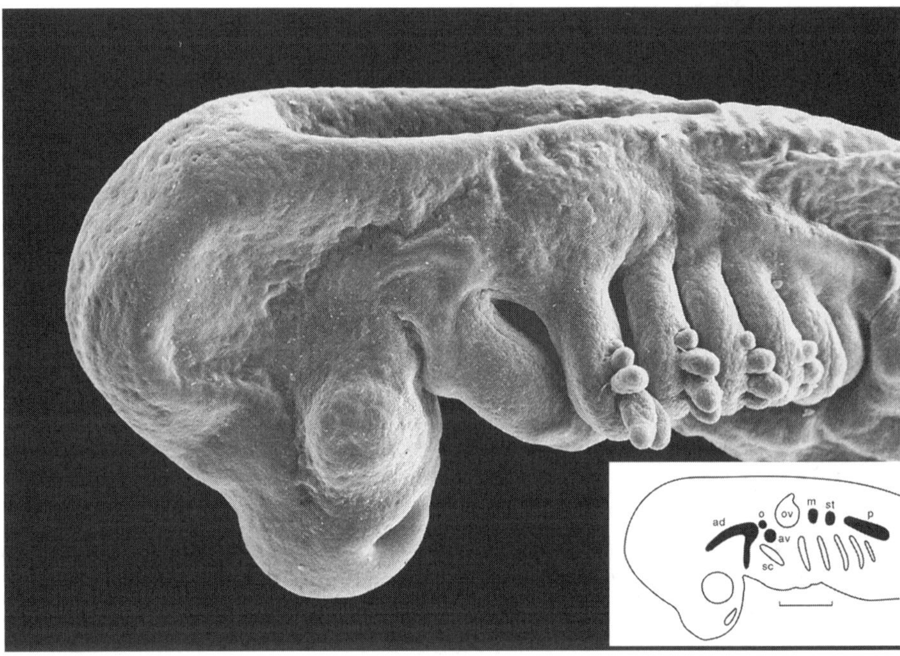

FIGURE 5.1. Scanning electron micrograph of the lateral surface of the head of a 17-day-old embryo of the clearnose skate, *Raja eglanteria*. The developing lateral line placodes can be recognized by their distinct hillock-like protrusions. *Raja eglanteria* requires approximately 12 weeks to develop, but the supra- and infraorbital sensory ridges of the anterodorsal placode are already beginning to elongate, and the posterior placode has begun to migrate. ad, Anterodorsal lateral line placode; av, anteroventral lateral line placode; m, middle lateral line placode; o, otic lateral line placode; ov, octaval (otic) vesicle; p, posterior lateral line placode; sc, spiracular cleft; st, supratemporal lateral line placode. Bar scale equals 500 μm. (From Northcutt 1997.)

Primitively, there are three preotic and three postotic lateral line placodes (Fig. 5.1). Each placode is named after the lateral line nerve that arises from that placode. Most amphibians appear to have lost the otic lateral line placode but retain the other five placodes, and each placode, with the exception of the posterior lateral line placode, develops in the same manner (Stone 1922; Northcutt et al. 1994; Schlosser and Northcutt 2000; Schlosser 2002b). In axolotls, each lateral line placode consists of tall columnar epithelium (Fig. 5.2A) and is easily distinguished from the adjacent ectoderm (Fig. 5.3A). Shortly after formation, the basement membrane beneath each placode is disrupted, and the neuroblasts that will form the sensory ganglia of the lateral line nerves are generated (Fig. 5.3B). As the neuroblasts differentiate, their axons enter the medulla, and their peripheral processes come into contact with the overlying placodes. At this

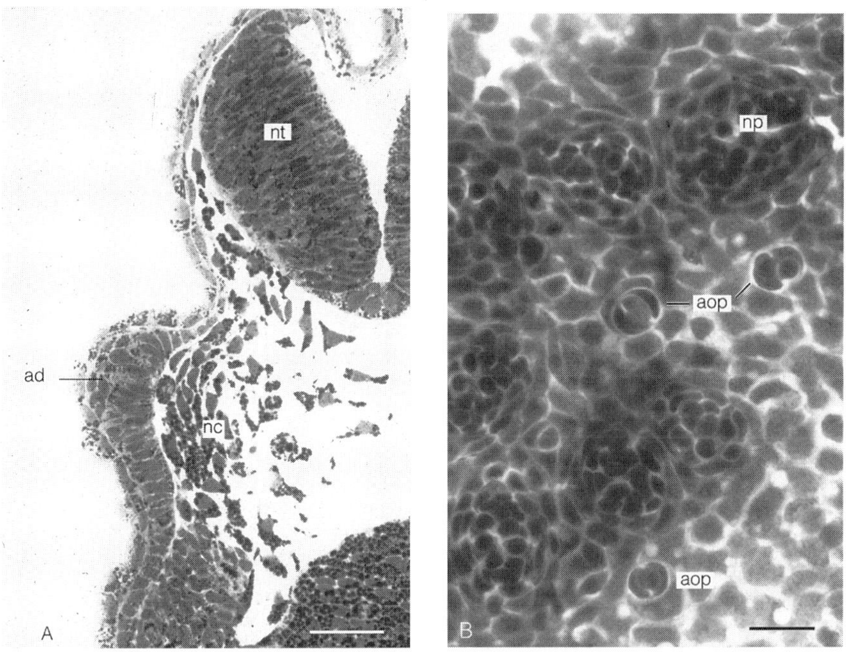

FIGURE 5.2. Photomicrographs of a transverse section (**A**) and a flat mount of head ectoderm (**B**), showing two stages in the development of a lateral line placode in an axolotl. (**A**) The transverse section through the rostral segment of the anterodorsal (ad) lateral line placode at stage 35 illustrates the distinct columnar epithelial cells that constitute a placode and the close relationship of the placode to migrating neural crest cells (nc), as they stream ventrally around the lateral edge of the neural tube (nt). (**B**) The flat mount of the cephalic ectoderm at stage 42 shows a segment of the supraorbital sensory ridge with ampullary organ primordia (aop) forming on the edge of the sensory ridge and neuromast promordia (np) forming more centrally. Bar scales equal 100 μm in **A** and 20 μm in **B**.

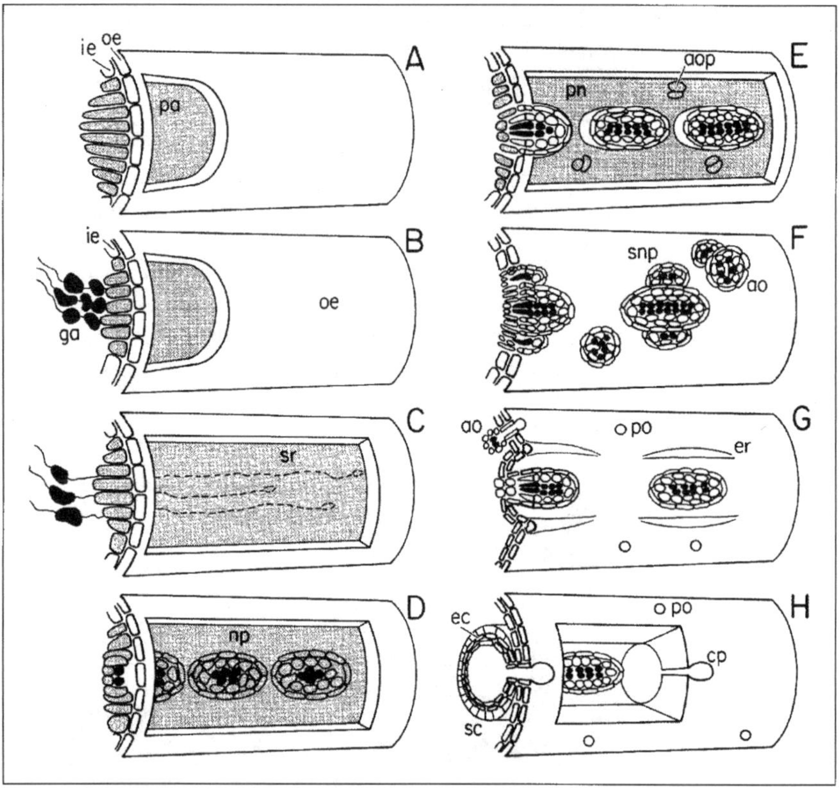

FIGURE 5.3. Stages in the development of a primitive lateral line placode. ao, ampullary organ; aop, ampullary organ primordium; cp, canal pore; ec, epithelial canal; er, ectodermal ridge; ga, ganglionic cells of lateral line nerve; ie, inner layer of ectoderm; np, primary neuromast primordium; oe, outer layer of ectoderm; pa, placode; *pn*, primary neuromast; po, ampullary pore; sc, secondary connective tissue canal; snp, secondary neuromast primordum; sr, sensory ridge. (From Northcutt et al. 1994.)

point, the placodes begin to elongate as a result of increased mitotic activity (Winklbauer and Hausen 1983), and each placode forms one or more sensory ridges (Fig. 5.3C). The posterior lateral line placode, however, is an exception, as it actively begins to migrate down the trunk, leaving neuromast primordia in its wake (Stone 1922; Metcalfe 1985).

Initially, each sensory ridge consists of morphologically uniform cells that subsequently begin to differentiate and form neuromast primordia (Figs. 5.2B, 5.3D). These primordia form only within the central zones of the sensory ridges, whereas ampullary primordia begin to form within the lateral zones of the sensory ridges (Fig. 5.2B) a short time later (Fig. 5.2E). There is no evidence that ampullary primordia bud off of neuromast primordia; rather, each type of receptor appears to arise from different cell lineages (Northcutt et al. 1994). Until this time, both types of receptor primordia are overlain by an outer layer of general ectoderm (Fig. 5.3D), which now begins to retreat, exposing the developing neuromast primordia (Fig. 5.3E). In axolotls and in all other living amphibians, the developing neuromasts form superficial lines, which remain on the surface (Fig. 5.3F). In most jawed fishes, however, ectodermal ridges form adjacent to the neuromasts, partially surrounding them in open grooves (Fig. 5.3G) that may subsequently fuse above the receptors (Allis 1889; Webb 1989b), enclosing them within an epithelial canal (Fig. 5.3H). These canals are usually invested subsequently by cartilage or bone that may be of neural crest origin.

After forming within the lateral zones of the sensory ridges (Fig. 5.3E), the ampullary primordia are also exposed by the retreating ectoderm that initially overlies the sensory ridges. In axolotls, however, ampullary primordia differentiate more slowly than neuromast primordia, and the ampullary organs erupt approximately 7 days later than neuromasts, at which point the larval axolotls begin to feed.

Most of the same stages in the development of lateral line placodes have been documented in those nonteleost fishes that possess primitive ampullary organs. Lateral line placodes have been recognized in elasmobranchs (Holmgren 1940; Northcutt 1997; Northcutt et al. 1997); chondrosteans (Bemis and Northcutt 2004; Gibbs and Northcutt 2004), gars (Landacre and Conger 1913), bowfins (Beckwith 1907), and lungfishes (Pehrson 1949). Similarly, sensory ridges have been documented in elasmobranchs (Johnson 1917; Ruud 1920; Disler 1977; Northcutt et al. 1997), sturgeons (Disler 1971; Gibbs and Northcutt 2004), paddlefishes (Bemis and Northcutt 2004), gars and bowfins (Allis 1889; Beckwith 1907; Landacre and Conger 1913), and lungfishes (Pehrson 1949). The origin of ampullary primordia within the lateral zones of sensory ridges has also been documented in elasmobranchs (Ruud 1920; Disler 1977; Northcutt et al. 1997) and sturgeons and paddlefishes (Disler 1971; Bemis and Northcutt 2004; Gibbs and Northcutt 2004).

Although ampullary fields usually occur in pairs, because they arise from within the lateral zones of a placode, the size of the ampullary fields varies greatly among amphibians and non-teleost fishes. In axolotls and other salamanders, the ampullary fields are never more than three or four receptors in width and usually occur along the entire rostrocaudal axis of a given sensory ridge. In elasmobranchs (Fig. 5.4), sturgeons, and paddlefishes, however, the ampullary fields are generally wider and may or may not extend the entire rostrocaudal length of a sensory ridge. The largest ampullary fields occur in sturgeons (Gibbs and Northcutt 2004) and paddlefishes (Nikolskaya 1983; Bemis and Northcutt 2004), where the ampullary fields of a single sensory ridge may be hundreds of times larger than the central zone. Paddlefishes appear to represent the extreme: their ampullary fields, which arise from the anteroventral lateral line placode, cover the entire cheek and operculum.

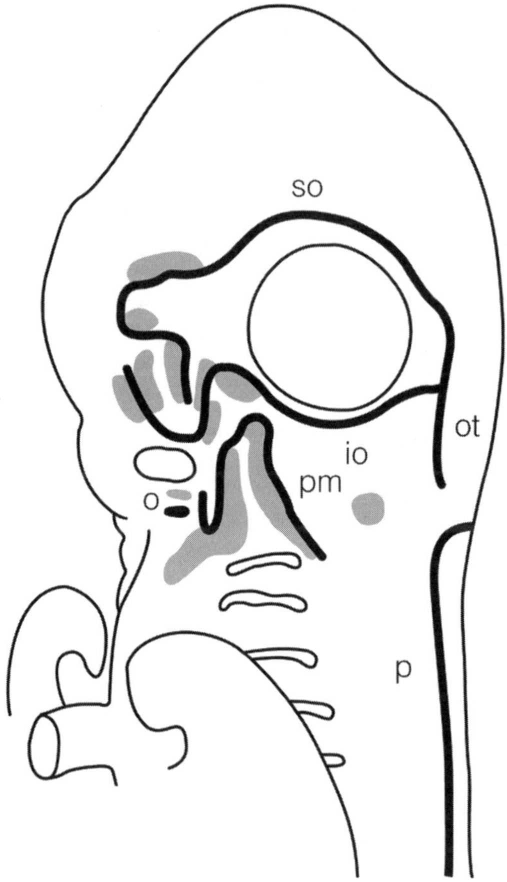

FIGURE 5.4. Distribution of fields of ampullary organ primordia (*gray*) adjacent to developing lines of neuromasts in a skate, *Raja.* io, infraorbital; o, oral; ot, otic; p, posterior; pm, preoperculomandibular; so, supraorbital sensory ridges. (Redrawn from Disler 1977.)

2.2 Development of Derived (Teleost) Electroreceptors

The earliest study on the development of teleost electroreceptors (Kirschbaum and Denizot 1975) focused on a gymnotid, *Eigenmannia*, and a mormyrid, *Marcusenius.* The authors noted that tuberous organs are present in the larvae of both genera 4 days after hatching, but they did not comment on the possible origin of these receptors. One of the authors noted in a subsequent paper (Denizot and Libouban 1985), however, that the sensory cells of the tuberous organs in another mormyrid, *Brienomyrus*, completely degenerated following transection of the lateral line nerve that innervates these organs, and it was therefore claimed that new sensory cells were subsequently generated from the accessory cells of the old organs, presumably induced by transection of the afferent nerve.

Bensouilah et al. (2002) described the development of larval tuberous electroreceptors in three additional mormyrid genera: *Campylomormyrus*, *Mormyrus*, and *Pollimyrus*. These authors report that in 12-day-old larvae the primordia of tuberous organs consist of a single sensory cell, located within an intraepidermal cavity on a platform of accessory cells. Although these larvae were too old for their embryonic origin to be determined, the authors did note that there appear to be two types of larval tuberous organs and that both types degenerate at the same time that the larval electric organ degenerates. Thus it appears that different classes of electroreceptors develop concurrently in embryonic and larval stages of mormyrid fishes. The ability of lateral line fibers to regenerate and possibly induce new electroreceptors (Roth 1986, 1993, 1994) has been interpreted as evidence that lateral line nerves induce electroreceptors from general ectoderm. To date, the most compelling evidence to support this interpretation comes from Bever and Borgens (1991a), who grafted electroreceptor-free tissue onto the anal fin and observed that electroreceptors form within the graft after the latter is invaded by lateral line fibers. Although it is possible that the invading lateral line fibers did induce electroreceptors from the ectoderm of the graft, it is also possible that stem cells from electroreceptors adjacent to the graft were attracted by chemical signals emanating from the fibers invading the graft, and that these stem cells invaded the graft. In this context, it is known that new electroreceptors do arise by budding off of older receptors in the gymnotid *Sternopygus* (Zakon 1984).

A series of developmental studies by Vischer and colleagues (Vischer 1989b, 1995; Vischer et al. 1989) avoids the problems inherent in regeneration studies. In 1989, Vischer published the first detailed study of the development of teleost electroreceptors in the gymnotid, *Eigenmannia*. Vischer noted that most of the lines of neuromasts have already formed by day 4 (following spawning) and that the first electroreceptor primordia form on the lateral edges of the neuromast lines. The first differentiated tuberous organs occur on the head on day 7, and the first differentiated ampullary organs occur on the head a day later. Ampullary and tuberous organs begin to occur on the trunk on day 8 and develop in a rostrocaudal direction. Of particular interest, Vischer observed that the first electroreceptors develop along the dorsal trunk line prior to the appearance of the first neuromasts, and he concluded, as did Northcutt et al. (1994), studying axolotls, that there is no evidence that electroreceptor primordia arise from developing neuromasts.

Given the data available to him in 1989, Vischer was unable to determine whether electroreceptor primordia arose from lateral line placodes or general ectoderm, but in a subsequent paper (Vischer et al., 1989) he and his colleagues suggested that the fibers of the lateral line nerves might induce both electroreceptors and neuromasts, because the nerves could be visualized prior to the receptors. In a final paper, Vischer (1995) concluded that electroreceptor primordia must be induced from general ectoderm by lateral line fibers, since the first cells of electroreceptor primordium are located within the stratum germinativum of the epidermis. Vischer did not indicate, however, whether the receptor primordia arise from the stratum germinativum of the general ectoderm or from that of the sensory ridges. In fact, it is impossible to determine from this work whether or not sensory ridges exist in *Eigenmannia,* as they are not described in any of Vischer's papers. In this context, Vischer (1989a) was unable to identify any lateral line placodes with the exception of the posterior placode as it migrated down the trunk.

Only two studies (Srivastava and Seal 1981; Northcutt 2003) suggest that teleost electroreceptors may arise from lateral line placodes. In studying the development of a number of Indian catfishes, Srivastava and Seal noted a similarity in the development of ampullary and neuromast primordia and suggested that they were homologous and arose from the same embryonic source. These workers, however, did not present evidence indicating whether the lateral line system in their fishes develops from placodes or whether both types of receptors arise from sensory ridges. With the exception of the posterior lateral line placode (Metcalfe 1985; Vischer 1989a), it was unclear until recently whether the cephalic lateral lines in teleosts develop from placodes. None of the early studies on the development of the lateral line system of teleosts (reviewed in Northcutt 2003) identified lateral line placodes, and they were first identified by Sahly et al. (1999) when they were visualized in zebrafish with the use of *eya1* gene expression.

The development of the lateral line system in the channel catfish, *Ictalurus nebulosus*, was described by this author (Northcutt 2003), beginning with the formation of the lateral line placodes. The development of this catfish has been staged (Fig. 5.5A). Hatching (stage 43) normally occurs 5 days following spawning, and the yolk-sac larvae begin to feed (stage 53) at approximately 10 days. By stage 24 (Fig. 5.5B) in channel catfish, one can recognize the preotic lateral line placodes: anterodorsal, anteroventral, and otic. The postotic placodes, i.e., the middle and posterior, cannot be recognized until stage 28. Channel catfish lack a supratemporal placode and, accordingly, fail to form a supratemporal commissure. Neurogenesis has begun in the anterodorsal placode by stage 24 but will not begin in the anteroventral and otic placodes until stage 28. Neurogenesis appears to continue in most placodes until stage 30, and each placode, with the exception of the posterior placode, begins to elongate and form sensory ridges by stage 34 (Fig. 5.5B). Although the posterior placode does not form a distinct sensory ridge, it begins to migrate onto the trunk during stage 34. The sensory ridges of the other placodes are well developed by stage 36, and neuromast primordia begin to form during stage 37. At this time, the sensory ridges are approximately 20 cells in width; the neuromast primordia are some 9 to 10 cells in width and occupy a central zone within each sensory ridge. Neuromasts begin to erupt at stage 41, and by stage 45 the lateral lines of the yolk-sac larvae consist of well developed superficial neuromasts.

Ampullary organ primordia can be recognized by stage 43 (hatching) within the lateral zones of the supraorbital and infraorbital sensory ridges. Similar primordia in the otic and middle lateral line placodes can be recognized a short time later (stage 44/45). The development of trunk electroreceptors was not

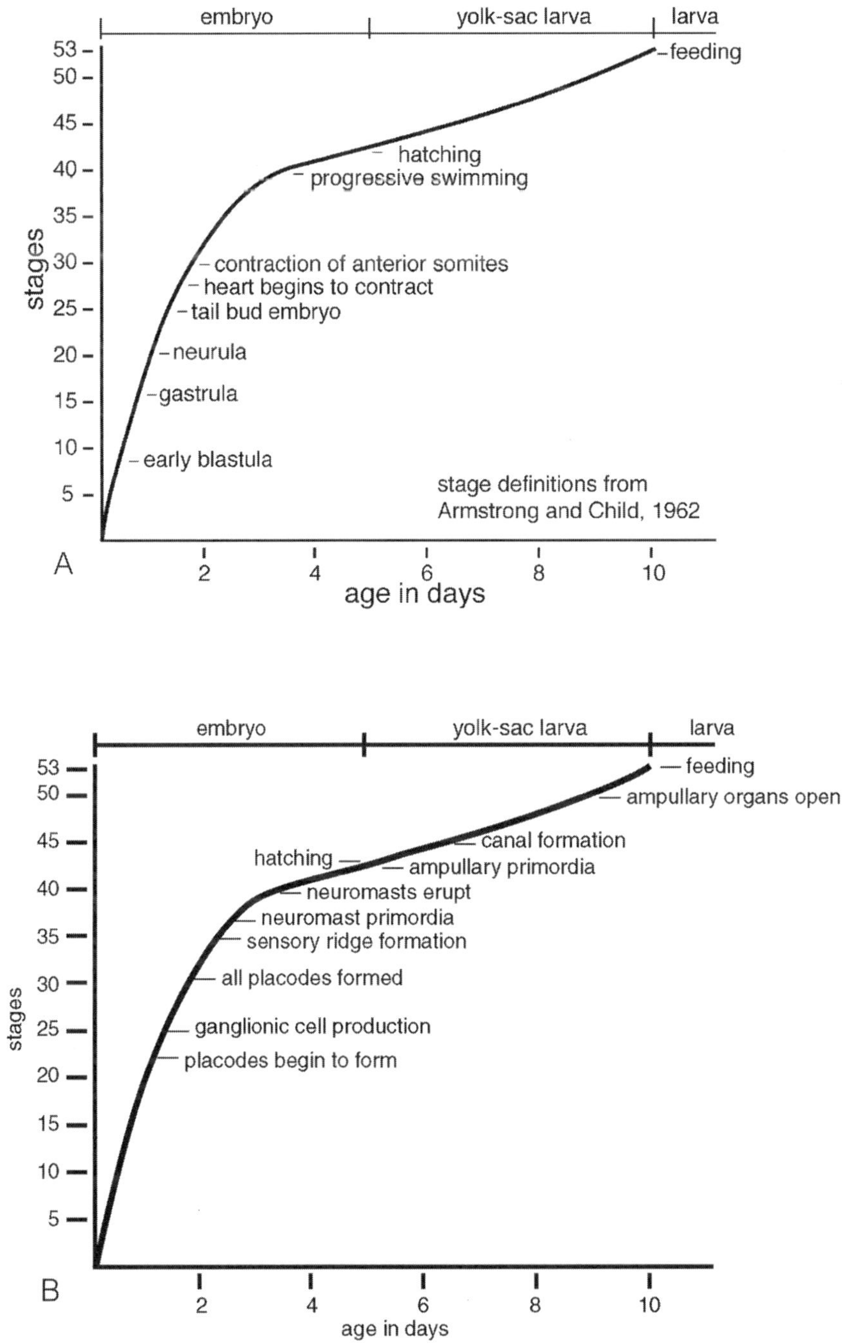

FIGURE 5.5. Timing of some major events in the development of the channel catfish (**A**) compared to the timing of major events in the development of lateral line placodes (**B**) excluding the posterior placode. (From Northcutt 2003.)

described in Northcutt (2003), but ampullary organ primordia on the head begin to open at stage 49, and by stage 51 (Fig. 5.6) a number of ampullary organ primordia and mature ampullary organs occur adjacent to each of the neuromast lines.

Although the relative timing of receptor development in *Eigenmannia* and *Ictalurus* is similar, reports by Vischer (1995) and Northcutt (2003) clearly differ regarding the site of origin of the electroreceptors. Resolution of these differences will require experiments in which one or more lateral line placodes are labeled with a cell marker such as DiI. If electroreceptor primordia incorporate the dye, it can be safely assumed that the primordia have a placodal origin. This type of experiment should be performed in gymnotids as well as catfishes, as it is possible that ampullary organs in both groups arise from lateral line placodes but that gymnotid tuberous organs have a different embryonic origin.

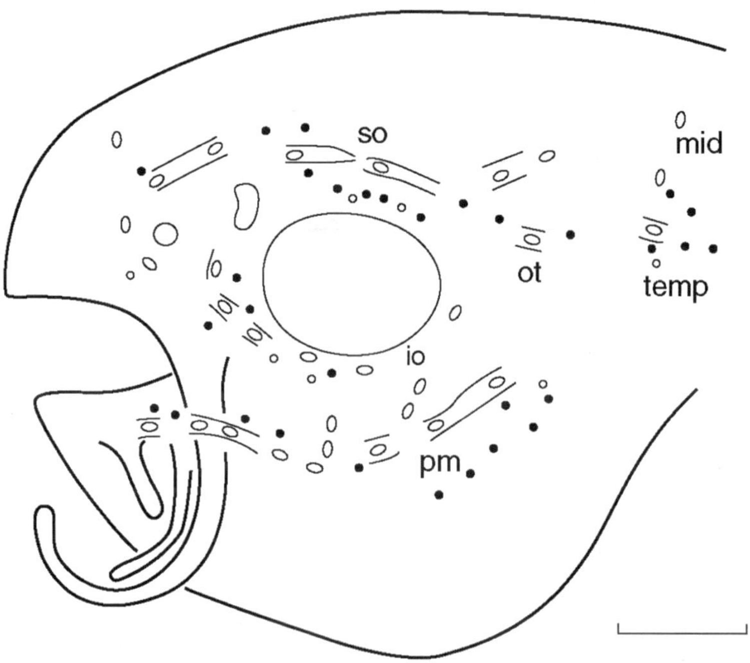

FIGURE 5.6. A map of the flattened head ectoderm from a stage 51 channel catfish yolk-sac larva indicating the positions of erupted neuromasts indicated by *open ovals*, which are beginning to be enclosed in ectodermal grooves indicated by the *parallel lines*. Mature ampullary organs indicated by *solid circles* and ampullary organ primordial indicated by *open circles* occur immediately adjacent to the developing lines of neuromasts. io, infraorbital; mid, middle; ot, otic; pm, preoperculomandibular; so, supraorbital; temp, temporal lines. Bar scale equals 500 μm. (From Northcutt 2003.)

2.3 Genetic Basis for Receptor Induction and Morphogenesis

There is growing evidence (reviewed in Baker and Bronner-Fraser 2001; Schlosser 2002a) that lateral line placodes (as well as all other placodes) arise from a common preplacodal ectodermal field at the neural plate border. If this is the case, then the first step in the induction of lateral line placodes is thus the formation of the preplacodal field. It is possible that many of the tissues and signals that induce neural crest also act to induce a preplacodal field. If so, then induction of the preplacodal field may involve signals from nonaxial mesoderm, as well as interactions with the adjacent neural crest and more lateral general ectoderm. In this context, it should be noted that a number of transcription factors, such as *Sox-2* and *Sox-3*, *Xiro-1-3*, *Sex 3*, and *Msx-1*, are expressed in a horseshoe-shaped ectodermal field surrounding the neural plate, and these may very well constitute the preplacodal field. The precise role of these transcription factors, however, is at present unclear.

All placodes express a number of genes (*Pax*, *Six*, *Eya*, and *Dach*), which probably form a regulatory network that specifies different groups of placodes (Baker and Bronner-Fraser 2001; Schlosser 2002a). One of the last steps in the specification of individual lateral line placodes may involve *Hox* genes (Northcutt 1996), as *Hoxb-3* is known to be expressed in the middle lateral line placode of axolotls (Metscher et al. 1997), and application of exogenous retinoic acid, a potent inhibitor of Hox gene activity, to late gastrula and early neurula axolotls repatterns the lateral line placodes (Gibbs and Northcutt 2004). Normally, all lateral line placodes, with the exception of the posterior placode, generate ampullary organs as well as neuromasts. As the concentration of retinoic acid is increased, however, fewer and fewer placodes generate ampullary organs, and at the highest concentration, no ampullary organs are generated at all, only neuromasts. This strongly suggests that retinoic acid, acting via the *Hox* genes, repatterns all placodes in these axolotls to a posterior placodal fate.

At present, nothing is known about the genetic basis of polarity determination (ganglionic cell production versus receptor production) of individual placodes or the patterning of their receptors. Since neuromasts form within the central placodal zone, whereas ampullary organs form within the lateral flanking zones, it is possible that lateral inhibitory signaling involving *Notch* and *Delta* patterns those placodes that generate ampullary organs and neuromasts.

At the moment, nothing is known about the genetic basis of electroreceptor development in teleosts, and little progress is likely to be made until the identification of the tissue(s) generating these receptors. Given the number of anatomical and functional similarities between non-teleost and teleost electroreceptors, however, it is likely that the receptors in both groups share a large part of the same developmental network. There are sufficient differences, however, also to suggest that the reevolution of electroreception in teleosts involves more than a single gene.

3. Development of Central Electroreceptive Nuclei and Pathways

Although primitive and derived electroreceptors are innervated by homologous lateral line nerves, their primary targets in the medulla are not homologous (McCormick 1982; Bullock et al. 1983; Bullock and Heiligenberg 1986). The primary medullary target of primitive electroreceptors is the dorsal octavolateralis nucleus (Bodznick and Boord 1986; Fritzsch and Münz 1986; Northcutt 1986; Ronan 1986; Bell and Maler, Chapter 4), whereas the primary medullary target of derived (teleost) electroreceptors is the the electroreceptive lateral line lobe (ELL) (Bell and Szabo 1986; Braford 1986; Carr and Maler 1986; Finger 1986). Amazingly, there are no developmental studies on the dorsal octavolateralis nucleus or on any of the higher centers that process electroreception in lampreys, chondrichthians, non-teleost bony fishes, or amphibians. This is due, in part, to the fact that most developmental studies on these taxa predate the discovery of electroreception, but there was also a subsequent lack of interest in the development of these centers. On the other hand, there are a limited number of developmental studies on electroreceptive centers of catfishes (Lannoo and Lannoo 1996) and gymnotiform fishes (Leyhausen et al. 1987; Lannoo et al. 1990, 1992). There appears to be only a single developmental study (Haugedé-Carré et al. 1977) of several of the electroreceptive centers in the medulla and cerebellum of a mormyrid species, *Pollimyrus*, and no developmental studies of the electroreceptive centers in either *Gymnarchus* or *Xenomystus*.

3.1 Development of Electroreceptive Centers in Catfishes

Lannoo and Lannoo (1996) described the development of the ELL in the channel catfish (*Ictalurus punctatus*) with reference to the onset of three behaviors: undulatory swimming, schooling, and feeding. These workers, as in Northcutt (2003), staged their embryos and larvae by modifying the normal tables of Armstrong and Child (1962) for *Ictalurus nebulosus* (Fig. 5.5). According to Lannoo and Lannoo, the central projections of the lateral line nerves that include electroreceptive fibers enter the medulla by stage 32. Thus lateral line fibers enter the medulla in the channel catfish, as in axolotls (Fig. 5.3), prior to the formation of the ampullary organs (Northcutt 2003). At the time that the developing lateral line fibers enter the medulla, the latter consists of a solid mass of neuroblasts and neurons, and a distinct ELL primordium, located at the extreme lateral edge of the fourth ventricle, can not be recognized until stage 33 or 34. During stages 36 through 46, the cells of the ELL primordium begin to differentiate, and a cerebellar crest can be recognized. By stage 49, a developing ELL can be distinguished from the more medial mechanoreceptive nucleus, and it forms a distinct lobe by stage 53 when feeding commences. Thus the maturation of cephalic ampullary organs (Northcutt 2003) and the ELL (Lannoo and Lannoo 1996) in the channel catfish appears to be timed so that both the receptors and their primary medullar center have developed just prior to the onset of feeding. Although the onset of undulatory swimming and schooling do not appear to be related to the development of the electrosensory system, they could be related instead to the development of the mechanoreceptive lateral line system (Fig. 5.5), which occurs much earlier (Northcutt 2003).

3.2 Development of Electroreceptive Centers in a Gymnotiform (Eigenmannia)

Because a normal staging table does not exist, the development of the central electroreceptive centers in the gymnotiform *Eigenmannia,* like the development of its electroreceptors, is described with reference to post-spawning ages (Vischer 1989a).

All electroreceptors in *Eigenmannia* are innervated by rami of the anterodorsal and anteroventral lateral line nerves (Northcutt and Vischer 1988), whose ganglia begin to differentiate by day 2 (Vischer et al. 1989). The initial peripheral rami also begin to form at this time, whereas the centrally directed fibers of these rami, unlike those in axolotls and catfishes, do not form until approximately day 5. In all of these taxa, however, the electroreceptive fibers of the lateral line nerves enter the medulla prior to the formation of the electroreceptors.

The germinal cells of the ELL in *Eigenmannia* arise dorsally in an alar periventricular zone, immediately behind the developing cerebellum, and migrate caudally and laterally within the developing medulla. These cells establish two ELL germinal zones: a medial zone, which forms the medial segment of the ELL, which is innervated by afferent fibers of the ampullary organs; and a lateral zone, which forms the lateral segment of the ELL, which is innervated by afferent fibers of the tuberous organs. Both germinal zones can be recognized by day 5, and the primary electrosensory fibers of the lateral line nerves enter both germinal zones by day 6. The ELL grows both laterally and caudally, differentiating in a rostrocaudal direction. By day 7 or 8, dorsal and ventral germinal subzones can be recognized in both the medial and lateral segments of the ELL. The dorsal subzones give rise to granular cell laminae, whereas the ventral subzones give rise to the pyramidal cell lamina. At present, the origin of the spherical cells of the ELL is unclear. By day 14, however, the granular cell and pyramidal cell laminae begin to be separated by a plexiform lamina formed by the ascending pyramidal cell axons, and the ELL begins to take on its mature laminar appearance.

Although there have been no developmental genetic studies of the ELL in any electroreceptive teleost, Lannoo et al. (1992) noted that the antibody anti-zebrin II recognizes the developing pyramidal cells in the medial ampullary processing segment of the ELL of *Eigenmannia.* The more lateral tuberous processing segments of ELL were unreactive, but the pyramidal cells of the mechanoreceptive medial octavolateralis nucleus were recognized by this antibody. Lannoo et al. (1992) suggested that the presence of anti-zebrin II–positive pyramidal cells in both the medial octavolateralis nucleus and the ampullary processing segment of the ELL might indicate that ampullary organs, but not tuberous organs, may have evolved from neuromasts. If this were the case, however, one might expect to see a close relationship between developing ampullary organs and neuromasts, but this does not appear to be the case.

Lannoo et al. (1990) provide less detail on the development of the torus semicircularis and optic tectum than they do on the ELL. They do note, however, that both the torus and tectum continue to grow by adding cells caudally. Because the ELL also grows caudally, the primary electroreceptive afferents must also shift caudally to maintain the somatotopic maps that are established in the ELL. This somatotopy is also maintained in the torus and tectum, but the mechanisms for this are unknown. Leyhausen et al. (1987) studied the relative growth of different parts of the brain in a related gymnotid, *Apteronotus*. These authors noted that the telencephalon grew at a constant rate, whereas the growth of the midbrain and hindbrain varied. The midbrain grew faster than the rest of the brain initially, but later it grew more slowly. The hindbrain was the opposite, initially growing more slowly than the rest of the brain, and later growing faster. These data suggest that midbrain centers mature earlier than hindbrain centers, which provide electroreceptive input to the midbrain. It is possible, therefore, that these relative growth rates are linked to the timing of the differentiation of their respective sense organs. If so, the retina and optic nerve projections to the midbrain should form before the electroreceptors and their related hindbrain centers.

3.3 Development of Electroreceptive Centers in a Mormyrid (Pollimyrus)

Mormyrid fishes are noted for their remarkable cerebellar hypertrophy (Bell and Szabo 1986; Bell and Maler, Chapter 4). As in other fishes, their cerebellum is divided into a corpus and a valvula, but in mormyrids the corpus is expanded in a rostrocaudal direction, forming three distinct lobes (C1–C3). This feature reaches an extreme in the genus *Gnathonemus*, but most of our information on development comes from *Pollimyrus,* which displays more moderate hypertrophy. In addition to lobes C1–C3, there are two caudally directed extensions of the cerebellum, the caudal lobe and the paired eminentiae granulares, which cover the dorsal and lateral surfaces of the enlarged lateral line lobes, respectively. The valvula is also enlarged in mormyrids. In most teleosts, this rostral continuation of the corpus extends into (and largely fills) the mesencephalic ventricle. In *Pollimyrus*, however, the valvula continues to enlarge, extending dorsally and rostrally, to arch over the surface of the optic tectum, and caudally to cover much of the corpus.

The development of the corpus and valvula, and that of numerous related structures, was described in *Pollimyrus* by Haugedé-Carré et al. (1977). Like that of most other electroreceptive teleost fishes, the ontogeny of *Pollimyrus* has not been staged. Unfortunately, Haugedé-Carré and colleagues provide few details regarding its ontogenetic stages. They do note that coordinated swimming occurs in 40-day-old larvae and that these fish are mature in 9 months. Otherwise, no other clues are provided regarding its general development.

Both the corpus and caudal lobe can be recognized initially in 6-day-old larvae and the eminentia granularis in 8-day-old larvae. The lateral line lobe also begins to develop in 8-day-old larvae, but the valvula and the second lobe of the corpus do not begin to develop until the larvae are 11 days old. The valvula completely fills the mesencephalic ventricle in 29-day-old larvae, and all cerebellar centers, with the exception of the valvula, are well developed in 38-day-old larvae. Development of the valvula is much slower, and its adult configuration is not reached until the larvae are approximately 6 months old.

4. Summary

Ontogenetic studies provide a key to understanding the organization of different classes of sensory receptors, their distribution on the body, and their innervation. This is particularly true for the lateral line system of fishes and amphibians, as one class of sensory receptors, electroreceptors, was lost and then regained at a later time in two groups of teleost fishes. There are extensive descriptive and experimental studies documenting that primitive electroreceptors, that is, nonteleost ampullary organs, arise from lateral line placodes that generate one or more lines of neuromasts and the lateral line nerve that innervates these receptors. Ampullary organ and neuromast primordia arise from lateral and central zones, respectively, of sensory ridges, which form by the elongation of the lateral line placodes. In contrast to primitive electroreceptors, the embryonic origin of the reevolved ampullary organs and the newly evolved tuberous electroreceptors in teleosts is uncertain. It has been claimed that ampullary organs in catfishes develop from lateral line placodes, whereas ampullary and tuberous organs in gymnotids are claimed to develop from general ectoderm. Clearly, experimental studies are needed to resolve these different interpretations.

Almost nothing is known regarding the genetic basis for the induction and morphogenesis of either primitive or derived (teleost) electroreceptors. Similarly, there are no existing descriptions of the development of the central electroreceptive centers in taxa with primitive electroreceptors, and very little is known regarding the development of these centers and pathways in electroreceptive teleosts.

Acknowledgments. I thank Ted Bullock and Carl Hopkins for inviting me to write this review. Jo Griffith, Preston Holmes, and Mary Sue Northcutt generated the composite figures. Sue Commerford and Mariola Milik provided general and technical support, as well as word processing and literature retrieval, respectively. Finally, Mary Sue Northcutt contributed to many phases of the research and preparation of the manuscript.

References

Allis EP (1889) The anatomy and development of the lateral line system in *Amia calva.* J Morphol 2:463–566.

Armstrong PB, Child JS (1962) Stages in the Development of *Ictalurus Nebulosus*. Syracuse, NY: Syracuse University Press.

Baker CVH, Bronner-Fraser M (2001) Vertebrate cranial placodes. I. Embryonic induction. Dev Biol 232:1–61.

Beckwith CJ (1907) The early development of the lateral line system of *Amia calva.* Biol Bull 14:23–28.

Bell CC, Szabo T (1986) Electroreception in mormyrid fish: central anatomy. In: Bullock TH, Heiligenberg W (eds), Electroreception. New York: John Wiley & Sons, pp. 375–422.

Bemis WE, Northcutt RG (2004) Development of the lateral line system of the paddlefish *Polyodon spathula* (Actinopterygii: Acipenseriformes). J Morphol 260:277.

Bensouilah M, Schugardt C, Roesler R, Kirschbaum F, Denizot J-P (2002) Larval electroreceptors in the epidermis of mormyrid fish: I. Tuberous organs of type A and B. J Comp Neurol 447:309–322.

Bever MM, Borgens RB (1991a) The regeneration of electroreceptors in *Kryptopterus*. J Comp Neurol 309:200–217.

Bever MM, Borgens RB (1991b) Patterning in the regeneration of electroreceptors in the fin of *Kryptopterus*. J Comp Neurol 309:218–230.

Bodznick D, Boord RL (1986) Electroreception in chondrichthyes: central anatomy and physiology. In: Bullock TH, Heiligenberg W (eds), Electroreception. New York: John Wiley & Sons, pp. 225–256.

Braford MR Jr. (1986) African knifefishes: the xenomystines. In: Bullock TH, Heiligenberg W (eds), Electroreception. New York: John Wiley & Sons, pp. 453–464.

Bullock TH, Heiligenberg W (eds) (1986) Electroreception. New York: John Wiley & Sons.

Bullock TH, Bodznick DA, Northcutt RG (1983) The phylogenetic distribution of electroreception: evidence for convergent evolution of a primitive vertebrate sense modality. Brain Res Rev 6:25–46.

Carr CE, Maler L (1986) Electroreception in gymnotiform fish: central anatomy and physiology. In: Bullock TH, Heiligenberg W (eds), Electroreception. New York: John Wiley & Sons, pp. 319–374.

Coombs S, Janssen J, Webb JF (1988) Diversity of lateral line systems: evolutionary and functional considerations. In: Atema J, Fay RR, Popper AN, Tavolga WN (eds), Sensory Biology of Aquatic Animals. New York: Springer-Verlag, pp. 553–593.

Coombs S, Görner P, Münz H (eds) (1989) The Mechanosensory Lateral Line: Neurobiology and Evolution. New York: Springer-Verlag.

Denizot JP, Libouban S (1985) New formation of sensory cells in the tuberous organ (electroreceptor) of *Brienomyrus niger* (Mormyridae) induced by transection of afferent nerve. Int J Dev Neurosci 3:323–330.

Disler NN (1971) Lateral Line Sense Organs and Their Importance in Fish Behavior (trans Mills H, Yariv M), Mills H (ed). Jerusalem Keter Press.

Disler NN (1977) The Lateral Line System Sense Organs of Sharks. Moscow: Science Publications.

Finger TE (1986) Electroreception in catfish: behavior, anatomy, and electrophysiology.

In: Bullock TH, Heiligenberg W (eds), Electroreception. New York: John Wiley & Sons, pp. 287–317.

Fritzsch B, Bolz D (1986) On the development of electroreceptive ampullary organs of *Triturus alpestris* (Amphibia: Urodela). Amphibia Reptilia 7:1–9.

Fritzsch B, Münz H (1986) Electroreception in amphibians. In: Bullock TH, Heiligenberg W (eds), Electroreception. New York: John Wiley & Sons, pp. 483–496.

Gibbs MA, Northcutt RG (2004) Development of the lateral line system in the shovelnose sturgeon. Brain Behav Evol 64:70–84.

Haugedé-Carré F, Kirschbaum F, Szabo T (1977) On the development of the gigantocerebellum in the mormyrid fish *Pollimyrus (Marcusenius) isidori*. Neursci Lett 6: 209–213.

Holmgren N (1940) Studies on the head in fishes; embryological, morphological, and phylogenetical researches. Acta Zool 21:51–267.

Johnson SE (1917) Structure and development of the sense organs of the lateral canal system of Selachians (*Mustelus canis* and *Squalus acanthias*). J Comp Neurol 28:1–74.

Kirschbaum F, Denizot J-P (1975) Sur la différenciation des électrorécepteurs chez *Marcusenius* sp. (Mormyrides) et *Eigenmannia virescens* (Gymnotides), poissons électriques à faible décharge. CR Acad Sci Paris 281:419–422.

Landacre FL, Conger AC (1913) The origin of the lateral line primordia in *Lepidosteus osseus*. J Comp Neurol 23:575–633.

Lannoo MJ, Lannoo SJ (1996) Development of the electrosensory lateral line lobe in the channel catfish, *Ictalurus punctatus*, with reference to the onset of swimming and feeding behaviors. Mar Fresh Behav Physiol 28:45–53.

Lannoo MJ, Vischer HA, Maler L (1990) Development of the electrosensory nervous system of *Eigenmannia* (Gymnotiformes): II. The electrosensory lateral line lobe, midbrain, and cerebellum. J Comp Neurol 294:37–58.

Lannoo MJ, Maler L, Hawkes R (1992) Zebrin II distinguishes the ampullary organ receptive map from the tuberous organ receptive maps during development in the teleost electrosensory lateral line lobe. Brain Res 586:176–180.

Leyhausen C, Kirschbaum F, Szabo T, Erdelen M (1987) Differential growth in the brain of the weakly electric fish, *Apteronotus leptorhynchus* (Gynotiformes), during ontogenesis. Brain Behav Evol 30:230–248.

McCormick CA (1982) The organization of the octavolateralis area in actinopterygian fishes: a new interpretation. J Morphol 171:159–181.

Metcalfe WK (1985) Sensory neuron growth cones comigrate with posterior lateral line primordial cells in zebrafish. J Comp Neurol 238:218–224.

Metscher BD, Northcutt RG, Gardiner DM, Bryant SV (1997) Homeobox genes in axolotl lateral line placodes and neuromasts. Dev Genes Evol 207:287–295.

New JG (1997) The evolution of vertebrate electrosensory systems. Brain Behav Evol 50:244–252.

Nikolskaya MP (1983) Development of the ampullar electroreceptor system in the ontogeny of *Polyodon* and Acipenseridae. Akad Nauk USSR 268:474–477.

Northcutt RG (1986) Electroreception in nonteleost bony fishes. In: Bullock TH, Heiligenberg W (eds), Electroreception. New York: John Wiley & Sons, pp. 257–285.

Northcutt RG (1989) The phylogenetic distribution and innervation of craniate mechanoreceptive lateral lines. In: Coombs S, Görner P, Münz H (eds), The Mechanosensory Lateral Line: Neurobiology and Evolution. New York: Springer-Verlag, pp. 17–78.

Northcutt RG (1997) Evolution of gnathostome lateral line ontogenies. Brain Behav Evol 50:25–37.

Northcutt RG (2003) Development of the lateral line system in the channel catfish. In: Browman HI, Skiftesvik AB (eds), The Big Fish Bang. Proceedings of the 26th Annual Larval Fish Conference. Bergen, Norway: Institute of Marine Research, pp. 137–159.

Northcutt RG, Bemis WE (1993) Cranial nerves of the coelacanth, *Latimeria chalumnae* [Osteichthyes: Sarcoptergii: Actinistia], and comparisons with other Craniata. Brain Behav Evol 42(S1):1–76.

Northcutt RG, Brandle K (1995) Development of branchiomeric and lateral line nerves in the axolotl. J Comp Neurol 355:427–454.

Northcutt RG, Vischer HA (1988) *Eigenmannia* possesses autapomorphic rami of the anterior lateral line nerves. Soc Neurosci Abstr 14:54.

Northcutt RG, Catania KC, Criley BB (1994) Development of lateral line organs in the axolotl. J Comp Neurol 340:480–514.

Northcutt RG, Brandle K, Fritzsch B (1995) Electroreceptors and mechanosensory lateral line organs arise from single placodes in axolotls. Dev Biol 168:358–373.

Northcutt RG, Holmes PH, Catania KC, Luer CA (1997) Lateral line system development in an elasmobranch, *Raja eglanteria*. Soc Neurosci Abstr 23:2380.

Pehrson T (1949) The ontogeny of the lateral line system in the head of dipnoans. Acta Zool 30:153–182.

Piotrowski T, Northcutt RG (1996) The cranial nerves of the Senegal bichir, *Polypterus senegalus* [Osteichthyes: Actinopterygii: Caldistia]. Brain Behav Evol 47:55–102.

Ronan M (1986) Electroreception in cyclostomes. In: Bullock TH, Heiligenberg W (eds), Electroreception. New York: John Wiley & Sons, pp. 209–224.

Roth A (1986) Afferent fibers induce electroreceptors in the skin of fish. Naturwissenschaften 73:264–266.

Roth A (1993) Regenerative outgrowth and distribution of the electroreceptive nerve fibers in the catfish *Kryptopterus*. J Comp Neurol 328:473–484.

Roth A (1994) Development of the electrosensory system. Naturwissenschaften 81:269–272.

Ruud G (1920) Uber Hautsinnesorgane bei Spinax niger Bon. Zool Jahrb Anat Ontogen 41:459–546.

Sahly I, Andermann P, Petit C (1999) The zebrafish *eya1* gene and its expression pattern during embryogenesis. Dev Genes Evol 209:399–410.

Schlosser G (2002a) Development and evolution of lateral line placodes in amphibians. I. Development. Zoology 105:119–146.

Schlosser G (2002b) Development and evolution of lateral line placodes in amphibians. II. Evolutionary diversification. Zoology 105:177–193.

Schlosser G, Northcutt RG (2000) Development of neurogenic placodes in *Xenopus laevis*. J Comp Neurol 418:121–146.

Song J, Northcutt RG (1991) The primary projections of the lateral line nerves of the Florida gar, *Lepisosteus platyrhincus*. Brain Behav Evol 37:38–63.

Srivastava CBL, Seal M (1981) Electroreceptors in Indian catfish teleosts. Adv Physiol 31:1–11.

Stone LS (1922) Experiments on the development of the cranial ganglia and the lateral line sense organs in *Amblystoma punctatum*. J Exp Zool 35:421–496.

Vischer HA (1989a) The development of lateral-line receptors in *Eigenmannia* (Teleostei,

Gymnotiformes). I. The mechanoreceptive lateral-line system. Brain Behav Evol 33: 205–222.
Vischer HA (1989b) The development of lateral-line receptors in *Eigenmannia* (Teleostei, Gymnotiformes). II. The electroreceptive lateral-line system. Brain Behav Evol 33: 223–236.
Vischer HA (1995) Electroreceptor development in the electric fish *Eigenmannia*: a histological and ultrastructural study. J Comp Neurol 360:81–100.
Vischer HA, Lannoo MJ, Heiligenberg W (1989) Development of the electrosensory nervous system in *Eigenmannia* (Gymnotiformes): I. The peripheral nervous system. J Comp Neurol 290:16–40.
Webb JF (1989a) Developmental constraints and evolution of the lateral line system in teleost fishes. In: Coombs S, Görner P, Münz H (eds), The Mechanosensory Lateral Line: Neurobiology and Evolution. New York: Springer-Verlag, pp. 79–97.
Webb JF (1989b) Gross morphology and evolution of the mechanoreceptive lateral-line system in teleost fishes. Brain Behav Evol 33:34–53.
Winklbauer R, Hausen P (1983) Development of the lateral line system in *Xenopus laevis*. I. Normal development and cell movement in the supraorbital system. J Embryol Exp Morphol 76:265–281.
Zakon HH (1984) Postembryonic changes in the peripheral electrosensory system of a weakly electric fish: addition of receptor organs with age. J Comp Neurol 228:557–570.
Zakon HH (1986) The electroreceptive periphery. In: Bullock TH, Heiligenberg W (eds), Electroreception. New York: John Wiley & Sons, pp. 103–156.

6

The Physiology of Low-Frequency Electrosensory Systems

David Bodznick and John C. Montgomery

1. Introduction

The ancestor to both jawed and jawless vertebrates had dermal electroreceptor organs to sense DC and low-frequency electric fields. Thus a low-frequency electrosense was among the first octavolateral sensory systems and is likely as old as vertebrates. Then as now, the natural stimuli would be local dipole fields generated by other animals and more uniform and large-scale fields created by ocean currents, the animal's own movements in the earth's magnetic field, and by electrochemical sources such as temperature or salinity transitions. Low-frequency electroreception allows vertebrate predators to find their prey at night or buried in the sand, can play a role in their social interactions, and may guide migrations and home range orientation.

The ancestral vertebrate electrosense was passed down to most major taxa of living fishes and aquatic amphibians (Bullock et al. 1983). It was lost in the radiation of fish leading to the teleosts, and again when vertebrates became fully terrestrial. Low-frequency electroreceptors subsequently reevolved two different times among teleosts and yet again in egg-laying mammals (Monotremata: platypus and echidnas) (Table 6.1) (Bullock et al. 1983; Zupanc and Bullock, Chapter 2). The ampullary electrosensory systems of fish and amphibians are modified hair cell senses with strong similarities to other octavolateral senses, particularly the lateral line and audition. The electroreceptors of monotremes, on the other hand, are simple trigeminal nerve endings in mucus glands on the animal's bill. Nonetheless, all of these systems share a selective sensitivity to low-frequency stimuli and all act primarily as passive sensors of natural electric fields from sources other than electric organs.

In this chapter we consider the neurophysiological properties of the low-frequency electrosensory systems. The focus is physiology but as a basis for this we first briefly consider the behaviorally relevant properties of natural electric fields that must be encoded by the receptors, and we summarize the basics of the electroreceptor anatomy. More details and references for the anatomy

TABLE 6.1. Evolutionarily distinct low-frequency electroreceptor systems.

Electroreceptors	Major taxa	Innervation	Excitatory polarity
I. Lorenzinian ampullary organs[a]	Petromyzontidae (lampreys)[b] Chondrichthyes (cartilaginous fish) Chondrostei (sturgeons, paddlefish) Sarcopterygii (lungfish, coelocanth) Amphibia (larval and aquatic adult urodeles, apodans)	Lateral line nerves	Cathodal (lumen negative)
Teleost ampullary organs			
II. In Osteoglossids[c]	Mormyriformes (African electric fish) Xenomystinae (African notopterids)	Lateral line nerves	Anodal (lumen positive)
III. In Ostariophysans[c]	Gymnotiformes (S. American electric fish) Siluriformes (catfishes, worldwide)	Lateral line nerves	Anodal
IV. Trigeminal electroreceptors	Monotremata (platypus, echidnas)	Trigeminal nerve	Cathodal

[a] After the ampullae of Lorenzini of elasmobranchs.

[b] The end bud electroreceptors of lampreys lack the canal and ampulla structures for which ampullary electroreceptors are named, but they are homologous with the other receptors of this group as evidenced by both physiological and other morphological traits.

[c] The phylogenetic relatedness of mormyrids and xenomystids on the one hand and gymnotids and silurids on the other suggest that low-frequency electroreceptors may have evolved just once in osteoglossid fishes and once again in ostariophysans.

and behavior are found in other chapters of this volume (Jørgensen, Chapter 3; Wilkens, Chapters 9).

Electroreceptors are sensors of potential and in this way they are no different from neurons and other excitable cells sensitive to transmembrane potentials. What sets low-frequency electroreceptors apart, however, is the phenomenal microvolt sensitivity of many of them, particularly in marine species. Mechanisms within the brain further enhance this sensitivity and selectively remove noise to improve signal detection. We give special attention here to the physiological mechanisms of sensitivity and noise reduction in low-frequency electrosensory systems.

1.1 Behaviorally Important Physical Properties of Natural Electric Fields

Natural electric fields in aquatic environments are very weak. Measured values for uniform fields in the oceans are typically less than 0.5 μV/cm, and calculated values for fields induced by a fish's swimming in the earth's magnetic field are similar. The potentials directly adjacent to organisms in the sea are generally 1 to 200 μV. The potential gradients of these local fields fall off with the cube of distance, so they are ordinarily much less than 1 μV/cm within 10 to 20 cm. Accordingly, there is a great premium on the ability to detect weak electric fields in order to detect predator or prey, at the greatest possible distances. Thresholds of behavioral responses extend down to less than 10 nV/cm in marine elasmobranchs, and the saltwater catfish, *Plotosus*, is nearly as sensitive (Kalmijn 1974; Obara and Sugawara 1984; Kalmijn 1988). Because of the lower shunt conductance of the water, natural electric fields may be at least 10 times stronger in freshwater than in marine environments, but still are very weak (Peters and Bretschneider 1972). Freshwater fish, and amphibians are generally one to two orders of magnitude less sensitive than marine fish, and this is probably a limit of useful sensitivity set by the higher ambient noise levels in freshwater. Behavioral thresholds for platypus appear to be 20 to 200 μV/cm (Pettigrew 1999).

In uniform electric fields, the field *intensity* and *orientation*, including *polarity*, can contain useful information about the relative velocity and direction of an ocean current. Similar cues generated by the animal's movement in the earth's magnetic field can indicate compass heading and velocity (Paulin 1995). For more local bioelectric fields the spatial distribution of these parameters becomes important. When localizing dipole fields of prey, predator, or conspecifics, directional information is available at close range in the steep fall off of the potential gradient away from the source. At greater distances, locating a dipole source may require a more complicated spatial analysis of the convergence or divergence of the curved dipole field lines. Alternatively, simpler localization strategies might be used such as aligning with, or maintaining a constant angle to, the direction of the curved field lines during approach (Schluger and Hopkins 1987; Kalmijn 1988).

Notably absent from this short list of behaviorally important features are precise stimulus timing and frequency. Electric fields propagate with nearly infinite speed and so they are present throughout their full extent instantaneously. There is no significant stimulus conduction time and no propagation-dependent changes in the fields. Precise time coding is thus apparently not important for low-frequency electroreception in any species, with the possible exception of platypus, in which comparison of onset times of electric and hydrodynamic stimuli might be used for distance measurements (Pettigrew 1999). Similarly, there is little to indicate that the particular frequencies of electric field stimuli within the detectable low-frequency range have any significance for behavior. In experimental tests, electroreceptive predators readily attack artificial dipoles without regard to the frequencies of the stimuli as long as they are within their sensitive range. Also, as noted, the lack of propagation of electric fields means that there are no changes in stimulus frequency composition with distance such as those that are important characteristics of sound or surface waves for audition and lateral line.

Hence it appears that the tasks for the low-frequency electroreceptors reduce to detecting extremely weak fields and to encoding the local intensity, orientation, and polarity of those fields. These, in combination with the wide spatial distribution of the receptors over the animal's body surface, appear to be all that are required for most behaviors dependent on low-frequency electroreception.

1.2 Basics of Receptor Anatomy

The anatomy of low-frequency electroreceptors is described in detail elsewhere (Jørgensen, Chapter 3; Zakon 1986), so it is necessary to describe only their most basic features here. Lorenzinian and teleost ampullary organs are like all other octavolateral receptors in that their receptor cells are embedded among support cells in a sensory epithelium, which lines the ampulla-like chamber at the base of each receptor organ. The receptor cells, which number from tens to thousands depending on taxa and habitat (higher in marine than freshwater species), have an apical cilium or microvilli and they communicate with from one to several primary afferent fibers through ribbon synapses on their basal face. Tight junctions between receptors and support cells place the apical and basal surfaces of the receptor cells in electrically and chemically distinct compartments. The apical cell surface looks at the water outside the animal through the low resistance of a very conductive but well-insulated jelly-filled canal. The base of the receptor faces the subdermal interior of the animal. Unlike most other octavolateral receptors, electroreceptors receive no efferent innervation.

The canals in Lorenzinian and teleost ampullary receptors are long in marine species and short, just traversing the skin, in freshwater species. This difference is related to the higher conductance of seawater and the lower skin resistance of marine versus freshwater forms. In marine fishes large-scale fields can pervade the internal tissues and so long canals are required to permit sampling of a significant portion of shallow uniform voltage gradients. The long canals also permit the clustering of the ampullae from many different receptors, with skin pores widely distributed across the body surface, in a single location so that they share a common internal reference potential. The high skin resistances in freshwater species effectively exclude external fields and receptors need only measure the potential drop across the skin referenced to the nearly isopotential body interior.

In contrast to the octavolateral electroreceptor systems of fish and amphibians, the electroreceptors of monotremes lack a proper sensory epithelium and voltage sensing is accomplished by the afferent nerve endings themselves. These free endings are found in a packed cell layer at the base of mucous sensory glands on the bill.

2. Receptor Physiology

2.1 Frequency Sensitivity

All low-frequency electroreceptors exhibit similar broad low-pass frequency response functions. The Lorenzinian and teleost ampullary receptors respond to field modulation rates from less than 0.1 Hz up to 10 to 50 Hz depending on species (Bullock 1982; Zakon 1986). For most species highest sensitivity is in the 1 to 10 Hz range. Platypus electroreceptors have greatest sensitivity from 50 to 100 Hz, and this may be a specialization for detecting muscle potentials created by the fast tail flicks of the freshwater shrimp they eat (Gregory et al. 1989). There is no evidence of differently tuned receptors in a single animal, and the individual receptors transmit only limited frequency information in the temporal modulation of their spiking. None of the low-frequency electroreceptors are truly DC sensitive, a property that is essential for their normal operation on a background of the animal's own substantial DC fields. But electroreceptive animals can sense external DC sources during the animal's own movements through these fields.

Although frequency discrimination may not be very important for low-frequency electroreception, a good match of the sensitive bandwidth of the receptors to the frequencies of natural stimuli has obvious significance for high sensitivity. Platypus electroreceptors may be an example of this, as noted earlier. The peak sensitivity and bandwidth of the electroreceptors in rays and skates is a good match to the electric fields of conspecifics. Furthermore, in stingrays, *Dasyatis sabina*, the frequency response functions change through the reproductive season in wild caught animals and in response to exogenous testosterone treatment in the laboratory, in a way that is consistent with their use of the sense in mating interactions (Sisneros and Tricas 2000, 2002).

2.2 Receptor Coding of Electric Field Intensity, Orientation, *and* Polarity

As Figure 6.1 illustrates, low frequency electroreceptor afferents have regular resting discharge rates that are increased or decreased, depending on *polarity*, by potential changes in the water outside the receptor pore (Obara and Sugawara 1984; Bennett and Obara 1986; Zakon 1986; Gregory et al. 1988). Lorenzinian ampullary organs are excited by a cathodal stimulus polarity, that is, when the pore and lumen of the organ is made negative relative to the basal face within the body. Anodal stimuli are inhibitiory. In contrast, the electroreceptors in both ostariophysine and osteoglossid teleosts are excited by the anodal stimulus polarity. The firing rate in each of the systems linearly encodes stimulus *intensity* over an operating range that extends two to three orders of magnitude. Furthermore, in each case this linear intensity–response function is smoothly continuous through zero so that the receptors operate in push/pull fashion like

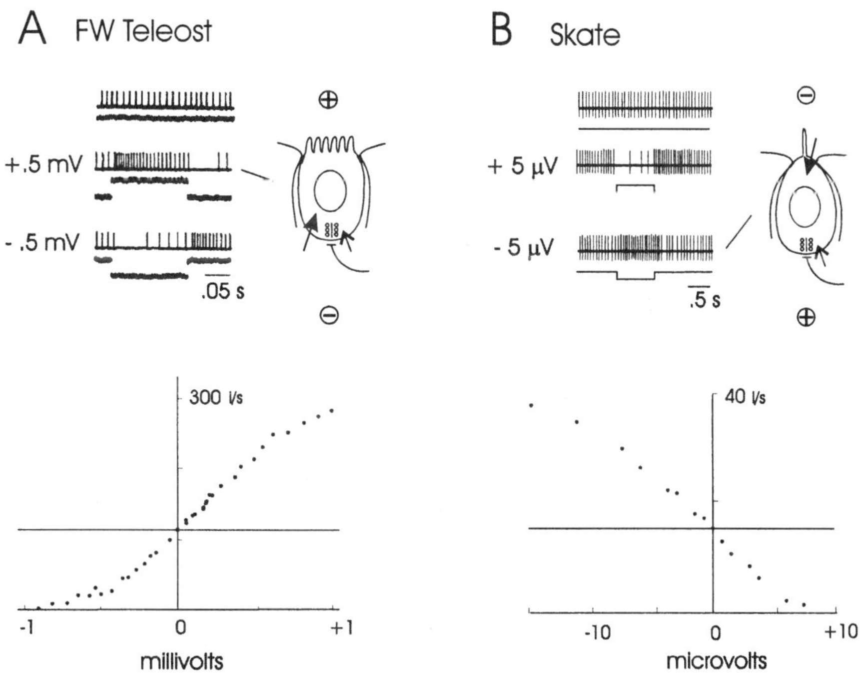

FIGURE 6.1. Responses of low-frequency electroreceptors in a freshwater teleost fish (*Gymnotus*, ostariophysan) and a marine skate (*Raja*, Lorenzinian ampullae). (**A**) *Top*: Teleost electroreceptors are excited by anodal stimuli, that is, positive potentials outside the receptor relative to the base, which depolarize the basal membrane of the receptor cells. Cathodal stimuli inhibit. The basal membrane contains a voltage-gated Ca conductance (*large arrow on diagram at right*) and acts as both voltage sensor and transmitter secreting surface. The apical surface is passive and low resistance. *Bottom*: Intensity–response function of the *Gymnotus* ampullary electroreceptor afferent. *Gymnotus* data are replotted from Bennett (1968). (**B**) *Top*: Lorenzinian ampullary electroreceptors are excited by cathodal and inhibited by anodal stimuli. These polarities depolarize and hyperpolarize, respectively, the apical receptor membrane which contains a voltage-gated Ca conductance (*large arrow on diagram*) and acts as the voltage sensor. This receptor current of the apical face then depolarizes the basal surface, which contains its own Ca conductance (*small arrow*) for transmitter secretion. *Bottom*: Intensity–response function of the skate electroreceptor afferent (Bodznick, unpublished). Note the linearity of the response through zero in both teleost and Lorinzinian receptors, and also the greater sensitivity typical of ampullary electroreceptors in marine versus freshwater species.

other hair cell sensors. All ampullary electroreceptors are very sensitive, with the receptors of marine species being about 10 times as sensitive as those of most freshwater species. The latter is true in both non-teleost and teleost electrosensory systems; sensitivity is related to habitat, not phylogeny. In the most sensitive marine elasmobranchs measurable changes in electroreceptor firing are seen with intensities of less than 1 μV at the receptor opening relative to the receptor base inside the body. The electroreceptor afferents of platypus are significantly less sensitive with measurable responses only with stimuli over a millivolt.

The ability to detect the *orientation* of a uniform electric field vector or the distribution of field orientations across the body is dependent on the directional sensitivity of the individual receptor organs and their locations on the body (Bodznick and Boord 1986). For marine fishes in which large-scale fields to a great extent pervade the fish's internal tissues, each receptor is most sensitive to a field aligned with the axis between receptor pore and the location of the ampulla within the body and is insensitive to fields orthogonal to this. In between these the potential across the receptors will be approximately given by the vector dot product of the **E** field ($\vec{E}$) and the vector given by the receptor length and orientation ($\vec{R}$).

$$\text{Effective stimulus} = \vec{E} \cdot \vec{R} = |\mathbf{E}||\mathbf{R}| \cos \Theta,$$

where Θ is the angle between the **E** field and receptor axis.

Similarly, in freshwater species each receptor is most sensitive to electric fields aligned with the axis between the skin pore and the field location corresponding to the isopotential point within the body. Receptors on head and tail are most sensitive to longitudinally oriented fields and those at mid-body levels are most sensitive to transverse fields.

2.3 Receptor Mechanisms

How do electroreceptors work? Most of what we know about ampullary electroreceptor mechanisms comes from studies on two marine fishes, a skate, *Raja*, with Lorenzinian ampullae, and a catfish, Plotosus, with teleost (ostariophysan) ampullary organs. These have been studied because of their extreme sensitivity, but similar mechanisms likely operate in the less sensitive freshwater ampullary systems. The receptors in *Raja* and *Plotosus* are independently evolved, so it is not surprising that there are differences in their means of achieving high sensitivity. Inaccessibility of the receptor cells has largely prevented intracellular studies, but the structure of ampullary organs in marine species has made it possible to apply current and voltage-clamp techniques to study the massed epithelial receptor currents in whole organs in situ or isolated from the animal. Receptor output can be measured simultaneously by recordings of the afferent nerves.

In ampullary electroreceptors sensory transduction is direct and accomplished by voltage-gated Ca channels in the receptor cell membranes. It is natural to think that extraordinary receptor sensitivity must require extraordinarily sensitive ion channels, but the pharmacology and electrical properties of summed receptor currents indicate that they are very similar to N and L type Ca channels (Sugawara 1989b; Lu and Fishman 1995). Instead, the high sensitivity of the electroreceptors appears to derive at least in part from (1) the elimination of response threshold by maintaining the receptor cells in a partially activated state or oscillating even in the absence of a stimulus, (2) DC synapses that also lack a threshold and steadily release transmitter as a function of membrane potential, and (3) a very large convergence ratio of receptor cells to afferent fibers. The structure of the receptor organs also plays a role by ensuring that nearly all the available stimulus potential is brought to bear directly across the sensory epithelium. The long canals in marine forms contribute particularly to sensitivity in shallow voltage gradients.

In mechanosensitive hair cells mechanoelectrical transduction takes place in the apical surface of the receptor cell and the resulting receptor potential acts on the basal membrane to modulate transmitter release. Teleost ampullary electroreceptors apparently evolved from lateral line mechanoreceptors and in each case the receptor cell basal membrane remains the voltage sensor while the apical membrane is passive and low resistance. In *Plotosus* and probably other teleosts an electrogenic Na–K pump in the basal surface of the receptor epithelium provides a steady outward bias current across the sensory epithelium at rest (Sugawara 1989a). This current is set at a level to partially activate a noninactivating Ca conductance in the basal receptor cell membrane, and voltage-clamp measures show that the receptor epithelium sits in the negative slope conductance region of its summed I–V curve even at rest. The Ca entry causes a steady release of an excitatory transmitter, similar to glutamate, onto the afferent nerve fibers. Although the afferent fibers in catfish and other ampullary receptors exhibit intrinsic pacemaker activity (Struik et al. 2002), under normal conditions their firing depends on this continuous synaptic input from the receptor cells.

Several thousand receptor cells synapse onto 5 to 10 afferent fibers in each ampullary organ. Lumen positive (anodal) sensory stimuli directly depolarize the basal face of the receptor cells, further increasing Ca conductance, transmitter release, and the afferent firing rate. Cathodal stimuli reduce ongoing transmitter release and inhibit (defacilitate) afferent firing. This accounts for the anodal polarity preference of teleost receptors. The receptor's sensitivity and linearity, in part at least, come from the continuous partial activation of the Ca conductance, so that even the most minute potential change across the basal membrane can influence the percentage of open Ca channels and the rate of transmitter release. The ribbon synapses apparently operate as DC synapses here as they do in other hair cells and retinal photoreceptors. The amplification achieved by the regenerative nature of the Ca conductance is limited by the heavy loading of the receptors down the low resistance of the receptor canal, which prevents full regenerative activity with all but very strong stimuli. The synaptic transfer function, that is, the efficiency of transmitter release and/or postsynaptic sensitivity, may contribute to amplification and sensitivity, but it has not yet been possible to measure this. Finally, the large convergence of receptor cells to afferent fibers improves the sensitivity to coherent external stimuli. For further references and details see (Obara and Sugawara 1984; Bennett and Obara 1986; Sugawara and Obara 1989).

In the ampullary receptors of *Raja* and other non-teleosts similar themes are applied in different ways. The major difference is that in the Loreznninian ampullae, the apical receptor cell membrane is the active voltage sensor and contains voltage-gated Ca channels. As in *Plotosus*, a bias current across the sensory epithelium partially activates this conductance even in the absence of a stimulus so that the sensory epithelium is held in the negative slope conductance region of the summed receptor *I–V* curve. This is evidenced by the fact that an excitatory cathodal sensory stimulus, which depolarizes the apical receptor, leads to an inward receptor current across the sensory epithelium, opposite to that which would be created by this voltage drop across a passive resistance. This inward receptor current is driven not by the voltage of the stimulus but rather by the net inward electrochemical gradient on Ca. The inward current across the apical membrane of the receptor cells flows outward across their basal membrane depolarizing it and triggering Ca influx and steady transmitter release at rest. As with *Plotosus*, sensitivity of the receptors in skates derives in part from the spontaneous activity in the receptor cells and the large convergence ratio of receptors to afferents. But placing the active sensory surface in the apical receptor membrane and separate from the basal transmitter secreting membrane provides an additional level of amplification and sensitivity not possible in *Plotosus*. There are two different models for how this might work.

The work of Bennett, Obara, and Clusin (Obara and Bennett 1972; Clusin and Bennett 1979) indicates that the apical membrane of the individual receptor cells is capable of full regenerative Ca spikes, which are evidenced by oscillations that they record in massed receptor currents. The oscillations appear to result from interactions of the Ca conductance with a Ca-activated K conductance in apical and/or basal membranes. These conductances and resulting oscillations appear very similar to those previously found in other hair cell receptors. Indeed they are the basis for the electrical tuning of hair cells first discovered in tuberous electroreceptors (see Zakon 1986) and later found to play a role in vestibular and auditory receptor tuning. In the skate electroreceptors the oscillations are thought to have nothing to do with tuning but everything to do with sensitivity. By this model the bias current across the sensory epithelium depolarizes the receptor cells so that they are each poised at threshold for a Ca spike and occasionally exceeding it. At rest a fraction of the receptor cells are spiking at any given moment and these are responsible for the baseline transmitter release and afferent firing. A cathodal stimulus, which depolarizes the apical face of the receptors, increases the proportion of active receptor cells; anodal stimuli decrease the proportion. Thus, maximum sensitivity is achieved by a positive feedback cycle poised at threshold. It is the very nonlinearity of the response that is responsible for the amplification of the receptor current.

Nevertheless, the intensity–response function of the afferent fibers is linear for most natural stimulus intensities. The linearity according to this model stems from the probabilistic nature of the response of the individual receptor cells and can be the case only if the individual receptor cells operate independently and oscillate asynchronously.

Lu and Fishman have proposed a somewhat different model (Lu and Fishman 1994, 1995). By an electrical measure termed the complex admittance of the voltage-clamped sensory epithelium, they confirmed that the epithelium at rest is maintained in a partially activated condition by a bias current that may be supplied by Na–K and Na–Ca exchangers that they identified on the basal surface of the receptor cells. The Lu and Fishman model differs from that of Bennett, Obara, and Clusin in that the individual receptor cells are proposed to each be maintained in the same partially activated condition but prevented from full regenerative activity because of loading by the canal and seawater. The linearity of the intensity response function of the receptor afferents is a property of each individual receptor cell rather than of just the receptor cell population. The enhanced sensitivity in this case results from having a sustained negative resistance of the apical receptor membrane in series with a significant positive resistance of the basal, transmitter secreting membrane. In theory, depending on the relative values of the negative and positive resistances, a very significant current amplification can be achieved.

Kalmijn (2003) recently presented arguments that support the Lu and Fishman model, and he described a method of intracellular recording from the receptor cells that may in the future allow discrimination between the two models. Recent studies on the paddlefish, *Polyodon*, a freshwater fish with Lorenzinian ampullae, provide evidence that in the intact fish there are continuous independent oscillations from epithelium and afferent nerve fibers (Neiman and Russell 2004). The epithelial oscillations are consistent with the receptor cells being a population of independent oscillators, and the fact that these recordings are from intact fish in the absence of stimulation may lend support for the Bennett, Obara, and Clusin model.

Nothing is known so far of the receptor mechanisms in platypus electroreceptors.

2.4 Receptors and Noise

High sensitivity is useful only if accompanied by mechanisms to limit noise. In the end the best measure of receptor sensitivity may be the signal-to-noise ratio (S:N) of the receptor response. The effects of internal noise, by which we mean the noise created by the receptor cells themselves as a result of random fluctuations in membrane potential, transmitter release, and the like, can be reduced significantly by convergence. A large number (N) of receptor cells converging onto a single afferent fiber can provide a $\sqrt{N}$ improvement in the ratio of external signal, which will be synchronous among receptor cells, to the asynchronous internal noise.

Unfortunately, simple receptor convergence will not reduce the response to external noise, which generally will affect all of the receptor cells of a single electrosensory organ in the same way. One very common source of external noise for ampullary electroreceptors is the animal's own bioelectric fields. Osmoregulatory ion pumping and electrochemical differences between animal and water create standing DC potentials across the body surface, and these DC potentials also give rise to low-frequency potential modulations during ventilatory and other movements (Bodznick et al. 1992). The standing DC fields can be ignored by the receptors, which remarkably can accommodate to strong DC fields without the loss of incremental sensitivity for microvolt stimuli (Murray 1965; Bodznick et al. 1993). However, in the absence of efferent innervation in electroreceptors, the effects of ventilatory and other movement-related external noise can be removed only within the brain, as discussed below.

2.5 Noise as Aid to Sensitivity?

It has been suggested that by the theory of stochastic resonance external or internal noise can in some special circumstances actually improve signal detection, in particular by boosting a periodic signal to levels exceeding threshold for detection. There is so far no direct physiological evidence of this in the low-frequency electrosense, and it is more difficult to understand in a system in which threshold may not exist (but see Bezrukov and Vodyanoy 1997). Nevertheless behavioral studies have suggested that it may occur in the paddlefish electrosense (Russell et al. 1999; Wilkens, Chapter 9).

3. Central Processing of Low-Frequency Electrosensory Information

Both the phylogenetically ancient ampullae of Lorenzini and the teleost ampullary electroreceptor system are octavolateralis senses, as noted earlier. As such, they share developmental and anatomical similarities with other hair cell–based sensory systems that include the mechanosensory lateral line and the sense of hearing. These developmental and anatomical similarities extend to the central projections and central processing of information from these senses. Like the lateral line, both these ampullary systems project to specialized hindbrain nuclei that are cerebellar-like. From here, the major ascending pathways include crossed lemniscal projections to nuclei adjacent to the lateral lemniscus and onward to the contralateral mesencephalon (for a discussion of these central pathways in elasmobranchs, chondrosteans, and teleost fishes, see Hofmann et al. 2002). Also characteristic of these octavolateralis senses is their ability to provide information on the location of stimuli external to the animal itself and the "map-like" representation of this spatial information within the brain. The primary functions of the hindbrain nuclei of these sensory pathways are to remove ambiguities in the sensory flow, cancel self-generated noise, and in some cases transform the sensory input into a form suitable for the generation of the maps at higher brain levels. Based on analogy with other senses, it is also likely that components of sensory inflow are segregated into separate streams that are processed independently. The electrosensory system of platypus is independently evolved from a cutaneous receptor precursor and so does not share the same central affinities as the other electrosensory systems. Little is known of the central processing of electrosensory information in platypus at lower brain levels, although the central cortical maps have been well described (Iggo et al. 1992; Krubitzer et al. 1995).

3.1 Hindbrain Contributions to Electrosensory Processing

Lorenzinian electroreceptors project to a distinctive hindbrain nucleus, called the dorsal octavolateralis nucleus (or dorsal nucleus). The recognition that the presence of a dorsal nucleus was diagnostic for electoreception was in fact instrumental in charting the phylogenetic distribution of electroreception and the ensuing discovery that electroreception was a phylogenetically ancient sense (Bullock et al. 1983). Teleost ampullary receptors project to a hindbrain nucleus known as the electrosensory lateral line lobe (ELL), and although the ELL is not homologous with the dorsal nucleus it is derived from the comparable lateral line nucleus and does share many similarities (Mongomery et al. 1995). In both the dorsal nucleus and the ELL the principal neurons (termed ascending efferent neurons [AENs] in non-teleosts and pyramidal, crest, or ganglion cells in teleosts) occupy a zone of the nucleus immediately beneath a distinctive molecular layer. This molecular layer matches that found in the cerebellum and is the reason that the dorsal nucleus and ELL are known as cerebellar-like structures. As in the cerebellum, the molecular layer is made up of a large number of parallel fibers and interneurons known as stellate cells. Spiny dendrites of the principal cells in each case penetrate the molecular layer and receive excitatory inputs from parallel fibers and inhibitory inputs from stellate cells. There are also differences between dorsal nucleus and ELL. Whereas AENs receive direct electrosensory input onto their ventral dendrites, electrosensory information from primary afferents reaches some of the principal neurons of ELL through interneurons in the ventral neuropil of the lobe. In addition, AENs form the main projection neurons of the dorsal nucleus, but the output of some of the principal cells in mormyrids is relayed via other neurons (Meek et al. 1996).

As noted earlier, two of the primary functions of the dorsal nucleus and the ELL are to remove ambiguities and to cancel self-generated noise in the electrosensory inflow—in effect to improve the S:N. In the dorsal nucleus, the receptive field structure of secondary neurons suppresses inputs that are common mode to all the receptors, and in both the dorsal nucleus and ELL, the molecular layer cap of the nuclei provides a substrate for an adaptive filter that cancels self-generated noise.

3.2 Receptive Field Structure and Common-Mode Suppression

Like many aquatic animals, elasmobranch fishes pump ions to maintain ion and water balance. In the process, they generate electrical fields within the body and the surrounding water (Bodznick et al. 1992). These fields are often modulated by ventilation, since the opening and closing of the mouth and gill slits changes the resistance of different pathways between the current source and sink. Elasmobranchs use their electrosensory system to detect the bioelectric fields of other animals, but these external fields must be detected over the top of the bioelectrical fields they are producing themselves. The extent of the problem is illustrated in Figure 6.2 which also provides a clear demonstration that this problem is solved in the hindbrain. Recordings from primary electrosensory afferent neurons reveal that they are driven over most of their dynamic range by the animal's own ventilation. This ventilatory reafference, as it is known, is many times stronger than the modulation generated by the biologically important external stimuli. So how are the biologically important signals detected against such a strong background noise generated by the animal's own ventilation? In part, the answer is that the biological signals are represented differentially across the receptor array, whereas the ventilatory reafference is uniform, or "common-mode," across all receptors. This means that by having a balance of excitatory and inhibitory inputs from different parts of the receptor array, the second-order cells (AENs) can cancel common-mode noise while preserving their sensitivity to external stimuli. The spatial extent of the excitatory and inhibitory component of the input to AENs is referred to as their receptive field structure. This has been studied in some detail in the little skate (*Raja erinacea*) (Bodznick et al. 1992) and the carpet shark (*Cephaloscyllium isabella*) (Bodznick and Montgomery 1992). Typically AENs have a focal excitatory receptive field and a diffuse inhibitory receptive field. Electrophysiological evidence points to the inhibitory receptive field being mediated via inhibitory interneurons of the ventral neuropil of the dorsal nucleus (Montgomery and Bodznick 1993) (Fig. 6.3).

Receptive field structure can have a major impact on stimulus encoding in the nervous system. For example, the well-known on-center/off-surround receptive field structure of retinal ganglion cells reduces their response to uniform illumination (a common-mode input in this case), but enhances sensitivity to contrast and movement. Within the elasmobranch electrosense, the typical focal excitatory and diffuse inhibitory field structure cancels the common-mode ventilation input, and reduces sensitivity to uniform electric fields, while preserving sensitivity to local dipole fields such as those produced by prey. A minority of AENs have receptive fields that consist of focal excitatory and focal inhibitory fields; these will still effect common-mode suppression, but depending on the spacing of the two focal regions, could enhance sensitivity to uniform fields or to dipoles of a particular size. It is unknown if encoding of these stimulus features is different in different elasmobranch species, or if this encoding is

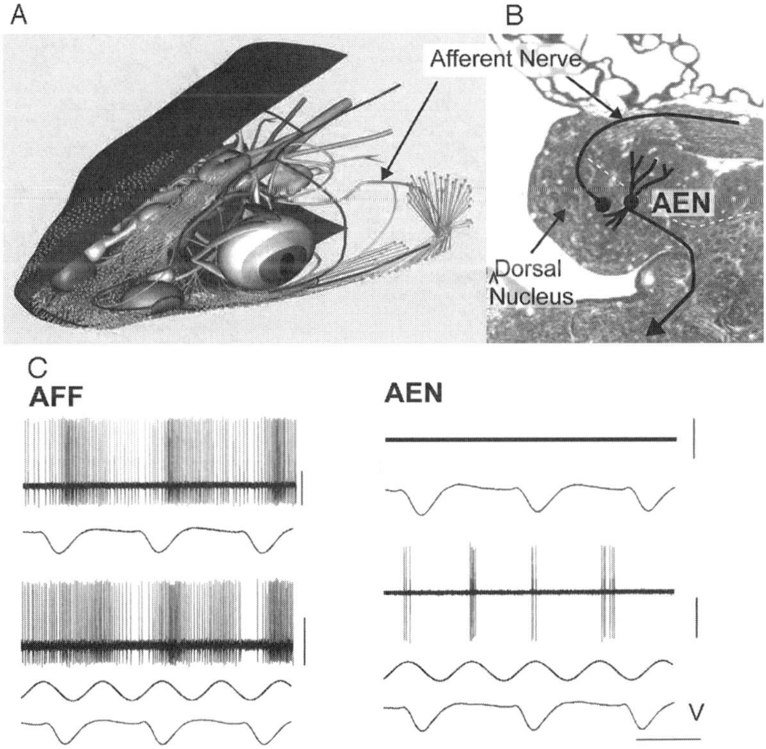

FIGURE 6.2. Representation of the electrosensory system of the dogfish, projection of the afferent nerves to the dorsal octavolateralis nucleus (or dorsal nucleus) of the hindbrain, and the main contribution of the dorsal nucleucs to electrosensory signal processing. (**A**) Anatomically realistic representation of the sense organs and brain of the dogfish. (Courtesy of Drs. Rachel Berquist and Michael Paulin.) Electrosensory pores are shown as dots on the skin on the animal's right side. The cut-away representation on the left side shows the electrosensory canals leading the ampullary clusters, and the main electrosensory nerves which project to the dorsal nucleus. (**B**) Cross section of the dorsal nucleus with a sketch of a single afferent electrosensory fiber entering in the nerve and contacting the ventral dendrites of an ascending efferent neuron (AEN). These are the projection neurons of the dorsal nucleus, and the axon of the AEN is shown leaving the dorsal nucleus and heading toward the midline. (**C**) Comparison of afferent (AFF) and AEN activity in relation to the animal's own ventilation and an extrinsic electrosensory stimulus. *Top two traces* for both AFF and AEN represent spontaneous spike activity and record of the animal's ventilation respectively. Note that the AFF is strongly spontaneously active and strongly driven by the animal's ventilation. The AEN is silent and unaffected by ventilation. The lower three traces for AFF and AEN represent the spike activity in response to a 1-Hz, 2-μV extrinsic electrosensory stimulus, the representation of the stimulus itself, and the record of ventilation. In this case note the strong response of the AEN to the extrinsic stimulus. In combination, these traces show that signal processing within the AEN is able to remove sensory reafference generated by the animal's own movement, while retaining sensitivity to external stimuli. Calibrations: Vertical—200 μV; horizontal—1 s.

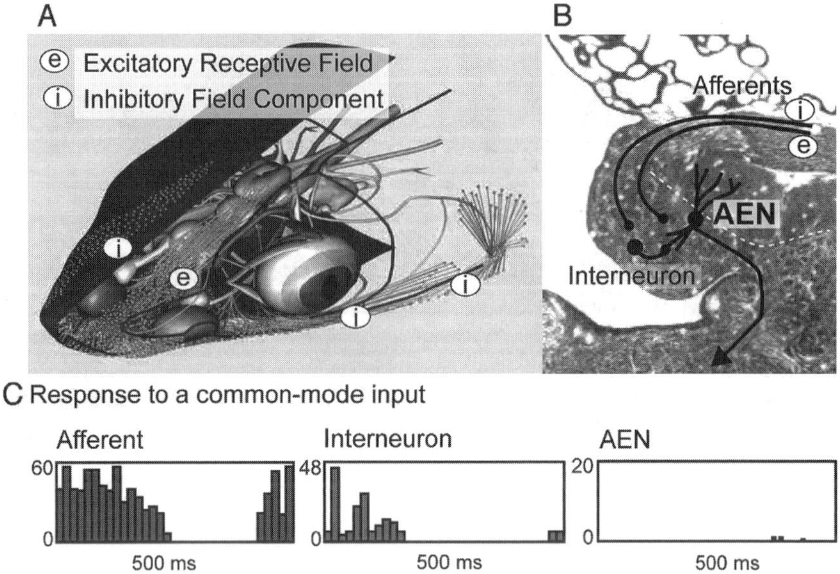

FIGURE 6.3. Receptive field structure of ascending efferent neurons (AENs) plays a role in removing common-mode sensory reafference. (**A**) Representation of the brain and sensory organs of the dogfish. (Courtesy of Drs. Rachel Bergquist and Michael Paulin.) The excitatory and inhibitory components of a particular AEN are shown as *circles* on the surface of the skin. Note that there is a single excitatory receptive field location and multiple inhibitory components. (From Bodznick and Montgomery 1992.) (**B**) Cross section of the dorsal nucleus with a sketch illustrating the principle that afferents from the excitatory receptive field activate AENs directly, whereas afferents from the inhibitory receptive field components activate the AEN via inhibitory interneurons in the ventral neuropil of the dorsal nucleus. (**C**) Illustrates spike histogram records of the response to a common-mode stimulus of an afferent fiber, an interneuron, and an AEN. Note that both the afferent and the interneuron are driven strongly by the common-mode stimulus, whereas the AEN is not. (From Montgomery and Bodznick 1993.)

preserved and segregated into separate sensory stream analysis within the central nervous system.

3.3 Hindbrain Adaptive Filters

The ampullary system of mormyrid weakly electric fish exhibits a very interesting special case of self-generated noise within a low-frequency electrosensory system. In this case, the animal's own electric organ discharge (EOD) can strongly activate the ampullary receptors. In a classic paper Bell (1982) showed that this form of reafference is cancelled within the ELL by a modifiable form of efference copy. In these experiments, the EOD was pharmacologically silenced, and the EOD command signal used to synchronize an external stimulus with the animal's intended, or fictive, EOD. In recordings from the principal neurons known as medium ganglion cells (MG), stimuli of this sort were progressively suppressed by the formation of a negative sensory image, or modifiable efference copy. The negative sensory image was revealed when the stimulus was turned off. In essence this system forms an adaptive filter that cancels self-generated noise generated by the animal's own EOD.

Within the elasmobranch electrosense it may seem as though common-mode suppression is sufficient to clean the system of self-generated noise. However, it was recognized early on that other forms of self-generated noise would not be common mode (Montgomery 1984). In addition, modeling studies Nelson and Paulin (1995) showed that the additional interneuron interposed on the inhibitory pathway to the AEN would slightly distort the timing of common-mode input, resulting in a residue of ventilation response within the AEN. These problems could be overcome by the addition of an adaptive filter to the common-mode mechanism. The presence of such an adaptive filter was demonstrated by recording from an AEN and pairing an external stimulus with the animal's own breathing (Montgomery and Bodznick 1994). As in the mormyrid example given earlier, a stimulus paired to ventilation is progressively suppressed by the formation of a negative sensory image. The negative image is again revealed when the stimulus is withheld (Fig. 6.4).

In both the mormyrid and the elasmobranch, the neuronal substrate for the adaptive filter is the molecular layer, and more precisely, synaptic plasticity at the inputs to the dorsal dendrites of the MG cells and the AENs (Bell et al. 1997a, 1999; Bodznick et al. 1999). The current thinking is that dendritic spikes in these dendrites drive an anti-Hebbian learning rule, such that each spike reduces the gain of coincidentally active parallel fibers. A converse rule increments the gain of active parallel fiber inputs in the absence of activity in the postsynaptic MG cell or AEN (Bell et al. 1993; Montgomery and Bodznick 1994). In vivo electrophysiology (Bodznick et al. 1999), in vitro slice experimentation (Bell et al. 1997b), and modeling studies (Nelson and Paulin 1995; Bodznick et al. 1999; Roberts and Bell 2000) all provide additional evidence that these learning rules are in operation and that they can effectively generate the negative sensory image required for cancellation. In the in vivo studies Bodznick et al. (1999) have shown that either pairing an intracellular current pulse within the AEN with ventilation, or pairing parallel fiber stimulation with an external stimulus, is effective at generating the predicted learning and negative sensory image. Within slices of the mormyrid ELL, Bell and co-workers (Bell et al. 1997b) have been able to determine precisely the temporal specificity of the learning rules. The dendritic spike has to occur within 50 ms of the parallel fiber excitatory postsynaptic potential (EPSP) to produce the gain reduction of the first anti-Hebbian learning rule. The contribution of modeling studies has been to show that using these learning rules even relatively arbitrary parallel fiber inputs can generate an appropriate negative sensory image. In addition, the modeling shows that plasticity at inhibitory synapses could also make an effective contribution to the function of these adaptive filters.

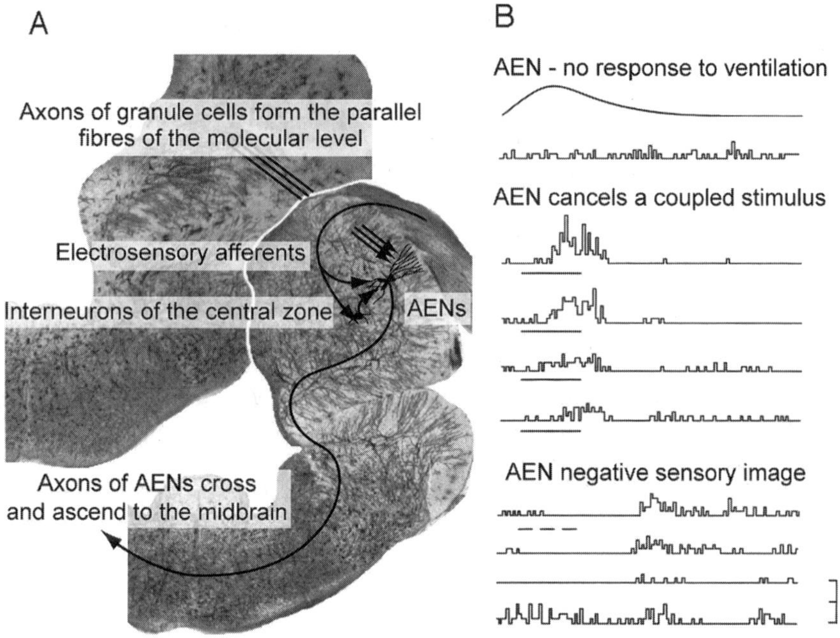

FIGURE 6.4. The molecular layer of the dorsal nucleus provides a substrate for an adaptive filter to cancel AEN activity associated with the animal's own movements. (**A**) The neural substrate for the adaptive filter is the molecular layer of the dorsal nucleus. Parallel fibers formed from granule cell axons make up the molecular layer and synapse with the extensive molecular layer dendrites of the AENs. This figure is made up of two Golgi-stained sections from the chimerid *Callorhynchus milli*. The more anterior section shows the cerebellar auricles which are the source of the granule cell axons making up the molecular layer. The more caudal section is at the level of the dorsal nucleus and shows a sketch of the afferent innervation of an AEN and the projection of its axon to higher brain areas. (**B**) A series of spike histograms illustrating the adaptive filter. The top two traces show a strain gauge record of the animal's ventilation and below that a spike histogram showing the low spontaneous activity of an AEN and its lack of response to ventilation. The next four traces show the effect of coupling an extrinsic stimulus to ventilation. Initially the AEN responds strongly with a period of excitation followed by inhibition. Over time the response decreases. The respective traces were taken immediately after coupling, and then at 2.5, 7.5, and 10 minutes. The final four traces show the negative sensory image revealed when the previously coupled stimulus is turned off. Initially there is a strong negative sensory image which decays over time. The respective traces represent 0, 5, 7.5, and 20 minutes after the stimulus was turned off. Vertical calibration is 10 counts per bin; each spike histogram record is 2 s long.

The same adaptive filter mechanisms are employed within the high-frequency electrosense of teleosts and it has also been explored extensively by Bastian in gymnotid electric fish (see Bastian and Zakon, Chapter 8).

3.4 Organization of Spatial Information Within the Brain

The distribution of electroreceptors on the surface of the body provides a spatial array of receptors intrinsically capable of representing stimulus location. The preservation of this peripheral spatial information is evident in the elasmobranch electrosensory system both at the level of the hindbrain and the midbrain. The dorsal nucleus of the hindbrain has a "receptotopic organization" in which the receptor clusters are represented at distinct locations in the nucleus and there is an orderly spatial representation of the layout of the receptor positions on the body surface (Bodznick and Schmidt 1984). In the midbrain tectum, there is a general "spatiotopic" map of the animal's dorsal hemisphere, where electrosensory information is mapped in alignment with visual input and where the region of space near the horizon is greatly overrepresented (Bodznick 1991). The ventral surface of the body, which is outside the visual field, is mapped in an adjacent brain area known as the lateral mesencephalic nucleus (see also Schweitzer 1986).

In platypus, there is an elaborate representation of the electrosensory surface of the bill in S1 neocortex (Pettigrew 1999). The detailed topographic representation of the bill surface and alternating bands of mechanoreceptive and electroreceptive information in stripe-like bands within this map have led to the hypothesis that this representation is important for locating an electrogenic target. The proposal is that reconstruction of stimulus field lines over the surface of the bill is necessary for determining source direction, and mechanosensory and electrosensory timing comparisons are important for determining source distance.

4. Summary

Of four different low-frequency electrosensory systems, three are octavolateral senses in fish or amphibians and the fourth is a trigeminal electrosense in monotremes. The electroreceptors in each of these systems operate in push/pull fashion to signal local electric field intensity, orientation, and polarity. Receptor mechanisms have not been studied in monotremes, but in ampullary electroreceptors, microvolt sensitivity is achieved with ordinary voltage-gated Ca channels by taking advantage of oscillating systems or the sensitivity inherent in a positive feedback system poised at threshold or partially activated. In teleosts a bias current across the receptor epithelium holds the basal membrane of the receptor cells in a partially activated state but they are prevented from full regenerative spiking by a well-adjusted resistive load. Even vanishingly small potentials can increase or decrease the proportion of active Ca channels and the level of ongoing transmitter release at DC synapses between the receptor cells and afferent fibers. Thus, there is in theory no response threshold. A large convergence of receptor cells to afferents further enhances sensitivity for signal over receptor noise. Similar mechanisms are responsible for sensitivity in the Lorenzinian ampullary receptors of non-teleosts except that it is the apical membrane of the receptor cells that is the voltage sensor. Positive feedback here amplifies receptor currents that then cause larger potential changes as they flow outward across the basal transmitter secreting membrane. In the Lorenzinian ampulla it is uncertain whether the individual receptor cells are partially activated but held below threshold for spiking by the resistive load as in teleosts, or whether they are each poised at threshold and occasionally or even regularly firing regenerative Ca spikes independently and with a probability determined by the stimulus potential. In either case, the firing rate of the afferents reflects the summed activity in the population of receptor cells and is linear with stimulus intensity for small stimuli through zero amplitude; the change is superimposed on the afferent's tonic discharge. Finally, the apical versus basal surface of the voltage sensor in non-teleost versus teleost electroreceptors accounts for the opposite stimulus polarity preference in the electroreceptors of the two.

For the ampullary systems, at the first processing station within the hindbrain sensitivity is further enhanced and noise is reduced through the convergence of inputs and the composite receptive fields of the principal neurons. In addition, an adaptive filter, which is implemented by cerebellar-like circuits and synaptic plasticity, uses feedforward and feedback signals to eliminate predictable electrosensory inputs including those created by the animal's own movements. Nothing is known of hindbrain processes in the monotreme electrosense and very little is known above medullary levels in any of the four systems except that spatial information is preserved in well-ordered topographic maps in the midbrain and telencephalon, and that multimodal integration occurs at both levels.

Neuroethologists often study unusual animals with highly specialized sensory systems because their extreme specializations are generally based on fundamental mechanisms pushed to their limits, and being extreme, these mechanisms are more easily identified and understood. Within the auditory world, bats and barn owls are ready examples. Because natural electric fields in aquatic environments are so very weak, the low-frequency electrosensory systems that evolved to detect them are among the most highly specialized systems for extreme sensitivity and for noise suppression, and so it is in this realm that they may have most to offer. We have described progress made so far in understanding extreme sensitivity and noise reduction mechanisms in both the receptors and in the medulla of the ampullary systems. Key components in each case are the control systems that permit the neurons and receptors to be held at their point of highest sensitivity. In the medulla the feedforward and feedback inputs to the principal neurons through the molecular layer are finely adjusted in an ongoing way to eliminate predictable signals and self-generated noise, while maintaining the neurons at their most sensitive level for responses to novel stimuli. In this case it appears that homeostatic control is achieved through conventional synaptic plasticity mechanisms directed by the activity of the principal neurons themselves. In a similar way it appears that control mechanisms are acting to keep the receptors poised at their most sensitive levels, seemingly on the brink of instability. In both Lorenzinian and teleost ampullary systems the receptor organs are actively maintained in the negative slope conductance region of their *I–V* curves by a bias current and an appropriate shunt conductance. Discovering the homeostatic mechanisms responsible for achieving this critical balance in a controlled way is a challenge for future work.

References

Bell CC (1982) Properties of a modifiable efference copy in an electric fish. J Neurophysiol 47:1043–1056.

Bell CC, Caputi A, Grant K, Serrier J (1993) Storage of a sensory pattern by anti-Hebbian synaptic plasticity in an electric fish. Proc Natl Acad Sci USA 90:4650–4654.

Bell C, Bodznick D, Montgomery J, Bastian J (1997a) The generation and subtraction of sensory expectations within cerebellum-like structures. Brain Behav Evol 50:17–31.

Bell CC, Han VZ, Sugawara Y, Grant K (1997b) Synaptic plasticity in a cerebellum-like structure depends on temporal order. Nature 387:278–281.

Bell CC, Han VZ, Sugawara Y, Grant K (1999) Synaptic plasticity in the mormyrid electrosensory lobe. J Exp Biol 202 (Pt 10):1339–1347.

Bennett MVL, Obara S (1986) Ionic mechanisms and pharmacology of electroreceptors. In: Bullock TH, Heiligenberg W (eds), Electroreception. New York: John Wiley & Sons, pp. 157–181.

Bezrukov SM, Vodyanoy I (1997) Stochastic resonance in non-dynamical systems without response thresholds. Nature 385:319–321.

Bodznick D (1991) Elasmobranch vision: multimodal integration in the brain. J Exp Zool Suppl 5:108–116.

Bodznick D, Boord RL (1986) Electroreception. New York: Springer-Verlag.

Bodznick D, Montgomery JC (1992) Suppression of ventilatory reafference in the elasmobranch electrosensory system: medullary neuron receptive fields support a common mode rejection mechanism. J Exp Biol 171:127–137.

Bodznick D, Schmidt AW (1984) Somatotopy within the medullary electrosensory nucleus of the little skate, *Raja erinacea*. J Comp Neurol 225:581–590.

Bodznick D, Montgomery JC, Bradley DJ (1992) Suppression of common mode signals within the electrosensory system of the little skate *Raja erinacea*. J Exp Biol 17:107–125.

Bodznick D, Hjelmstad G, Bennett MVL (1993) Accommodation to maintained stimuli in the ampullae of Lorenzini: how an electroreceptive fish achieves sensitivity in a noisy world. Jpn J Physiol 43:S231–237.

Bodznick D, Montgomery JC, Carey M (1999) Adaptive mechanisms in the elasmobranch hindbrain. J Exp Biol 202:1357–1364.

Bullock TH (1982) Electroreception. In: Cowan MW (ed), Annual Review of Neuroscience, Vol. 5. Palo Alto, CA: Annual Reviews, pp. 121–170.

Bullock TH, Bodznick DA, Northcutt RG (1983) The phylogenetic distribution of electroreception: evidence for convergent evolution of a primitive vertebrate sense modality. Brain Res 287:25–46.

Clusin WT, Bennett MVL (1979) The ionic basis of oscillatory responses of skate electroreceptors. J Gen Physiol 73:703–723.

Gregory JE, Iggo A, McIntyre AK, Proske U (1988) Receptors in the bill of the platypus. J Physiol 400:349–366.

Gregory JE, Iggo A, McIntyre AK, Proske U (1989) Responses of electroreceptors in the platypus bill to steady and alternating potentials. J Physiol 408:391–404.

Hofmann MH, Wojtenek W, Wilkens LA (2002) Central organization of the electrosensory system in the paddlefish (*Polyodon spathula*). J Comp Neurol 446:25–36.

Iggo A, Gregory JE, Proske U (1992) The central projection of electrosensory information in the platypus. J Physiol 447:449–465.

Kalmijn AJ (1974) The detection of electric fields from inanimate and animate sources other than electric organs. In: Fessard A (ed), Handbook of Sensory Physiology, Vol. III/3. Berlin: Springer-Verlag, pp. 148–200.

Kalmijn AJ (1988) Detection of weak electric fields. In: Atema J, Fay RR, Popper AN, Tavolga WN (eds), Sensory Biology of Aquatic Animals. New York: Springer-Verlag, pp. 151–186.

Kalmijn AJ (2003) Graded positive feedback in elasmobranch ampullae of Lorenzini. AIP Conf Proc 665:133–141.

Krubitzer L, Manger P, Pettigrew J, Calford M (1995) Organization of somatosensory cortex in monotremes: in search of the prototypical plan. J Comp Neurol 351:261–306.

Lu J, Fishman HM (1994) Interaction of apical and basal membrane ion channels underlies electroreception in ampullary epithelia of skates. Biophys J 67:1525–1533.

Lu J, Fishman HM (1995) Ion channels and transporters in the electroreceptive ampullary epithelium from skates. Biophys J 69:2467–2475.

Meek J, Grant K, Sugawara Y, Hafmans TG, Veron M, Denizot JP (1996) Interneurons of the ganglionic layer in the mormyrid electrosensory lateral line lobe: morphology, immunohistochemistry, and synaptology. J Comp Neurol 375:43–65.

Montgomery J, Coombs S, Conley R, Bodznick D (1995) Hindbrain sensory processing in lateral line, electrosensory, and auditory systems: a comparative overview of anatomical and functional similarities. Audit Neurosci 1:207–231.

Montgomery JC (1984) Noise cancellation in the electrosensory system of the thornback ray: common mode rejection of input produced by the animal's own ventilatory movement. J Comp Physiol A 155:103–111.

Montgomery JC, Bodznick D (1993) Hindbrain circuitry mediating common mode suppression of ventilatory reafference in the electrosensory system of the little skate *Raja erinacea*. J Exp Biol 183:203–215.

Montgomery JC, Bodznick D (1994) An adaptive filter that cancels self-induced noise in the electrosensory and lateral line mechanosensory systems of fish. Neurosci Lett 174:145–148.

Murray RW (1965) Electroreceptor mechanisms: the relation of impulse frequency to stimulus strength and responses to pulsed stimuli in the ampullae of Lorenzini of elasmobranchs. J Physiol 180:592–606.

Neiman AB, Russell DF (2004) Two distinct types of noisy oscillators in electroreceptors of paddlefish. J Neurophysiol 92:492–509.

Nelson ME, Paulin MG (1995) Neural simulations of adaptive reafference suppression in the elasmobranch electrosensory system. J Comp Physiol A 177:723–736.
Obara S, Bennett MVL (1972) Mode of operation of ampullae of Lorenzini of the skate, *Raja*. J Gen Physiol 60:534–557.
Obara S, Sugawara Y (1984) Electroreceptor mechanisms in teleost and non-teleost fishes. In: Bolis L, Keynes RD, Maddrell SHP (eds), Comparative Physiology of Sensory Systems. New York: Cambridge University Press, pp. 509–523.
Paulin MG (1995) Electroreception and the compass sense of sharks. J Theor Biol 174: 325–339.
Peters RC, Bretschneider F (1972) Electric phenomena in the habitat of the catfish *Ictalurus nebulosus* LeS. J Comp Neurol 81:345–362.
Pettigrew JD (1999) Electroreception in monotremes. J Exp Biol 202 (Pt 10):1447–1454.
Roberts PD, Bell CC (2000) Computational consequences of temporally asymmetric learning rules: II. Sensory image cancellation. J Comput Neurosci 9:67–83.
Russell DF, Wilkens LA, Moss F (1999) Use of behavioural stochastic resonance by paddle fish for feeding. Nature 402:291–294.
Schluger JH, Hopkins CD (1987) Electric fish approach stationary signal sources by following electric current lines. J Exp Biol 130:359–367.
Schweitzer J (1986) Functional organization of the electroreceptive midbrain in an elasmobranch (*Platyrhinoidis triseriata*). A single-unit study. J Comp Physiol A 158:43–58.
Sisneros JA, Tricas TC (2000) Androgen-induced changes in the response dynamics of ampullary electrosensory primary afferent neurons. J Neurosci 20:8586–8595.
Sisneros JA, Tricas TC (2002) Neuroethology and life history adaptations of the elasmobranch electric sense. J Physiol Paris 96:379–389.
Struik ML, Bretschneider F, Peters RC (2002) Spontaneous nerve activity and sensitivity in catfish ampullary electroreceptor organs after tetanus toxin application. Pflugers Arch 443:903–907.
Sugawara Y (1989a) Electrogenic Na-K pump at the basal face of the sensory epithelium in the *Plotosus* electroreceptor. J Comp Physiol A 164:589–596.
Sugawara Y (1989b) Two Ca current components of the receptor current in the electroreceptors of the marine catfish *Plotosus*. J Gen Physiol 93:365–380.
Sugawara Y, Obara S (1989) Receptor Ca current and Ca-gated K current in tonic electroreceptors of the marine catfish *Plotosus*. J Gen Physiol 93:343–364.
Zakon HH (1986) The electroreceptive periphery. In: Bullock DTH, Heiligenberg W (eds), Electroreception. New York: John Wiley & Sons, pp. 103–156.

7

Physiology of Tuberous Electrosensory Systems

MASASHI KAWASAKI

1. Introduction

Tuberous electrosensory systems are highly specialized sensory systems that appear to have evolved exclusively for processing the electric organ discharge (EOD) signals generated by electric fishes. They are unique to the two orders of freshwater electric fishes that use the electric signals for communication and electrolocation. The Gymnotiformes from South America and the Mormyriformes from Africa emit EODs from electric organs in their tails, and they detect and analyze distortions in their EODs, including changes in amplitude, waveform, and time delay, caused by objects in the surrounding environment. These fish also generate EODs for social communication, a process in which the tuberous receptors sense and analyze the discharges of other individuals. The fish in the mormyriform family, Mormyridae, have two types of tuberous organs, of which one, the knollenorgan, is used solely for communication. All others have two types of tuberous receptors that are shared for both behavioral functions.

Tuberous electroreceptors, which are named for their tuberlike anatomical arrangement in the skin (see Jørgensen, Chapter 3) respond only to high-frequency components of EODs, from tens of Hz to more than 1 kHz. They are insensitive to the constant, DC signals or low-frequency signals that activate ampullary electroreceptors so effectively. Ampullary electroreceptors, which are named for the ampullary canal leading from receptors to the exterior surface, are found in mormyriform and gymnotiform fishes, in catfishes, and in many species of non-teleost fishes (see Bodznick and Montgomery, Chapter 6).

Because tuberous electroreceptors are used primarily for sensing EODs, the general difficulty of identifying relevant stimuli for this system is much easier than it is for other sensory systems. EODs are relatively simple signals, and by describing them, we can precisely know the relevant stimuli for which the entire system has evolved. EODs are easily recorded, analyzed, and altered according to specific parameters of amplitude, time, and waveform for use as artificial stimuli. The relevance of the altered artificial stimuli can be examined with behavioral experiments or with physiological studies of the central nervous system (CNS). This neuroethological approach has been successful in demonstrating the significance of the temporal structure of EOD pulses in communication as well as physiological mechanisms for detection of species-specific EODs of mormyrid fishes (Hopkins and Bass 1981; Amagai 1998; Amagai et al. 1998; Friedman and Hopkins 1998) (see Sections 2.2.2 and 3.3.3).

Studies of tuberous electroreceptor systems have provided an opportunity for comparative studies—another hallmark of neuroethological research. Tuberous electroreceptors emerged independently in the two main groups of freshwater electric fishes. None of the common ancestors of these two groups is capable of either electrogenesis or electroreception. Yet, within each group, there is a proliferation of species with varied electric signals, behavioral functions, and physiological designs. By comparing tuberous systems within these two groups and between the groups, we gain insights into evolution of receptor design both within a lineage and between lineages evolving independently. Electric fishes provide one of the most successful subjects for comparative physiology of sensory systems (see Bullock and Hopkins, Chapters 1 and 15).

This review begins with a description of the physiological types of tuberous electroreceptors with references to their behavioral significance (see Section 2). It continues (see Section 3) with major structures in the CNS having distinct physiological and behavioral functions, including more complex functions such as descending control of afferent information, and ends (see Section 4) with a description of interactions between the tuberous sensory system and the electromotor corollary discharge system.

2. Receptor Physiology

2.1 General

After the electrolocation capabilities of weakly electric fish were discovered using behavioral methods (Lissmann 1958; Lissmann and Machin 1958), the tuberous electroreceptors were identified electrophysiologically (Bullock et al. 1961; Fessard and Szabo 1961). Szabo (1965) then showed that tuberous electroreceptor cells are positioned at the base of the tuberlike capsule, which is created by an invagination of the epidermis (Fig. 7.1). The receptor cells expose a large part of their outer-face membrane to the lumen of the tubular capsule, which is indirectly connected to the external medium. A remarkable morphological specialization is the microvilli on the outer face of the sensory cells, which are believed to act as a coupling capacitor for the high-pass characteristic of tuberous receptors (Bennett 1967). Only a small part of the inner face of the sensory cells is anchored to the supporting cells at the base. There the sensory cells make a synaptic connection to afferent fibers. The outer face is thought to act as passive resistance because of its relatively large surface area (low membrane resistance), while the smaller inner surface through which current is funneled contributes to transduction (Bennett 1971). Accordingly, the effective

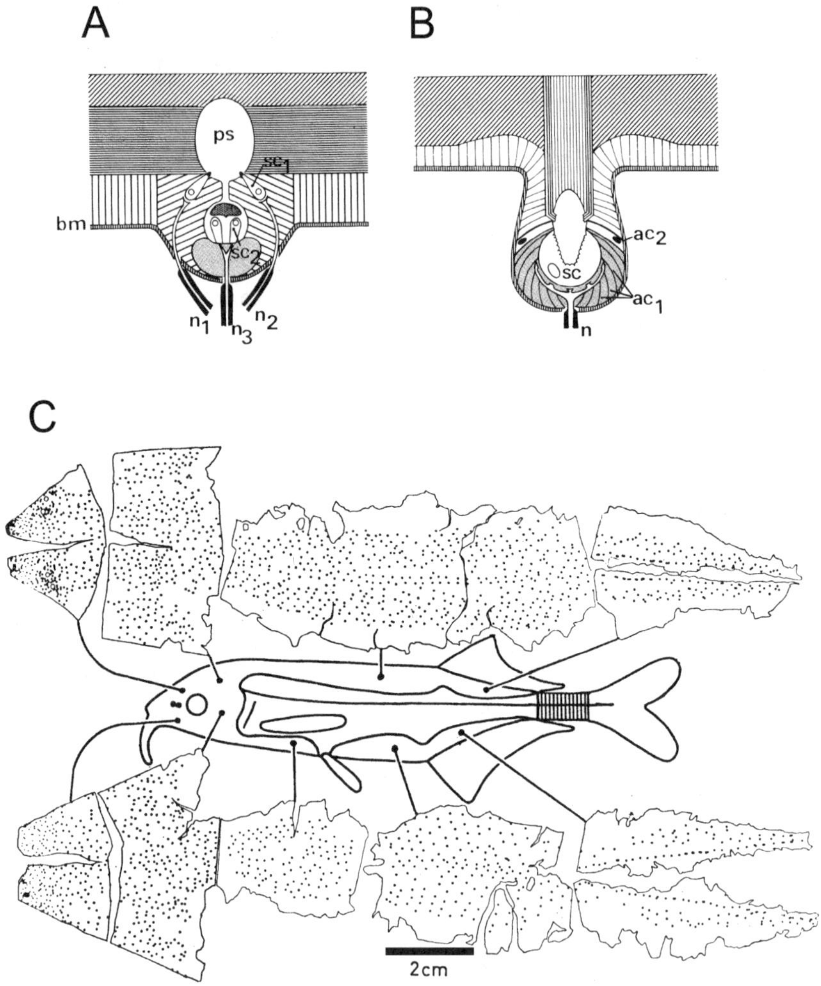

FIGURE 7.1. Schematic drawings of tuberous electroreceptors of a mormyrid fish (**A**) and of *Gymnarchus* (**B**). bm, basement membrane; n, afferent nerve fiber; sc, sensory cells. (**C**) Distribution of mormyromast electroreceptors on flattened skin of *Gnathonemus*. (**A** and **B** from Szabo 1974; **C** from Harder 1968.)

current for transduction is inward current. When current flows into the fish's body, the inner surface of the sensory cells is depolarized. Although all tuberous receptor types follow the same basic physiological design, considerable anatomical variation exists among different types of tuberous receptors in various species (Szabo 1974) (see Jørgensen, Chapter 3).

Tuberous electroreceptor cells are innervated by primary afferent fibers whose somata are found in the lateral line nerve ganglion. Each fiber innervates from one to tens of closely clustered electroreceptor organs. The spatial receptive field of a single afferent fiber is large, reflecting the fact that in electrolocation objects do not form sharp images on the body surface as do optical images on the retina. Spatial receptive fields for single afferent fibers have been studied with electrolocation targets by Heiligenberg (1977) and Bastian (1986a). For comprehensive reviews of electroreceptor physiology see Bennett (1971), Szabo and Fessard (1974), and Zakon (1986).

2.2 Physiological Types

Using artificial electrical stimuli that mimic a fish's own EODs, spike trains in primary afferent fibers from electroreceptors can be studied to determine their responsiveness to signal amplitude, time, frequency, and waveform. For example, some fibers respond to amplitude changes of EOD mimics with a variable number of spikes (amplitude coders), while others respond to an EOD with a single spike (time coders). Unlike the auditory system of birds and mammals, in which time and amplitude features of stimuli are derived postsynaptically from a single receptor type in the periphery, electroreceptors exhibit remarkable specializations for encoding amplitude versus time even at the level of the sensory receptors in the skin for all electric fish species. All weakly electric fish species possess at least one type of electroreceptor suited for encoding amplitude information and one for encoding temporal. The amplitude of a signal conveys information about size, distance, and resistivity of electrolocation objects, while temporal features of stimuli convey information about capacitance of objects and the waveform of EODs. Both parameters may be affected jointly when the fish's own EODs are electrically mixed with those of other individuals in social situations.

2.2.1 Amplitude Coding Tuberous Electroreceptors

Electrolocation targets with lower and higher electrical resistivity than the surrounding water create larger and smaller amplitudes respectively in the EOD feedback signal at the skin area near the target, and this local amplitude is encoded as the probability or frequency of action potentials (amplitude-rate coding) by amplitude-coding electroreceptors. In wave-type gymnotiform fishes, which generate quasi-sinusoidal continuous discharges, amplitude-coding afferent fibers fire action potentials in a probabilistic manner in which there are more action potentials for larger amplitudes and fewer action potentials for smaller amplitudes (Hagiwara and Morita 1963) (Fig. 7.2B). *Gymnarchus,* a wave-type African electric fish, also possesses a similar amplitude-coding electroreceptor subtype (Bullock et al. 1975). In pulse-type gymnotiform fishes, which generate short pulses with relatively longer interpulse intervals, burst-duration coder fire bursts of action potentials of varying duration to represent the local amplitude of the EOD (Bastian 1976) (Fig. 7.3). Pulse-type mormyrid electric fishes also possess mormyromasts, a functionally similar type of electroreceptor organ that

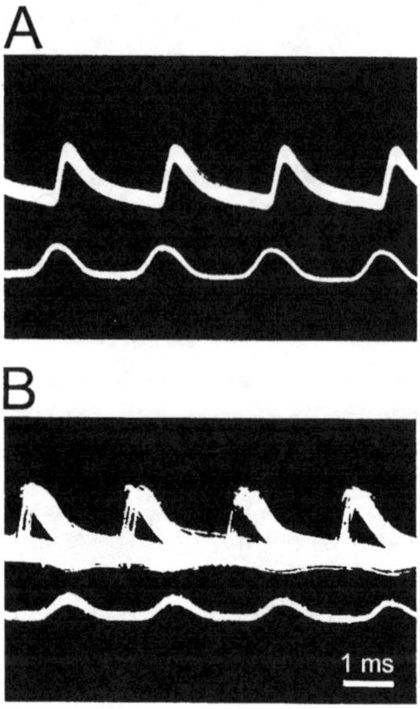

FIGURE 7.2. Time-coding, or T units (**A**) and amplitude-coding, or P units (**B**), recorded in the anterior lateral line nerve in *Eigenmannia*. *Upper traces*: nerve spikes; *lower traces:* EODs that acted as stimuli. Note that the T units fire for every cycle of the stimulus while P units fire in a probabilistic manner. (From Hagiwara and Morita 1963.) Calibration = 1 ms.

generates more action potentials for larger stimulus amplitude. Mormyromasts are unique for having two distinct types of sensory cells (A cells and B cells) located in different chambers (outer and inner) in the receptor organ. A and B cells are innervated by physiologically distinct afferent fibers, type A and type B respectively (Bell et al. 1989; Bell 1990). These fibers, like burst-duration coders in gymnotiform pulse fishes, fire with shorter latencies and larger numbers of action potentials for larger signal amplitudes (Bennett 1965; Szabo and Hagiwara 1967; Szabo and Fessard 1974; Bell 1990). Physiological properties of type A and type B fibers differ in threshold, maximum spike numbers, amplitude-rate curves, and frequency tuning (Bell 1990). In a modeling study by Shuai et al. (1998), the physiological differences between A and B cells could be explained from the known anatomical differences in the receptor cells combined with differences in membrane conductances. Type B fibers, but not type A, are sensitive to subtle variation in the waveform of EOD, suggesting functions other than simple amplitude coding (von der Emde and Bleckmann 1997) (see Section 2.2.2). Reflecting these physiological differences, type A and B afferent nerve fibers terminate in different areas in the brain (Bell et al. 1989).

Mormyromast electroreceptors may encode the amplitude of a stimulus by time or latency of the first action potential relative to the pulse-type EODs (Szabo and Hagiwara 1967). Latency of the first spike varies from 1.5 ms to 12 ms according to EOD amplitude. Since these fish have access to the electromotor corollary discharge signals for EOD generation, the latency of first spike times could be compared with the corollary discharge (see Section 4).

Latency coding of amplitude has been suggested by a behavioral study in a mormyrid fish (Hall et al. 1995).

2.2.2 Time-Coding Tuberous Electroreceptors

Tonic Time (Phase) Coder[1]

An interesting class of electroreceptor afferents was found in 1965 by Bullock and Chichibu (1965) in a wave-type gymnotiform fish, *Sternopygus,* which continuously generates EODs at approximately 100 Hz. Each cycle of the EOD generated one action potential in these afferent fibers; thus the afferent fibers continuously fire at the same frequency as the EOD in a time-locked (phase-locked) fashion. Since the spike frequency is constant and is the same as that of the constant EODs, they argued that in these fibers only the time of each afferent spike may be used to represent sensory information. Similar time-coding afferents that continuously fire one-to-one to EOD cycles were later found in all wave-type electric fishes (Scheich et al. 1973; Bullock et al. 1975) (Fig. 7.2A). The precision of time locking is often very high—on the order of 10^{-5}s for a stimulus period of 2.5 to 3×10^{-3}s (Rose and Heiligenberg 1985b; Carr et al. 1986a; Guo and Kawasaki 1997). The time-coding tuberous afferents of wave-type electric fishes, carrying sensory information purely by times of action potentials, represent the clearest form of time coding of all sensory systems.

Time coding by these electroreceptor afferents is important for at least two behavioral functions, the jamming avoidance response (JAR) (see Section 3.3) and electrolocation of capacitive objects. JARs require time differences at different electroreceptors that are created by interference of the fish's own EODs by a neighboring fish's EODs. Central mechanisms for detecting time differences have been found in *Eigenmannia* and *Gymnarchus* (see Section 3.2). Electric fish are capable of electrolocating objects not only by their resistivity but also by their electric capacitance (Meyer 1982; von der Emde and Ringer 1992; von der Emde 1993, 1997, 1998). Natural objects such as leaves and small organisms have capacitance that introduces local delays into the feedback signal. Scheich et al. (1973) demonstrated that an electrolocation target with electrical capacitance of 1 nF caused a significant delay in time-coding afferents.

Pulse Marker

Pulse-type gymnotiform fishes have a type of time-coding electroreceptor called the pulse marker (M type) that produces one spike per EOD pulse over a wide range of stimulus amplitudes (Szabo and Fessard 1974; Bastian 1976)

---

1. The term "phase" is often used in the literature to indicate a "time" in cyclic firing of neurons or EODs of wave-type electric fishes. "Time," instead of "phase" is used throughout this chapter for both wave- and pulse-type fishes for consistency. "Time comparison" or "time-locked" in this review is synonymous with "phase comparison" or "phase-locked" that may be found in the literature.

(Fig. 7.3A1). The afferent spikes from this type of electroreceptor also convey precise times of the EOD to the CNS by time coding.

Although the times of action potentials are accurately locked to stimulus times at a given amplitude, both the tonic time coders and pulse markers show varying degrees of amplitude-dependent latency changes as originally shown by Bullock and Chichibu (1965) (Bastian 1976; Guo and Kawasaki 1997), thereby creating a time-amplitude ambiguity.

Waveform Sensitivity in Tuberous Receptors

Both type A and type B mormyromast electroreceptors generate more spikes in response to stronger stimuli, as mentioned in Section 2.2.1. Type B afferents are, however, also sensitive to the waveform of the EOD. Capacitive loads by electrolocation objects can introduce delays in the EOD feedback signal as well as distortions in the EOD waveform owing to frequency-dependent phase shifts of high-frequency EOD components (von der Emde 1997) (Fig. 7.4). Some pulse-type mormyrid fishes have evolved very short pulses (approximately a few hundred microseconds) (Hopkins 1986b) that contain strong high-frequency components, probably to improve electrolocation capabilities by capacitance-induced waveform changes (von der Emde and Ringer 1992).

Mormyrid fishes (*Gnathonemus petersii*, *Pollomyrus isidori*, *Mormyrus rume jubelini*, and *Mormyrus caschive*) were trained, in behavioral experiments, to

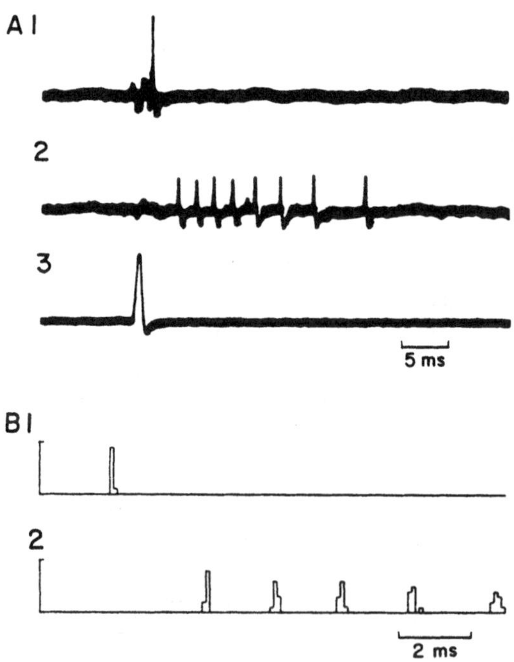

FIGURE 7.3. Tuberous receptors of a pulse-type gymnotiform fish, *Hypopomus*. (**A1**) Pulse marker, (**A2**) burst-duration coder, (**A3**) an EOD pulse. (**B**) Poststimulus time histograms of 10 responses of a pulse marker (**B1**) and a burst-duration coder (**B2**). (From Bastian 1976.)

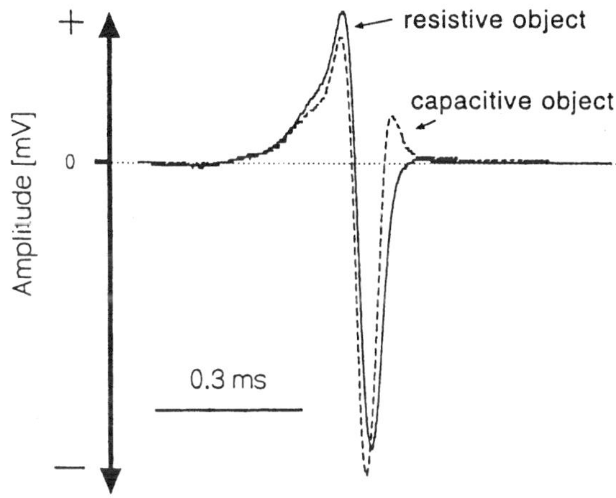

FIGURE 7.4. An EOD pulse recorded near a *Gnathonemus petersii* with a resistive or capacitive object (1 nF) present. The two objects had identical absolute impedance creating identical peak-to-peak amplitude of EOD signals. Note different waveforms. (From von der Emde 1997.)

discriminate capacitive objects (von der Emde 1992; von der Emde and Ringer 1992). In physiological studies, von der Emde and Bleckmann (1997) discovered that type B afferents from the mormyromast electroreceptors are sensitive to the waveform of EODs (Fig. 7.5). The afferents showed stronger responses to distorted waveform than to undistorted normal EODs—that is, *inverse waveform tuning*. The selectivity of type B fibers to waveforms has been successfully modeled (Shuai et al. 1999).

The sensitivity of gymnotiform pulse marker receptors to the waveform of the EOD pulse has been demonstrated by a behavioral study in which the fish changed their EOD frequencies in response to waveform distortion of stimulus pulses (Heiligenberg and Altes 1978).

Knollenorgan

Mormyrid fishes possess a unique type of time-coding electroreceptor organ, the knollenorgan, which responds to both the fish's own and other individuals' EODs. Its specialization for electrocommunication has been shown by the receptor's physiological properties and central processing of afferent signals from it. The response is usually one spike per EOD pulse stimulation for a wide range of amplitudes. The afferent signal generated by the fish's own EODs, however, is blocked in the first central relay nucleus, the nucleus of the electrosensory lateral line lobe (nELL) in the medulla, by electric organ corollary discharges (EOCD) from the animal's own electromotor system (for EOCD see Section 4). Thus, the spikes evoked by the fish's own EODs are blocked and only afferent signals generated as a result of EODs of other individuals are conveyed to higher brain centers. Hopkins and Bass (1981) demonstrated that males of a mormyrid fish discriminate between male and female EODs and between normal female and time reversed or phase-shifted female EODs. This study suggested that mormyrid fish discriminate the waveform of an EOD by

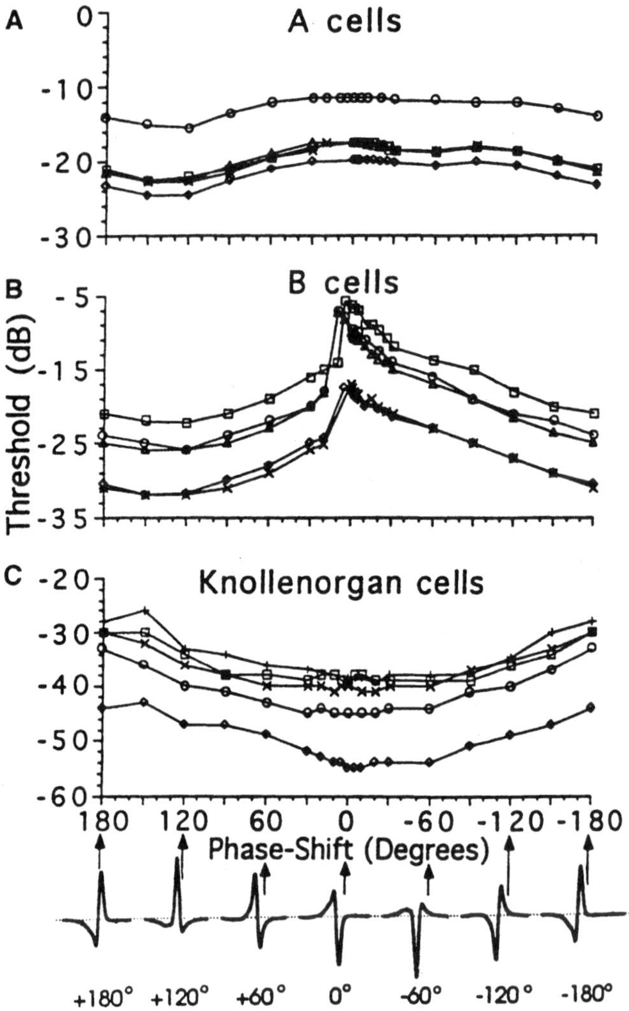

FIGURE 7.5. Tuning curves of three receptor types in *Gnathonemus* for phase shifting of frequency components in an EOD pulse (five different fibers in each type). Type B fibers (B cells) are sensitive to distorted wave forms, whereas other types of receptors are relatively immune to waveform distortion. (From von der Emde and Bleckmann 1997.)

its temporal rather than spectral components. The waveform and duration of EODs are different among species of mormyrids and serve for species recognition and ethological isolation (Hopkins 1986a,b) (see Hopkins, Chapter 10).

Like pulse marker afferents in pulse-type gymnotiform fishes, knollenorgans fire a time-locked action potential in response to an EOD. The threshold is low and the amplitude-dependent latency shift is small (Bell 1990) (Fig. 7.6C). How

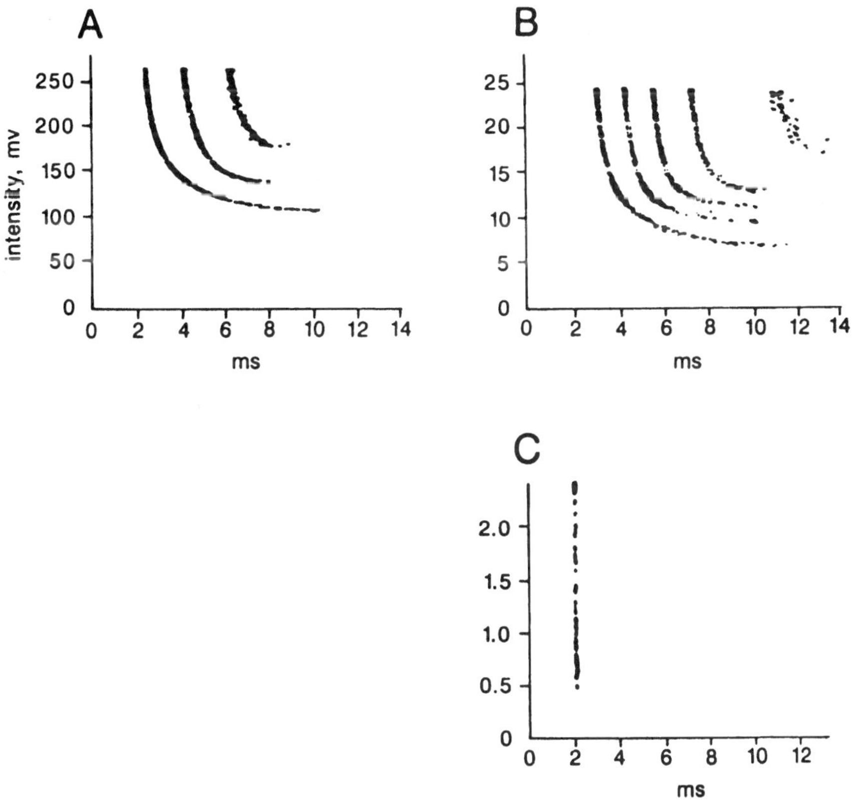

FIGURE 7.6. Raster displays of responses of primary afferent fibers in *Gnathonemus* in response to increasingly stronger stimulus amplitude. *Abscissa:* latency, *ordinate:* stimulus intensity. (**A**, **B**) Different types of mormyromast receptors. (**C**) Knollenorgan. Note different intensity scales and constant latency of knollenorgan responses. (From Bell 1990.)

can a single action potential ever encode waveform information from an external source where there is no absolute time reference? Hopkins (1986b) proposed that the waveform is encoded by an ensemble code of action potentials from different knollenorgan afferents from different parts of the body that respond to different phases of an EOD pulse (Hopkins 1986b). Central physiological mechanisms for decoding the waveform information from knollenorgan afferents have been discovered (see Section 3.3.3).

2.3 Frequency Tuning and Receptor Oscillation

Tuberous electroreceptors in general are sharply tuned to a particular frequency and show V-shaped threshold tuning curves similar to those seen in auditory neurons. The best frequencies for firing are closely matched to the EOD frequency of the species (or the individual). Examples from gymnotiform wave species are shown in Figure 7.7. Matching of frequency between EOD and receptor tuning is also observed in pulse species in which the frequency components within a pulse rather than its repetition frequency are important. Fourier analysis of EOD pulses shows broad spectra with peaks roughly matched to the best frequency of receptor tuning in gymnotiform pulse species (Bastian 1976, 1977) and in mormyromasts and knollenorgans of mormyrids (Hopkins 1981).

Some tuberous electroreceptors show damped oscillation after a brief artificial stimulation in the absence of the EOD (Bennett 1971). In an extreme case in *Gymnarchus*, all tuberous electroreceptors on the skin continuously oscillate in synchrony with no stimulus at all, generating an electric field of significant strength around the fish (Kawasaki 2001). This oscillation is observed when *Gymnarchus* interrupts its EOD as a natural submissive behavior for a prolonged time (minutes) (Hopkins 1974). The frequency of oscillation is close to that of the EOD prior to interruption. Physiological recording during the interruption showed that time-coding afferent neurons continued firing as if they were stimulated by the fish's own EODs.

The mechanisms of frequency tuning and oscillation at basic EOD rate in tuberous electroreceptors are not understood. They may involve the anatomical structure of receptor organs, membrane channel properties of receptor neurons, properties of the receptor-afferent synapses, and coupling between receptors and afferent fibers (Bennett 1965, 1967, 1971). Hair cells of lower vertebrates (amphibians, reptiles, and birds) show electrical tuning and behave like tuberous electroreceptors (Crawford and Fettiplace 1981). The tuning mechanisms are being discovered in these animals (Art et al. 1993; Wu et al. 1995; Ramanathan et al. 1999; Smotherman and Narins 1999). Electroreceptor tuning may change during sexual maturation in cases where there are sex-associated changes in the EOD (Zakon and Meyer 1983; Keller et al. 1986; Zakon et al. 1990) (see Bastian and Zakon, Chapter 8) and in response to administered hormones (Bass and Hopkins 1984).

Because ampullary electroreceptors are tuned to low frequencies (see Bodznic and Montgomery, Chapter 6), they are insensitive to EOD frequencies in wave-type electric fishes. During social communication displays, however, when the EOD frequency is rapidly modulated either during courtship or agonistic displays, ampullary receptors may respond to the low-frequency components of the discharge during the modulation. (Naruse and Kawasaki 1998). A series of EOD interruptions (chirps), which occur during courtship and aggression in a wave-type gymnotiform fish, *Eigenmannia*, also stimulate ampullary electroreceptors because the steady-state voltage during the interruptions is nonzero and deviated from the mean voltage during EOD (Metzner and Heiligenberg 1991).

For a comprehensive review of the frequency tuning of electroreceptors, see Zakon (1986) (also see Bastian and Zakon, Chapter 8).

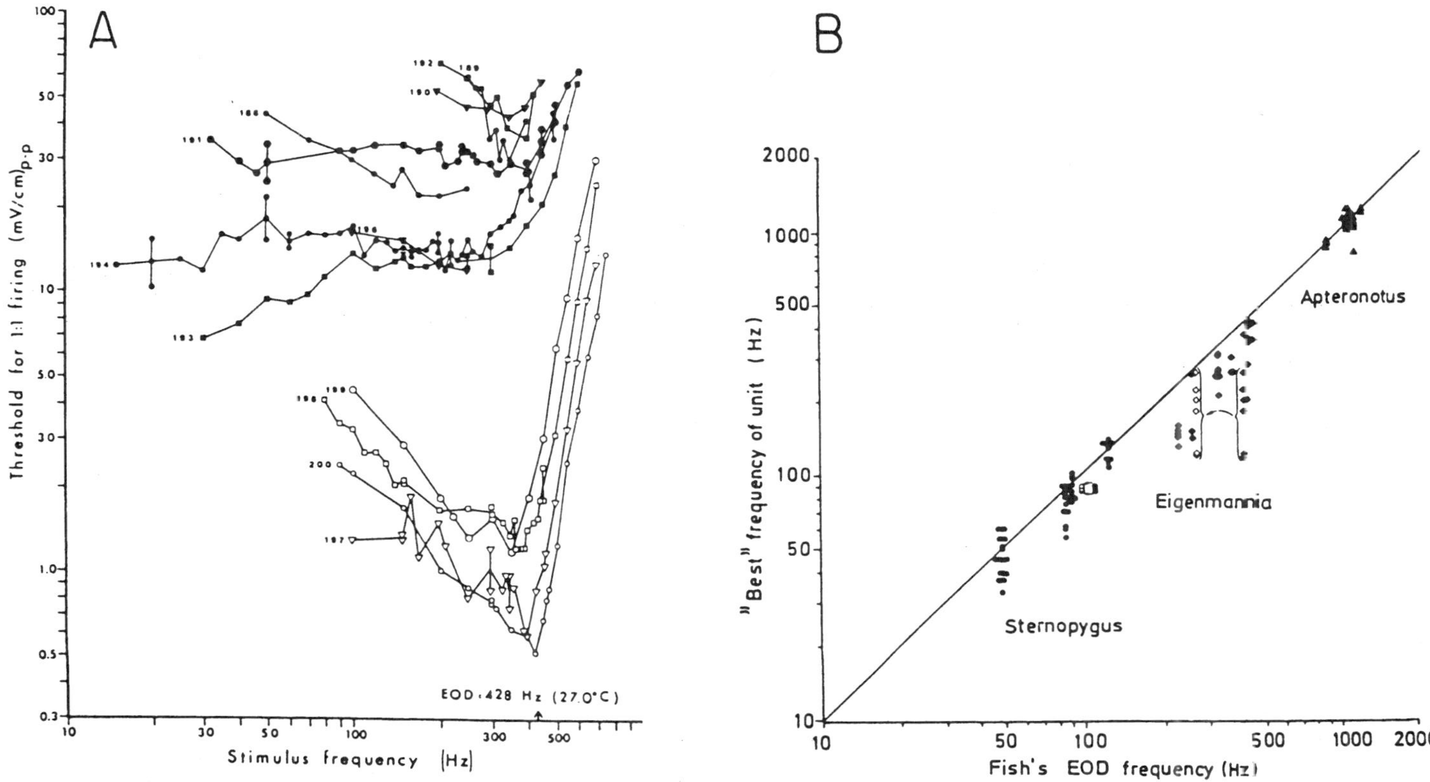

FIGURE 7.7. (**A**) Frequency tuning curves of tuberous receptors in *Eigenmannia* (from one individual). *Filled circle*: amplitude-coding P-type afferent; *open circles*: time-coding T-afferent. (**B**) Relationship between individuals' EOD frequency and best frequencies for tuberous receptors in three genera of gymnotiform wave-type fishes (Hopkins 1976).

3. Central Physiology

3.1 Overview of Central Physiology

In all gymnotiforms and mormyriforms primary electrosensory afferents enter the brain via the lateral line nerve and terminate in the hindbrain electrosensory lateral line lobe (ELL). The ELL consists of several different zones (or segments) that receive inputs from different types of primary afferent fibers. Figure 7.8 shows for three different genera of fishes the functional connections of different types of tuberous afferents to the ELL and connections between the ELL and higher midbrain centers. Amplitude information undergoes substantial processing in the ELL in all species (see Section 3.2). For time processing, time differences between receptor afferents from different areas of the skin carry behaviorally important information (see Section 3.3). Time differences are detected in the ELL in *Gymnarchus* (see Section 3.3.2) and in the midbrain in gymnotiform and mormyrid fishes (see Sections 3.3.1 and 3.3.3). Midbrain nuclei project back to the ELL probably in all species, and this descending projection strongly affects ELL neurons (see Sections 3.2.1 and 4). In mormyrids, the electromotor system sends corollary discharge signals to the ELL via several relay nuclei to modify sensory processing in ELL (see Section 4). The physiology of amplitude and time processing pathways in the three groups of fishes is described in the sections that follow (see Bell and Maler, Chapter 4 for corresponding central anatomy).

3.2 Amplitude Processing Pathway

3.2.1 The ELL in Gymnotiformes

The amplitude-coding afferents in wave-type gymnotiform fishes terminate in the deep layers of the ELL cortex. This is a well-studied structure, both anatomically and physiologically, and much is known about the mechanisms of amplitude processing (Maler et al. 1981; Carr and Maler 1986; Berman and Maler 1999).

In the ELL, the amplitude-coding afferent terminals form excitatory synaptic connections with basilar dendrites of basilar pyramidal cells. The afferent also forms excitatory synaptic connections with the granule cells, which in turn form inhibitory connections with non-basilar pyramidal cells. Enger and Szabo (1965) were the first to record from neurons in the ELL and show a variety of responses to metal and plastic electrolocation targets moved past the skin surface. A later study by Bastian (1981a) showed two physiological types, E cells and I cells, that responded respectively either to an increase or decrease in stimulus amplitude. These two physiological types were later confirmed to be basilar and non-basilar pyramidal cells by intracellular recording and labeling (Saunders and Bastian 1984). Both E and I cells responded to a narrower range of amplitude modulation frequencies than do primary afferent fibers (Bastian 1981a,b, 1986a).

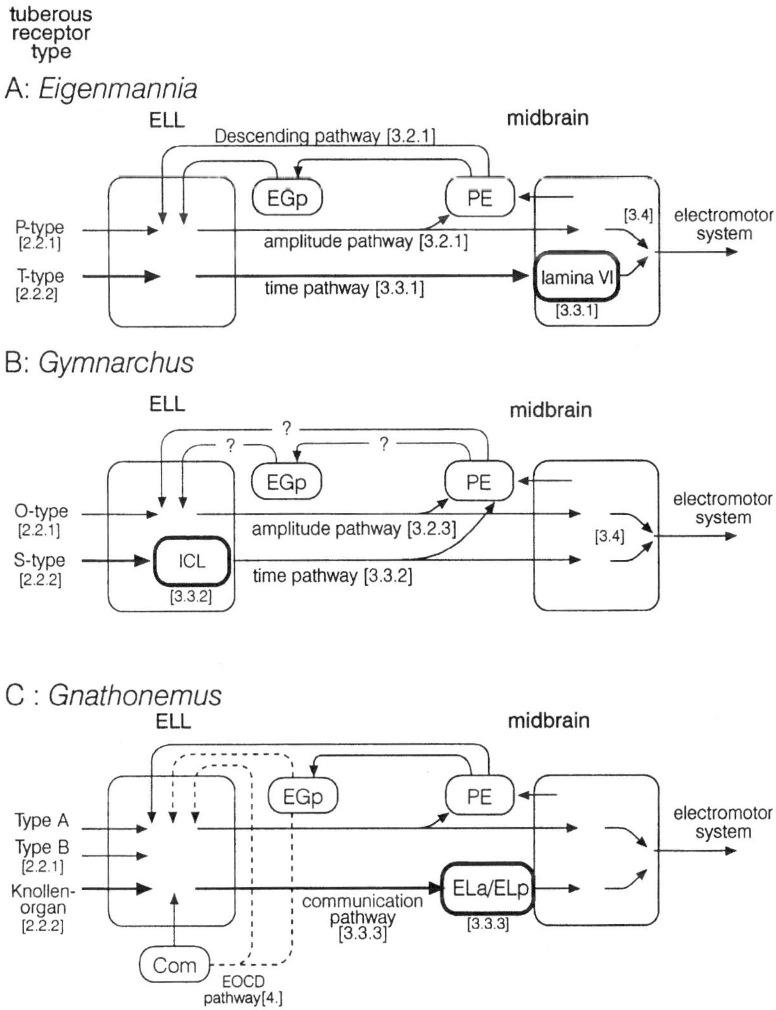

FIGURE 7.8. Major functional pathways of tuberous information in three genera of electric fishes, (**A**) *Eigenmannia* (Gymnotiformes), (**B**) *Gymnarchus* (Mormyriformes), and (**C**) *Gnathonemus* (Mormyriformes). Electroreceptor types (*left columns*) are associated with central pathways (*remaining figure*). Numbers in *square brackets* indicate section numbers in this chapter where description is found. *Thick arrows* indicate time pathways in which time information is conveyed by temporal code. Areas surrounded by a thick envelope perform time comparison. Electric organ corollary discharge (EOCD) pathway (*dashed lines*) is highly abbreviated with many nuclei omitted for brevity. *Lines with a question mark* indicate presumed pathways. Com, command nucleus; EGp, eminentia granularis posterior; ELa/ELp, anterior/posterior exterolateral toral nucleus; ELL, electrosensory lateral line lobe; PE, preeminentialis nucleus.

Recently developed statistical methods have been applied to quantitative analyses of the spike trains in afferent fibers and these ELL neurons (Gabbiani and Metzner 1999). Time-varying amplitude information (dynamic changes in amplitude) is conveyed by a series of spikes of amplitude-coding afferents in *Eigenmannia* (P-type afferents) to pyramidal cells which extract behaviorally important features. In these analyses, the amount of time-varying amplitude information was quantitatively analyzed in the spike trains in P-type afferents and pyramidal cells by the stimulus estimation method (Poor 1994). P-type afferents were found to accurately convey temporal signals, while pyramidal cells extract features such as increases and decreases in amplitude (Gabbiani et al. 1996; Wessel et al. 1996; Metzner et al. 1998). ELL cells with concentric receptive fields, excitatory center-inhibitory surround, were analyzed by the stimulus estimation method (Bastian et al. 2002). Stimuli applied only to the center of the receptive field are encoded with high efficiency, while larger stimuli covering the entire receptive field fail to evoke responses with the same high degree of temporal precision.

Heiligenberg and Dye (1982) discovered that each P-type amplitude-coding afferent fiber trifurcates into the centromedial, centrolateral, and lateral segments of the ELL. Moreover, neurons in each segment are topographically organized to form a spatial map of the electrosensory periphery (Carr et al. 1982). The three maps are physiologically and anatomically different (Shumway 1989a, b) and are flanked by still another map for the low-frequency ampullary system. Neurons in the lateral map have larger, coarser receptive fields. They tend to be the most sensitive of the three maps, and respond to high frequencies of amplitude modulation. The neurons in the centromedial map have the smallest, most precise receptive fields, with large inhibitory surrounds, and respond to only to low-frequency amplitude modulations. The neurons in the centrolateral zone show intermediate spatial and temporal characteristics. γ-Aminobutyric acid (GABA) plays a critical role in shaping both temporal and spatial characteristics of ELL neurons (Shumway and Maler 1989). In addition, the rates of convergence (the number of afferent fibers that an ELL cell receives) are different between maps. These physiological differences among segments must reflect differences in their electrolocation functions, such as temporal and spatial processing. By selectively lesioning the maps, Metzner and Juranek (1997a,b) demonstrated different functions for the maps. The lateral map in *Eigenmannia* and *Apteronotus* was especially important in evoking "chirping" responses, while the centromedial map was important in the jamming avoidance response (see Section 3.4).

Recently it has been discovered that the pyramidal cells in the ELL of *Apteronotus* tend to fire at a particular oscillatory frequency (approximately 30 Hz) only when amplitude modulation of sensory signal is applied "globally" covering nearly the entire fish. Oscillations like these are important in the function of visual, somatosensory, and olfactory systems (Gray et al. 1989; Laurent 1996; Murthy and Fetz 1996). Global modulation of the EOD occurs during communication because the EODs of a distant neighboring fish tend to modulate feedback EODs over a large area of the skin surface. Physiological experiments and modeling have shown that the oscillatory response is due to a recurrent descending inhibitory influence on the ELL neurons (see Section 3.2.2). When amplitude modulation is applied only locally, mimicking an electrolocation target, the same physiological preparation and the model tend not to oscillate (Doiron et al. 2003). Chacron et al. (2003) has further demonstrated that frequency tuning of pyramidal cells changes according to the size of a stimulus area where amplitude modulation is applied. The neurons' frequency tuning switches from low modulation rates to high when the area of amplitude modulation extends from the central to peripheral receptive field.

3.2.2 Descending Control of the ELL

The ELL of gymnotiform fishes receives not only afferent inputs but also massive descending inputs from the nucleus of praeeminentialis (PE), a midbrain structure (Fig. 7.9) (Sas and Maler 1983) (see Bell and Maler, Chapter 4). The PE sends descending output via two different paths: stellate cells in the PE

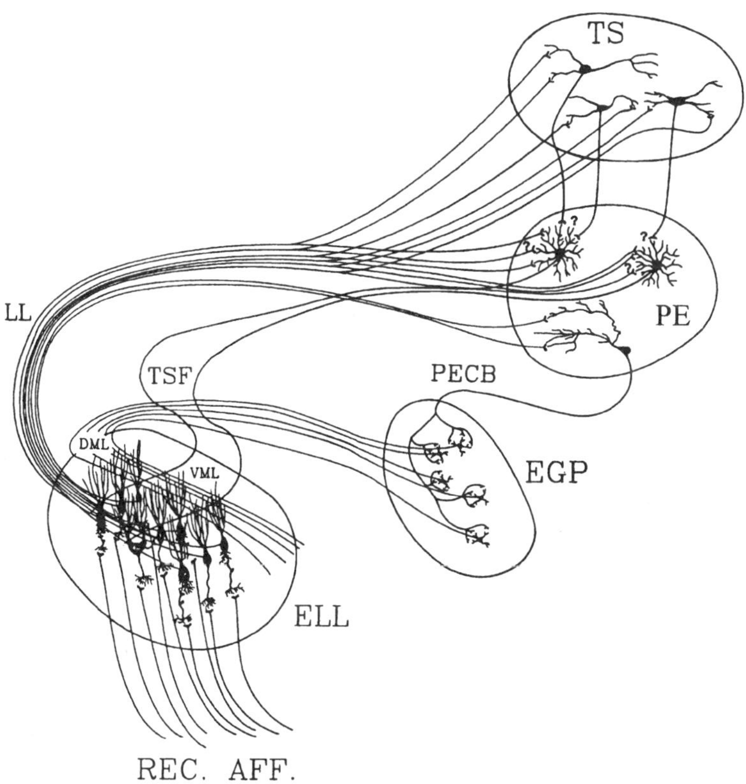

FIGURE 7.9. Pyramidal cells in ELL project to the torus semicircularis (TS) and nucleus preeminentialis (PE) in the midbrain. The PE projects back to the ELL directly and indirectly via the eminentia granularis. (From Bratton and Bastian 1990.)

directly project to the ELL, whereas the multipolar cells project to a relay nucleus, the eminentia granularis posterior (EGp), that in turn projects to the ELL (Sas and Maler 1983). Bastian (1986b, c) showed by lesion experiments in *Apteronotus* that removal of the descending inputs results in a large increase of amplitude gain and the size of the receptive field of ELL neurons, suggesting that the descending inputs serve as gain control mechanisms. Differences in physiological properties between these cell types were studied by intracellular labeling and recording techniques (Bastian and Bratton 1990; Bratton and Bastian 1990). The multipolar cells respond in a tonic fashion to amplitude changes of prolonged duration (minutes), suggesting their involvement in gain control. No other known neurons in the CNS show such nonadapting steady responses to amplitude changes. The multipolar neurons send descending input to the dorsal molecular layer of the ELL via the eminentia granularis (see Bell and Maler, Chapter 4). The stellate neurons, however, do not show steady responses to prolonged change in amplitude but respond better to a small moving electrolocation object. The stellate cells directly project to the ventral molecular layer of the ELL. Bratton and Bastian (1990) suggested that the direct pathway from the PE to the ELL may enhance the sensitivity of ELL neurons to a local and novel stimulus and serve as a "sensory searchlight" (Crick 1984). The cellular properties that support the searchlight hypothesis in ELL neurons, such as GABA-mediated inhibition, voltage-dependent excitatory postsynaptic potentials (EPSPs), dendritic spike bursts, and frequency-dependent synaptic facilitation have been extensively studied in in vivo and in vitro experiments (Berman et al. 1995, 1997, 2001; Berman and Maler 1998a–c, 1999; Doiron et al. 2001).

The effects of the feedback pathway on ELL neurons in *Apteronotus* is highly plastic. ELL neurons adapt to repeated sensory stimulation if it is time locked to the animal's own acts, and this plastic change in ELL neurons is mediated by the feedback pathway described earlier (Bastian 1996a,b, 1998) (also see Bastian and Zakon, Chapter 8).

As mentioned, a burst-like firing pattern, rather than individual spikes, in ELL neurons may be important in conveying sensory information (Gabbiani et al. 1996; Wessel et al. 1996; Metzner et al. 1998). The bursting properties of ELL neurons have been analyzed in vitro (Mathieson and Maler 1988; Turner et al. 1994, 1996; Lemon and Turner 2000) and in vivo (Bastian and Nguyenkim 2001). These studies show that tetrodotoxin (TTX)-sensitive sodium channels exist in the soma and the dorsal dendrite but not the basal dendrite of the basilar and non-basilar pyramidal neurons in the ELL; onset and termination of bursts depend on the interaction of dendritic and somatic after-potentials, and bursting characteristics are different in the three different tuberous maps of the ELL. In their in vivo study, Bastian and Nguyenkim (2001) showed that spontaneous bursting characteristics of ELL neurons are greatly affected by glutamatergic synaptic blockers to apical dendrites of the neurons, suggesting that the bursting characteristics are under control of the descending inputs.

3.2.3 Torus Semicircularis in Gymnotiformes

The torus semicircularis (hereafter "torus") is a huge, unpaired tegmental structure in the midbrain. The torus receives inputs from basilar and non-basilar pyramidal cells in the ELL through terminals in layers III, V, VII, and VIII (Carr et al. 1981). Extensive physiological studies on selectivity for amplitude modulation have been conducted in the torus. Scheich and Bullock (1974) first discovered neurons that selectively respond to amplitude modulation. After a demonstration that the jamming avoidance response (JAR) of *Eigenmannia* requires both amplitude and time (phase) modulation be coordinated with each other in time (Heiligenberg et al. 1978; Heiligenberg and Bastian 1980) (see Section 3.4), Bastian and Heiligenberg (1980a) discovered neurons in the torus that were sensitive either to amplitude or to time (phase) modulation alone or to both in combination. As in the ELL, the amplitude-sensitive neurons in the torus respond to each cycle of periodic amplitude modulation with a burst. This time-locking of bursting is an important feature for the JAR, which requires temporal analysis of amplitude and time (phase) modulation. These neurons were studied physiologically and anatomically with intracellular recording and labeling, and were shown to exist in layers V, VII, and VIII (Rose and Heiligenberg 1985a). They project to various parts of the brain, such as the optic tectum, the lateral mesencephalic area, and the nucleus praeeminentialis as well as other layers within the torus, indicating their involvement not only in JAR but also in other behavioral functions. Whereas amplitude-sensitive neurons in the ELL respond to a wide range of amplitude modulations (1 to more than 100 Hz) (Bastian 1981b; Shumway 1989a), most neurons in the torus respond to slow modulations (approximately 4 Hz) which most strongly elicit the JAR (Rose and Heiligenberg 1986a).

How does this behaviorally relevant selectivity for slow modulations emerge in the torus? To answer this question, Rose and his colleagues investigated toral neurons with their pioneering in vivo whole-cell recording technique that provided stable recording of high-quality intracellular potentials (Rose and Fortune 1996). In a series of studies, Rose and his colleagues showed three physiological mechanisms that contribute to the emergence of a low-pass tuning of the torus neurons to amplitude modulation frequency. The first mechanism is based on electrical cable properties of dendritic spines. Rose and Call (1992, 1993) found that low-pass characteristics correlate with the density of dendritic spines in torus neurons. Neurons with a large number of dendritic spines showed large postsynaptic potentials to low-frequency amplitude modulation (approximately 4 Hz) but not to high-frequency modulations (10 to 20 Hz). The dendritic spines have been suggested to act as a low-pass filter (Koch and Poggio 1983; Fortune and Rose 1996). The second mechanism is attributable to voltage-dependent conductances in the torus neurons (Fortune and Rose 1997b). Steady current injection into torus neurons, which shifted the working range of membrane voltage for postsynaptic potentials, altered their frequency responses. The third mechanism is frequency-dependent synaptic depression at input synapses of to-

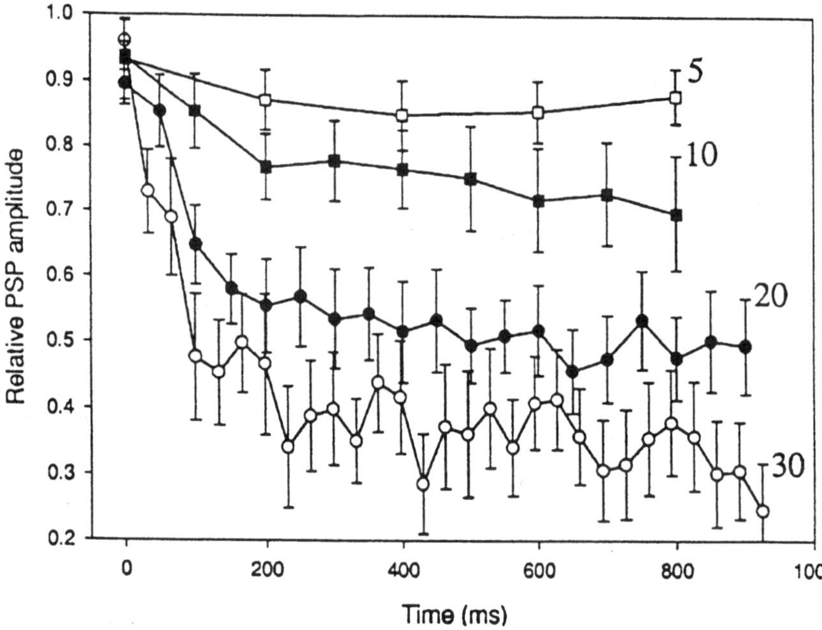

FIGURE 7.10. Time course of postsynaptic potential reduction to successive amplitude modulation in a neuron in the torus semicircularis in *Eigenmannia*. (From Rose and Fortune 1999.)

rus neurons (Rose and Fortune 1999; Fortune and Rose 2000). The responsiveness of torus neurons to high-frequency amplitude modulation shows rapid depression (time constant = a few hundred milliseconds) with rapid recovery (in tens of milliseconds), whereas neurons persistently respond to continuous low-frequency modulation (Fig. 7.10). This frequency-dependent depression in torus neurons parallels the behavioral selectivity (Rose and Fortune 1999). Intermittent bursts of high rates of amplitude modulation evoke stronger frequency-shifting behavioral responses (Takizawa et al. 1999) than do continuous rapid amplitude modulation. The depression appears to occur within the torus, since responses evoked by electrical stimulation of input fibers to the torus similarly show stimulation frequency-dependent depression (Fortune and Rose 2000). Similar temporal selectivity is found in the low-frequency ampullary electrosensory system in the same fish (Fortune and Rose 1997a).

3.2.4 Amplitude Processing in *Gymnarchus*

The amplitude-coding afferent fibers in *Gymnarchus* (see Section 2.2.1) terminate in the dorsal and medial zones of the ELL (Kawasaki and Guo 1996). There, in vivo whole-cell recording revealed neurons that respond to amplitude changes in electrosensory stimuli (Kawasaki and Guo 1998). As in *Eigenman-*

nia, E and I type neurons respond respectively to increase and decrease of sensory stimulus amplitude. Morphology of basal dendrites was different for those two groups. They also show large dorsal dendrites, indicating that they receive descending inputs from higher centers (see Section 3.2.2), as do the Gymnotiformes. These ELL neurons project to the midbrain nuclei N. praeeminentialis and the lateral toral nucleus. In the toral nucleus, neurons are also amplitude-sensitive (see Section 3.3.2). In the context of the JAR (see Section 3.4), *Gymnarchus* is capable of detecting a modulation as little as 0.02% of the peak-peak amplitude (Guo and Kawasaki 1997). Correspondingly, some central neurons exhibit responses to amplitude modulation of less than 0.5% (Kawasaki and Guo 1998). The amplitude processing system of *Gymnarchus* is remarkably similar to that of *Eigenmannia* despite its independent evolution (see Section 3.3.4).

3.2.5 The ELL in Mormyrids

The amplitude-coding afferent fibers type A and type B (see Section 2.2.1) respectively project to the medial and dorsolateral zones in the ELL. In mormyrid fish, afferent information is heavily modified by the electric organ corollary discharges (EOCD), which are an efference copy signal generated by the electromotor system for EOD. The efference copy signal reaches the ELL for strong modification of sensory signals (see Section 4). In physiological studies of the ELL in mormyrids, much focus has been given to the interaction of EOCD and afferent signals rather than afferent response properties to various stimulus parameters. An exception to this is that neurons in the dorsolateral zone were found to be sensitive to the waveform of EOD much as in the type B afferent fibers (von der Emde and Bell 1994). As in gymnotiform fishes, the ELL receives descending inputs from the PE that contain both sensory and EOCD components (Bell et al. 1992; von der Emde and Bell 1996). A more detailed description of ELL neurons in mormyrids is provided in the context of EOCD gating in Section 4.

3.3 Time Processing Pathways

As mentioned in Section 2.2.2, time-coding primary afferent fibers convey information about precise occurrence times of sensory signals by their times of spikes (time-locking). To be useful, spike time information may need to be compared with a time reference signal. Since electrolocation signals are the results of the animal's own EODs, the obvious time reference signal would be the electromotor command signal that triggers the firing of EODs. Times of afferent spikes may be changed as a result of capacitance of objects or another fish's EOD as mentioned in Section 2.2.2. Surprisingly, however, in all electric fish species so far examined, the electromotor command signal is not used to decode time information in the time-coding afferent signal (with the possible exception of mormyromast information in mormyrids—see Section 4). Instead, time-locked spikes are compared between afferent signals originating from different areas on the body much as in binaural time comparison in the vertebrate auditory system for purposes of determining interaural time differences. Time comparison circuits have been found in three electric fish species with distinct phylogenetic relations (Hopkins 1995). The time scale involved in these systems is remarkably short—on the order of microseconds. This high degree of temporal precision and repeatability requires specializations in synaptic transmission and in physiological and anatomical properties of time-coding neurons as seen in the time-coding pathways of the vertebrate auditory system (Irvine 1992; Trussell 1999; Carr et al. 2001).

3.3.1 Time-Coding System in *Eigenmannia*

In *Eigenmannia* (see Section 2.2.2), time-coding electroreceptor afferents fire one action potential per EOD cycle time-locked to near the zero crossing of the EOD waveform, thereby conveying behaviorally significant local timing information to the brain (Fig. 7.2A). Electroreceptors at different locations on the body surface receive electrosensory feedback of the EOD at slightly different times. When the EOD is jammed by a second EOD stimulus, the feedback signal from the fish's own EODs is summed with the jamming EOD to produce an amplitude- and time-modulated "beat." Time and amplitude modulations are periodic, and the magnitude (depth) of the cyclic time modulation differs for different body locations. The time disparity thus created in the firing of T-type electroreceptors from different locations (Heiligenberg 1991a) was demonstrated to be one of two essential cues for the JAR (Heiligenberg et al. 1978). Although the magnitude of the time disparity that evokes strong JARs is on the order of 10^{-4}s, the behavioral threshold value is very small (approximately 300 ns) (Rose and Heiligenberg 1985b). This acuity represents one of the highest time resolutions found in the nervous system of any animal (Carr 1993; Simmons et al. 1995; Guo and Kawasaki 1997). As mentioned in Section 2.2.2, the capacitive component of nearby objects may also change times of the EOD feedback signal, again creating time disparity between electrosensory signals at different body locations (Rose and Heiligenberg 1986a). The significance of the time disparity was demonstrated also by a behavioral study in which a fish was trained to discriminate between resistive and capacitive objects (von der Emde 1998).

The time-coding, T-type afferent enters the brain via the anterior lateral line nerve and terminates on the soma of the spherical cells in all three topographical maps in the ELL (see Section 3.2.1). Unlike amplitude information, time-locked firing of time-coding afferents does not undergo substantial processing in the ELL, but the jitter of afferent signals may be reduced by convergent innervation of many afferents on spherical cells (Maler et al. 1981). The synaptic connection from the afferents to the spherical cells is electrical, and the spherical cells simply fire a single action potential in response to coherent inputs from the afferents. Spherical cells project to the lamina VI of the torus semicircularis in the midbrain, where they form synapses on the soma of giant cells and the dendrite of the small cells (Fig. 7.11) (Carr et al. 1986b). The giant cells also

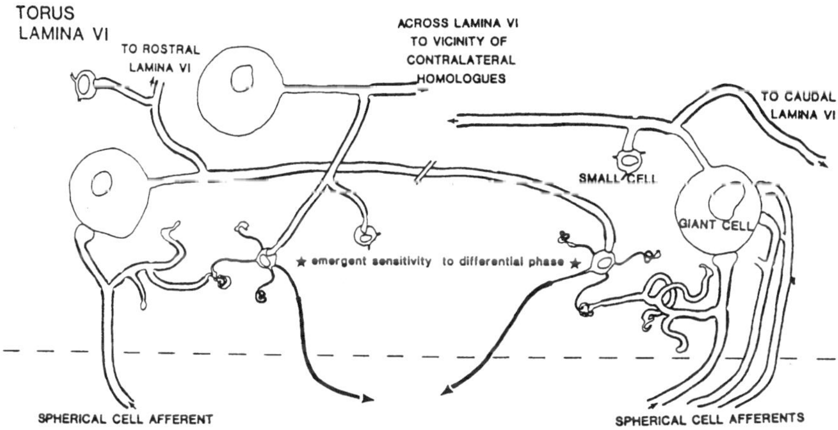

FIGURE 7.11. Time comparison circuit in the midbrain of *Eigenmannia*. The time comparator, the small cells, receive time-locked inputs from spherical cells on their dendrites and from giant cells on their soma. (From Carr et al. 1986b.)

fire one spike in response to inputs from spherical cells, again in a time-locked manner. Carr et al. (1986a) compared the time accuracy of firing (jitter) between T-type afferents and time-locked neurons in the torus semicircularis and found that the central neurons exhibit smaller standard deviation in jitter than the afferents (approximately 11 μs compared to approximately 30 μs). This improved time accuracy is correlated with the anatomical convergence of neurons at the somata of the spherical and giant cells (Maler et al. 1981; Carr et al. 1986b; Fig.4.10 in Heiligenberg 1991a).

Heiligenberg and Rose (1985) made intracellular recordings from and labeled neurons (small cells) in the torus that responded to time disparities between time-locked inputs from both giant cells in the torus and spherical cells in ELL. Other types of neurons in different layers of the torus also responded specifically to time disparities between electrosensory signals applied at different body locations (Bastian and Heiligenberg 1980b; Rose and Heiligenberg 1985a). Rose and Heiligenberg (1986b) measured sensitivity of the time disparity of neurons in the torus and found that threshold sensitivity ranged from 6.4 μs to 165 μs. The highest single neuron sensitivity to time disparity (approximately 1 μs) has been recorded in the prepacemaker nucleus (Kawasaki et al. 1988b), which is a premotor nucleus that controls the frequency of EOD in the JAR.

In pulse-type gymnotiform fish, time-coding pulse marker afferents terminate on spherical cells in the ELL as in wave-type gymnotiform fishes (Szabo 1967; Castello et al. 1998). These spherical cells project to the magnocellular mesencephalic nucleus of the midbrain, where the time-locked field potential can be recorded as in layer VI of wave-type gymnotiform fishes (Réthelyi and Szabo 1973). Time comparison presumably occurs in the nucleus, but to date no neurons or behaviors have been demonstrated for time comparison in pulse-type gymnotiform fishes.

3.3.2 Time-Coding System in *Gymnarchus*

Although belonging to phylogenetically separate lineages *Gymnarchus* and *Eigenmannia* both emit wave-type EODs and exhibit JARs that are qualitatively identical (Bullock et al. 1975) (see Section 3.4) using the same complex computational algorithms (Kawasaki 1993) involving time disparity between different body locations as an essential cue. Both *Eigenmannia* and *Gymnarchus* show temporal hyperacuity with sensitivity to time differences on the order of 10^{-6}s between sensory signals (Guo and Kawasaki 1997; Kawasaki 1997).

In spite of the similarity of computational rules, the neural circuits are highly divergent. In *Gymnarchus,* the time comparison circuit was found in the hindbrain, not in the midbrain as in *Eigenmannia* (Kawasaki 1996). S-type afferent fibers of *Gymnarchus* respond in time-locked fashion (Bullock et al. 1975) much like the time-coding T afferents in *Eigenmannia.* The S afferents project to the inner cell layer (ICL) of the medial zone of the ELL where they give rise to collaterals that project to the soma of adendritic cells, the giant cells (Bass and Hopkins 1982; Kawasaki and Guo 1996). These giant cells fire in time-locked manner to every cycle of the EOD signal. Giant cells project bilaterally to the ICL, allowing for convergence between time-locked spikes from S-type afferent fibers and from the giant cells on the same, or opposite side of the body surface. It is here that detection of time disparity is expected (Fig. 7.12).

Matsushita and Kawasaki (2004) discovered a postsynaptic neuron type (the ovoidal cell) that receives inputs from the S-type afferents and giant cells and presumably detects time disparity between their firings. A giant cell terminal forms a large single synapse that embraces the soma of the ovoidal cell and covers approximately 85% of the somatic membrane surface with no gap. The large continuous synaptic cleft may lower the voltage inside the synaptic cleft when inward (excitatory) postsynaptic current occurs. The lowered voltage, in turn, ephaptically depolarizes presynaptic potential, creating positive feedback to the presynaptic terminal for fast signal transmission. The ovoidal cells transmit synaptic output via their dendrites to pyramidal cells in the ICL (Fig. 7.13). These cells show sensitivity to time disparities on the order of 10^{-6}s and project to the midbrain (Kawasaki and Guo 1996, 1998).

Since time-locked inputs to the ovoidal cells vary only on the order of 10^{-5}s we may ask how such accurate times arise in the time comparator neurons as postsynaptic potentials, and how do the comparator neurons distinguish these very small time differences? In modeling approaches, Kashimori et al. (2001) and Takagi and Kawasaki (2003) have shown that time disparity on the order of microseconds between two synaptic potentials results in firing of the model neuron even when the rise time of synaptic potentials had much longer time course than the time disparity. Kashimori et al. (2001) also proposed an additional model in which arrays of less accurate time comparators converge to

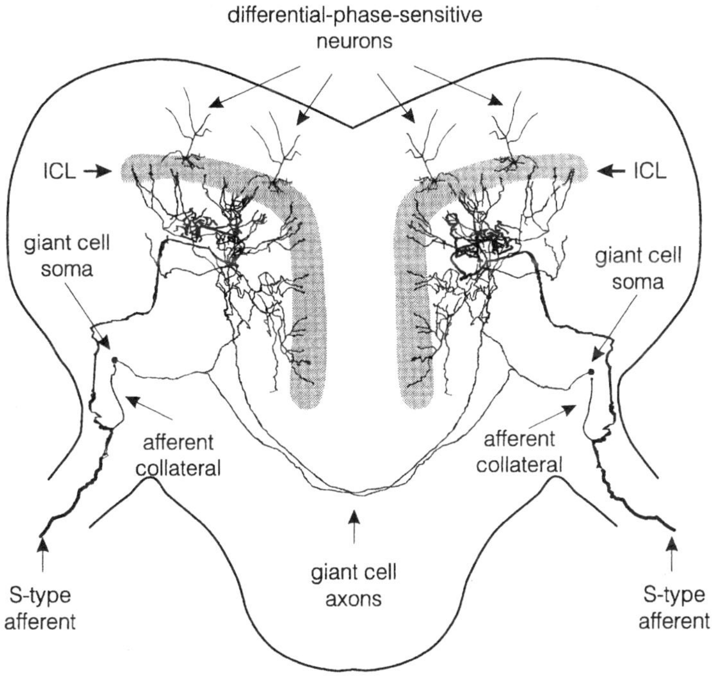

FIGURE 7.12. Time comparison circuit in *Gymnarchus*. Time-locked S-type afferents terminate in the inner cell layer (ICL) of the ELL, which sends a collateral to the soma of the giant cell. Giant cells bilaterally send axon terminals to ICL. These time-locked neurons terminate on ovoidal cells (not shown) in the ICL. Differential-phase-sensitive (time-sensitive) neurons respond to time differences between time-locked inputs. (From Guo and Kawasaki 1997.)

produce more accurate neurons. The essential convergent rule in this model is that axons of time comparators with different time preferences terminate at different dendritic sites of the output neuron.

3.3.3 Time-Coding System in *Brienomyrus*

The knollenorgan electrosensory system of mormyrid pulse fishes has been an exceptional subject for the neurobiological study of central circuits dedicated to animal communication (Amagai 1998; Amagai et al. 1998; Friedman and Hopkins 1998). Behavioral studies demonstrated that a pulse-type mormyrid electric fish, *Brienomyrus*, uses pulse duration (or waveform information) for species and sex recognition (Hopkins and Bass 1981; Hopkins 1986a,b).

The knollenorgan fires a single action potential either to the up- or down-edge of a square pulse or a natural EOD, depending on the geometrical relationship between the electroreceptor and a neighboring fish. If, for example, the electric current generated by a neighbor's EOD penetrates from the right to

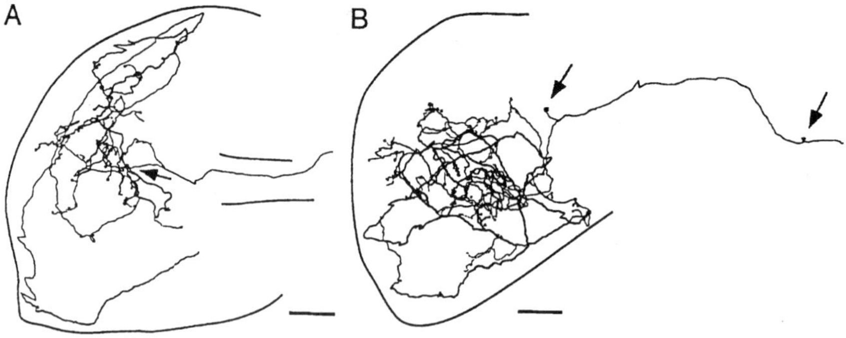

FIGURE 7.13. Axon terminals of an nELL neuron in *Brienomyrus*. Intracellular injection and labeling revealed a large terminal field in the ELa. Calibration = 100 μm. Terminals onto large cells are indicated by an *arrow*. (From Friedman and Hopkins 1998.)

the left side of a fish, knollenorgans in the two sides are stimulated by different strokes (up or down edges) of the EOD pulses, and thus the time difference between action potentials from the two sides encode the duration of the EOD pulse.

On entering the brain, knollenorgan afferents terminate on adendritic somata of neurons in the nELL, which receives massive inhibitory inputs from the EOCD system (see Section 4) timed to intercept and inhibit knollenorgan input from the fish's own EOD. As a consequence, the only afferent signals that get through to higher centers are those evoked by EODs of other individuals. As in the responses of knollenorgan afferents (Fig. 7.6C), the firing of nELL neurons is also tightly time-locked to the afferent signal. The nELL neurons bilaterally project to a midbrain structure, the nucleus exterolateralis pars anterior (ELa), where the projecting axons of nELL neurons recursively bifurcate to synapse onto numerous small cells and several large cells. The large cells again show time-locked responses and project to the small cells (Mugnaini and Maler 1987a; Friedman and Hopkins 1998).

The projection pattern of the nELL axons and the large cells suggests that the small cells compare firing times of these input neurons and detect pulse duration or waveform information. Friedman and Hopkins (1998) showed that axons of large cells in ELa have an extensive arborization (Fig. 7.13). The axons may act as a delay line that adjusts arrival times of action potentials to time-comparing small cells. Although intracellular recording from and labeling of the small cells have not been successful in the ELa, Amagai (1998) characterized neurons in the adjacent structure, the nucleus exterolateralis pars posterior (ELp), to which small cells project. There, type I neurons monotonically increase firing frequency with increased pulse duration, whereas type II neurons are tuned to a specific duration of pulses. Amagai (1998) proposed that the monotonic response of the type I neuron is created at the small cell in the ELa by convergent projection of time-locked signals and that type I neurons with different cutoff pulse durations converge onto type II cells in the ELp to create the selectivity to a pulse duration. The pulse durations that evoke strongest responses in type II neurons range from approximately 100 μs to approximately 10 ms, which corresponds to naturally occurring EOD pulse durations.

Mugnaini and Maler (1987a) demonstrated GABAergic synapses of the large cells (i.e., interstitial cells) to the small cells by immunocytochemistry. Application of picrotoxin to the ELa and ELp resulted in alteration of type A cell responses, consistent with Amagai's proposed model for temporal selectivity of the small cells (Amagai 1998; Amagai et al. 1998).

3.3.4 Comparison of Time-Coding Systems

The time-coding systems of *Eigenmannia*, *Gymnarchus*, and *Brienomyrus* exhibit striking similarities despite their distant phylogenetic relationship (Table 7.1). *Eigenmannia* belongs to the Gymnotiformes, a relatively new group of teleost, but *Gymnarchus* and *Brienomyrus* belong to the Osteoglossomorpha, one of the oldest teleost lineages. These two groups lack common electric fish ancestors and are believed to have evolved electromotor and electrosensory systems independently. Their time systems' shared features include (1) large-diameter and fast-conducting axons in the time pathways; (2) adendritic somata

TABLE 7.1. Comparison of time-coding systems.

Order	Gymnotiformes	Mormyriformes	
genus	*Eigenmannia*	*Gymnarchus*	*Brienomyrus*
Shared by all species	• Thick fast-conducting primary afferents (large diameter, heavy myelination) • Thick fast-conducting secondary neurons (spherical cells, giant cells, nELL) • Time-locked fibers ramify to give direct and indirect inputs to time comparators • Mixed synapses at comparators • Mixed synapses at relay synapses • Adendritic somata at relay synapses (spherical, giants, large) • Accurate firing of time-locked cells • Small size of time comparator		
Involvement of GABA in comparison	Unknown	Unknown	Present
Time scale involved	10^{-6} to 10^{-3}	10^{-6} to 10^{-3}	10^{-4} to 10^{-2}
Site of time comparison	Midbrain (torus)	Hindbrain (ELL)	Midbrain (ELa)
Primary function	JAR, electrolocation	JAR, electrolocation	Species and sex recognition
Comparator structure	Fused at midline	Bilateral	Bilateral

in time-locked neurons; and (3) existence of mixed synapses (electrical and chemical) both at time-conserving synapses that conserve spike time sequence and at time-comparing synapses that drive time comparator postsynaptic neurons. Fast conduction is believed to contribute to accurate conduction of time-locked firing with minimal jitter. Fast-conducting thick axons are also found in time-locked auditory systems in birds and mammals. Adentritic postsynaptic cells are probably advantageous for fast electrotonic propagation and integration of synaptic potentials (Carr 1993; Carr and Friedman 1999). Another striking similarity is that time-locked axons bifurcate into direct and indirect paths to provide inputs to time comparators—the time comparator neurons (small cells in *Eigenmannia* and *Brienomyrus* and ovoidal cells in *Gymnarchus*) receive inputs from the indirect and direct paths for time comparison. This pattern of projection is different from that in the mammalian auditory system in which the neurons in the medial superior olive, the first site of binaural convergence, receive direct excitatory inputs from the anteroventral cochlea nucleus for time comparison (Joris et al. 1998; Grothe 2000).

Perhaps the most remarkable difference among the groups of electric fishes is their anatomical sites of time comparison. Time comparison circuits of similar organization occur in the midbrain in *Eigenmannia* but in the hindbrain in *Gymnarchus* (Kawasaki 1996). Despite a much closer phylogenetic relationship (see Chapter 13), the time comparison circuit of *Gymnarchus* is in the hindbrain and that of *Brienomyrus* is in the midbrain. The commissural connection by time-locked giant cell axons occurs in the hindbrain in *Gymnarchus*, but *Brienomyrus* lacks commissural connections in the hindbrain for the knollenorgan time system; instead, *Brienomyrus*'s time-locked neurons in the hindbrain (nELL neurons) project bilaterally, allowing bilateral time comparison in the midbrain. The apparent lack of a homologous relationship between the time circuits in *Gymnarchus* and *Brienomyrus* invites further comparative studies of other pulse-type mormyrid electric fishes that exhibit a diverse range of behaviors and pulse durations.

3.4 Convergence of Amplitude and Time Information for the JAR

As described in Sections 3.2 and 3.3, amplitude and time information are sampled by different classes of electroreceptors and processed independently by separate channels in the brain in all electric fish species. The amplitude and time information converge, in higher brain centers to control behaviors that require combinatorial information about amplitude and time. The JAR of *Eigenmannia* and *Gymnarchus* requires such combination and is one of the best understood behaviors in terms of its underlying neuronal mechanisms. Heiligenberg (1991b) has provided a comprehensive review of the physiology of the tuberous system in *Eigenmannia* and Kawasaki (1993, 1996) has provided further comparative discussion of the JAR. These fishes emit EODs of an individually fixed constant frequency (250 to 500 Hz). When neighboring two fish have similar EOD frequencies (within approximately 1 to 8 Hz), they shift their

EOD frequencies away from each other to create a larger frequency difference in order to avoid mutual jamming of their electrolocation systems. In this behavior, a fish determines whether a neighbor's EOD frequency is higher or lower than its own by examining the temporal pattern of modulation occurring in the mixture of both EODs. Temporal modulation occurs both in the amplitude and time of zero crossing in the EOD mixture of the two fish. Neither amplitude nor zero crossing time is sufficient to unambiguously determine the jamming frequency of the neighbor, because the modulation frequency is determined by the *absolute* difference frequency between the two encountering fish. The time relationship between amplitude and time modulations is, however, different for a higher and a lower frequency neighbor. This has been illustrated using phase plots of time versus amplitude throughout the beat cycle in which positive and negative difference frequencies have been compared (Heiligenberg 1991a).

Behavioral experiments demonstrated that combined evaluation of amplitude and time modulation is essential for performing the JAR (Heiligenberg et al. 1978; Heiligenberg and Bastian 1980; Heiligenberg 1991a; Kawasaki 1993). Although many neurons in the ELL and the torus semicircularis in the midbrain are sensitive to either amplitude or time modulation exclusively, Rose and Heiligenberg (1985a, 1986a) discovered neurons in *Eigenmannia* that respond specifically to a temporal combination of the two in the torus. These neurons were found to project to various parts of the diencephalon and midbrain, where neurons again show sensitivity to a temporal combination of amplitude and time modulation (Kawasaki et al. 1988b; Keller 1988; Rose et al. 1988). These interface neurons exist in nuclei that are known to control the frequency of the pacemaker nucleus for JAR—the nucleus of electrosensorius and the prepacemaker nucleus (Kawasaki and Heiligenberg 1988; Kawasaki et al. 1988a; Keller et al. 1990; Heiligenberg et al. 1991; Keller et al. 1991; Metzner 1993).

The presence of midbrain neurons that selectively respond to temporal patterns of amplitude and time modulations in the gymnotiform *Sternopygus*, a species that does not show any JAR, suggests that combinatorial sensitivity is useful not only for JAR but must also be useful for other functions such as electrolocation (Rose et al. 1987).

A similar combination of parallel and hierarchical processing of amplitude and time information is found in *Gymnarchus*. Whereas amplitude- and time-sensitive neurons in the ELL do not discriminate between neighbors with higher and lower EOD frequencies (Kawasaki and Guo 1996, 1998), most neurons in the midbrain are sensitive to a combination of amplitude and time modulation (Kawasaki and Guo 2002; Carlson and Kawasaki 2004).

4. EOCD Gating of Afferent Signals in Mormyrid Fishes

Afferent signals arise in sensory organs not only by stimuli from the external world but also from stimulation by the animal's own motor acts. It is essential for sensory systems to distinguish between these two types of afferent signals for appropriate recognition and sensory guidance of behaviors. The need for physiological mechanisms that can make such a discrimination has long been suggested (Grüsser 1995). Centrally generated motor command signals that drive motor effectors such as electric organ or musculature may be copied and sent to sensory systems to modify sensory processing of afferent signals. This "copy signal" was termed *efference copy* by von Holst and *corollary discharge* by Sperry (Sperry 1950; von Holst and Mittelstaedt 1950). Interaction between these centrally generated signals and afferent signals is best understood in pulse-type mormyrid electric fishes. The afferent signal generated by the animal's own activity is termed *reafference*, and the signal generated by external stimuli is called *exafference*. How can the CNS distinguish between them? In physiological experiments on electric fishes, the cholinergic synaptic blocker curare silences the EOD by blocking synapses at the electric organ. In such EOD-silenced preparations, motor command signals to the electric organ—called command discharge in pulse-type fishes—still operate normally and can be recorded at the tail. Using the command discharges as a time reference, artificial electrosensory stimuli, in place of silenced EODs, can be delivered at natural times to mimic reafference or with a delay to mimic exafference. Using a curarized preparation of *Gnathonemus*, Bennett and Steinbach (1969) first recorded EOCDs in the brain of *Gnathonemus*. The EOCD was time locked to command discharges and occurred in the absence of any afferent signal. It was recorded in many areas throughout the brain including the ELL and the valvula cerebelli. They further demonstrated that afferent field potentials generated by nerve stimulation in a midbrain structure (the ELa) were blocked if the stimulation was given approximately 5 ms after an EOCD. The EOCD-related inhibition is very precise in time (lasting 2 to 3 ms) and complete (Russell and Bell 1978). Zipser and Bennett (1976b) recorded intracellularly from neurons in the nELL and showed the same type of time-locked inhibition by EOCD. In these experiments, afferent responses were of knollenorgan origin. The role of EOCD inhibition on knollenorgan-originated afferents is clearly to block reafferent signals caused by the fish's own EOD, but the short period of inhibition permits afferent signals caused by other fish to reach higher centers for communication (see Section 3.3.3).

In addition to the knollenorgan, mormyromast electroreceptors are stimulated by both the animal's own EOD and external stimuli (Bell and Russell 1978). Reafferent signals that are associated with the animal's own EODs should be useful in electrolocation, but afferent signals generated by another fish's EOD (exafference) may add only confusion. To elucidate how exafference and reafference are distinguished, Zipser and Bennett (1976a, b) showed that neurons in the mormyromast region of the ELL respond with subthreshold EPSPs to either EOCD or afferent stimulation. Spiking resulted when they coincide in time, indicating that these neurons act as an EOCD gate for reafference. The role of this gating in active electrolocation was suggested by behavioral experiments by Heiligenberg (1976). When artificial and external EOD pulses were given at the exact moment of the fish's own EOD, its electrolocation performance was disrupted, but noncoincidental pulses were ineffective. Behavioral experiments by Meyer and Bell (1983) showed that the sensory gate is activated immediately after an EOCD for a period of approximately 30 ms in *Gnathonemus* or for approximately 10 ms in *Pollimyrus*, a fish with a much narrower EOD pulse.

Interaction of the EOCD and afferent signals in the tuberous (medial) zone in the ELL has been further studied. EOCD-related field potential occurs in all layers in the ELL (for discussion of layers see Bell and Maler, Chapter 4). The potential in the granule cell layer is earliest and strongest, suggesting that EOCD reaches this layer first. Sensory evoked field potential is also large, suggesting that EOCD-afferent interaction occurs in this layer (Bell et al. 1992). By extracellular recording, Bell et al. (1992) found several neuron types in the medial zone that respond to EOCD alone and with a paired afferent stimulation. E cells responded to a sensory stimulus, and this response is strongly facilitated by the EOCD when the stimulus is given immediately after the EOCD. There is little or no response to the EOCD in the absence of sensory stimulation. I cells showed strong responses to EOCD alone, but responses were inhibited when sensory stimulation was paired. Responses to paired stimuli were highly plastic (see Bastian and Zakon, Chapter 8 for plastic responses). Recently, Mohr et al. (2003a) identified the most extensive list of cell types in the medial zone with intracellular labeling and recording. In this in vivo study, most neuron types showed nonlinear interaction between EOCD and sensory stimulation. Mohr et al. (2003b) also stimulated different relay stations for EOCD (see below) to distinguish the origins of EOCD-driven potentials in each cell type. Together with previous in vitro (Bell et al. 1997; Han et al. 1999; Sugawara et al. 1999; Han and Bell 2002) and anatomical works (Meek et al. 1999, 2001), the basic circuit for EOCD gating in the ELL is beginning to be revealed. These studies also significantly contribute to an understanding of plasticity in the ELL (see Bastian and Zakon, Chapter 8).

Elaborate EOCD pathways have been found in *Gnathonemus* (Bell et al. 1981, 1983, 1995; Bell and von der Emde 1995) (see Fig. 4.10 in Chapter 4). EOCDs originate from the command nucleus in the medulla, which determines the precise moment of an EOD (see Macadar et al., Chapter 14 for the EOD motor system). Neurons in the command nucleus, while sending their axons to the medial relay nucleus for EODs, send their collaterals to the bulbar command-associated nucleus (BCA) for EOCDs. BCA in turn bilaterally project to two nuclei, the mesencephalic command-associated nucleus (MCA) and the paratrigeminal command-associated nucleus (PCA). Field potentials recorded in these nuclei are precisely time locked to the command signal. The PCA projects to the outer two thirds of the molecular layer of the ELL via EGp. The EOCD mediated carried by this PCA pathway provides the ELL with EOD times for gating information with strong plasticity (Bell et al. 1995) (see Bastian and Zakon, Chapter 8). The physiological role of the MCA was examined by recording, stimulation, and lesion experiments (Bell et al. 1995). Lesioning the MCA eliminates a large part of field potential in various layers in the ELL. Before reaching the ELL, EOCD signals were passed from the MCA to a structure near the lateral lemniscus. Neurons in this juxtalemniscal region project to the nELL, where inhibition of knollenorgan afferents takes place (Mugnaini and Maler 1987a,b; Bell and Grant 1989). The juxtalemniscal region also projects to an anteroventral margin of the ELL, the medial juxtalobar nucleus. Recording, stimulation, and lesion experiments by Bell and von der Emde (1995) suggested that the EOCD via the medial juxtalobar nucleus excites granule cells in the ELL for gating electrosensory information. High temporal precision of the juxtalobar EOCD neurons (approximately 50 μs) suggested that EOCD may be used in the ELL to decode amplitude-dependent latency shift of afferent spikes for amplitude coding. As mentioned in Section 2.2.1, some tuberous afferent fibers show amplitude-dependent latency shifts. The use of time code for amplitude coding has been indicated by a behavioral study (Hall et al. 1995).

The ampullary region of the ELL is also affected by EOCDs and shows the same plasticity as in the mormyromast region. Bell (1981) showed that in neurons in the ampullary region of the ELL, pairing of the EOCD and a low-frequency signal for several minutes resulted in adaptation to the stimulus, and a reverse polarity response emerged after the paired stimulation was removed. See Chapters 4 and 8 for further discussion of plasticity.

Gymnarchus, the only wave-discharging species in Africa, is the only mormyriform without an EOCD system (Kawasaki 1993, 1994). The computational rules used for JAR of *Gymnarchus* do not take advantage of the timing pulse from the EOCD that is used by the mormyrids. Instead, the computational rules seem to parallel those of a wave gymnotiform, *Eigenmannia*, and the anatomy strongly suggests independent and parallel evolution to solve a similar problem.

5. Summary

Tuberous electrosensory system has two major behavioral functions, electrolocation and electrocommunication, which are unique to gymnotiform and mormyriform electric fishes that emit EODs with high-frequency components. This chapter presented an abundance of information about how peripheral and central physiological mechanisms carry out these behavioral functions. Electrolocation involves identification of stimulus features such as location (direction and distance), size, and electrical properties (resistance and capacitance) of objects. These features are detected and analyzed by two separate tuberous pathways specialized for amplitude and time processing. The parallel processing of amplitude and time in the hindbrain converges in the midbrain to produce neurons with more complex response properties. Electrocommunication for sex recognition in pulse-type mormyriforms relies on time-coding of waveform (pulse duration) of their EODs. Pulse duration is sampled by a type of time-coding electroreceptors (knollenorgan) that are specialized for communication and is decoded in the midbrain.

Understanding of the tuberous electrosensory system of electric fishes has been greatly advanced as a result of the interdisciplinary approach involving ethological, anatomical, physiological, and computational methods. Future studies with these methods should continue to provide complementary data for better understanding of behavioral mechanisms.

Acknowledgments. The author thanks Drs. Curtis Bell, Theodore Bullock, Carl Hopkins, and Art Popper for critically reading the manuscript.

References

Amagai S (1998) Time coding in the midbrain of mormyrid electric fish. II. Stimulus selectivity in the nucleus exterolateralis pars posterior. J Comp Physiol A 182:131–143.

Amagai S, Friedman MA, Hopkins CD (1998) Time coding in the midbrain of mormyrid electric fish. I. Physiology and anatomy of cells in the nucleus exterolateralis pars anterior. J Comp Physiol 182:115–130.

Art JJ, Fettiplace R, Wu YC (1993) The effects of low calcium on the voltage-dependent conductances involved in tuning of turtle hair cells. J Physiol 470:109–126.

Bass AH, Hopkins CD (1982) Comparative aspects of brain organization of an African "wave" electric fish, *Gymnarchus niloticus*. J Morphol 174:313–334.

Bass AH, Hopkins CD (1984) Shifts in frequency tuning of electroreceptors in androgen-treated mormyrid fish. J Comp Physiol 155:713–724.

Bastian J (1976) Frequency response characteristics of electroreceptors in weakly electric fish (Gymnotoidei) with a pulse discharge. J Comp Physiol 112:165–180.

Bastian J (1977) Variations in the frequency response of electroreceptors dependent on receptor location in weakly electric fish (Gymnotoidei) with a pulse discharge. J Comp Physiol 121:53–64.

Bastian J (1981a) Electrolocation. I. How the electroreceptors of *Apteronotus albifrons* code for moving objects and other electrical stimuli. J Comp Physiol 144:465–479.

Bastian J (1981b) Electrolocation. II. The effects of moving objects and other electrical stimuli on the activities of two categories of posterior lateral line lobe cells in *Apteronotus albifrons*. J Comp Physiol 144:481–494.

Bastian J (1986a) Electrolocation. Behavior, anatomy, and physiology. In: Bullock TH, Heiligenberg W (eds), Electroreception. New York: John Wiley & Sons, pp. 577–612.

Bastian J (1986b) Gain control in the electrosensory system: a role for the descending projections to the electrosensory lateral line lobe. J Comp Physiol 158:505–515.

Bastian J (1986c) Gain control in the electrosensory system mediated by descending inputs to the electrosensory lateral line lobe. J Neurosci 6:553–562.

Bastian J (1996a) Plasticity in an electrosensory system. I. General features of a dynamic sensory filter. J Neurophysiol 76:2483–2496.

Bastian J (1996b) Plasticity in an electrosensory system. II. Postsynaptic events associated with a dynamic sensory filter. J Neurophysiol 76:2497–2507.

Bastian J (1998) Plasticity in an electrosensory system. III. Contrasting properties of spatially segregated dendritic inputs. J Neurophysiol 79:1839–1857.

Bastian J, Bratton B (1990) Descending control of electroreception. I. Properties of nucleus praeeminentialis neurons projecting indirectly to the electrosensory lateral line lobe. J Neurosci 10:1226–1240.

Bastian J, Heiligenberg W (1980a) Neural correlates of the jamming avoidance response in *Eigenmannia*. J Comp Physiol 136:135–152.

Bastian J, Heiligenberg W (1980b) Phase-sensitive midbrain neurons in *Eigenmannia*: neural correlates of the jamming avoidance response. Science 209:828–831.

Bastian J, Nguyenkim J (2001) Dendritic modulation of burst-like firing in sensory neurons. J Neurophysiol 85:10–22.

Bastian J, Chacron MJ, Maler L (2002) Receptive field organization determines pyramidal cell stimulus-encoding capability and spatial stimulus selectivity. J Neurosci 22:4577–4590.

Bell CC (1981) An efference copy which is modified by reafferent input. Science 214: 450–452.

Bell CC (1990) Mormyromast electroreceptor organs and their afferent fibers in mormyrid fish. III. Physiological differences between two morphological types of fibers. J Neurophysiol 63:319–332.

Bell CC, Grant K (1989) Corollary discharge inhibition and preservation of temporal information in a sensory nucleus of mormyrid electric fish. J Neurosci 9:1029–1044.

Bell CC, Russell CJ (1978) Effect of electric organ discharge on ampullary receptors in a mormyrid. Brain Res 145:85–96.

Bell CC, von der Emde G (1995) Electric organ corollary discharge pathways in mormyrid fish: II. The medial juxtalobar nucleus. J Comp Physiol 177:463–479.

Bell CC, Finger TE, Russell CJ (1981) Central connections of the posterior lateral line lobe in mormyrid fish. Exp Brain Res 42:9–22.

Bell CC, Libouban S, Szabo T (1983) Pathways of the electric organ discharge command and its corollary discharges in mormyrid fish. J Comp Neurol 216:327–338.

Bell CC, Zakon H, Finger TE (1989) Mormyromast electroreceptor organs and their afferent fibers in mormyrid fish: I. Morphology. J Comp Neurol 286:391–407.

Bell CC, Grant K, Serrier J (1992) Sensory processing and corollary discharge effects in the mormyromast regions of the mormyrid electrosensory lobe. I. Field potentials, cellular activity in associated structures. J Neurophysiol 68:843–858.

Bell CC, Dunn K, Hall C, Capti A (1995) Electric organ corollary discharge pathways in mormyrid fish. I. The mesencephalic command associated nucleus. J Comp Physiol 177:449–462.

Bell CC, Han VZ, Sugawara Y, Grant K (1997) Synaptic plasticity in a cerebellum-like structure depends on temporal order. Nature 387:278–281.

Bennett MVL (1965) Electroreceptors in mormyrids. Cold Spring Harbor Symp Quant Biol 30:245–262.

Bennett MVL (1967) Mechanisms of electroreception. In Cahn P (ed), Lateral Line Detectors. Bloomington: Indiana University Press, pp. 313–393.

Bennett MVL (1971) Electroreception. In Hoar WS, Randall DJ (eds), Fish Physiology, Vol. 5: Sensory Systems and Electric Organs. New York: Academic Press, pp. 493–574.

Bennett MVL, Steinbach AB (1969) Influence of electric organ control system of electrosensory afferent pathways in mormyrids. In: Llinas R (ed), Neurobiology of Cerebellar Evolution and Development. Chicago: American Medical Association, pp. 207–214.

Berman NJ, Maler L (1998a) Inhibition evoked from primary afferents in the electrosensory lateral line lobe of the weakly electric fish (*Apteronotus leptorhynchus*). J Neurophysiol 80:3173–3196.

Berman NJ, Maler L (1998b) Interaction of GABAB-mediated inhibition with voltage-gated currents of pyramidal cells: computational mechanism of a sensory searchlight. J Neurophysiol 80:3197–3213.

Berman NJ, Maler L (1998c) Distal versus proximal inhibitory shaping of feedback excitation in the electrosensory lateral line lobe: implications for sensory filtering. J Neurophysiol 80:3214–3232.

Berman NJ, Maler L (1999) Neural architecture of the electrosensory lateral line lobe: adaptations for coincidence detection, a sensory searchlight and frequency-dependent adaptive filtering. J Exp Biol 202:1243–1253.

Berman NJ, Hincke MT, Maler L (1995) Inositol 1,4,5-trisphosphate receptor localization in the brain of a weakly electric fish (*Apteronotus leptorhynchus*) with emphasis on the electrosensory system. J Comp Neurol 361:512–524.

Berman NJ, Plant J, Turner RW, Maler L (1997) Excitatory amino acid receptors at a feedback pathway in the electrosensory system: implications for the searchlight hypothesis. J Neurophysiol 78:1869–1881.

Berman NJ, Dunn RJ, Maler L (2001) Function of NMDA receptors and persistent sodium channels in a feedback pathway of the electrosensory system. J Neurophysiol 86:1612–1621.

Bratton B, Bastian J (1990) Descending control of electroreception. II. Properties of nucleus praeeminentialis neurons projecting directly to the electrosensory lateral line lobe. J Neurosci 10:1241–1253.

Bullock TH, Chichibu S (1965) Further analysis of sensory coding in electroreceptors of electric fish. Proc Natl Acad Sci USA 54:422–429.

Bullock TH, Hagiwara S, Kusano K, Negishi K (1961) Evidence for a category of electroreceptors in the lateral line of gymnotid fishes. Science 134:1426–1427.

Bullock TH, Behrend K, Heiligenberg W (1975) Comparison of the jamming avoidance responses in Gymnotoid and Gymnarchid electric fish: a case of convergent evolution of behavior and its sensory basis. J Comp Physiol 103:97–121.

Carlson BA, Kawasaki M (2004) Nonlinear response properties of combination-sensitive electrosensory neurons in the midbrain of *Gymnarchus niloticus*. J Neurosci 24(37): 8039–8048.

Carr C, Maler L (1986) Electroreception in gymnotiform fish. Central anatomy and physiology. Bullock TH, Heiligenberg W. (eds), Electroreception. New York: John Wiley & Sons

Carr CE (1993) Processing of temporal information in the brain. Annu Rev Neurosci 16:223–243.

Carr CE, Friedman MA (1999) Evolution of time coding systems. Neural Comput 11: 1–20.

Carr CE, Maler L, Heiligenberg W, Sas E (1981) Laminar organization of the afferent and efferent systems of the torus semicircularis of Gymnotiform fish: morphological substrates for parallel processing in the electrosensory system. J Comp Neurol 203: 649–670.

Carr CE, Maler L, Sas E (1982) Peripheral organization and central projections of the electrosensory nerves in gymnotiform fish. J Comp Neurol 211:139–153.

Carr CE, Heiligenberg W, Rose GJ (1986a) A time-comparison circuit in the electric fish midbrain. I. Behavior and physiology. J Neurosci 6:107–119.

Carr CE, Maler L, Taylor B (1986b) A time-comparison circuit in the electric fish midbrain. II. Functional morphology. J Neurosci 6:1372–1383.

Carr CE, Soares D, Parameshwaran S, Perney T (2001) Evolution and development of time coding systems. Curr Opin Neurobiol 11:727–733.

Castello ME, Caputi A, Trujillo-Cenoz O (1998) Structural and functional aspects of the fast electrosensory pathway in the electrosensory lateral line lobe of the pulse fish *Gymnotus carapo*. J Comp Neurol 401:549–563.

Chacron MJ, Doiron B, Maler L, Longtin A, Bastian J (2003) Non-classical receptive field mediates switch in a sensory neuron's frequency tuning. Nature 423:77–81.

Crawford AC, Fettiplace R (1981) An electrical tuning mechanism in turtle cochlear hair cells. J Physiol 312:377–422.

Crick F (1984) Function of the thalamic reticular complex:the searchlight hypothesis. Proc Natl Acad Sci USA 81:4586–4590.

Doiron B, Longtin A, Berman N, Maler L (2001) Subtractive and divisive inhibition: effect of voltage-dependent inhibitory conductances and noise. Neural Comput 13: 227–248.

Doiron B, Chacron MJ, Maler L, Longtin A, Bastian J (2003) Inhibitory feedback required for network oscillatory responses to communication but not prey stimuli. Nature 421:539–543.

Enger P, Szabo T (1965) Activity of central neurons involved in electroreception in some weakly electric fish (Gymnotidae). J Neurophysiol 28:800–818.

Fessard A, Szabo T (1961) Mise évidence d'un récepteur sensible à l'électricité dans la peau d'un mormyre. C. rendu hebd. Séanc Acad Sci Paris 253:1859–1860.

Fortune ES, Rose GJ (1996) Temporal filtering properties of midbrain neurons in an electric fish: implications for the function of dendritic spines. Neurosci Res 26:89–94.

Fortune ES, Rose GJ (1997a) Temporal filtering properties of ampullary electrosensory neurons in the torus semicircularis of *Eigenmannia*: evolutionary and computational implications. Brain Behav Evol 49:312–323.

Fortune ES, Rose GJ (1997b) Passive and active membrane properties contribute to the temporal filtering properties of midbrain neurons in vivo. J Neurosci 17:3815–3825.

Fortune ES, Rose GJ (2000) Short-term synaptic plasticity contributes to the temporal filtering of electrosensory information. J Neurosci 20:7122–7130.

Friedman MA, Hopkins CD (1998) Neural substrates for species recognition in the time-coding electrosensory pathway of mormyrid electric fish. J Neurosci 18:1171–1185.

Gabbiani F, Metzner W (1999) Encoding and processing of sensory information in neuronal spike trains. J Exp Biol 202:1267–1279.

Gabbiani F, Metzner W, Wessel R, Koch C (1996) From stimulus encoding to feature extraction in weakly electric fish. Nature 384:564–567.

Gray CM, Konig P, Engel AK, Singer W (1989) Oscillatory responses in cat visual cortex exhibit inter-columnar synchronization which reflects global stimulus properties. Nature 338:334–337.

Grothe B (2000) The evolution of temporal processing in the medial superior olive, an auditory brainstem structure. Prog Neurobiol 61:581–610.

Grüsser OJ (1995) On the history of the ideas of efference copy and reafference. Clio Med 33:35–55.

Guo Y-X, Kawasaki M (1997) Representation of accurate temporal information in the electrosensory system of the African electric fish, *Gymnarchus niloticus*. J Neurosci 17:1761–1768.

Hagiwara S, Morita H (1963) Coding mechanisms of electroreceptor fibers in some electric fish. J Neurophysiol 26:551–567.

Hall C, Bell CC, Zelick R (1995) Behavioral evidence of a latency code for simulus intensity in mormyrid electric fish. J Comp Physiol 177:29–39.

Han VZ, Bell CC (2002) Cell morphology and local circuitry of the central lobes of the mormyrid cerebellum. Ann NY Acad Sci 978:515–516.

Han VZ, Bell CC, Grant K, Sugawara Y (1999) Mormyrid electrosensory lobe in vitro: morphology of cells and circuits. J Comp Neurol 404:359–374.

Harder W (1968) Die beziehungen zwischen elektrorezeptoren, elektrischem organ, seitenlinienorganen und nervensystem bei den Mormyridae (Teleostei, Pisces). Z Vergl Physiol 59:272–318.

Heiligenberg W (1976) Electrolocation and jamming avoidance in the mormyrid fish *Brienomyrus*. J Comp Physiol 109:357–372.

Heiligenberg W (1977) Principles of Electrolocation and Jamming Avoidance in Electric Fish—A Neuroethological Approach. Berlin: Springer-Verlag.

Heiligenberg W (1991a) Neural nets in electric fish. Cambridge, MA: MIT Press.

Heiligenberg W (1991b) The neural basis of behavior: a neuroethological view. Annu Rev Neurosci 14:247–267.

Heiligenberg W, Altes RA (1978) Phase sensitivity in electroreception. Science 199: 1001–1004.

Heiligenberg W, Bastian J (1980) The control of *Eigenmannia*'s pacemaker by distributed evaluation of electroreceptive afferences. J Comp Physiol 136:113–133.

Heiligenberg W, Dye J (1982) Labeling of electroreceptive afferents in gymnotid fish by intracellular injection of HRP: the mystery of multiple maps. J Comp Physiol 148: 287–296.

Heiligenberg W, Baker C, Matsubara J (1978) The jamming avoidance response in *Eigenmannia* revisited: the structure of a neuronal democracy. J Comp Physiol 127:267–286.

Heiligenberg W, Rose GJ (1985) Phase and amplitude computations in the midbrain of an electric fish: intracellular studies of neurons participating in the jamming avoidance response of *Eigenmannia*. J Neurosci 5:515–531.

Heiligenberg W, Keller CH, Metzner W, Kawasaki M (1991) Structure and function of neurons in the complex of the nucleus electrosensorius of the gymnotiform fish *Eigenmannia*: detection and processing of electric signals in social communication. J Comp Physiol 169:151–164.

Hopkins CD (1974) Electric communication in fish. Am Sci 62:426–437.

Hopkins CD (1976) Stimulus filtering and electroreception: tuberous electroreceptors in three species of gymnotoid fish. J Comp Physiol 111:171–207.

Hopkins CD (1981) On the diversity of electric signals in a community of mormyrid electric fish in West Africa. Am Zool 21:211–222.

Hopkins CD (1986a) Temporal structure of non-propagated electric communication signals. Brain Behav Evol 28:43–59.

Hopkins CD (1986b) Behavior of mormyridae. In: Bullock TH, Heiligenberg W (eds), Electroreception. New York: John Wiley & Sons, pp. 527–576.

Hopkins CD (1995) Convergent designs for electrogenesis and electroreception. Curr Opin Neurobiol 5:769–777.

Hopkins CD, Bass AH (1981) Temporal coding of species recognition signals in an electric fish. Science 212:85–87.

Irvine DR (1992) Physiology of auditory brainstem. In: Popper AN, Fay RR (eds), The Mammalian Auditory Pathway: Neurophysiology. New York: Springer-Verlag, pp. 153–231.

Joris PX, Smith PH, Yin TC (1998) Coincidence detection in the auditory system: 50 years after Jeffress. Neuron 21:1235–1238.

Kashimori Y, Inoue S, Kambara T (2001) A neural mechanism of hyperaccurate detection of phase advance and delay in the jamming avoidance response of weakly electric fish. Biol Cyber 85:117–131.

Kawasaki M (1993) Independently evolved jamming avoidance responses employ identical computational algorithms: a behavioral study of the African electric fish, *Gymnarchus niloticus*. Comp Physiol A 173:9–22.

Kawasaki M (1994) The African wave-type electric fish, *Gymnarchus niloticus*, lacks corollary discharge mechanisms for electrosensory gating. J Comp Physiol A 174: 133–144.

Kawasaki M (1996) Comparative analysis of the jamming avoidance response in African and South American wave-type fishes. Biol Bull 191:103–108.

Kawasaki M (1997) Sensory hyperacuity in the jamming avoidance response of weakly electric fish. Curr Opin Neurobiol 7:473–479.

Kawasaki M (2001) Cutaneous electrical oscillation in a weakly electric fish, *Gymnarchus niloticus*. J Comp Physiol A 187:597–604.

Kawasaki M, Guo Y-X (1996) Neuronal circuitry for comparison of timing in the electrosensory lateral line lobe of the African wave-type electric fish *Gymnarchus niloticus*. J Neurosci 16:380–391.

Kawasaki M, Guo Y-X (1998) Parallel projection of amplitude and phase information from the hindbrain to the midbrain of the African electric fish *Gymnarchus niloticus*. J Neurosci 18:7599–7611.

Kawasaki M, Guo Y-X (2002) Emergence of temporal-pattern sensitive neurons in the midbrain of weakly electric fish *Gymnarchus niloticus*. J Physiol Paris 96:531–537.

Kawasaki M, Heiligenberg W (1988) Individual prepacemaker neurons can modulate the pacemaker cycle of the gymnotiform electric fish, *Eigenmannia*. J Comp Physiol 162: 13–21.

Kawasaki M, Maler L, Rose GJ, Heiligenberg W (1988a) Anatomical and functional organization of the prepacemaker nucleus in gymnotiform electric fish: the accommodation of two behaviors in one nucleus. J Comp Neurol 276:113–131.

Kawasaki M, Rose GJ, Heiligenberg W (1988b) Temporal hyperacuity in single neurons of electric fish. Nature 336:173–176.

Keller CH (1988) Stimulus discrimination in the diencephalon of *Eigenmannia*: the emergence and sharpening of a sensory filter. J Comp Physiol 162:747–757.

Keller CH, Zakon HH, Sanchez DY (1986) Evidence for a direct effect of androgens upon electroreceptor tuning. J Comp Physiol 158:301–310.

Keller CH, Maler L, Heiligenberg W (1990) Structural and functional organization of a diencephalic sensory-motor interface in the gymnotiform fish, *Eigenmannia*. J Comp Neurol 293:347–376.

Keller CH, Kawasaki M, Heiligenberg W (1991) The control of pacemaker modulations for social communication in the weakly electric fish *Sternopygus*. J Comp Physiol 169:441–450.

Koch C, Poggio T (1983) A theoretical analysis of electrical properties of spines. Proc R Soc Lond B Biol Sci 218:455–477.

Laurent G (1996) Dynamical representation of odors by oscillating and evolving neural assemblies. Trends Neurosci 19:489–496.

Lemon N, Turner RW (2000) Conditional spike backpropagation generates burst discharge in a sensory neuron. J Neurophysiol 84:1519–1530.

Lissmann HW (1958) On the function and evolution of electric organs in fish. J Exp Biol 35:156–191.

Lissmann HW, Machin KE (1958) The mechanism of object location in *Gymnarchus niloticus* and similar fish. J Exp Biol 35:451–486.

Maler L, Sas E, Rogers J (1981) The cytology of the posterior lateral line lobe of high frequency weakly electric fish (Gymnotidae). Dendritic differentiation and synaptic specificity in a simple cortex. J Comp Neurol 195:87–139.

Mathieson WB, Maler L (1988) Morphological and electrophysiological properties of a novel in vitro preparation: the electrosensory lateral line lobe brain slice. J Comp Physiol 163:489–506.

Matsushita A, Kawasaki M (2004) Unitary giant synapses embracing a single neuron at the convergent site of time-coding pathways of an electric fish, *Gymnarchus niloticus*. J Comp Neurol 472(2):140–155.

Meek J, Grant K, Bell CC (1999) Structural organization of the mormyrid electrosensory lateral line lobe. J Exp Biol 202:1291–1300.

Meek J, Hafmans TG, Han V, Bell CC, Grant K (2001) Myelinated dendrites in the mormyrid electrosensory lobe. J Comp Neurol 431:255–275.

Metzner W (1993) The jamming avoidance response in *Eigenmannia* is controlled by two separate motor pathways. J Neurosci 13:1862–1878.

Metzner W, Heiligenberg W (1991) The coding of signals in the electric communication of the gymnotiform fish *Eigenmannia*: from electroreceptors to neurons in the torus semicircularis of the midbrain. J Comp Physiol 169:135–150.

Metzner W, Juranek J (1997a) A sensory brain map for each behavior? Proc Natl Acad Sci USA 94(26):14798–14803.

Metzner W, Juranek J (1997b) A method to biotinylate and histochemically visualize ibotenic acid for pharmacological inactivation studies. J Neurosci Methods 76(2):143–150.

Metzner W, Koch C, Wessel R, Gabbiani F (1998) Feature extraction by burst-like spike patterns in multiple sensory maps. J Neurosci 18:2283–2300.

Meyer J (1982) Behavioral response of weakly electric fish to complex impedances. J Comp Physiol 145:459–470.

Meyer JH, Bell CC (1983) Sensory gating by a corollary discharge mechanism. J Comp Physiol 151:401–406.

Mohr C, Roberts PD, Bell CC (2003a) Cells of the mormyromast region of the mormyrid electrosensory lobe: I. Responses to the electric organ corollary disharge and to electrosensory stimuli. J Neurophysiol 90:1193–1210.

Mohr C, Roberts PD, Bell CC (2003b) Cells of the mormyromast region of the mormyrid electrosensory lobe: II. Responsese to input from central sources. J Neurophysiol 90: 1211–1223.

Mugnaini E, Maler L (1987a) Cytology and immunocytochemistry of the nucleus exterolateralis anterior of the mormyrid brain: possible role of GABAergic synapses in temporal analysis. Anat Embryol (Berl) 176:313–336.

Mugnaini E, Maler L (1987b) Cytology and immunocytochemistry of the nucleus of the lateral line lobe in the electric fish *Gnathonemus petersii* (Mormyridae): evidence suggesting that GABAergic synapses mediate an inhibitory corollary discharge. Synapse 1:32–56.

Murthy VN, Fetz EE (1996) Synchronization of neurons during local field potential oscillations in sensorimotor cortex of awake monkeys. J Neurophysiol 76:3968–3982.

Naruse M, Kawasaki M (1998) Possible involvement of the ampullary electroreceptor system in detection of frequency-modulated electrocommunication signals in Eigenmannia. J Comp Physiol 183:543–552.

Poor H (1994) An introduction to signal detection and estimation. New York: Springer-Verlag.

Ramanathan K, Michael TH, Jiang GJ, Hiel H, Fuchs PA (1999) A molecular mechanism for electrical tuning of cochlear hair cells. Science 283:215–217.

Réthelyi M, Szabo T (1973) A particular nucleus in the mesencephalon of weakly electric fish, *Gymnotus carapo* (Gymnotidae, Pisces). Exp Brain Res 17:229–241.

Rose GJ, Call SJ (1992) Evidence for the role of dendritic spines in the temporal filtering properties of neurons: the decoding problem and beyond. Proc Natl Acad Sci USA 89:9662–9665.

Rose GJ, Call SJ (1993) Temporal filtering properties of midbrain neurons in an electric fish: implications for the function of dendritic spines. J Neurosci 13:1178–1189.

Rose GJ, Fortune ES (1996) New techniques for making whole-cell recordings from CNS neurons in vivo. Neurosci Res 26:89–94.

Rose GJ, Fortune ES (1999) Frequency-dependent PSP depression contributes to low-pass temporal filtering in *Eigenmannia*. J Neurosci 19:7629–7639.

Rose GJ, Heiligenberg W (1985a) Structure and function of electrosensory neurons in the torus semicircularis of *Eigenmannia*: morphological correlates of phase and amplitude sensitivity. J Neurosci 5:2269–2280.

Rose GJ, Heiligenberg W (1985b) Temporal hyperacuity in the electric sense of fish. Nature 318:178–180.

Rose GJ, Heiligenberg W (1986a) Neural coding of difference frequencies in the midbrain of the electric fish *Eigenmannia*: reading the sense of rotation in an amplitude-phase plane. J Comp Physiol 158:613–624.

Rose GJ, Heiligenberg W (1986b) Limits of phase and amplitude sensitivity in the torus semicircularis of *Eigenmannia*. J Comp Physiol 159:813–822.

Rose GJ, Keller CH, Heiligenberg W (1987) "Ancestral" neural mechanisms of electrolocation suggest a substrate for the evolution of the jamming avoidance response. J Comp Physiol 160:491–500.

Rose GJ, Kawasaki M, Heiligenberg W (1988) "Recognition units" at the top of a neuronal hierarchy? Prepacemaker neurons in *Eigenmannia* code the sign of frequency differences unambiguously. J Comp Physiol 162:759–772.

Russell CJ, Bell CC (1978) Neuronal responses to electrosensory input in mormyrid valvula cerebelli. J Neurophysiol 41:1495–1510.

Sas E, Maler L (1983) The nucleus praeeminentials: a Golgi study of a feedback center in the electrosensory system of gymnotiform fish. J Comp Neurol 221:127–144.

Saunders J, Bastian J (1984) The physiology and morphology of two types of electrosensory neurons in the weakly electric fish *Apteronotus leptorhynchus*. J Comp Physiol 154:199–209.

Scheich H, Bullock TH (1974) The detection of electric fields from electric organs. In: Fessard A (ed), Handbook of Sensory Physiology, Vol. III/3: Electroreceptors and Other Specialized Receptors in Lower Vertebrates, Berlin: Springer-Verlag, pp. 201–256.

Scheich H, Bullock TH, Hamstra RH (1973) Coding properties of two classes of afferent nerve fibers: high-frequency electroreceptors in the electric fish, *Eigenmannia*. J Neurophysiol 36:39–60.

Shuai J, Kashimori Y, Kambara T (1998) Electroreceptor model of the weakly electric fish *Gnathonemus petersii*. I. The model and the origin of differences between A- and B-receptors. Biophys J 75:1712–1726.

Shuai J, Kashimori Y, Hoshino O, Kambara T, von der Emde G (1999) Electroreceptor model of weakly electric fish *Gnathonemus petersii*: II. Cellular origin of inverse waveform tuning. Biophys J 76:3012–3025.

Shumway CA (1989a) Multiple electrosensory maps in the medulla of weakly electric gymnotiform fish. I. Physiological differences. J Neurosci 9:4388–4399.

Shumway CA (1989b) Multiple electrosensory maps in the medulla of weakly electric gymnotiform fish. II. Anatomical differences. J Neurosci 9:4400–4415.

Shumway CA, Maler L (1989) GABAergic inhibition shapes temporal and spatial response properties of pyramidal cells in the electrosensory lateral line lobe of gymnotiform fish. J Comp Physiol 164:391–407.

Simmons JA, Saillant PA, Ferragamo MJ, Haresign T, Dear SP, Fritz J, McMullen TA (1995) Auditory computations for biosonar target imaging in bats. In: Hawkins HL, McMullen TA, Popper AN, Fay RR (eds), Auditory Computation. New York: Springer-Verlag, pp. 401–468.

Smotherman MS, Narins PM (1999) The electrical properties of auditory hair cells in the frog amphibian papilla. J Neurosci 19:5275–5292.

Sperry R (1950) Neural basis of the spontaneous optokinetic response produced by visual inversion. J Comp Phyiol Psychol 43:482–489.

Sugawara Y, Grant K, Han V, Bell CC (1999) Physiology of electrosensory lateral line lobe neurons in *Gnathonemus petersii*. J Exp Biol 202:1301–1309.

Szabo T (1965) Sense organs of the lateral line system in some electric fish of the Gymnotidae, Mormyridae and Gymnarchidae. J Morphol 117:229–250.

Szabo T (1967) Activity of peripheral and central neurones involved in electroreception In: Cahn PH (ed), Lateral Line Detectors. Bloomington, IN: Indiana University Press, pp. 295–311.

Szabo T (1974) Anatomy of the specialized lateral line organs of electroreception, In: Fessard A (ed), Handbook of Sensory Physiology, Vol. III/3: Electroreceptors and Other Specialized Receptors in Lower Vertebrates. Berlin: Springer-Verlag, pp. 13–58.

Szabo T, Fessard A (1974) Physiology of electroreceptors. In: Fessard A (ed), Handbook of Sensory Physiology, Vol. III/3: Electroreceptors and Other Specialized Receptors in Lower Vertebrates. Berlin: Springer-Verlag, pp. 59–124.

Szabo T, Hagiwara S (1967) A latency change mechanisms involved in sensory coding of electric fish (mormyrids). Physiol Behav 2:331–335.

Takagi H, Kawasaki M (2003) Modeling of time disparity detection by the Hodgkin-Huxley equations. J Comp Physiol A Neuroethol Sens Neural Behav Physiol 189: 257–262.

Takizawa Y, Rose GJ, Kawasaki M (1999) Resolving competing theories for control of the jamming avoidance response: the role of amplitude modulations in electric organ discharge decelerations. J Exp Biol 202:1377–1386.

Trussell LO (1999) Synaptic mechanisms for coding timing in auditory neurons. Annu Rev Physiol 61:477–496.

Turner RW, Maler L, Deerinck T, Levinson SR, Ellisman MH (1994) TTX-sensitive dendritic sodium channels underlie oscillatory discharge in a vertebrate sensory neuron. J Neurosci 14:6453–6471.

Turner RW, Plant JR, Maler L (1996) Oscillatory and burst discharge across electrosensory topographic maps. J Neurophysiol 76:2364–2382.

von der Emde G (1992) Electrolocation of capacitive objects in four species of pulse-type weakly electric fish. II. Electric signalling behaviour. Ethology 92:177–192.

von der Emde G (1993) Capacitance discrimination in electrolocating weakly electric pulse-fish. Naturwiss 80:231–233.

von der Emde G (1997) Electroreception. In: Evans D (ed), The Physiology of Fishes Boca Raton: CRC Press, pp. 313–343.

von der Emde G (1998) Capacitance detection in the wave-type electric fish *Eigenmannia* during active electrolocation. J Comp Physiol 182:217–224.

von der Emde G, Bell CC (1994) Responses of cells in the mormyrid electrosensory lobe to EODs with distorted wave-forms: implications for capacitance detection. J Comp Physiol 175:83–93.

von der Emde G, Bell CC (1996) The nucleus preeminentialis of mormyrid fish, a center for recurrent electrosensory feedback: I. Electrosensory and corollary discharge responses. J Neurophysiol 76:1581–1596.

von der Emde G, Bleckmann H (1997) Waveform tuning of electroreceptor cells in the weakly electric fish, *Gnathonemus petersii.* J Comp Physiol 181:511–524.

von der Emde G, Ringer T (1992) Electrolocation of capacitive objects in four species of pulse-type weakly electric fish. I. Discrimination performance. Ethology 91:326–338.

von Holst E, Mittelstaedt H (1950) Das Reafferenz-Prinzip. Naturwiss 37:464–476.

Wessel W, Koch C, Gabbiani F (1996) Coding of time-varying electric field amplitude modulations in a wave-type electric fish. J Neurophysiol 75:2280–2293.

Wu YC, Art JJ, Goodman MB, Fettiplace R (1995) A kinetic description of the calcium-activated potassium channel and its application to electrical tuning of hair cells. Prog Biophys Mol Biol 63:131–158.

Zakon H (1986) The electroreceptive periphery. In: Bullock TH (ed), Electroreception. New York: John Wiley & Sons, pp. 103–156.

Zakon HH, Meyer JH (1983) Plasticity of electroreceptor tuning in the weakly electric fish, *Sternopygus dariensis.* J Comp Physiol 153:477–487.

Zakon HH, Yan HY, Thomas P (1990) Human chorionic gonadotropin-induced shifts in the electrosensory system of the weakly electric fish, *Sternopygus.* J Neurobiol 21: 826–833.

Zipser B, Bennett MVL (1976a) Responses of cells of the posterior lateral line lobe to activation of electroreceptors in a mormyrid fish. J Neurophysiol 39:693–712.

Zipser B, Bennett MVL (1976b) Interaction of electrosensory and electromotor signals in lateral line lobe of a mormyrid fish. J Neurophysiol 39:713–721.

8

Plasticity of Sense Organs and Brain

JOSEPH BASTIAN AND HAROLD H. ZAKON

1. Introduction

Plasticity of neural function is a dominant theme in contemporary neuroscience. Nervous systems and the behaviors that they control need to change as organisms grow and reproductive states change. Environments change predictably (i.e., seasonally) as well as unpredictably, and an animal's physiological requirements fluctuate over a wide range of time scales. Plasticity is a broad term that covers many phenomena that enable an organism to cope with such changes. In this chapter we first cover plasticity of the electrosensory periphery. Here, the major plastic changes are the growth-dependent addition of sensory elements and restructuring of afferent axon terminals to accommodate this process. Changes in the sensitivity and coding properties of receptors result from these maturational processes. In addition, there are seasonal hormonally driven changes in the tuning of the electroreceptors that accompany changes in reproductive status. The second portion of this chapter focuses on the plasticity of central electrosensory networks. The consequences of rapid changes in neural function that apparently occur without changes in synaptic strength as well as the effects of short-term (facilitation and depression) and long-term potentiation and depression-like (LTP- and LTD-like) changes are described. These changes in central nervous system function result in the adaptive reconfiguration of circuits that optimize the extraction of behaviorally meaningful information from the sensory inflow.

2. Electroreceptor Plasticity

Many poikilothermic vertebrates continue to grow throughout their lifetimes. This growth is often accompanied by the addition of sensory receptors to various sensory organs, such as photoreceptors in the retina and hair cells in the inner ear and lateral line (Lombarte and Popper 1994; Higgs et al. 2002). Receptor cell addition and, as a consequence of this, remodeling of the sensory organs and their afferent terminals, occurs in the electroreceptors of elasmobranchs and teleosts.

2.1 Receptor Cell Addition

The new electroreceptors originate from the single-celled epithelial or support cell layer in which they are embedded much as the support cells in the inner ear are believed to generate new hair cells. New electroreceptors are added at a slow but constant rate. The best-studied examples of electroreceptor addition are in the ampullary and tuberous receptor organs of gymnotiforms (Zakon 1984, 1987; Vischer 1995). The former are specialized for the detection of low-frequency signals, usually of environmental origin, while the latter are specialized for detection of high-frequency signals caused by an animal's own, or conspecifics', electric organ discharges (EODs; see Jørgensen, Chapter 3; Bodznick and Montgomery, Chapter 6; Kawasaki, Chapter 7). Examples of the proliferation of ampullary receptor organs as a function of fish size are shown in Figure 8.1. In the electroreceptor organs of these fish receptor cells are added at a low rate until a critical number is reached or until the organ grows to a specific size, and then the organ divides into two daughter organs. The observation that some organs are in the process of dividing and others are not, even when innervated by the same afferent terminal, suggests it is unlikely that signals from the terminal trigger organ division. The most extreme cases reported of receptor cell addition and subsequent division of organs occur in the ampullary system of gymnotiforms in which a single initial organ can divide into more than 30 organs, and the nerve terminal changes from a small thin terminal with a few boutons to one with many tens of branches and a very extensive field of boutons. While the number of organs per afferent fiber increases with fish size, the number of afferent fibers remains constant throughout the fish's life (Carr et al. 1982; Zakon 1984).

The functional consequence of receptor cell addition is an increase in the sensitivity of the receptor organs. In the gymnotiform fish *Sternopygus*, sensitivity is increased about 1 log unit as fish increase threefold in length (Sanchez and Zakon 1990). However, there appears to be differential addition of organs to some receptors and less so to others so that a bimodal distribution eventually develops. Intracellular recordings of afferent axons indicate that the receptor afferent that innervates clusters of organs are the T-type receptors, whereas those that innervate only one or two organs are P type. T-type organs are more sharply tuned and more sensitive. Similarly in the ampullary system of catfish, the number of organs per afferent fiber increases as fish grow and the increased number of organs is correlated with increase in sensitivity and sharper tuning (Peters and van Ieperen 1989; Peters et al. 1997).

Although it is not yet known whether the number of receptor cells per organ increases in elasmobranchs as it does for auditory afferents (Corwin 1983), the sensitivity of their organs does increase as they age (Sisneros et al. 1998; Sisneros and Tricas 2002).

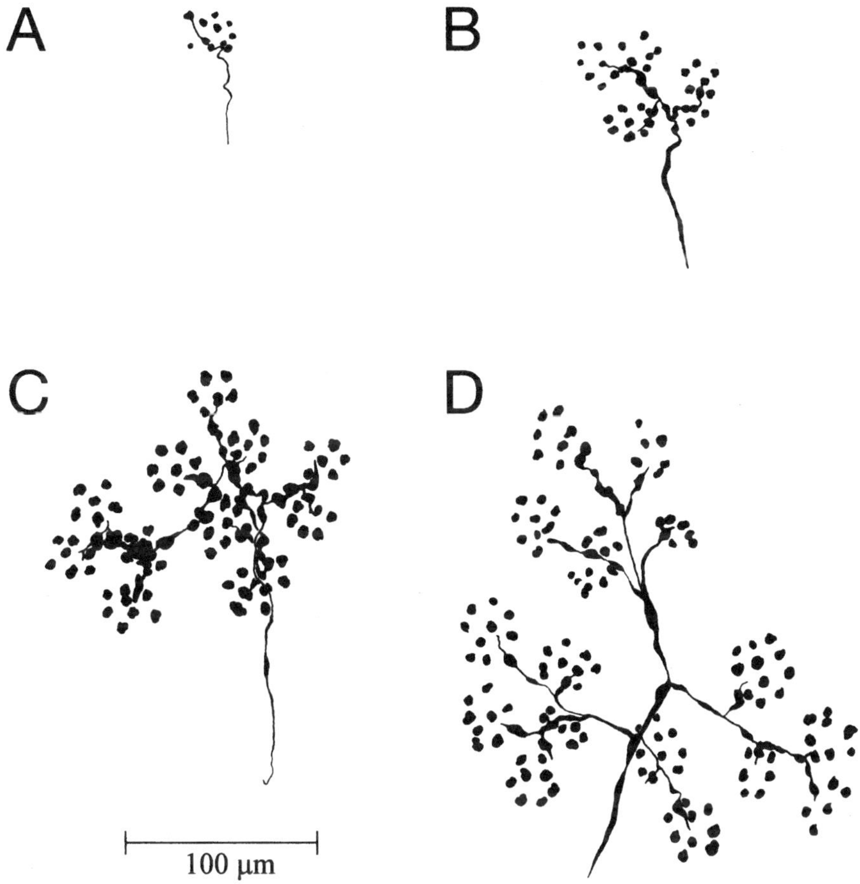

FIGURE 8.1. Addition of ampullary organs and growth of the ampullary axon terminals in fish of increasing size: (**A**) 4.1 cm, (**B**) 12.5 cm, (**C**) 18.0 cm, (**D**) 26.5 cm. (Reprinted with permission from Zakon 1984.)

2.2 Degeneration and Regeneration of Electroreceptor Organs

Electroreceptor cells depend on their innervation for trophic support. When an organ is denervated and the nerve prevented from returning, the receptor cells die within one to a few days, and within weeks the whole organ may disappear. The absence of the nerve or the appearance of degenerating receptor cells triggers the support cells to begin dividing and generating new sensory cells (Weisleder et al. 1994, 1996; Bensouilah and Denizot 1994). The new sensory cells differentiate but they in turn die without innervation. The fact that denervated tuberous organs generate tuberous receptor cells indicates that the determination of receptor cell fate (tuberous versus ampullary versus lateral line hair cell) is not dictated by the nerve. Ampullary organs can be forced to be innervated by tuberous afferents (Zakon et al. 1998). When this occurs, the ampullary receptor cells die but then reappear on reinnervation. This result suggests that the ampullary receptor cells can be supported by the trophic effects of a tuberous axon, and furthermore, that the identity of the cell is independent of the identity of the innervating afferent fiber (see Section 2.3).

Receptor organs regenerate if they are injured. If a patch of skin is removed the skin regenerates and new organs appear within a few weeks. The precise origin of the new organs is still not known, but it is likely that they are the progeny of support cells of intact organs surrounding the lesion. Support cells may be stimulated to divide by the injury and their progeny then migrate into the newly developing epithelium.

A nice set of studies has been done on the ampullary electroreceptors on the fin in the glass catfish (*Kyptopterus*) (Bever and Borgens 1991a,b). In this species, a row of receptor organs is located in the proximal portion of the fin whereas the distal part is free of electroreceptors. When a piece of the fin is rotated and sutured back into the fin, afferent fibers grow past the former distal domain, even though it is now closer, and continue to grow to the site of their original location even though it is now farther. There they induce a new organ at the site of the organ that had degenerated on denervation. In a separate experiment, they demonstrated that a new organ formed even when a nerve was forced to innervate a skin transplant that was initially free of electroreceptors (Bever and Borgens 1991b). Their interpretation of this was that the afferent induced a new organ from the epidermal cells of the skin. Another possibility is that support cells from old organs migrated into the skin along with the nerve. As in gymnotiforms, the source of new receptor cells is still unknown.

2.3 Plasticity of the Afferent Terminal

The afferent endings on the different types of electroreceptor cells differ. For example, ampullary organs are innervated by afferents that form a single large bouton at the base of each sensory cell whereas the so-called type I tuberous electroreceptor organ has a large calyx-like ending that envelops the organ. These are reminiscent of the bouton and calyceal terminals that innervate the hair cells of the vertebrate vestibular organs. This raises the question of how the afferent terminal type becomes matched to the class of sensory receptor. There are three logical possibilities to explain how this is accomplished: (1) Both the receptor cell identity and the identity of the afferent are fixed and they match via a recognition process; (2) the identity of the receptor is fixed, and that of the afferent fiber is plastic and its morphology is dictated by the type of receptor it innervates; or (3) the afferent fiber's identity is fixed and once it innervates a receptor cell it dictates the receptor cell's identity. These possibilities were tested in an experiment in which an ampullary organ was transplanted in place of a type I tuberous organ and forced to be innervated by the severed tuberous afferent (Zakon et al. 1998). In all cases, the tuberous afferent retained the distinctive calbindin label (ampullary afferents do not express calbindin) but showed an ampullary-like, large bouton type morphology. The results of this experiment support the idea that the receptor cell dictates the morphology of the terminal ending. This is in agreement with the observation that vestibular afferents that innervate both type I and type II hair cells in the mammalian vestibular organs show both calyceal and bouton endings appropriate to each cell. One additional point is that even though it is not known what trophic substance the afferent provides to the electroreceptor cells, ampullary receptor cells can be supported by tuberous afferents.

2.4 Hormone-Dependent Changes in Tuning

One of the intriguing aspects of tuberous receptors in gymnotiforms is that they are tuned to, that is they are most sensitive to, each fish's EOD frequency. Because EOD frequency is sexually dimorphic and individually distinct (i.e., individuals' EOD frequencies typically differ), the tuning of receptors is also sexually dimorphic and individually distinct. Furthermore, the tuning of electroreceptors is plastic and can be varied by treatment with hormones. For example, males of the gold-lined black knifefish *Sternopygus* discharge their electric organs at a lower frequency than females. In this species, treatment of fish with the androgens testosterone or dihydrotestosterone (DHT) gradually lowers EOD frequency. That is, androgens masculinize the fish. In parallel, the tuning of its tuberous receptors shifts downward maintaining the match to its EOD frequency (Meyer and Zakon 1982; Zakon and Meyer 1983).

How is receptor tuning kept in register with the EOD frequency when both are shifted with hormones? This is not so straightforward since the electrical activity of the electroreceptors and pacemaker neurons are different, and therefore likely possess different suits of ion channels. One hypothesis is that hormones act directly on the pacemaker nucleus and that the receptor tuning is then secondarily shifted because they are being stimulated by a different EOD frequency (activity dependence). This can be tested by measuring receptor sensitivity following spinal cord transection that severs the descending axons from the pacemaker to the spinal cord electromotoneurons and eliminates the EOD. When this is done, receptor tuning remains intact (suggesting that the tuning is not an ongoing result of electrical stimulation) and shifts downward by the appropriate amount following androgen treatment. Thus the receptors are not tuned by the ongoing EOD.

A second hypothesis is that the receptors may still be tuned by the EOD but might be sensitive to it only when hormone titers are high (hormone-induced plasticity). This hypothesis can be tested by treating spinalized fish with androgens in the presence of an artificial EOD that is gradually *increasing* in frequency. When this is done by wires implanted in their tails, the receptors shift downward rather than upward, again ignoring the electric field (Keller et al. 1986).

While the EOD is silenced in spinalized fish, the endogenously firing neurons of the pacemaker nucleus are still active. Thus, a third possibility is that the pacemaker nucleus somehow orchestrates the tuning of electroreceptors (central control of tuning). This possibility was eliminated when it was shown that receptors remain tuned and become shifted downward by androgens in fish whose pacemakers were removed by a chemical lesion (Ferrari and Zakon 1989). Furthermore, when a patch of skin is removed and new receptor cells regenerate in the new skin, they will become tuned to the appropriate frequency for that fish even in a fish with a pacemaker lesion.

Finally, it is possible that the hormones act only on the receptors and that as the tuning of the receptors shifts, the fish behaviorally adjusts its pacemaker to emit at a frequency to which its receptors are the most sensitive (behavioral sensitivity). This is not supported because when fish are treated with hormones, the EOD frequency shifts in advance of changes in receptor tuning. These experiments force us to conclude that hormones act directly on both the receptor cells as well as the pacemaker neurons. In support of this, tuberous electroreceptors label with an antibody to the androgen receptor (Zakon 2000).

There are also species differences in the hormone sensitivity of the electroreceptors of gymnotiforms. For example, brown ghosts (*Apteronotus leptorhynchus*) show reverse sexual dimorphism; that is, unlike other species of wave-type gymnotiforms, the males have higher EOD frequencies than the females. In this species, androgen implants raise EOD frequency and estrogen lowers it. As expected, androgens increase the best frequency of tuberous electroreceptors as well.

Tuberous electroreceptors evolved independently in the other group of teleosts with weak electric organs, the mormyriforms. The EODs of mormyrids are pulse-like and the receptors are more broadly tuned. Nevertheless, males generate broader pulses than females, which means that the peak frequency in the energy spectrum of their EODs is lower in frequency. Consequently, the tuning of their receptors is also at a lower frequency (Bass and Hopkins 1984). One major difference with wave-type gymnotiforms is that their electroreceptors apparently need an ongoing EOD in order to be shifted in their tuning by steroid hormones. However, only a single study has addressed this question and more work is needed to characterize this form of plasticity.

Surprisingly, plasticity of tuning also occurs in the ampullary receptors of elasmobranchs. Receptor tuning of the stingray (*Dasyatis sabina*) shows seasonal changes and androgenic modulation. As in elasmobranchs generally, the ampullary receptors of this species are fairly broadly tuned although they have a distinct best frequency (BF). The ampullary receptors of this species are best tuned around 2 to 4 Hz.

In this species there are two peaks of androgen during the breeding season. During the primary androgen increase in wild-caught males, the electrosensory primary afferent neurons showed an increase in discharge regularity, a downward shift in BF and passband, and a greater sensitivity to low-frequency stimuli from 0.01 to 4 Hz (Sisneros and Tricas 2000). Implants of the androgen DHT caused a similar downward shift in BF and increased sensitivity of the afferents to lower frequency stimuli especially in the range of 0.5 to 2 Hz (Fig. 8.2). The significance of this is that during the breeding season female stingrays enter shallow water first and bury themselves in the sand. The males appear later and cruise over the sand to find and dig up the females to mate with them. The males localize the females by sensing the electric fields produced by the female's rhythmic breathing which are dominated by frequencies to which the males are most sensitive (Tricas et al. 1995).

Hormone-dependent changes in receptor sensitivity may be a more general characteristic of acousticolateralis systems. Recent data on a nonelectric teleost,

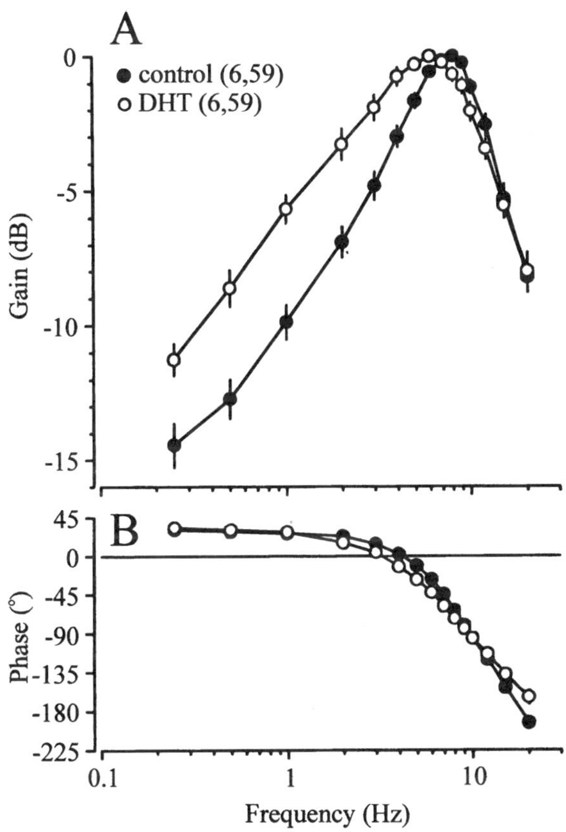

FIGURE 8.2. Androgens enhance low-frequency sensitivity of male skate electroreceptors. Bode plots of the frequency responses of electrosensory primary afferents recorded from adult male atlantic stingrays before and after DHT implants. (**A**) Response gain as a function of stimulus frequency. (**B**) Response phase as a function of stimulus frequency. The numbers of animals and electrosensory primary afferents tested are indicated in *parentheses*. (Reprinted with permission from Sisneros and Tricas 2000.)

the plainfin midshipman, show that the sensitivity of the auditory afferents from the sacculus also show seasonal variation in sensitivity (Sisneros and Bass 2003). This effect has recently been shown to be hormone-dependent (Sisneros et al. 2004).

3. Plasticity in the Central Nervous System

Here we focus on functional changes in central sensory neuron responsiveness and stimulus selectivity, or tuning, that occur in response to specific spatial and temporal stimulus characteristics. Developmental and seasonal changes in neural architecture are not discussed, although numerous studies of such changes have appeared (Zupanc and Heiligenberg 1989; Zupanc and Zupanc 1992; Zupanc et al. 1996; Zupanc 1999, 2001; Zupanc and Ott 1999; Clint and Zupanc 2001; Zupanc and Clint 2001). A variety of mechanisms are known to contribute to changes in sensory neuron function at various levels within nervous systems, including short-term forms of plasticity (e.g., facilitation and depression), long-term plasticity (LTP- and LTD-like mechanisms), and virtually instantaneous or switch-like changes mediated by modulatory influences of specific neural pathways. Examples of each of these mechanisms are described in the context of an important information processing task for these fish: the discrimination of signals due to electrolocation targets, such as prey, from communication signals and other patterns of electrosensory input that have the potential to mask electrolocation stimuli.

3.1 Discrimination of Electrolocation versus Electrocommunication Stimuli

The detection of prey as well as other targets in the environment and the analysis of communication signals generated by conspecifics are two principal functions of active electrosensory systems. The African mormyriforms and the South American gymnotiforms, the two major groups of active electrosensory fish, show many similarities in the general organization and function of their electrosensory systems (see Bell and Maler, Chapter 4; Bodznick and Montgomery, Chapter 6; Kawasaki, Chapter 7). Both have ampullary and tuberous receptors that are specialized for the detection of low-frequency signals, usually of environmental origin, and high-frequency signals caused by an animal's own, or conspecifics', EODs, respectively. The tuberous electroreceptors can be further subdivided into categories thought to serve primarily either communication or electrolocation functions (see Jørgensen, Chapter 3; Kawasaki, Chapter 7). Similarities are also seen centrally. The electrosensory lateral line lobe (ELL), the first-order nucleus, receives the receptor afferent projection as well as massive inputs descending from higher centers and, in both groups, the ELL is divided into separate subdivisions or somatotopic maps. The ampullary receptors project to a unique map while tuberous afferents terminate in three additional maps. In mormyrids, subtypes of tuberous receptor afferents project uniquely to different maps, which are clearly associated with processing information related to electrolocation versus electrocommunication signals (Bell 1989). In gymnotids there is also strong evidence that specific ELL maps are required for the performance of specific behaviors (Metzner and Juranek 1997). However, each tuberous receptor afferent branches, providing copies of the afferent input to each of the three tuberous-recipient regions, raising the possibility that there may be overlap in map functions (Heiligenberg and Dye 1982).

3.1.1 Corollary Discharge Gating of Receptor Afferent Input in Mormyrids

Perhaps the most significant difference in the designs of the mormyrid and gymnotid central electrosensory systems is in the presence of corollary discharge signals associated with the motor commands that control the EOD. Corollary discharge signals project to several electrosensory nuclei in mormyrids (Bennett and Steinbach 1969; Bell et al. 1983, 1995) but these signals are absent in gymnotids. In mormyrids, the corollary discharge provides a gating signal that, at the level of the ELL, facilitates the discrimination of afferent inputs caused by the animal's own discharge from those evoked by the discharges of other fish. Mormyromast electroreceptor afferents, those important for electrolocation, project to the medial and dorsolateral zones of the ELL (Bell et al. 1989). Neurons within this ELL subdivision also receive a corollary discharge input that facilitates responses to mormyromast activity occurring within a short time window (less than 30 to 40 ms; Bell and Grant 1992). Hence, the corollary discharge input to the mormyromast regions acts as a gate selectively facilitating sensory inputs that result from the animal's own discharge and are, therefore, likely to provide electrolocation information.

A different subdivision, the nucleus of the ELL, receives input from a separate tuberous receptor category, knollenorgans, which transmit precise temporal information about when EODs occur. Both an animal's own EOD as well as discharges of other fish activate knollenorgans, however, the corollary discharge input to the nucleus of the ELL is inhibitory (Zipser and Bennett 1976; Bell and Grant 1989). Hence, the corollary discharge input to the nucleus of the ELL also acts as a gate but its effect is to block responses to the animal's own EOD. Responses to the discharges of other fish are not blocked and are passed on to higher centers where sex-and species-specific information is extracted from measurements of the temporal characteristics of the electric organ discharge waveform (see Hopkins, Chapter 10; Hopkins and Bass, 1981; Amagai 1998; Amagai et al. 1998; Xu-Friedman and Hopkins 1999). The corollary discharge signals along with specializations of tuberous receptor subtypes result in a complete specialization of separate ELL subdivisions for processing electrolocation versus electrocommunication information.

3.1.2 Alternative Strategies in the Absence of Corollary Discharge Gating

Gymnotiform fish do not have corollary discharge signals associated with EOD production and each tuberous afferent trifurcates within the ELL contributing to three complete somatotopic maps of the sensory epithelium (Heiligenberg and Dye 1982). The ELL efferent neurons, pyramidal cells, receive receptor afferent inputs directly, or indirectly via inhibitory interneurons, and have typical antagonistic center surround receptive field structure (see Bell and Maler, Chapter 3; Kawasaki, Chapter 7; Maler 1979; Bastian 1981; Maler et al. 1981). On average, there are significant differences in the physiological properties of pyramidal cells from the different maps (Shumway 1989a,b; Turner et al. 1996), but there is also considerable variation among cells within the same map (Bastian and Courtright 1991; Bastian et al. 2002). The lack of a strict segregation of afferent input related to electrolocation versus electrocommunication in the gymnotid ELL raises the question of how these fish discriminate stimuli of these categories. Recent results indicate that switch-like changes in pyramidal cell amplitude modulation (AM) frequency tuning and firing statistics may contribute to this discrimination (Chacron et al. 2003; Doiron et al. 2003).

Electrosensory stimuli generated by small prey such as Daphnia cause spatially restricted or *local stimuli* that influence receptor afferents within relatively small (5 to 20 mm diameter) regions of the skin (Fig. 8.3A). Based on the velocity of the fish relative to the prey at the time of detection, Nelson and MacIver (1999) determined that the electrosensory stimulus was dominated by low temporal frequencies having a bandwidth ranging from 1 to about 20 Hz. Electrocommunication behaviors occur when the distance separating two or more fish allows each to sense the discharges of others. Each fish is "immersed" in the EOD field of its neighbor and this results in stimulation of large regions of the body surface referred to as *global stimulation.* In the case of just two fish (Fig. 8.3A), each receives a signal that is the sum of both discharges and this composite signal "beats." That is, the composite signal will be amplitude and phase modulated at a frequency determined by the difference between the two discharge frequencies. Low beat frequencies evoke the well-known jamming avoidance response (JAR) (see Kawasaki, Chapter 7; Hopkins, Chapter 10; Watanabe and Takeda 1963; Bullock et al. 1972a,b; Heiligenberg 1977, 1991), while a wider range of beat frequencies, up to at least 200 Hz, evoke "chirps" (Larimer and MacDonald 1968; Bullock 1969; Hopkins 1974; Hagedorn and Heiligenberg 1985; Zupanc and Maler 1993; Dulka and Maler 1994; Engler et al. 2000; Bastian et al. 2001; Engler and Zupanc 2001). Not only are chirps evoked by high-frequency AM stimuli, but the behavior itself, a rapid upward frequency modulation of the EOD, provides a brief AM dominated by frequencies from 50 to 100 Hz (Zupanc and Maler 1993). Hence, as indicated in Figure 8.1A, electrolocation and at least some communication signals have very different spatiotemporal signatures: local and low-frequency versus global and high-frequency, respectively.

Pyramidal cells switch their response properties contingent on these spatiotemporal signatures. The AM frequency tuning curves of Figure 8.3B summarize ELL pyramidal cell responses to sinusoidal EOD amplitude modulations (SAMs) of various frequencies presented with prey-like and communication-like stimulus geometries. Responses to stimuli presented with prey-like geometry

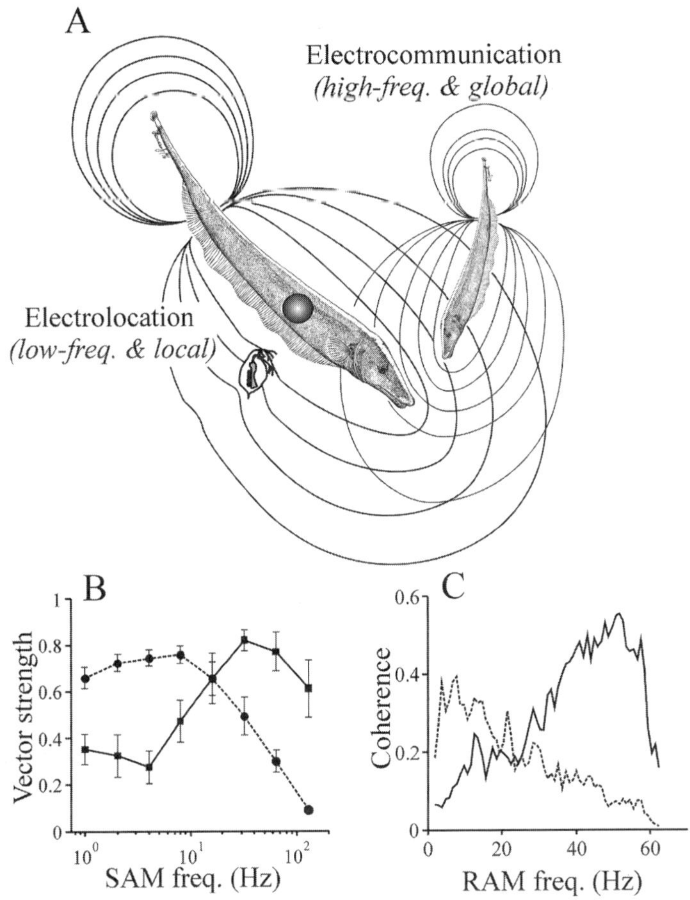

FIGURE 8.3. Electrolocation and electrocommunication stimuli generally have different spatial extents that control frequency tuning of neurons within the first-order electrosensory nucleus. (**A**) Illustrations of the wave-type weakly electric fish *Apteronotus leptorhynchus* showing the spatial characteristics of electrolocation versus electrocommunication stimuli. Prey such as *Daphnia* cause distortions of the electric organ discharge field (*curved lines*) that are localized to small regions of the body surface (*gray spherical region*) and are dominated by low frequencies. During electrocommunication each fish is effectively "immersed" in the EOD field of its neighbor, hence large regions of the sensory epithelium are stimulated. (**B**) Electrosensory lateral line lobe pyramidal cells are tuned to lower-frequency sinusoidal amplitude modulations (SAMs) of the electric organ discharge when the stimulus is applied locally (*dashed line*). Tuning shifts to higher frequencies when SAMs of the same amplitude are presented globally (*solid line*). Tuning is quantified as the vector strength which ranges from 0, when the spike train is random relative to the SAM, to 1 when the spikes are perfectly phase locked to the SAM. Error bars indicate ± 1.0 SEM, $n = 6$ cells). (**C**) Analysis of the coherence of spike trains and random amplitude modulations (RAMS, bandwidth 0 to 60 Hz) reveal the same shift in tuning when stimuli are presented locally (*dashed line*) versus globally (*solid line*). (**B** and **C** reprinted with permission from Chacron et al. 2003.)

peaked at SAM frequencies less than 10 Hz (dashed line); however, responses to the same AMs presented with communication-like geometry peaked between 40 and 60 Hz (solid line). Random amplitude modulations (RAMs) were also used in these studies, and Figure 8.3C shows plots of stimulus-spike train coherence, demonstrating that similar shifts in pyramidal cell tuning occur under these conditions. Information-theoretic measures (coding fraction and mutual information; Rieke et al. 1997; Gabbiani and Koch 1998; Borst and Theunissen 1999) can be used to quantify the ELL neurons' ability to transmit information about the random stimuli. Earlier use of these measures demonstrated that tuberous electroreceptor afferents perform exceptionally well, typically encoding from 40% to 60% of the time-varying amplitude of a globally presented stimulus (coding fractions from 0.4 to 0.6; Wessel et al. 1996). Paradoxically, under similar stimulus conditions, the encoding ability of ELL single pyramidal cells were found to be quite poor (typically encoding less than 20% of the stimulus), and even when pairs of pyramidal cells were considered, encoding ability did not approach that of single receptor afferents. However, under these conditions, the ELL cells performed well as feature detectors, indicating the time of occurrence specific stimulus events with brief bursts of action potentials (Gabbiani et al. 1996; Metzner et al. 1998; Gabbiani and Metzner 1999; Krahe et al. 2002).

The ability of ELL pyramidal cells to transmit detailed information about a stimulus depends on stimulus geometry as well as stimulus bandwidth. When random stimuli are used that match the spatiotemporal signatures of natural stimulus categories, *local low-frequency* and *global high-frequency*, increases in coding fraction and mutual information rates of from 100% to 300% are seen compared to other spatiotemporal combinations (Bastian et al. 2002; Chacron et al. 2003). The striking increases in encoding ability seen when stimuli mimic naturally occurring configurations suggest that certain behaviors require detailed information about time-varying EOD amplitude modulations. The use of random stimuli also revealed that pyramidal cells change the patterning of their spike trains given specific stimulus configurations. High-frequency global stimuli, which have characteristics in common with electrocommunication signals, cause pyramidal cells to adopt an oscillatory bursting firing pattern, and diffuse inhibitory electrosensory feedback from higher centers is necessary for this shift to bursty firing (Doiron et al. 2003). This bursty behavior shows similarities to pyramidal cell firing studied in in vitro preparations and may reflect biophysical specializations including cyclic somatic depolarization due to backpropagation of dendritic spikes (Turner et al. 1994, 1996; Turner and Maler 1999; Doiron et al. 2001, 2002). Hence, changes in pyramidal cell AM tuning, stimulus encoding ability, and spike train statistics all occur contingent upon specific spatiotemporal characteristics of the stimulus. One or more of these changes may contribute to the fish's ability to differentiate between electrolocation and electrocommunication stimuli.

3.1.3 Nonclassical Receptive Field Stimulation Contributes to Pyramidal Cell Tuning Shifts

The switch in pyramidal cell properties occurs immediately on changes in stimulus geometry, suggesting that experience-dependent synaptic plasticity is not involved. Rather, differences in the areas of the sensory epithelium stimulated by prey-like versus communication-like stimuli determine the cell's behavior by recruiting different constellations of synaptic input. This idea is supported by the results of experiments in which regions of the fish were electrically isolated via an insulating partition allowing selective stimulation of different body regions. The low-frequency preference of pyramidal cells, seen with local stimuli, was abolished if the same stimulus pattern was simultaneously applied to a cell's receptive field center and to larger regions of the sensory epithelium distant from and not including the cell's classical center surround receptive field (Chacron et al. 2003). These larger regions comprise a "modulatory" or nonclassical receptive field (Vinje and Gallant 2000) since their stimulation alone does not influence pyramidal cell firing. Additional results from this study indicate that the nonclassical receptive field stimulation recruits inhibitory feedback descending from higher-order electrosensory regions that attenuates responses to low-frequency stimuli without interfering with responses to high frequencies.

Unlike mormyrid fish in which there is a strict segregation of neurons processing electrolocation and electrocommunication within separate ELL subdivisions, single ELL neurons of gymnotiform fish may process sensory information relevant to both behavioral categories. Thus, in gymnotids, discrimination of these stimulus categories may require that populations of higher-order cells be capable of selectively responding to the bursty spike trains of pyramidal cells encoding communication stimuli while other populations selectively respond to the more regular spike trains of perhaps these same pyramidal cells as they process electrolocation stimuli.

4. Low-Pass Filtering Improves Electrolocation in the Presence of Interfering Stimuli

The ELL efferent neurons project via the lateral lemniscus to the nucleus praeeminentialis, a structure principally involved in feedback control of the ELL (Maler et al. 1982; Sas and Maler 1983, 1987; Bastian and Bratton 1990; Bratton and Bastian 1990), and the midbrain torus semicircularis which is the homolog of the inferior colliculus (see Bell and Maler, Chapter 4; Carr et al. 1981; Carr and Maler 1985, 1986). Electrosensory neurons within the torus can be categorized according to their temporal frequency response characteristics. Most preferentially respond to low-frequency (2 to 8 Hz) EOD AMs although some cells are tuned to higher frequencies (Partridge et al. 1981; Rose and Heiligenberg 1986). Low-frequency AMs due to the presence of electrolocation targets, as well as the continuous low-frequency beats due to the presence of fish with slightly different EOD frequencies, are well matched to the tuning characteristics of these torus cells, suggesting that the beats might mask electrolocation signals. Behavioral studies verified that electrolocation performance is compromised in the presence of low-frequency beats (Heiligenberg 1973, 1977, 1991). When the beat frequency is less than about 15 Hz the fish execute the JAR (Watanabe and Takeda 1963; Bullock et al. 1972a,b; Heiligenberg 1991; Metzner 1999). During this behavior each fish alters its EOD frequency in the direction that enlarges the frequency difference, thereby increasing the beat rate to higher values. The deleterious effects of beats on electrolocation disappear when the beat frequency rises above about 15 Hz (Heiligenberg 1973). Paradoxically, the jamming avoidance response shifts the beat frequency to higher values that are *better* matched to the pyramidal cells' tuning for such global stimuli (Fig. 8.3B, C). Despite this, electrolocation performance improves. This suggests that filtering mechanisms at higher processing stages must preferentially pass responses to low-frequency AMs associated with electrolocation while attenuating responses to the high-frequency AMs or beats.

4.1 Short-Term Synaptic Plasticity Contributes to Temporal Filtering in the Electrosensory Midbrain

In an elegant series of studies, employing in vivo patch recording and clever stimulation techniques, Rose and collaborators described the AM frequency tuning of torus neurons and identified characteristics of toral cells that correlate with this tuning. Intracellular recording and single cell staining experiments showed that cells most sensitive to low-frequency AMs, low-pass cells, typically possess dendrites heavily invested with dendritic spines, while cells tuned to high-frequency AMs have smooth dendrites. Neurons with spiny dendrites were also able to encode the time course of the AM accurately as changes in a cell's membrane potential whereas those with smooth dendrites were not (Rose and Call 1992, 1993). An example of a low-pass neuron's responses to AMs or beats swept from high to low frequencies is shown in Figure 8.4A1. Postsynaptic potentials grow as AM frequency falls below about 10 Hz and plateau for AMs less than about 6 Hz. In addition, adding high-frequency AMs to ongoing low-frequency AMs did not result in attenuation of the low-frequency responsiveness (compare Fig. 8.4A2, A3). This is a critical feature of the filtering mechanism in that it allows these cells to maintain responsiveness to low frequencies while attenuating responses to higher-frequency components of the complex stimuli that result from electrolocation targets mixed with the discharges of neighboring fish. The biophysical properties of cells with spiny dendrites (the probable long membrane time constant of the spine itself), the presence of voltage-dependent conductances that preferentially amplify postsynaptic potentials evoked by low-frequency inputs, as well as gain control mechanisms all probably contribute to the tuning characteristics of these low-pass neurons (Rose and Call 1993; Rose et al. 1994; Fortune and Rose 1997; Rose and Fortune 1999).

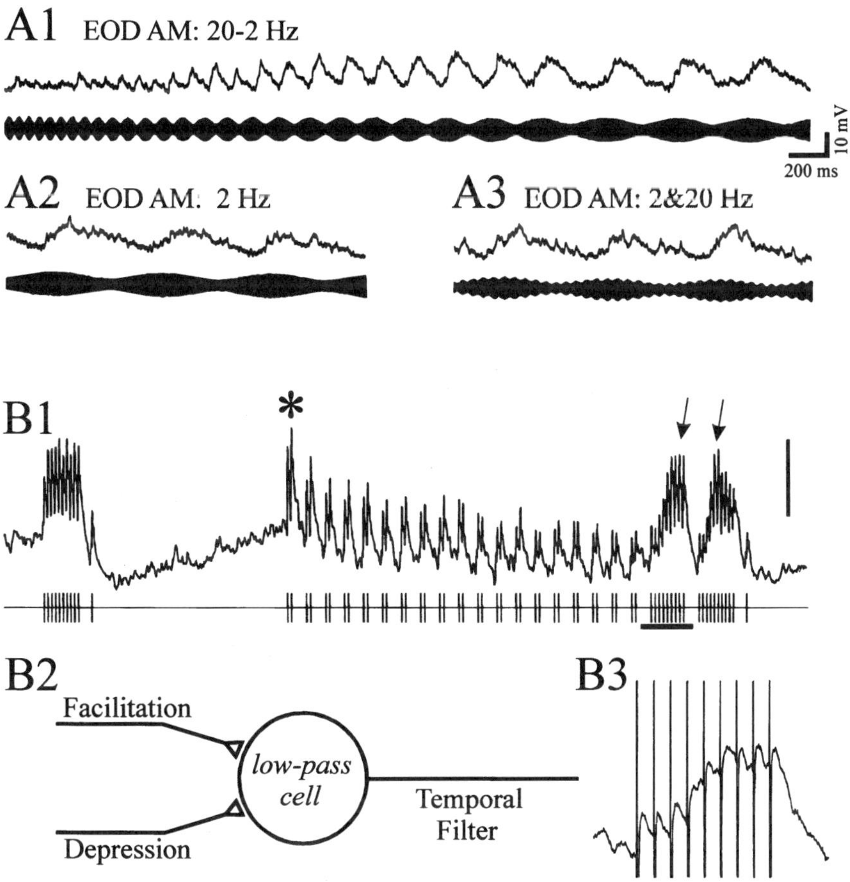

FIGURE 8.4. Examples of low-pass filtering of SAMs resulting from beats that simulate the interaction of the EODs of two fish having different discharge frequencies. (**A1**) Postsynaptic potentials (*upper trace*) recorded from low-pass torus cells of *Eigenmannia* accurately track only the low-frequency AMs when the beat rate is swept from 20 to 2Hz. (**A2**) Membrane potential of the same cell during stimulation with 2 Hz beats. (**A3**) Adding 20-Hz beats to the 2-Hz beats does not degrade responses to the low-frequency AM. (**B1**) Membrane potential changes evoked by different temporal patterns of electrical stimuli applied to the lateral lemniscus of *Eigenmannia*. Bursts of stimuli (10 stimuli, 10-ms spacing), which mimic responses to low-frequency AMs, evoke facilitating EPSPs, but pairs of stimuli (10-ms spacing) repeated at 50 Hz initially evoke large depolarization (*asterisk*) but then depressing EPSPs. Mimics of low-frequency sensory input applied 10 ms after a train of depressing stimuli still show facilitation (*arrows*). (**B2**) Model proposed by Fortune and Rose (2001) to illustrate how the interaction of facilitating and depressing synaptic inputs can confer low-pass properties on the postsynaptic cell. (**B3**) Responses to the burst of lateral lemniscus stimuli underlined in (**B1**) with an expanded time scale. (Reprinted with permission from Rose et al. 1994 [**A1 to A3**] and Fortune and Rose 2001 [**B1–B3**].)

4.1.1 Spike Train Temporal Characteristics Selectively Control Facilitation and Depression to Enhance Low-Pass Filtering

More recent studies demonstrate that neither biophysical properties nor the presence of voltage-dependent conductances *fully* account for the observed quality of low-pass filtering and that synaptic depression is also a major contributor to the low-frequency selectivity of many torus cells. Torus afferents were directly activated by electrical stimulation of the lateral lemniscus. This technique bypasses processing in the ELL, eliminating its potential contribution to the filtering process. Patterns of lateral lemniscus stimuli that mimicked the activity of ELL output cells evoked by beats of different frequencies were used and for many low-pass cells obvious synaptic depression occurred with mimics of responses to high- but not low-frequency beats (Fortune and Rose 2000). Hence, these low-pass cells could represent a neural correlate of the animals' ability to maintain accurate electrolocation behavior in the presence of high-frequency beats. However, low-pass filtering based on synaptic depression alone cannot explain the ability of these cells to maintain sensitivity to low-frequency stimuli as demonstrated in Fig. 8.4A3. Once a synapse has depressed it should be unable to contribute to any additional responses whether evoked by low- or high-frequency stimuli.

Fortune and Rose (2000, 2001) demonstrated that many low-pass neurons are capable of responding with either depressing or facilitating postsynaptic potentials contingent on the temporal pattern of lateral lemniscus stimulation. An example of this phenomenon is shown in Figure 8.4B1. The upper trace shows a low-pass cell's membrane potential and the lower indicates the pattern of lateral lemniscus stimulation. First, a mimic of ELL efferent activity due to low-frequency AMs (burst of 10-stimuli with 10 ms interpulse interval) was applied and the cell responded with a large depolarization due to summation of facilitating excitatory postsynaptic potentials (EPSPs). Following this, a train of pulse-pairs (10-ms ipi and 20-Hz repetition rate) mimicking the periodic bouts of high-frequency activity due to 20-Hz beats was applied and the EPSPs depressed as occurs with actual 20-Hz beat stimulation. However, despite the fact that synaptic depression occurred, the cell remained capable of responding to mimics of low-frequency sensory inputs. This was demonstrated by applying additional bursts of 10 stimuli 50 ms after the last "high-frequency" stimulus pair (arrows). The first of these responses is shown with an expanded time scale in Figure 8.4B3. Thus, depending on the temporal pattern of afferent input either net depression or facilitation can occur enabling the cell to maintain responses to low-frequency stimuli while the effects of high-frequency AMs are attenuated.

A simple network model suggested by Fortune and Rose (2001) to explain these results is shown in Fig. 8.4B2. Two classes of presynaptic inputs to the torus cells are proposed; one displays short-term depression while the other facilitates with repetitive activity. These two classes of inputs could represent separate classes of ELL efferents preferentially responsive to high- and low-frequency AMs, respectively. Individual high-frequency AMs, perhaps produced by chirps, would evoke brief bouts of activity and large EPSPs as seen in the initial responses to the high-frequency stimulus (Fig. 8.4B1, asterisk). Continuous high-frequency AMs, however, would result in depression at this synapse. Low-frequency stimuli preferentially activate the facilitating input, resulting in large summating EPSPs independent of the status of the depressing input. As suggested by Fortune and Rose (2001), the fact that pyramidal cells from different ELL maps show different AM tuning (Shumway 1989b) raises the possibility that the facilitating and depressing torus inputs could be associated with the efferents of specific maps.

In addition, the changes in ELL pyramidal cell tuning and spike train statistics that occur contingent on switching between local and global stimuli could also contribute to the mechanism outlined in Figure 8.4B3. The transition to oscillatory bursting seen when pyramidal cells receive global stimuli not only increases the probability of high-frequency spike doublets (approximately 10 ms interspike intervals) but also results in oscillatory firing with a periodicity at about 30 Hz (Doiron et al. 2003). Thus, pyramidal cell axon terminals that tend to depress are most likely to do so in response to higher-frequency global stimuli.

5. Long-Term Plasticity and Adaptive Filtering of Predictable Sensory Inputs

Many sensory systems are sensitive to reafferent patterns of input, those resulting from an animal's own activity, as well as to stimuli exclusively from the external world. Animals with active sensory systems, such as weakly electric fish, are obviously sensitive to reafferent electrosensory inputs since changes in posture, as occur during locomotion and as the animals explore their environment, alter the spatial relationship between the electric organ and the receptor array. This results in large changes in the EOD voltage over various regions of the body (Heiligenberg 1975, 1977; Assad et al. 1999) and receptor afferents respond strongly to such reafferent inputs (Bastian 1995). Although most sharks and rays do not produce electric organ discharges, the extraordinarily high sensitivity of their ampullary receptors (Kalmijn 1988) results in these being driven by modulations of the animal's own bioelectric fields as a result of gill movements and locomotor patterns (Montgomery 1984; New and Bodznick 1990; Bodznick et al. 1999; Montgomery and Bodznick 1999). Reafferent sensory inputs are likely to carry little useful information and are also potentially disruptive in that they can mask other important signals. However, reafferent sensory inputs are also predictable given appropriate proprioceptive information or corollary discharges of the relevant motor commands. Remarkably efficient mechanisms that rely on this predictability have been discovered that selectively cancel reafferent responses but preserve sensitivity to novel stimuli.

5.1 Cerebellar-Like Organization Is Associated with Adaptive Filtering

Electroreceptor, normal lateral line, and auditory afferents project to cerebellum-like structures that share a common general design (Fig. 8.5) across species (Montgomery et al. 1995; Bell et al. 1997a; Bell 2002). Receptor afferents project either directly, or indirectly via interneurons, to a population of cells variously referred to as ascending efferent neurons (AENs) in elasmobranchs, pyramidal cells in gymnotiform fish, or principal cells in mormyriforms. For simplicity, *principal cells* will be used to refer generally to these neurons. Arrays of principal cells not only receive the receptor afferent projection, forming somatotopic map(s) of the body surface, but also receive inputs from huge numbers of parallel fibers via the cells' apical dendrites. The parallel fibers, axons of typical cerebellar granule cells, convey proprioceptive information, corollary discharges of motor commands, and descending or feedback inputs of the same

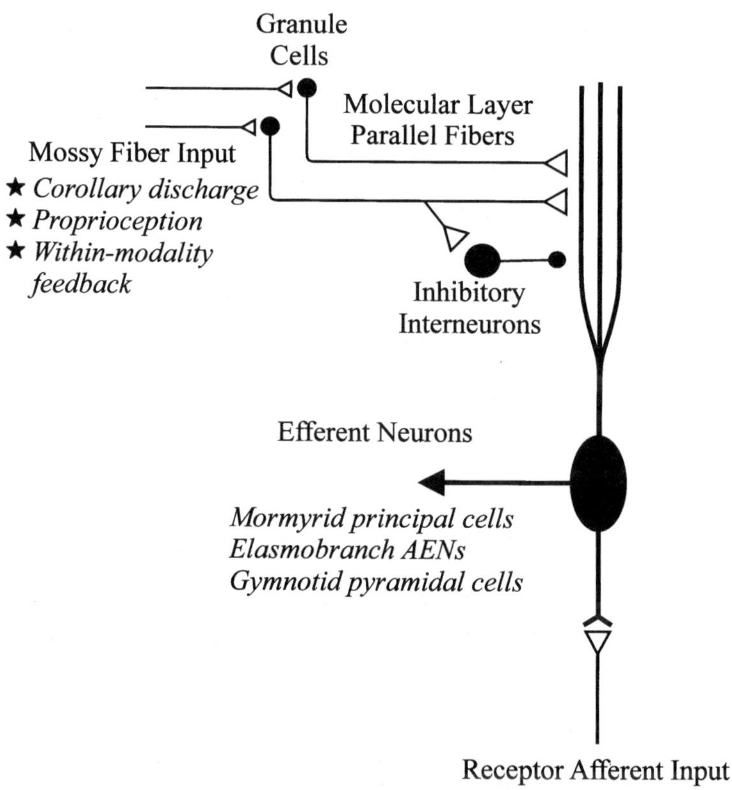

FIGURE 8.5. Diagram of the cerebellar-like circuitry characteristic of sensory processing regions known to be capable of adaptively filtering predictable inputs. See text for explanation.

modality (e.g., electrosensory information) to the principal cells (Bodznick and Boord 1986; Schmidt and Bodznick 1987; Bell et al. 1992; Conley and Bodznick 1994; Bastian 1995; Hjelmstadt et al. 1996).

The constellation of sensory and corollary discharge signals carried by the parallel fibers provides patterns of synaptic input to the principal cells' apical dendrites which can cancel responses to predictable afferent inputs. Cancellation occurs because the apical dendritic inputs cause a sequence of membrane potential changes that is the inverse, or *negative image*, of that expected as a result of the predictable afferent input. Hence, summation of the negative image inputs with receptor afferent inputs results in the cancellation of the predictable components of a stimulus without compromising the system's sensitivity to novel input patterns. Plasticity at the parallel fiber to pyramidal cell synapses ensures that the cancellation remains optimal even if characteristics of the predictable afference evolve (see Bell et al. 1997a; Bell 2001, 2002 for reviews).

This adaptive cancellation mechanism was initially described in studies of the ampullary, low-frequency sensitive, component of the mormyrid electrosensory system. Ampullary electroreceptors normally provide information about weak voltage sources extrinsic to the fish. Although ampullary receptors also respond to the fish's own brief EOD, these responses probably carry little useful information (Bell and Russell 1978). Principal cells of the mormyrid ELL receive ampullary afferents as well as electric organ corollary discharge (EOCD) signals, and Bell (1981, 1982, 1986a) showed that the EOCD provides a predictive signal that attenuates principal cell responses to ampullary input caused by the animal's own discharge. An example of an experiment demonstrating the adaptive nature of this cancellation mechanism is shown by the raster display of Figure 8.6 (Bell 1986a). In this experiment the fish's normal EOD was blocked with a curare-like drug; however, the EOD command could still be recorded at the level of the spinal cord (its timing is indicated by the arrow) and used to trigger stimuli applied to the ampullary receptors. The corollary discharge input to the ELL principal cells also remains intact in curarized fish and the uppermost segment of the raster (C alone, pre) shows that initially the cell was unresponsive to the corollary discharge alone. A brief inhibitory electrosensory stimulus was then presented at the time when the EOD would normally occur (vertical black line within the raster) so that the cell received paired corollary discharge inputs plus reduced receptor afferent input. Initially this caused strong inhibition followed by excitation, (C+S, initial) but with continuous presentation these responses decayed (C+S, late). Attenuation of the responses results from the formation of a negative image of the expected sensory input via adjustments of the synaptic strengths of corollary discharge inputs to the principal cells. The cell's response to the newly developed negative image is revealed when the electrosensory stimulus is removed (C alone, post). Despite the fact that the electrosensory stimulus is no longer being given, a pattern of excitation followed by inhibition is seen that is due to the altered synaptic strength of corollary discharge inputs to the cell.

More recent studies of several species of fish provided additional examples

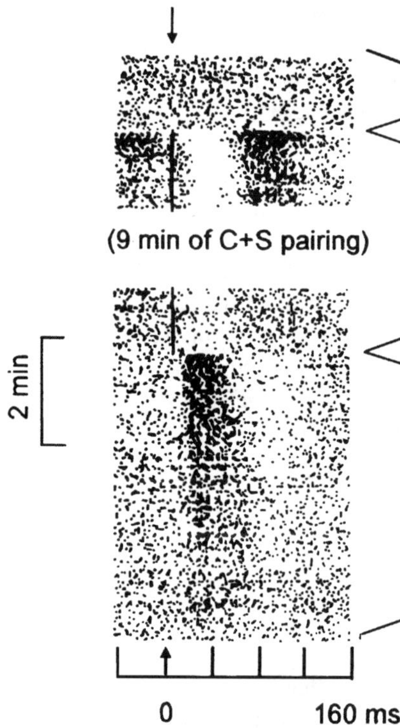

FIGURE 8.6. Raster display demonstrating the cancellation of receptor afferent input via the development of negative image responses by a principal cell within the ampullary receptor–recipient region of the mormyrid ELL. See text for explanation. (Reprinted with permission from Bell 1986a.)

of this adaptive filter. Gill movements of sharks and rays result in a continuous cyclic electrosensory input that strongly modulates ampullary afferents; however, the principal cells (ascending efferent neurons, AENs) of the first electrosensory processing station, the dorsal octavolateral nucleus (DON), are virtually insensitive to these inputs. Common-mode rejection via well-balanced antagonistic center-surround receptive field organization contributes to the principal cells' ability to reject these signals (Montgomery 1984; New and Bodznick 1990; Bodznick and Montgomery 1992; Bodznick et al. 1992; Montgomery and Bodznick 1993, 1999). Montgomery and Bodznick (1994) also demonstrated that an adaptive cancellation mechanism, virtually identical to that initially described by Bell (1981), was operating in this system as well. The apical dendrites of the DON principal cells also receive parallel fiber inputs that provide proprioceptive information related to body movements, electrosensory inputs descending from higher centers, as well as corollary discharges of motor commands (New and Bodznick 1990; Conley and Bodznick 1994; Hjelmstadt et al. 1996).

The parallel fibers provide the "raw data" for predicting the sensory consequences of these movements and plasticity at the principal cell apical dendritic synapses tunes the negative image to optimally cancel the predictable components of the stimulus. In addition, Montgomery and Bodznick (1994) formulated a set of learning rules, described in Section 5.2, that govern the formation and continuous updating of negative images.

Gymnotiform fish, such as shown in Fig. 8.3, often bend their body into an arc-like posture during exploratory behaviors. This results in large changes in the amplitude of the electric organ discharge at various sites on the body and electroreceptor afferents respond strongly to these (Bastian 1995, 1996a,b). However, ELL principal cells, which receive synaptic inputs from these afferents, are relatively insensitive to the electrosensory consequences of changes in posture. As described earlier, the ability of these pyramidal cells to preferentially reject the reafferent electrosensory inputs relies on the integration of a negative image input with the afferent signal to be cancelled. Proprioceptive information, electrosensory information descending from higher centers, and, possibly, corollary discharges of motor commands all contribute to the generation of the negative images (Sas and Maler 1983, 1987; Bastian and Bratton 1990; Bastian 1995, 1999). In addition to these examples from electrosensory systems, Montgomery and Bodznick (1999) demonstrated a similar cancellation mechanism in the medial octavolateral nucleus (normal lateral line processing region) of the scorpion fish and Bell et al. (1992) demonstrated similar adaptive cancellation in the mormyromast (electrolocation) region of the mormyrid ELL.

5.2 Anti-Hebbian Plasticity Contributes to Adaptive Filtering in Cerebellar-like Structures

The amazing similarity of the physiological characteristics of the cancellation mechanism plus the similar cerebellar-like organization of the brain areas involved suggests that a common cellular mechanism may underlie the formation of the negative image inputs, and Montgomery and Bodznick (1994) proposed a general model and set of learning rules to account for the adaptability of the negative images (Fig. 8.7). The principal cells receive afferent input consisting of a signal embedded in self-generated or otherwise predictable noise ($S + N_{sg}$). A negative image of the noise ($-N_{sg}$) is computed as a function of parallel fiber inputs carrying motor corollary discharge, proprioceptive, and electrosensory information. Integration of these signals by the principal cell results in the selective removal of the noise without compromising responses to the signal S. The adaptive component of the negative image is proposed to result from an anti-Hebbian form of plasticity where presynaptic activity paired with appropriately timed postsynaptic depolarization results in decreased excitatory synaptic strength. This plasticity is governed by two rules. First, parallel fibers active coincident with the postsynaptic target cell's activity leads to gain reduction, or depression, at those particular synapses. Second, parallel fibers active at times when the postsynaptic cell is inactive lead to increased gain at those

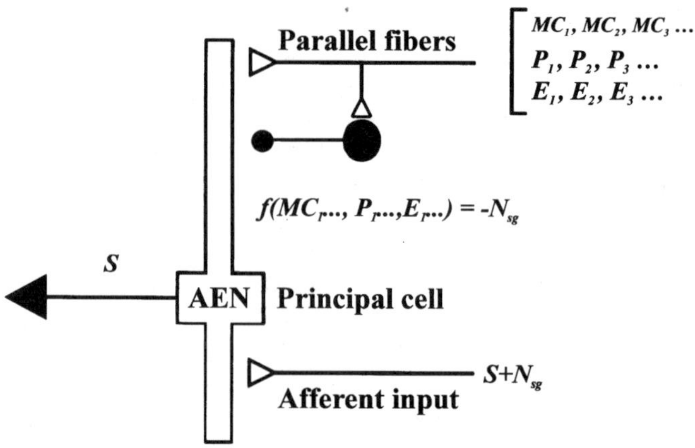

FIGURE 8.7. Summary of learning rules proposed by Montgomery and Bodznick (1994) governing changes in synaptic strength (anti-Hebbian plasticity) that optimize negative images of predictable inputs. See text for explanation. (Reprinted with permission from Montgomery and Bodznick 1994.)

particular synapses. Hence, the net effect of those parallel fibers carrying information related to the noise will be to provide *relative* hyperpolarization at times when the afferent input depolarizes the cell and increased depolarization at times when the afferent input evokes hyperpolarization. Inhibitory synaptic inputs are likely to contribute to the negative images; however, these must be governed by a complementary learning rule (Montgomery and Bodznick 1994). That is, inhibitory inputs active coincident with postsynaptic depolarization would be strengthened while those active coincident with postsynaptic hyperpolarization would be weakened. Modeling studies confirm that cancellation would be most efficient if the gain of both excitatory and inhibitory synapses were governed by these rules (Nelson and Paulin 1995).

That negative image formation involves changes in the gain of synaptic inputs to the principal cells has been verified in studies in which the cell's membrane potential is directly manipulated via intracellular current injection and paired with different predictive inputs—either EOD corollary discharge signals in mormyrids (Bell et al. 1993), gill movements in elasmobranchs (Bodznick et al. 1996), or with electrosensory stimulation in gymnotids (Bell et al. 1997a). Following the paired stimulation, negative images of the principal cells' responses to current injection appeared in response to the predictive inputs alone. The specific involvement of parallel fiber input in the generation of negative images has been demonstrated by pairing direct electrical stimulation of the parallel fibers with either sensory stimulation or current injection to the principal cells. Again, changes in gain of parallel fiber inputs occurred consistent with Montgomery and Bodznick's (1994) learning rules and the formation of negative images (Bastian 1998a; Bodznick et al. 1999). In gymnotids, a second purely electrosensory pathway, which provides feedback information to proximal regions of the principal cells' apical dendrites, has also been shown to participate in negative image formation (Bastian 1996b; Bastian et al. 2004). This suggests that some systems are capable of adaptively canceling predictable inputs using only information received via that same modality. In fact, in both elasmobranchs and gymnotids negative image responses will develop in response to cyclically predictable electrosensory inputs alone (Bastian 1995, 1996a,b; Bodznick et al. 1999).

The cellular mechanisms underlying the adaptive cancellation have been studied both in vivo as well as in vitro preparations of the gymnotid and mormyrid ELL. The similarity of the brain areas showing this plasticity to the cerebellum suggests that negative image plasticity may, at least in part, be due to a long-term depression-like (LTD-like) phenomenon at the parallel fiber to principal cell synapses. Negative images can be long term, lasting tens of minutes (Bell 1986a; Bastian 1996a), and in some cases for in excess of 3 hours (Bodznick et al. 1999). Like cerebellar LTD, the EPSP depression following parallel fiber activity paired with postsynaptic depolarization is Ca^{2+} dependent (Bastian 1998b; Han et al. 2000). However, unlike cerebellar LTD this depression requires *N*-methyl-*D*-aspartate (NMDA) receptor activation (Han et al. 2000; Bastian, unpublished observations). Depression at a subset of otherwise active parallel fiber synapses could result in the hyperpolarization needed to cancel afferent depolarization, and enhancement of concomitantly active inhibitory synapses would further strengthen the cancellation (Nelson and Paulin 1995). Conversely, the increased excitatory synaptic strength needed to cancel hyperpolarizing patterns of afferent input could simply result from relaxation of the LTD-like depression of excitatory synapses, however, the possibility also exists that the increased synaptic strength could reflect potentiation at excitatory synapses. In vitro studies of the mormyrid ELL (Bell et al. 1997b; Han et al. 2000) and of the gymnotid ELL (Wang and Maler 1997, 1998; Lewis and Maler 2002) have demonstrated potentiation at both the parallel fiber synapses and at synapses conveying descending electrosensory information directly to principal cells in gymnotids (Oswald et al. 2002). In all cases, however, the potentiation seems to be nonassociative; it does not require any temporal relationship between pre- and postsynaptic activity and it is likely to have a presynaptic locus. Both short-term potentiation (post-tetanic potentiation) as well as long-term potentiation have been observed in both mormyrids (Han et al. 2000) and gymnotids (Lewis and Maler, personal communication).

The timing of pre- and postsynaptic activity is critically important for modifications of the negative image in mormyrids. In this system principal cells produce two types of action potentials: low threshold small and narrow spikes and higher threshold broad spikes (Bell et al. 1993). The anti-Hebbian depression of parallel fiber EPSPs occurs only when the broad spike occurs within about 50 ms of the onset of the parallel fiber EPSP. All other temporal relationships between pre- and postsynaptic activity, as well as sufficiently frequent presynaptic activity alone, result in EPSP potentiation (Han et al. 2000). Thus, the gain of parallel fiber synapses changes in opposite directions depending on the temporal relationship between pre- and postsynaptic (broad spike) activity. Recent computational and modeling studies verified that this temporally asymmetric learning rule, established in in vitro studies, results in the formation of accurate and persistent negative images similar to those seen in the in vivo studies (Roberts 1999; Roberts and Bell 2000, 2002).

6. Summary

Structural and functional plasticity are ubiquitous features of nervous systems, yet the roles for the resulting changes in neuronal processing are often poorly understood and not clearly linked to specific aspects of the animal's behavior. Among the advantages associated with studying electroreceptive animals is the relative simplicity of the set of behaviorally relevant stimuli. This simplicity greatly facilitates the design of experiments capable of revealing neural processing algorithms that are actually involved in specific behaviors rather than simply enumerating physiological properties of various nervous system components. For example, tuning of the electroreceptors to the spectral characteristics of the animal's own discharge is an obvious advantage and it is not surprising that tuning is plastic given that the discharge itself changes as the animals mature and as their reproductive status changes. However, the mechanism by which the receptor tuning parallels EOD frequency changes in gymnotiforms is surprising and raises additional important questions. How do changing androgen titers alter EOD frequency and afferent tuning in parallel given that there is no obvious effect of either process on the other? Changes in the kinetics of ionic currents involved in the control of the EOD and in receptor transduction are undoubtedly involved, but the precise mechanisms, particularly at the receptor level, are not understood. Most studies of afferent plasticity have focused on gymnotiform fish with wave discharges. However, many gymnotiform fish produce pulse discharges. Not only are these also often sexually dimorphic (Hopkins 1999) but the spectral characteristics of the EOD can vary substantially at different sites on the body as does receptor tuning (Bastian 1977; Watson and Bastian 1979). If hormone-mediated tuning shifts also occur independent of the EOD in these animals then the hormone must have differential effects on receptor tuning contingent on the location or identity of the receptor cells. Alternatively, in gymnotiform pulse fish the tuning shift may require an ongoing EOD as in mormyrids which also produce a pulse discharge. The robust plasticity of electroreceptor tuning suggests the possibility that tuning shifts may also occur in hair cells of other octavolateral systems. In particular, the presence of electrical tuning of hair cells in lower vertebrates as well as the hormonally mediated seasonal alterations in vocalizations of many species suggests that changes in hair cell tuning and/or sensitivity may also occur as a function of reproductive status. Seasonal plasticity of the frequency sensitivity of the peripheral auditory system has very recently been demonstrated in a vocalizing fish (Sisneros and Bass 2003; Sisneros et al. 2004).

In addition to short and long-term forms of synaptic plasticity seen at different levels in central electrosensory systems, pyramidal cells within the gymnotiform ELL show a switch-like change in tuning and firing statistics contingent on the spatiotemporal signature of the electrosensory stimulus. Tuning shifts to low-frequency when the stimulus geometry mimics that generated by small prey and to high-frequency when the geometry is global, as in the case of communication signals. These shifts may facilitate the animal's ability to discriminate electrolocation from communication stimuli and low-pass torus cells are expected to respond best to pyramidal cell output when these cells process electrolocation stimuli. In addition, the pyramidal cells respond to repetitive global stimuli, such as high-frequency beats, with oscillatory burst-like patterns of activity. This raises the possibility that subsets of pyramidal cell axon terminals will depress, ensuring that the low-pass cells' responses to electrolocation stimuli are not masked. Thus, the midbrain filtering mechanism may, at least in part, rely on changes in the firing pattern of their afferent axons.

The similarities in the organization of the electrosensory lateral line lobes of weakly electric fish, the cerebellum, as well as other cerebellar-like structures, including the dorsal cochlear nucleus (DCN), suggest that such structures might perform similar functions. The idea that a principal role of sensory processing structures having this organization is to remove, or somehow correct for, movement-related or other predictable input patterns that might otherwise mask important stimulus features is particularly attractive and suggests a role for the proprioceptive inputs to the DCN principal cells (Young and Davis 2001). Pinna movements are obviously associated with auditory localization in many species including cats (Populin and Yin 1998). DCN principal cells not only are sensitive to auditory stimuli but also respond to somatosensory inputs related to pinna movements (Young et al. 1995; Kanold and Young 2001), and DCN lesions interfere with sound localization behavior (Sutherland et al. 1998; May 2000). Since the spectral cues exploited for sound source localization change as a function of pinna position (Young et al. 1996; Young and Davis 2001), either corrections for pinna position need to be incorporated into the analysis, or as suggested by Young and Davis (2001), the spectral changes due to pinna movement may provide information useful for optimizing pinna position for sound source identification. The precise role(s) of the DCN in auditory processing, however, remains to be determined.

One constant feature of the processing that occurs in the cerebellum-like structures, as well as the cerebellum itself, may be the potential to generate negative images of expected inputs. These typically appear within one to just a few minutes and can be revealed by appropriately designed physiological studies as described earlier, as well as with behavioral techniques. For example, recent studies of orientation responses in fish demonstrate that compensation for a predictable (conditioned) postural perturbation is associated with development of an after effect triggered by the conditioning cue alone (Montgomery et al. 2002). The after-effect is analogous to the negative image responses seen in physiological studies although involvement of cerebellum or cerebellar-like structures was not established. Humans learn to compensate for changes in motor performance due to predictable perturbations in sensory inflow. Following learning, removal of the perturbation results in the generation of after-effects that are also analogous to the negative images described above. Elegant studies comparing the compensation ability of normal subjects with that of subjects with cerebellar damage demonstrate that a functional cerebellum is necessary for this compensation and for the appearance of after-effects (Martin et al. 1996; Lang and Bastian 1999; Bastian 2002). Perhaps sound localization experiments employing reversible modification of the pinna and the position-dependent spectral cues may lead to the development of negative images or after-effects that provide insights as to the role of cerebellar-like structures in auditory function.

Acknowledgments. This work was supported by grants from the NIH to J.B. and H.Z.

References

Amagai S (1998) Time-coding in the midbrain of mormyrid electric fish. II. Stimulus selectivity in the nucleus exterolateralis pars posterior. J Comp Physiol A 182:131–143.

Amagai S, Friedman MA, Hopkins CD (1998) Time-coding in the midbrain of mormyrid electric fish. I. Physiology and anatomy of cells in the nucleus exterolateralis pars anterior. J Comp Physiol A 182:115–130.

Assad C, Rasnow B, Stoddard PK (1999) Electric organ discharges and electric images during electrolocation. J Exp Biol 202:1185–1193.

Bass AH, Hopkins CD (1984) Shifts of frequency tuning in electroreceptors in androgen-treated mormyrid fish. J Comp Physiol A 155:713–724.

Bastian AJ (2002) Cerebellar limb ataxia: abnormal control of self-generated and external forces. Ann NY Acad Sci 978:16–27.

Bastian J (1977) Variations in the frequency response of electroreceptors dependent on receptor location in a weakly electric fish (Gymnotoidei) with a pulse discharge. J Comp Physiol 121:53–64.

Bastian J (1981) Electrolocation. II. The effects of moving objects and other electrical stimuli on the activities of two categories of posterior lateral line lobe cells in *Apteronotus albifrons*. J Comp Physiol 144:481–494.

Bastian J (1995) Pyramidal-cell plasticity in weakly electric fish: a mechanism for attenuating responses to reafferent electrosensory inputs. J Comp Physiol A 176:63–78.

Bastian J (1996a) Plasticity in an electrosensory system. I. General features of a dynamic sensory filter. J Neurophysiol 76:2483–2496.

Bastian J (1996b) Plasticity in an electrosensory system. II. Postsynaptic events associated with a dynamic sensory filter. J Neurophysiol 76:2497–2507.

Bastian J (1998a) Plasticity in an electrosensory system. III. Contrasting properties of spatially segregated dendritic inputs. J Neurophysiol 79:1839–1857.
Bastian J (1998b) Modulation of calcium-dependent postsynaptic depression contributes to an adaptive sensory filter. J Neurophysiol 80:3352–3355.
Bastian J (1999) Plasticity of feedback inputs in the apteronotid electrosensory system. J Exp Biol 202:1327–1337.
Bastian J, Bratton B (1990) Descending control of electroreception. I. Properties of nucleus praeeminentialis neurons projecting indirectly to the electrosensory lateral line lobe. J Neurosci 10:1226–1240.
Bastian J, Courtright J (1991) Morphological correlates of pyramidal cell adaptation rate in the electrosensory lateral line lobe of weakly electric fish. J Comp Physiol 168: 393–407.
Bastian J, Schniederjan S, Nguyenkim J (2001) Arginine vasotocin modulates a sexually dimorphic communication behavior in the weakly electric fish *Apteronotus leptorhynchus*. J Exp Biol 204:1909–1923.
Bastian J, Chacron MJ, Maler L (2002) Receptive field organization determines pyramidal cell stimulus-encoding capability and spatial stimulus selectivity. J Neurosci 22:4577–4590.
Bastian J, Chacron MJ, Maler L (2004) Plastic and nonplastic pyramidal cells perform unique roles in a network capable of adaptive redundancy reduction. Neuron 41:767–779.
Bell CC (1981) An efference copy which is modified by reafferent input. Science 214: 450–453.
Bell CC (1982) Properties of a modifiable efference copy in an electric fish. J Neurophysiol 47:1043–1056.
Bell CC (1986a) Duration of plastic change in a modifiable efference copy. Brain Res 369:29–36.
Bell CC (1989) Sensory coding and corollary discharge effects in mormyrid electric fish. J Exp Biol 146:229–253.
Bell CC (2001) Memory-based expectations in electrosensory systems. Curr Opin Neurobiol 11:481–487.
Bell CC (2002) Evolution of cerebellum-like structures. Brain Behav Evol 59:312–326.
Bell CC, Grant K (1989) Corollary discharge inhibition and preservation of temporal information in a sensory nucleus of mormyrid electric fish. J Neurosci 9:1029–1044.
Bell CC, Grant K (1992) Corollary discharge effects and sensory processing in the mormyrid electrosensory lobe: II. Cell types and corollary discharge plasticity. J Neurophysiol 68:859–875.
Bell CC, Russell CJ (1978) Termination of electroreceptor and mechanical lateral line afferents in the mormyrid acousticolateral area. J Comp Neurol 182:367–382.
Bell CC, Libouban S, Szabo T (1983) Neural pathways related to the electric organ discharge command in mormyrid fish. J Comp Neurol 216:327–338.
Bell CC, Zakon H, Finger TE (1989) Mormyromast electroreceptor organs and their afferent fibers in mormyrid fish. I. Morphology. J Comp Neurol 286:391–407.
Bell CC, Grant K, Serrier J (1992) Corollary discharge effects and sensory processing in the mormyrid electrosensory lobe. I. Field potentials and cellular activity in associated structures. J Neurophysiol 68:843–858.
Bell CC, Caputi A, Grant K, Serrier J (1993) Storage of a sensory pattern by anti-Hebbian synaptic plasticity in an electric fish. Proc Natl Acad Sci USA 90:4650–4654.

Bell CC, Dunn K, Hall C, Caputi A (1995) Electric organ corollary discharge pathways in mormyrid fish. I. The mesencephalic command associated nucleus. J Comp Physiol A 177:449–462.

Bell CC, Bodznick D, Montgomery J, Bastian J (1997a) The generation and subtraction of sensory expectations within cerebellum-like structures. Brain Behav Evol 50(Suppl):17–31.

Bell CC, Han VZ, Sugawara Y, Grant K (1997b) Synaptic plasticity in a cerebellum-like structure depends on temporal order. Nature 387:278–281.

Bennett MVL, Steinbach AB (1969) Influence of electric organ control system on electrosensory afferent pathways in mormyrids. In: Llinas R (ed), Neurobiology of Cerebellar Evolution and Development. Chicago: American Medical Association, pp. 207–214.

Bensouilah M, Denizot JP (1994) Formation of new sensory cells in deafferented tuberous organs of the gymnotid fish *Eigenmannia virescens*. J Neurosci Res 39:545–555.

Bever MM, Borgens RB (1991a) The regeneration of electroreceptors in *Kryptoperus*. J Comp Neurol 309:200–217.

Bever MM, Borgens RB (1991b) Patterning in the regeneration of electroreceptors in the fin of *Kryptopterus*. J Comp Neurol 309:218–230.

Bodznick D, Boord RL (1986) Electroreception in chondrichthyes: central anatomy and physiology. In: Bullock TH, Heiligenberg W (eds), Electroreception. New York: John Wiley & Sons, pp. 225–256.

Bodznick D, Montgomery JC (1992) Suppression of ventilatory reafference in the elasmobranch electrosensory system: medullary neuron receptive fields support a common mode rejection mechanism. J Exp Biol 171:127–137.

Bodznick D, Montgomery JC, Bradley DJ (1992) Suppression of common mode signals within the electrosensory system of the little skate *Raja erinacea*. J Exp Biol 171: 127–138.

Bodznick D, Carey MR, Larner BW (1996) Locating the site of adaptive reafference suppression: evidence for a role of parallel fibers. Soc Neurosci Abstr 22:449.

Bodznick D, Montgomery JC, Carey M (1999) Adaptive mechanisms in the elasmobranch hindbrain. J Exp Biol 202:1357–1364.

Borst A, Theunissen FE (1999) Information theory and neural coding. Nat Neurosci 2: 947–957.

Bratton B, Bastian J (1990) Descending control of electroreception. II. Properties of nucleus praeeminentialis neurons projecting directly to the electrosensory lateral line lobe. J Neurosci 10:1241–1253.

Bullock TH (1969) Species differences in effect of electroreceptor input on electric organ pacemakers and other aspects of behavior in electric fish. Brain Behav Evol 2:85–118.

Bullock TH, Hamstra RH, Scheich H (1972a) The jamming avoidance response of high frequency electric fish. I. General features. J Comp Physiol 77:1–22.

Bullock TH, Hamstra RH, Scheich H (1972b) The jamming avoidance response of high frequency electric fish. II. Quantitative aspects. J Comp Physiol 77:23–48.

Carr CE, Maler L (1985) A Golgi study of the cell types of the dorsal torus semicircularis of the electric fish *Eigenmannia*: functional and morphological diversity in the midbrain. J Comp Neurol 235:207–240.

Carr CE, Maler L (1986) Electroreception in gymnotiform fish: central anatomy and physiology. In: Bullock TH, Heiligenberg W (eds), Electroreception. New York: John Wiley & Sons, pp. 319–374.

Carr CE, Maler L, Heiligenberg W, Sas E (1981) Laminar organization of the afferent and efferent systems of the torus semicircularis of gymnotiform fish: morphological substrates for parallel processing in the electrosensory system. J Comp Neurol 203: 649–670.

Carr CE, Maler L, Sas E (1982) Peripheral organization and central projections of the electrosensory nerves in gymnotiform fish. J Comp Neurol 211:139–153.

Chacron MJ, Doiron B, Maler L, Longtin A, Bastian J (2003) Frequency tuning of a sensory neuron mediated by non-classical receptive field. Nature 423:77–81.

Clint SC, Zupanc GKH (2001) Neuronal regeneration in the cerebellum of adult teleost fish, *Apteronotus leptorhynchus*: guidance of migrating young cells by radial glia. Dev Brain Res 130:15–23.

Conley RA, Bodznick D (1994) The cerebellar dorsal granular ridge in an elasmobranch has proprioceptive and electroreceptive representations and projects homotopically to the medullary electrosensory nucleus. J Comp Physiol A 174:707–721.

Corwin JT (1983) Postembryonic growth of the macula neglecta auditory detector in the ray, *Raja clavata:* continual increases in hair cell number neural convergence, and physiological sensitivity. J Comp Neurol 217:345–356.

Doiron B, Longtin A, Turner RW, Maler L (2001) Model of gamma frequency burst discharge generated by conditional backpropagation. J Neurophysiol 86:1523–1545.

Doiron B, Laing C, Longtin A, Maler L (2002) Ghostbursting: a novel neuronal burst mechanism. J Comput Neurosci 12:5–25.

Doiron B, Chacron MJ, Maler L, Longtin A, Bastian J (2003) Inhibitory feedback required for network burst responses to communication but not prey stimuli. Nature 421:539–543.

Dulka JG, Maler L (1994) Testosterone modulates female chirping behavior in the weakly electric fish, *Apteronotus leptorhynchus*. J Comp Physiol A 174:331–343.

Engler G, Zupanc GK (2001) Differential production of chirping behavior evoked by electrical stimulation of the weakly electric fish, *Apteronotus leptorhychus*. J Comp Physiol A 187:747–756.

Engler G, Fogarty CM, Banks JR, Zupanc GKH (2000) Spontaneous modulations of the electric organ discharge in the weakly electric fish, *Apteronotus leptorhychus*: a biophysical and behavioral analysis. J Comp Physiol A 186:645–660.

Ferrari MB, Zakon HH (1989) The medullary pacemaker nucleus is unnecessary for electroreceptor tuning plasticity in *Sternopygus*. J Neurosci 9:1354–1361.

Fortune ES, Rose GJ (1997) Passive and active membrane properties contribute to the temporal filtering properties of midbrain neurons *in vivo*. J Neurosci 17:3815–3825.

Fortune ES, Rose GJ (2000) Short-term synaptic plasticity contributes to the temporal filtering of electrosensory information. J Neurosci 20:7122–7130.

Fortune ES, Rose GJ (2001) Short-term synaptic plasticity as a temporal filter. Trends Neurosci 24:381–385.

Gabbiani F, Koch C (1998) Principles of spike train analysis. In: Koch C, Segev I (eds), Methods in Neuronal Modeling, 2nd ed. Cambridge, MA: MIT Press, pp. 313–360.

Gabbiani F, Metzner W (1999) Encoding and processing of sensory information in neural spike trains. J Exp Biol 202:1267–1279.

Gabbiani F, Metzner W, Wessel R, Koch C (1996) From stimulus encoding to feature extraction in weakly electric fish. Nature 384:564–567.

Hagedorn M, Heiligenberg W (1985) Court and spark: signals in the courting and mating of gymnotid fish. Anim Behav 33:254–265.

Han VZ, Grant K, Bell CC (2000) Reversible associative depression and nonassociative potentiation at a parallel fiber synapse. Neuron 27:611–622.

Heiligenberg W (1973) Electrolocation of objects in the electric fish *Eigenmannia* (Rhamphichthyidae, Gymnotoidei). J Comp Physiol 87:137–164.

Heiligenberg W (1975) Electrolocation and jamming avoidance in the electric fish *Gymnarchus niloticus*. J Comp Physiol 103:55–67.

Heiligenberg W (1977) Principles of electrolocation and jamming avoidance in electric fish. A neuroethological approach. In: Braitenberg V (ed), Studies of Brain Function, Vol. 1. New York: Springer-Verlag, pp. 1–85.

Heiligenberg W (1991) Neural Nets in Electric Fish. Cambridge, MA: MIT Press.

Heiligenberg W, Dye JC (1982) Labelling of electroreceptive afferents in gymnotoid fish by intracellular injection of HRP: the mystery of multiple maps. J Comp Physiol A 148:287–296.

Higgs DM, Souza MJ, Wilkins HR, Presson JC, Popper AN (2002) Age-and size-related changes in the inner ear and hearing ability of the adult zebrafish (*Danio rerio*). J Assoc Res Otolaryngol 3:174–184.

Hjelmstadt GO, Parks G, Bodznick D (1996) Motor corollary discharge activity and sensory responses related to ventilation in the skate vestibulolateral cerebellum: implications for electrosensory processing. J Exp Biol 199:673–681.

Hopkins CD (1974) Electric communication: functions in the social behavior of *Eigenmannia virescens*. Behavior 50:270–305.

Hopkins CD (1999) Design features for electric communication. J Exp Biol 202:1217–1228.

Hopkins CD, Bass AH (1981) Temporal coding of species recognition signals in an electric fish. Science 212:85–87.

Kalmijn AJ (1988) Detection of weak electric fields. In: Atema J, Fay RR, Popper AN, Tavolga WN (eds), Sensory Biology of Aquatic Animals. New York: Springer-Verlag, pp. 151–186.

Kanold PO, Young ED (2001) Proprioceptive information from the pinna provides somatosensory input to cat dorsal cochlear nucleus. J Neurosci 21:7848–7858.

Keller C, Zakon HH, Sanchez DY (1986) Evidence for a direct effect of androgens upon electroreceptor tuning. J Comp Physiol A 158:301–310.

Krahe R, Kreiman G, Gabbiani F, Koch C, Metzner W (2002) Stimulus encoding and feature extraction by multiple sensory neurons. J Neurosci 22:2374–2382.

Lang CE, Bastian AJ (1999) Cerebellar subjects show impaired adaptation of anticipatory EMG during catching. J Neurophysiol 82:2108–2119.

Larimer JL, MacDonald JA (1968) Sensory feedback from electroreceptors to electromotor pacemaker centers in gymnotids. Am J Physiol 214:1253–1261.

Lewis JE, Maler L (2002) Dynamics of electrosensory feedback: short-term plasticity and inhibition in a parallel fiber pathway. J Neurophysiol 88:1695–1706.

Lombarte A, Popper AN (1994) Quantitative analyses of postembryonic hair cell addition in the otolithic endorgans of the inner ear of the European hake, *Merluccius merluccius* (Gadiformes, Teleostei). J Comp Neurol 345:419–428.

Maler L (1979) The posterior lateral line lobe of certain gymnotoid fish: quantitative light microscopy. J Comp Neurol 345:323–363.

Maler L, Sas E, Rogers J (1981) The cytology of the posterior lateral line lobe of high frequency weakly electric fish (Gymnotidae): dendritic differentiation and synaptic specificity in a simple cortex. J Comp Neurol 195:87–139.

Maler L, Sas E, Carr CE, Matsubara J (1982) Efferent projections of the posterior lateral line lobe of gymnotiform fish. J Comp Neurol 211:154–164.

Martin TA, Keating JG, Goodkin HP, Bastian AJ, Thach WT (1996) Throwing while looking through prisms. I. Focal olivocerebellar lesions impair adaptation. Brain 119: 1183–1198.

May BJ (2000) Role of the dorsal cochlear nucleus in the sound localization behavior of cats. Hearing Res 148:74–87.

Metzner W (1999) Neural circuitry for communication and jamming avoidance in gymnotiform electric fish. J Exp Biol 202:1365–1375.

Metzner W, Juranek J (1997) A sensory brain map for each behavior? Proc Natl Acad Sci USA 94:14798–14803.

Metzner W, Koch C, Wessel R, Gabbiani F (1998) Feature extraction by burst-like spike patterns in multiple sensory maps. J Neurosci 18:2283–2300.

Meyer JH, Zakon HH (1982) Androgens alter the tuning of electroreceptors. Science 217:635 637.

Montgomery JC (1984) Noise cancellation in the electrosensory system of the thornback ray: common mode rejection of input produced by the animal's own ventilatory movement. J Comp Physiol A 155:103–111.

Montgomery JC, Bodznick D (1993) Hindbrain circuitry mediating common mode suppression of ventilatory reafference in the electrosensory system of the little skate *Raja erinacea*. J Exp Biol 183:203–215.

Montgomery JC, Bodznick D (1994) An adaptive filter that cancels self-induced noise in the electrosensory and lateral line mechanosensory systems of fish. Neurosci Lett 174:145–148.

Montgomery JC, Bodznick D (1999) Signals and noise in the elasmobranch electrosensory system. J Exp Biol 202:1349–1355.

Montgomery JC, Coombs S, Halstead MBD (1995) Biology of the mechanosensory lateral line in fishes. Rev Fish Bio Fisheries 5:399–416.

Montgomery JC, Carton G, Bodznick D (2002) Error-driven motor learning in fish. Biol Bull 203:238–239.

Nelson ME, MacIver MA (1999) Prey capture in the weakly electric fish *Apteronotus leptorhynchus* sensory acquisition strategies and electrosensory consequences. J Exp Biol 202:1195–1203.

Nelson ME, Paulin MG (1995) Neural stimulations of adaptive reafference suppression in the elasmobranch electrosensory system. J Comp Physiol A 177:723–736.

New JG, Bodznick D (1990) Medullary electrosensory processing in the little skate. II. Suppression of self-generated electrosensory interference during respiration. J Comp Physiol A 167:295–307.

Oswald AM, Lewis JE, Maler L (2002) Dynamically interacting processes underlie synaptic plasticity in a feedback pathway. J Neurophysiol 87:2450–2463.

Partridge BL, Heiligenberg W, Matsubara J (1981) The neural basis of a sensory filter in the jamming avoidance response: no grandmother cells in sight. J Comp Physiol 145:153–168.

Peters RC, van Ieperen S (1989) Resting discharge and sensitivity of ampullary electroreceptors in *Clarias gariepinus* related to convergence ratio during ontogeny. Brain Behav Evol 34:43–47.

Peters RC, Brans RJ, Bretschneider F, Versteeg E, Went A (1997) Converging electroreceptor cells improve sensitivity and tuning. Neurosci Lett 81:297–301.

Populin LC, Yin TCT (1998) Pinna movements of the cat during sound localization. J Neurosci 18:4233–4243.

Rieke F, Warland D, de Ruyter van Steveninck R, Bialek W (1997) Spikes: Exploring the Neural Code. Cambridge, MA: MIT Press.

Roberts PD (1999) Computational consequences of temporally asymmetric learning rules: I. Differential Hebbian learning. J Comp Neurosci 7:235–246.

Roberts PD, Bell CC (2000) Computational consequences of temporally asymmetric learning rules: II Sensory image cancellation. J Comput Neurosci 9:67–83.

Roberts PD, Bell CC (2002) Spike timing dependent synaptic plasticity in biological systems. Biol Cybern 87:392–403.

Rose GJ, Call SJ (1992) Evidence for the role of dendritic spines in the temporal filtering properties of neurons: the decoding problem. Proc Natl Acad Sci USA 89:9662–9665.

Rose GJ, Call SJ (1993) Temporal filtering properties of midbrain neurons in an electric fish: implications for the functions of dendritic spines. J Neurosci 13:1178–1189.

Rose GJ, Fortune ES (1999) Mechanisms for generating temporal filters in the electrosensory system. J Exp Biol 202:1281–1289.

Rose GJ, Heiligenberg W (1986) Limits of phase and amplitude sensitivity in the torus semicircularis of *Eigenmannia*. J Comp Physiol A 159:813–822.

Rose GJ, Etter N, Adler TB (1994) Responses of electrosensory neurons in the torus semicircularis of *Eigenmannia* to complex beat stimuli: testing hypotheses of temporal filtering. J Comp Physiol 175:467–474.

Sanchez DY, Zakon HH (1990) The effects of postembryonic cell addition on the response properties of electroreceptive afferents. J Neurosci 10:361–369.

Sas E, Maler L (1983) The nuclei praeeminentialis: a Golgi study of a feedback center in the electrosensory system of gymnotid fish. J Comp Neurol 221:127–144.

Sas E, Maler L (1987) The organization of afferent input to the caudal lobe of the cerebellum of the gymnotid *Apteronotus leptorhynchus*. Anat Embryol 177:55–79.

Schmidt AW, Bodznick D (1987) Afferent and efferent connections of the vestibulolateral cerebellum of the little skate, *Raja erinacea*. Brain Behav Evol 30:282–302.

Shumway CA (1989a) Multiple electrosensory maps in the medulla of weakly electric gymnotiform fish. I. Physiological differences. J Neurosci 9:4388–4399.

Shumway CA (1989b) Multiple electrosensory maps in the medulla of weakly electric gymnotiform fish. II. Anatomical differences. J Neurosci 9:4400–4415.

Sisneros JA, Bass AH (2003) Seasonal plasticity of peripheral auditory frequency sensitivity. J Neurosci 23:1049–1058.

Sisneros JA, Tricas TC (2000) Androgen-induced changes in the response dynamics of ampullary electrosensory primary afferent neurons. J Neurosci 20:8586–8595.

Sisneros JA, Tricas TC (2002) Ontogenetic changes in the response properties of the peripheral electrosensory system in the Atlantic stingray (*Dasyatis sabina*). Brain Behav Evol 59:130–140.

Sisneros JA, Tricas TC, Luer CA (1998) Response properties and biological function of the skate electrosensory system during ontogeny. J Comp Physiol A 183:87–99.

Sisneros JA, Forlano PM, Deitcher DL, Bass AH (2004) Steroid-dependent auditory plasticity leads to adaptive coupling of sender and receiver. Science 305:404–407.

Sutherland DP, Masterton RB, Glendenning KK (1998) Role of acoustic striae in hearing: reflexive responses to elevated sound-sources. Behav Brain Res 97:1–12.

Tricas TC, Michael SW, Sisneros JA (1995) Electrosensory optimization to conspecifics phasic signals for mating. Neurosci Lett 202:129–132.

Turner RW, Maler L (1999) Oscillatory and burst discharge in the Apteronotid electrosensory lateral line lobe. J Exp Biol 202:1255–1265.

Turner RW, Maler L, Deerinck T, Levinson SR, Ellisman MH (1994) TTX-sensitive dendritic sodium channels underlie oscillatory discharge in a vertebrate sensory neuron. J Neurosci 14:6453–6471.

Turner RW, Plant JR, Maler L (1996) Oscillatory and burst discharge across electrosensory topographic maps. J Neurophysiol 76:2364–2382.

Vinje W, Gallant JL (2000) Sparse coding and decorrelation in primary visual corterx during natural vision. Science 287:1273–1276.

Vischer H (1995) Electroreceptor deveopment in the electric fish *Eigenmannia*: a histological and ultrastructural study. J Comp Neurol 360:81–100.

Wang D, Maler L (1997) Plasticity of the direct feedback pathway in the electrosensory system of *Apteronotus leptorhynchus*. J Neurophysiol 78:1882–1889.

Wang D, Maler L (1998) Differential role of Ca^{2+}/calmodulin-dependent kinases in post-tetanic potentiation at input selective glutamatergic pathways. Proc Natl Acad Sci USA 95:7133–7138.

Watanabe A, Takeda K (1963) The change of discharge frequency by A.C. stimulus in a weakly electric fish. J Exp Biol 40:57–66.

Watson D, Bastian J (1979) Frequency response characteristics of electroreceptors in the weakly electric fish, *Gymnotus carapo*. J Comp Phsiol 134:191–202.

Weisleder P, Lu Y, Zakon HH (1994) Effects of denervation upon receptor cell survival and basal cell proliferation in tuberous electroreceptor organs of a weakly electric fish. J Comp Neurol 347:545–552.

Weisleder P, Lu Y, Zakon HH (1996) Tuberous electroreceptor organs form in denervated regenerating skin of a weakly electric fish. J Comp Neurol 367:563–574.

Wessel R, Koch C, Gabbiani F (1996) Coding of time-varying electric field amplitude modulations in a wave-type electric fish. J Neurophysiol 75:2280–2293.

Xu-Friedman MA, Hopkins CD (1999) Central mechanisms of temporal analysis in the knollenorgan pathway of mormyrid electric fish. J Exp Biol 202:1311–1318.

Young ED, Davis KA (2001) Circuitry and function of the dorsal cochlear nucleus. In: Oertel D, Fay RR, Popper AN (eds), Integrative Functions in the Mammalian Auditory Pathway. New York: Springer-Verlag, pp. 160–206.

Young ED, Nelken I, Conley RA (1995) Somatosensory effects on neurons in dorsal cochlear nucleus. J Neurophysiol 73:743–765.

Young ED, Rice JJ, Tong SC (1996) Effects of pinna position on head-related transfer functions in the cat. J Acoust Soc Am 99:3064–3076.

Zakon HH (1984) Postembryonic changes in the peripheral electrosensory system of a weakly electric fish: addition of receptor organs with age. J Comp Neurol 228:557–570.

Zakon HH (1987) Variation in the mode of receptor cell addition in the electrosensory system of gymnotiform fish. J Comp Neurol 262:195–214.

Zakon HH (2000) Sex steroids and weakly electric fish: a model system for activational mechanisms of hormone action. In: Matsumoto A (ed), Sexual Differentiation of the Brain. CRC Press, pp. 95–112.

Zakon HH, Meyer JH (1983) Plasticity of electroreceptor tuning in the weakly electric fish, *Sternopygus dariensis*. J Comp Physiol A 153:477–487.

Zakon HH, Lu Y, Weisleder P (1998) Sensory cells determine afferent terminal morphology in cross-innervated electroreceptor organs: implications for hair cells. J Neurosci 18:2581–2591.

Zipser B, Bennett MVL (1976) Interaction of electrosensory and electromotor signals in the lateral line lobe of a mormyrid fish. J Neurophysiol 39:713–721.

Zupanc GKH (1999) Neurogenesis, cell death and regeneration in the adult gymnotiform brain. J Exp Biol 202:1435–1446.

Zupanc GKH (2001) Distribution of radial glia in the central posterior/prepacemaker nucleus of weakly electric fish, *Apteronotus leptorhynchus*. In: Gottingen Neurobiology Report, Proceedings of the Fourth Meeting of the German Neuroscience Society 2001, Vol. 2. Stuttgart/New York: George Thieme Verlag, p. 976.

Zupanc GKH, Clint SC (2001) Radial glia-mediated up-regulation of somatostatin in the regenerating adult fish brain. Neurosci Lett 309:149–152.

Zupanc GKH, Heiligenberg W (1989) Sexual maturity-dependent changes in neuronal morphology in the prepacemaker nucleus of adult weakly electric knifefish, *Eigenmannia*. J Neurosci 9:3816–3827.

Zupanc GKH, Maler L (1993) Evoked chirping in the weakly electric fish, *Apteronotus leptorhynchus*: a quantitative biophysical analysis. Can J Zool 71:2301–2310.

Zupanc GKH, Ott R (1999) Cell proliferation after lesions in the cerebellum of adult teleost fish: time course, origin, and type of new cells produced. Exp Neurol 160:78–87.

Zupanc GKH, Zupanc MM (1992) Birth and migration of neurons in the central posterior/prepacemaker nucleus during adulthood in weakly electric knifefish (*Eigenmannia* sp.). Proc Natl Acad Sci USA 89:9539–9543.

Zupanc GKH, Horschke I, Ott R, Rascher GB (1996) Postembryonic development of the cerebellum in gymnotiform fish. J Comp Neurol 370:443–464.

9

Behavior of Animals with Passive, Low-Frequency Electrosensory Systems

LON A. WILKENS AND MICHAEL H. HOFMANN

1. Introduction

The electrosense and associated behaviors appeared early in the evolutionary history of vertebrates as an ampullary sensory system with electroreceptors sensitive to weak, low-frequency electrical signals (less than 1 $\mu V\ cm^{-1}$, less than 50 Hz) in the aquatic environment. The ampullary organs are apparently ancestral to jawed fish (Fig. 9.1), the electrosense having appeared 400 million years earlier in the agnathans, represented by homologous epidermal end bud electroreceptors in lampreys (Ronan 1986). Jawed fish consist of the cartilaginous elasmobranchs and chimeras and the bony fishes. The bony fishes can be further divided into ray- and lobe-finned sister groups.

Ampullary organs, with receptors featuring an apical kinocilium and cathodal excitability, and derived from ectodermal placodes along with mechanosensory neuromasts (Northcutt 1986; see also Jørgensen, Chapter 3), are present in all extant cartilaginous and bony fishes except for the Neopterygii (gars, bowfin, and teleosts; Fig. 9.1). Thus, having disappeared in the late Paleozoic prior to the emergence of teleosts; the electrosense is absent in the overwhelming majority of fish. Nevertheless, the electrosense is a cardinal feature of all chondrichthyan and non-neopterygian bony fish. The list of the early aquatic vertebrates featuring the archetypal ampullary electrosense includes lampreys, elasmobranchs and chimeras, bichirs and chondrosteans, lungfish, coelacanths, and the more aquatic urodeles as well as limbless caecilian amphibians.

Within this diverse group of animals the electrosense is considered *passive*, sensing only exafferent signals (i.e., external electric fields from either animate or inanimate sources). With the exception of skates (Rajidae) and rays (e.g., *Torpedo*) that deliver either weak or strong and aggressive electric discharges, respectively (Bass 1986; Macadar et al., Chapter 14), none of the fish and amphibians with the plesiomorphic ampullary system produce the ongoing electric signals that characterize *active* electrosensory fish. The active electrosense, in which fish monitor their own regular or continuous electrical discharges, exists in only two groups of unrelated teleosts, the weakly electric gymnotiforms and

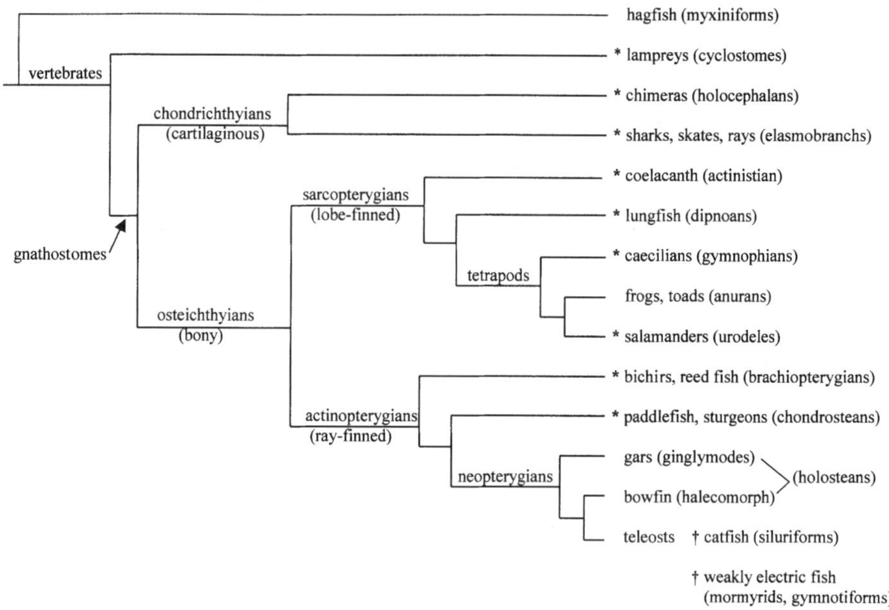

FIGURE 9.1. Cladogram of the early aquatic vertebrates, fish, and amphibians, showing groups in possession of Lorenzinian ampullary organs (*), and those with nonhomologous teleost ampullae (†). The cladogram is adapted from various sources including Ladich and Popper (2004), Bullock et al. (1983), and Northcutt (1986).

momyrids. This active electrosense is mediated by a different class of high-pass tuberous receptors tuned to their high-frequency reafferent signals (Kawasaki, Chapter 7).

Low-frequency ampullary electroreceptors are also found in teleosts, including the weakly electric fish, but these receptor organs differ from those of non-teleosts in having apical microvilli and anodal excitability. Teleost ampullary receptors are the apparent result of convergent trends in the evolution of analogous receptors in more recent fishes (properties of ampullae are reviewed by Bullock et al. 1983; Zakon 1986). Active electroreception is the subject of chapters by Hopkins (Chapter 10) and Nelson (Chapter 11), although passive, ampullary electrolocation in weakly electric fish is discussed as well by Hopkins. Electrosensory behaviors, such as the proposed compass sense in elasmobranchs (Kalmijn 1974; Paulin 1995), are "active," but only in the sense that they are based on the swimming motions of the fish (see Section 4.2.2).

In addition, the electrosense was reacquired independently in siluriforms (catfish) and in a single mammalian taxon (Monotremata) that includes the platypus and echidnas, two unrelated vertebrate groups that also utilize the conductive properties of the aquatic, or in the case of echidnas, a moist environment. Each of these involves passive mechanisms. In the lone terrestrial example, the echidna has adapted the monotreme electrosense as a possible mechanism for feeding (Manger et al. 1997). Unlike fish and amphibian electroreceptors, the monotreme electrosensory organs are derived from mucous glands in the skin.

Despite the wide distribution of passive electrosensory systems, there are notable gaps in our understanding of the behaviors and therefore the biological functions served by the electrosense. Behavioral correlates of the electric sense in groups such as lampreys, chimeras, sturgeons, coelacanths, and lungfish are not well investigated. In this chapter we review known electrosensory behaviors, including the responses of catfish where electrosensitivity was initially described, the classic prey detection behaviors in sharks and rays, more recent descriptions of stingray mating and paddlefish planktivory, the convergent rostral electrosense of the platypus, amphibian prey catching, and electronavigation in elasmobranchs. Convincing evidence exists concerning the role of the passive electrosense in feeding behaviors and its sensitivity to weak electric fields. Less explored is its role in orientation and navigation in response to geoelectric fields and competing theories remain for using the earth's magnetic field for electronavigation.

2. Low-Frequency Electric Signals in the Aquatic Environment

The aquatic environment provides many sources of electric fields. They range from local fields produced by animals to large-scale, quasi-uniform fields produced by ocean currents, boundary layer effects, or the earth's magnetic field. According to their origin electric fields are frequently divided into (1) animate electric fields, (2) fields caused by physicochemical effects, and (3) electric fields induced by magnetic induction.

Before reviewing some data on electric field intensities, it must be emphasized that accurate measurement of DC or AC fields is very difficult and comparisons made between different authors and techniques is problematic. Aside from technical problems regarding electrode polarization and drift, the exact location of both the recording electrode, as well as the reference electrode, is very important. Because of the weak fields, noise in the electrode and amplifier usually requires placement of the electrode as close as possible to the source. The actual field values at some distance relative to the location of a possible electroreceptor are not only much weaker (by the third power of the distance in the far field), but also depend on the size and orientation of the often-multipolar sources. Even if distance, orientation, and source amplitude are kept constant, the size of a dipole alone has profound effects on the field amplitude at a distance. Extreme caution is advised for attempts to compare source field intensities with receptor threshold values.

2.1 Fields from Animate Sources

Soon after the discovery that sharks and catfish avoid the electric fields of metal rods in the water by means of their ampullary electroreceptors (Parker and van Heusen 1917; Dijkgraaf and Kalmijn 1962), the search for behaviorally relevant stimuli began. It was already established that animals as well as plants show steady electrical potentials (Burr and Northrop 1939). Kalmijn (1966) provided the first behavioral evidence that sharks could sense the weak electric fields of a flatfish buried in the sand. Later, he measured the electric fields of many aquatic animals and also gave more detailed information about the frequency composition of these bioelectric fields (Kalmijn 1972).

All aquatic animals show a pattern of DC field potentials around the body; in fish it is generally strongest in the head and gill region. It is strongest in teleosts (up to 500 μV, measured 1 mm above the skin; Kalmijn 1974). Potentials are about 10 times weaker in elasmobranchs. The DC potentials of crustaceans are in the same range as those of elasmobranchs, but potentials of more than 1 mV were found in wounded specimens (Kalmijn 1972, 1974). Potentials up to 1 mV measured directly at the carapace were also recorded from the freshwater plankton, *Daphnia*, decreasing with the power of three or more as distance is increased (Wojtenek et al. 2001b). Many other invertebrates have potentials only on the order of 10 μV, values that also depend greatly on the distance of the electrode from the source. The sources of these DC potentials are not well understood. The potentials persist if the animal is anesthetized and thus are not dependent on muscle contractions (Kalmijn 1972). In catfish and trout, Roth (1972) found negative DC potential fields around the mouth and positive ones at the gill openings. With one electrode 1 to 3 mm above the skin and a remote reference electrode, he recorded potentials up to 3.5 mV from the trout and 1.5 mV from catfish. The trunk and tail did not produce measurable potentials. Peters and Bretschneider (1972), also using catfish, recorded bioelectric fields that were largest within 5 mm of the gill openings (800 μV, Fig. 9.2). However, in contrast to Roth (1972), Peters and Bretschneider (1972) reported negative values around the gills and positive ones at the mouth; the trunk and tail also showed positive electric fields.

Respiratory movements often modulate the DC potentials. These modulations can be as large as the DC potential itself, but "normally presented themselves more as a 'ripple' on a relatively high dc level" (Kalmijn 1974). Respiratory potentials are probably not caused by the contraction of muscles since passive movements of the gill covers in anesthetized animals cause modulations similar to the ones observed in awake animals. The frequency of respiratory potentials is usually in the range of 0.5 to a few Hertz. Low-frequency potentials of invertebrates are rarely correlated with respiratory movement. Taylor et al. (1992) described the AC fields of a variety of platypus prey items. Swimming earthworms showed a 3 Hz rhythmic field with an amplitude of 3 μV cm^{-1}. Leeches can reach up to 210 μV cm^{-1} at a peak frequency of 6.2 Hz. The fields of decapod crustaceans are characterized by their tail flick with resulting

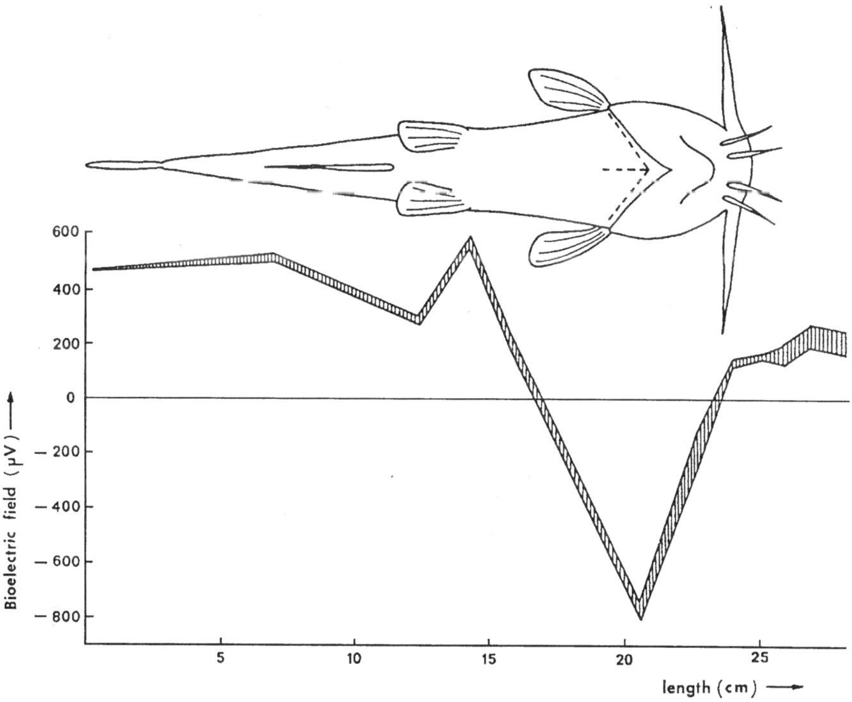

FIGURE 9.2. Bioelectric field of the catfish (*Ictalururs nebulosus*). The potentials were measured 5 mm from the ventral side of the animal with respect to a 1 m distant reference electrode. The thickness of the line indicates the amplitude of the respiratory potential. (From Peters and Bretschneider 1972.)

amplitudes of about 150 μV cm^{-1} in the redclaw and freshwater shrimp and peak frequencies of 4.5 and 8.8 Hz, respectively. An exception is the atyid shrimp, which generates potentials up to 1900 μV cm^{-1} with a broad frequency peak between 1.6 and 3.1 Hz. Several insects were also investigated. Many showed only weak or no potentials, but some reach amplitudes up to 800 μV cm^{-1}. The frequency peaks were in the range of 5 Hz, but some were as high as 20 Hz. It is not clear whether the fields measured by Taylor et al. (1992) represent nerve or muscle potentials, or whether they are the result of secondary modulations of a DC field. Unfortunately, DC fields were apparently not measured by Taylor et al. (1992).

Higher-frequency signals (above 20 Hz) are produced by the direct action of nerves or muscles (Kalmijn 1974). Most passive electrosensory animals are insensitive in the frequency range of muscle potentials but coordinated synchronous contractions may dramatically increase the amplitude of the electric fields. For example, strong contractions cause potentials on the order of 10 to 20 μV in teleosts, even higher (100 μV) in tunicates and mollusks. Kleerekoper and

Sibakin (1956a,b, 1957) reported high-amplitude spike potentials from the lamprey (200 to 300 μV measured 1.5 to 2 cm over the eye). The source of these potentials was not determined but the authors suggested a possible role in active electroreception. Strong potentials are also associated with contractions of the sonic muscle of the sound producing toadfish *Opsanus* (Bennett 1970). He recorded electric fields of tenths of a millivolt and suggested that the sonic muscle spikes are a forerunner of electric organs. The toadfish is not electroreceptive, however, and the 100 to 200-Hz electrical potentials are probably only a by-product of the sound producing system.

The weakly electric fish produce greater amplitude electric fields by means of specially developed organs. Most of these fish emit either pulses or sinusoidal fields with a peak amplitude of a few volts. These fields are used by the active electrosensory fish to probe their environment or for communication and are discussed by Hopkins (Chapter 10). However, some passive electrosensory catfish are specialized to detect these fields and use this information to capture their prey (Westby 1988). A few animals produce strong electric discharges that are used to stun their prey. These discharges are not likely to be of any relevance for passive electroreceptive fish and are not discussed here.

2.2 Fields from Physicochemical Sources

Any boundary between chemically and physically different materials will generate electric potentials that may reach intensities far greater than those from animate sources. If two substances that differ in chemical composition, concentration, or temperature, come into contact, there is a current flow from one to the other. For example, a metal object put into water will attract or lose electrons depending on its electrochemical force relative to the force of the surrounding water. Since the positive charges of a metal cannot move, equilibrium is reached very soon and results in a steady DC potential without further current flow. If both positive and negative charges are free to flow, as is the case in two liquids, a steady current may flow. Both forms of electrochemical potential are very common in nature and form a rich electrical landscape. Any body of water is surrounded by boundaries with different kinds of solid material creating regional or global fields that are quite stable over time and could be used for orientation and navigation. In addition, vertical or horizontal stratifications of water masses add other components to the local electric fields. These are more variable in time and can change in a matter of days or hours. However, these changes frequently occur in a predictable way as water stratifications, such as temperature or salinity zones, and are subject to daily or seasonal variations. Of course, these zones also could be detected directly by measuring temperature or salinity, but the electric fields between and within zones could give additional directional cues based on their orientation and size.

There are several sources of electric potentials that contribute to characteristic electric fields in a river (Krajew 1957): (1) ion absorption at the riverbed, (2) water filtration through the riverbed, (3) the river current, and (4) temperature gradients. Each equates to an electric field intensity well above the thresholds of passive electrosensory animals. Krajew (1957) gives examples of filtration fields across a riverbed on the order of 10 to 20 mV m^{-1}, and flow fields along the river of 19 to 30 mV m^{-1}. The current fields are proportional to the flow velocity and thus change at any mechanical barrier along the river. Measuring the longitudinal component of the electric fields may give an animal some information on the flow speed and direction. This could be particularly important for animals that are not in tactile or visual contact with the riverbed. In the open ocean, similar mechanisms apply although the fields caused by ocean currents are more complex. Furthermore, salinity differences are probably more pronounced, particularly at river mouths and vertical currents. Organic matter within ocean layers can alter electric fields as well. Jumps in potential as a result of accumulation of phytoplankton and bacteria were reported by Bogorov et al. (1969).

Stable electric fields, which could guide animals through their aquatic environment, were measured in freshwater by Peters and Bretschneider (1972). They found local fields up to 15 mV m^{-1} that are quite stable over time. The potentials in salt water are much weaker (Pals and Schoenhage 1979; Pals et al. 1982a) and can be distinguished as either local or regional in nature. Local electric fields are associated with the substrate and measure up to 1.5 mV m^{-1} for mud or clay bottoms while weaker fields occur over sandy substrates. These are stable over time at any one location, although there is considerable variation between locations. Regional fields correlate strongly with tidal currents and were measured with wide-span electrodes 100 m apart. Pals et al. (1982a) suggested that an animal could use these fields to determine the direction and velocity of the water current carrying the animal. This is important information for an animal that has no tactile or visual contact with the seabed.

2.3 Fields Created by Magnetic Induction

Changes of or relative movement within a magnetic field creates an electric field perpendicular to the magnetic field lines. Since the earth has a magnetic field, any conducting medium moving through it induces an electric field. For an electrosensory animal two potential mechanisms are available to perceive the magnetic fields: a passive detection of gradients from ocean streams, such as the Gulf Stream, or an active detection of gradients set up by swimming movements. According to the *passive* model of electronavigation, an ocean stream subject to the vertical component (vector) of the Earth's magnetic field would generate a transverse flow of current at the surface and an ohmic voltage gradient that the fish could interpret as either upstream or downstream in direction. This would allow a fish to utilize or compensate for drift in navigation. Voltage gradients of 0.05 to 0.5 μV cm^{-1} have been measured in the surface waters of large-scale oceanic gyres, and 0.25 μV cm^{-1} gradients reported for strong tidal currents (data cited in Kalmijn 1974), both within the then accepted threshold sensitivity limits for sharks and rays (0.01 μV cm^{-1}).

The *active* swimming model is predicted to enable the fish to derive actual compass headings and position. Swimming in the horizontal component of the earth's magnetic field would generate currents that pass vertically through the fish—upward when swimming east (Fig. 9.3), downward when swimming west. Therefore, voltage gradients detected between ampullae on the dorsal and ventral surfaces could be interpreted to signal east–west compass headings. Swimming parallel to the magnetic field would generate no current flow and could be interpreted as north or south. Conversely, the earth's vertical magnetic component would affect the ampullae that extend laterally by generating a parallel or transverse current. A voltage gradient relative to the inclination of the vertical axis may give information on the corresponding latitude, thus providing the fish equivalent of global positioning. Moderate swimming speeds of 50 cm s^{-1} yield theoretical voltage gradients of 0.1 to 0.2 μV cm^{-1} (Kalmijn 1988), which are well within threshold sensitivity. Thus, a swimming fish probing the magnetic field by making frequent turns could develop an unambiguous sense of direction from the induced currents. The "compass sense" works only in a dynamical mode and requires the constant computation of sensory input with the speed and direction of the fish's own movement. For example, a shark swimming in a southwesterly direction would perceive an increase in voltage by turning toward the west, but this would also occur with an increase in swimming speed.

A modification of the electromagnetic induction theory has been proposed more recently (Paulin 1995) that addresses perceived weaknesses in the original theory. These include the ability of the animal to compensate for passive "drift potentials" when navigating using the active mode. In addition, a passively induced DC potential would be undetectable by the ampullae of Lorenzini since

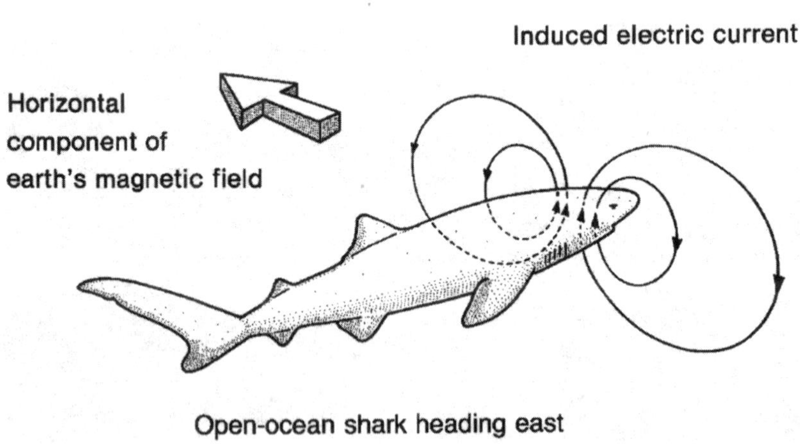

FIGURE 9.3. A shark swimming through the Earth's magnetic field. If it swims eastbound, the horizontal component of the magnetic field induces an electric field that flows ventrodorsally through the fish's body. (From Kalmijn 1974.)

these receptors adapt to a steady offset potential. It seems unlikely, however, that passive drift potentials would be purely DC since streams, eddies, and so forth are seldom uniform and nonuniformity is also characteristic of local magnetic fields (Klimely 1993). Also, ampullary receptors are responsive to very low frequencies of 0.1 Hz and less (Murray 1962; Tricas and New 1998; Hofmann and Wilkens, unpublished observations from paddlefish ampullae). A key feature of Paulin's theory requires that voltage modulations that occur during turns be compared with the waveform of head movement registered by the vestibular system. These learned cues would be independent of interference from environmental electric fields such as the drift currents. The lateral undulations of the head of a shark during swimming would induce continually varying currents and, as a result, register continual compass information. According to this theory, skates and rays would presumably benefit less since they swim without lateral undulations. Paulin's theory has not been tested behaviorally.

3. Electroreception in Cyclostomes

The earliest evolutionary existence of the electrosense in a living vertebrate occurs in lampreys. Electroreceptors are distributed along the entire length of the adult lamprey, are innervated by branches of the anterior lateral line nerve (ALLN) whose axons terminate in the dorsal (electrosensory) nucleus of the medulla (DON), and are excited by weak, low-frequency cathodal stimuli (Bodznick and Preston 1983; Ronan and Northcutt 1987). These features are homologous with the primitive non-teleost electrosensory system (Bullock et al. 1983), with the exception that lamprey electroreceptors are not restricted to the head, and suggest that electroreception was an ancestral feature predating the evolution of jawed fish. Hagfish lack any vestige of the electrosense and have either lost this sensory modality or, as seems more likely, represent a separate nonvertebrate lineage that predates the electrosense (Fig. 9.1). The more ancient cephalochordates also appear to lack the electric sense.

Despite the presence of electrosensors, no electrosensory behavior has yet been described in lampreys. Possible functions, as advanced by Bodznick and Preston (1983), include near-field detection of and orientation toward the DC or low-frequency electric fields of other (prey) fish, and use of the earth's geomagnetic fields as guidance for their long-distance migrations, the latter resulting from detection of electric fields induced by movements of the fish and/or oceanic currents (Kalmijn 1974). More recently, electrosensitivity has been demonstrated in larval lampreys (Ronan 1988). For the burrowing ammocete larvae, long-lived microphagus feeders, a potential electrosensory function might include predator detection and avoidance by cessation of physiological activities such as pharyngeal pumping that generates water currents, as has been shown for skates (Sisneros et al. 1998).

4. Electrolocation in Chondrichthyes

Passive electrosensory behaviors are well documented for cartilaginous fish, beginning with the classic experiments of Kalmijn (1971) in which he described the biological significance of the electrosense for feeding by sharks and rays. Earlier behavioral experiments (Dijkgraaf and Kalmijn 1962) had established that the dogfish, *Scyliorhinus canicula*, was sensitive to the galvanic currents associated with metal wires advanced to within a few centimeters of the head, with the ampullae of Lorenzini as the candidate sensory organs. Stimulation by metallic wires or rods, but not glass rods, evoked oriented escape reactions, as did electric field gradients of 1 μV cm^{-1} or more delivered by electrodes in the tanks. These experiments confirmed the historic account of electrosensitivity in fish by Parker and van Heusen (1917), who demonstrated a galvanic sensitivity to metallic rods in the catfish, followed by avoidance reactions to currents measuring around 1 μA. The biological relevance of avoiding metallic objects has little evolutionary significance aside from the aforementioned practicality, although there are implications for impacts from man-made structures placed in the aquatic environment (Gurgens et al. 2000).

4.1 Passive Electroreception in Holocephali

The chimeras or ratfish, although predating elasmobranchs in the fossil record by several hundred million years, nevertheless possess a seemingly equivalent electrosensory system of Lorenzinian ampullae. The electrosense has been studied in only a single species, *Hydrolagus colliei,* a bottom-dwelling fish from deeper subtidal waters along the eastern north Pacific coast. Fields et al. (1993) describe a set of ca. 500 ampullary pores on the head of *H. colliei* and tectal evoked potentials in response to 5 Hz cathodal stimuli of less than 1 μV cm^{-1}, but in the two animals available for this study no overt behavioral response to novel electrical stimuli was observed. To elicit an electrosensory response, a ratfish was subjected to associative conditioning using 10-μA stimulus pulses and prodding with a glass rod. After training, fish responded to the electrical stimulus with an escape response involving erection of the dorsal spine and an abrupt reversal in the direction of swimming. With decreasing intensity, response percentages dropped from 100% at 10 μA to 35% at 2 μA, with threshold potentials calculated to be in the range of 0.2 μV cm^{-1} or less.

Behavioral and physiological thresholds for the ratfish are approximately 100 times less sensitive than for sharks (Fields et al. 1993). Nonetheless, the electric fields of potential prey organisms are within the range of sensitivity of the ratfish (see Section 2), and the concentration of ampullae about the head and mouth of the ratfish suggests that the electrosense is used in feeding, for example, in searching for its typical prey of benthic invertebrates. The head also features prominent lateral line canals juxtaposed with the fields of ampullae, so a bimodal sensory function is likely. Unfortunately, the captive ratfish available for this study did not feed, making it impossible to test these ideas. Predator detection is also a possible function given that conditioning stimuli trigger spine erection and escape swimming.

4.2 Electrosensory Behavior in Elasmobranchs

All sharks, skates, and rays are electrosensory, and all examined to date appear to use their ampullary electrosense passively. Skates (Rajidae) have weak electric organs and generate electric organ discharges (EODs) during social interactions with other skates, and they discharge in response to experimental stimuli, both electrical and mechanical (Bratton and Ayers 1987; Sisneros et al. 1998). However, they do not appear to use EODs for active electrolocation. Only ampullae with canals and pores directed posteriorly respond to the self-generated EODs and this is relatively weak compared to their response to ventilatory reafference (New 1994).

4.2.1 Threshold Sensitivity

The role of the elasmobranch electrosense best characterized is in detecting local dipole fields such as those associated with potential prey. Early evidence for this assertion revolved around measuring threshold sensitivity in sharks and skates and equating this with the electric fields characteristic of their natural prey (summarized in Kalmijn 1974, 1978, 1988). In experiments using either DC or 5 to 8 Hz stimuli, variously as uniform or local dipole fields presented to quiescent sharks and skates, reflex responses were triggered at voltage gradients as low as 0.1 $\mu V\ cm^{-1}$. Sharks responded by eyelid closure and skates by respiratory phase shifts seen as short interruptions in rhythmic ventilatory movements of the spiracle valves. Stronger stimuli (1 to 10 $\mu V\ cm^{-1}$) triggered startle or avoidance responses in sharks. To evaluate threshold sensitivity under more natural conditions, Kalmijn recorded heartbeat signals from free-swimming sharks and skates. Bradycardia could be elicited in unconditioned skates (*Raja*) at a voltage gradient of 0.01 $\mu V\ cm^{-1}$ in uniform 5-Hz square-wave fields. This represented the greatest sensitivity to date for any aquatic organism, and fell well within the range of electric field strengths of known prey (10 to 500 μV close to the surfaces of marine invertebrates and fish, see Section 2).

Maximum sensitivity is lower yet by approximately an order of magnitude (1 to 5 $nV\ cm^{-1}$), as demonstrated in several behavioral contexts. In field experiments in which the orienting response of the dogfish *Mustelus canis* to a local dipole field was quantified, responses were observed at distances exceeding 38 cm that corresponded to a field strength conservatively estimated at approximately 5 $nV\ cm^{-1}$ (Kalmijn 1982). In a companion experiment in the laboratory, the stingray *Urolophus halleri* was successfully conditioned to orient toward and enter the correct compartment on the basis of detecting the polarity of a uniform electrical field, again at a field strength of 5 $nV\ cm^{-1}$. In other conditioning experiments nurse sharks (*Ginglymostoma cirratum*) were trained to detect steel balls whose electric fields were determined by measuring their dipole moments (Johnson et al. 1984). Average threshold sensitivity was 18.6 nV cm^{-1}, but this was lowered to 4.3 nV cm^{-1} in the presence of uniform background fields of 5 nV cm^{-1}. Thus, ambient field strengths equivalent to thresholds obtained for sharks in experiments at sea appear to enhance a behavior likened to feeding (Kalmijn 1988).

In more recent experiments confirming this exquisite sensitivity, Haine et al. (2001) report that both sharks and rays orient to dipole fields at distances equivalent to threshold strengths of 4 nV cm^{-1}, and Kajiura and Holland (2002) report that sharks respond to field strengths as low as 1 nV cm^{-1}. The latter study was designed expressly to evaluate whether the wider dimension of the head of hammerhead sharks confers greater sensitivity to electric fields. Sandbar (*Carcharhinus plumbeus*) and hammerhead (*Sphyrna lewini*) sharks of equivalent size were presented dipole fields within enclosures in their natural environment, but no differences in threshold sensitivity were observed. However, the "hammer" shape does appear to facilitate greater maneuverability in orienting to the stimulus, a benefit for the capture of prey bent on escape, and it may also widen the search area with respect to local dipole fields.

4.2.2 Prey Detection

Kalmijn (1971, 1982) performed a series of experiments that convincingly demonstrated a primary role for the electric sense in directing the final stage of the feeding attack. Equivalent results were obtained for both the skate (*Raja clavata* L.) and the shark (*Scyliorhinus canicula* L.) in carefully controlled laboratory experiments, although we will describe only the shark experiments. Sharks were allowed to search their tanks for small flatfish (the plaice, *Pleuronectes*) that buried themselves in the sand. Dispersing a few drops of fish juice (from a whiting) in the water heightened feeding behavior. The sharks were adept at locating the fish but only when passing within 15 cm or less of the hidden or nearly hidden prey (Fig. 9.4a). Following detection sharks made sudden turns toward the plaice, exposing it by removal of the sand and grasping and shaking the fish before swallowing it in smaller pieces.

To demonstrate that the electric field of the plaice was instrumental in guiding the attack a series of control feeding experiments were performed to eliminate the possibility that visual, chemical, or mechanical stimuli were involved. To this end a live plaice was encased in a seawater–agar chamber with inlet and outlet tubes to irrigate the fish that was then buried in the sand. The chamber was determined to be electrically transparent by recording electric field potentials across the agar walls, but it served to eliminate any chemical or mechanical signals from the plaice and hide it from view. Sharks aroused to feed initiated attacks similar to those of uncased fish by uncovering and then grasping the agar chamber as if attempting to extract the bait, but as before, only if passing within 15 cm. Attacks were directed specifically at the head end of the chamber (Fig. 9.4b), the region of the fish producing the largest potentials. If instead small pieces of whiting were placed in the chamber the shark now investigated

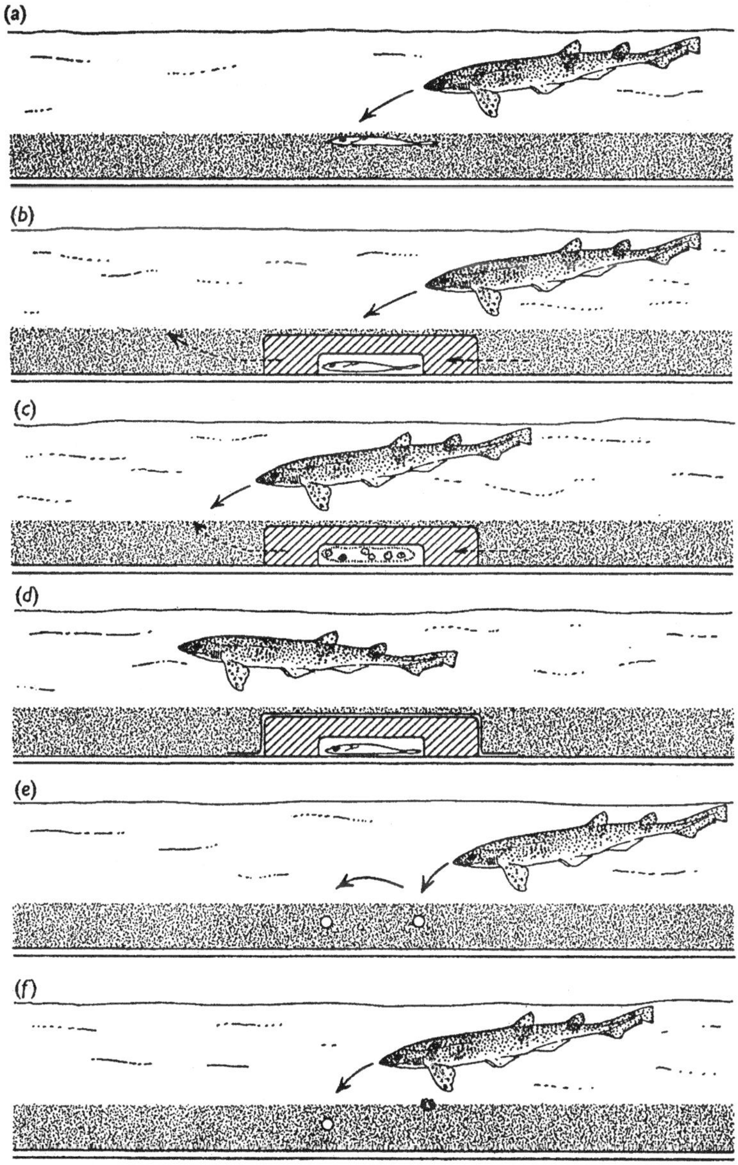

FIGURE 9.4. Feeding attacks by the shark on (**a**) a buried flatfish, (**b**) a flatfish encased in an agar chamber, (**c**) odor of fish pieces, (**e**) dipole electrodes, and (**f**) electrode versus bait. In (**d**) the agar chamber is insulated by plastic film. *Solid arrows* indicate point of attack, *broken arrows* seawater flow through the agar chamber. (From Kalmijn 1971. Reprinted with permission from The Company of Biologists, Ltd.)

the outlet tube, ignoring the chamber (Fig. 9.4c). When the chamber was insulated with a thin sheet of plastic before it was covered with sand the shark no longer appeared able to detect it (Fig. 9.4d). These simple but elegant experiments showed that sharks responded with normal feeding responses to sensory signals limited to the electric fields of its prey, thus clearly showing that the passive electrosense is an important sensory adaptation for feeding behavior. To reinforce this conclusion with a more direct test of the electrodetection hypothesis, electrodes that created dipole fields simulating the plaice were inserted into the chamber in place of the fish. The shark attacked the agar chamber similarly to the way it would if a live fish were present (Fig. 9.4e). In a final test, the shark was given a choice between fresh pieces of whiting and the artificial dipole field. Attacks were aimed at the electrodes, ignoring the bait (Fig. 9.4f), and further certify that the electrosense is the *dominant* near-field sensory modality used in directing the terminal predatory strike.

These two species, *Raja* and *Scyliorhinus*, are bottom-dwelling elasmobranchs where the electrosense would have the advantage of penetrating the substrate in search of prey, that is, if their predatory behavior in the natural environment is equivalent to that observed in the laboratory environment. A subsequent test of the electrosensory feeding hypothesis under natural environmental conditions has validated this prediction in another bottom-dwelling dogfish, *Mustelus canis*, and extends these conclusions to include the epipelagic blue shark, *Prionace glauca* (Kalmijn 1978, 1982) where prey must be captured in a three-dimensional environment. This experimental protocol utilized two sets of dipole electrodes, separated by an odor source to attract the sharks and lowered to substrate in shallow water or suspended 5 m below the ocean surface in deeper water. Current was supplied variously to one set of electrodes (8 μA DC, 2 to 5 cm tip separation), with the other set serving as a control. Sharks responding to the target were observed at night from a nonmetal inflatable raft under dimly lit conditions. Small dogfish oriented to the benthic target by making a sudden turn and invariably attacking the live pair of electrodes (Fig. 9.5A). The orientation response distances exceeded 15 cm in more than a third of the attacks, with the greatest (38 cm) corresponding to the oft-reported 5 nV cm^{-1} threshold sensitivity value. With both sets of dipoles active sharks preferentially attacked the closely spaced electrodes, and often avoided the electrodes with wider separation. Blue sharks approached the suspended target more cautiously but attacked the electrically active electrodes in 80% of the observed responses (Fig. 9.5B).

Field studies on two additional shark species, the white shark (*Carcharodon carcharias*) and the swell shark (*Cephaloscyllium ventriosum*), have produced similar results. White sharks attracted to horseflesh bait were presented a choice between baits with and without dipole electrodes delivering either pulsed or DC currents (2.2 V at the source). Although the results were reported as "cursory" in nature, the sharks nevertheless attacked the electrified baits three times more often (Tricas and McCosker 1984). Furthermore, these sight predators roll their

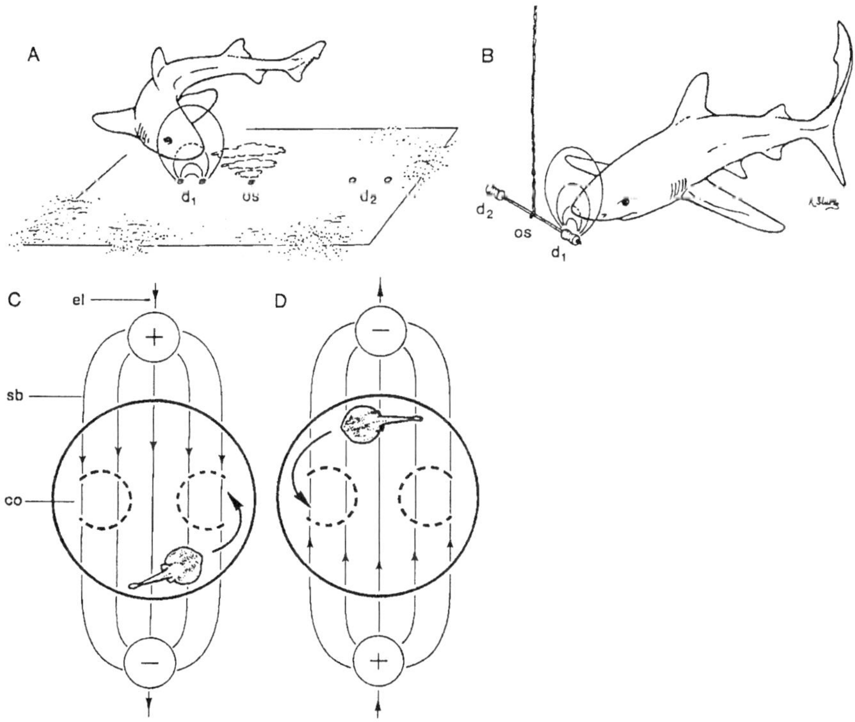

FIGURE 9.5. Open-water feeding attacks of (**A**) the shallow-water dogfish at the sea bed and (**B**) the oceanic blue shark in the water column. d_1, Active dipole; d_2, control dipole; *os*, odor source. (**C**, **D**) Orientation of the stingray to a 5 nV cm^{-1} uniform electric field, with the ray trained to enter the corral (co) on the left relative to electric field direction. el, electrodes; sb, salt bridges. (From Kalmijn 1982. Reprinted with permission from the American Association for the Advancement of Science.)

eyes back during the final moments of an attack, suggesting that they rely on the electrosense for near-field orientation to the prey while protecting their eyes (Tricas and McCosker 1984). In a more quantitative study the role of the electrosense in prey detection was tested in the swell shark (Tricas 1982), a bottom-feeding ambush predator that captures small fish such as blacksmith (*Chromis punctipinnis*) as they rest on exposed bottom at night. Blacksmith tethered to poles and presented to sharks by divers were readily taken by either a "yawn" or "gulp" capture sequence when brought within 5 cm of the head of the shark. Prey was not taken if sealed in plastic bags or if presented in a special chamber that was sealed with plastic film. Blacksmiths in the open chamber, or when enclosed by an agar lid, were attacked in 80% and 60% of the respective feeding trials, the latter eliminating all sensory cues except for the electric field of the blacksmith.

4.2.3 Predator Detection

Local dipole fields are as likely to provide information about predators as they are for prey. However, no studies have systematically examined whether the passive electrosense mediates locomotory escape or avoidance behaviors. Vision and lateral line senses are clearly important as early warning systems, and where turbidity is low and/or the predator is large these senses would appear to function at distances greater than would be possible for the electrosense. However, there are indications that electrical signals do convey information about which fish are wary. As noted early on, strong electric fields (1 $\mu V\ cm^{-1}$) and galvanic potentials trigger escape responses in sharks (Kalmijn 1974), and small sharks that attack closely spaced electrodes avoid those with wider spacing (Kalmijn 1982; Pals et al. 1982b). Sharks conditioned to orient to and grasp steel balls avoid the largest size, presumably because of its stronger electric field (Johnson et al. 1984). These somewhat anecdotal observations suggest that electric fields of greater intensity or dimensions may be interpreted as belonging to larger animals, and therefore threatening. It is equally plausible that stronger artificial electric fields do not accurately simulate the dipole characteristics of any organism and are simply avoided.

That the electric sense is used to detect and avoid predators is most clearly shown in experiments on embryos of the clearnose skate, *Raja eglanteria* (Sisneros et al. 1998). These skates are oviparous and deposit egg cases containing a single embryo on the benthic substrate. Toward the end of their 12-week embryonic development period skates are quite active and continuously ventilate the inside of the egg case by whipping the tail, a motion that produces a streaming current of water through the case. Embryos exhibit a freeze response to various external stimuli including weak electric fields (0.56 $\mu V\ cm^{-1}$) at 0.5 and 1.0 Hz, frequencies representative of the DC electric fields of large predators modulated by respiratory or locomotory (relative to the embryo) movements (Tricas et al. 1995). "Freezing" the ventilatory current eliminates a hydrodynamic signal easily detected by the lateral line receptors of potential predatory fish known to prey on skate embryos. Near-term embryos have functionally developed ampullary receptors with best frequency responses around 1 to 2 Hz and therefore are well equipped to detect predatory signals (Sisneros et al. 1998). This match between freeze response behavior, electroreceptor sensitivity, and predatory signal bandwidths is a clear indication that the passive electrosense is an important mechanism for detecting and avoiding potential predators.

4.2.4 Mating Behavior

As cited previously, skates discharge their electric organs when in the company of conspecifics (Bratton and Ayers 1987; Tricas et al. 1995; Sisneros et al. 1998) and their ampullary receptors have been shown to be sensitive to these signals (New 1994). However, no social interaction or behavioral response has been linked directly to the EOD. Rays, which lack electric organs, nonetheless rely heavily on electrosensory communication as they congregate in shallow waters during the mating season. Tricas et al. (1995) have provided convincing evidence that round stingray males (*Urolophus halleri*) utilize the substrate penetrating features of the electrosense described previously for elasmobranch feeding to locate females. Male rays cruise the shallow sand flats searching for reproductively active females that have buried themselves in the sand. Location of the female is signaled by a sharp turn toward the inconspicuous potential mate (Fig. 9.6A) after which the male excavates the female and begins courtship behavior. Female rays can also locate buried consexuals and position and bury themselves nearby. Detection of conspecifics relies on the bioelectric potentials that arise across the skin and gills (see Section 2). Tricas et al. (1995) have shown that these DC potentials are strongly modulated by movements of the gill slits and spiracles and that they have a frequency spectrum maximum at the 1 Hz ventilatory rate (Fig. 9.6B). These signals were recorded and played back to the wild population of rays in their natural environment. Playback fields (modulated by an 8 μA current) were delivered by dipole electrodes implanted in plastic scale models of female rays that were buried in the sand. Male rays showed distinct orientation reactions and approached to inspect the model (Fig. 9.6C). Females also responded, often burying near the model. Recordings of the stingray electroreceptors again showed a close match to the bioelectric signals with a peak sensitivity of 1 to 2 Hz. These experiments demonstrate clearly that the passive electrosense is used as a mechanism of communication in social behavior, in addition to its role in feeding and predator detection.

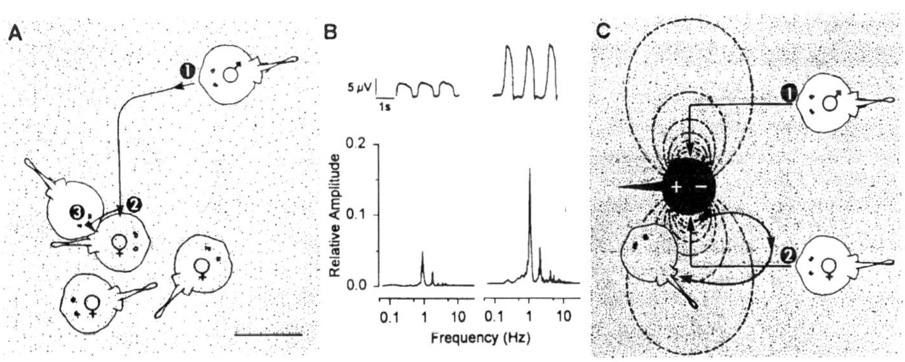

FIGURE 9.6. (**A**) Orientation responses of round stingrays in the wild. Male search path (1) shows an abrupt turn toward and (2, 3) inspection of disc margin of a buried female. (**B**) Modulated respiratory potentials of female ray recorded dorsally at the spiracle (*top left*) and ventrally at the gill slits (*top right*). Frequency spectrum (*bottom traces*) shows maximum power at 1 Hz. (**C**) Orientation responses to simulated respiratory potentials from electrodes in a buried plastic ray model. Males (1) and females (2) orient toward and investigate the model. Females often bury close by (cf. in **A**). (From Tricas et al. 1995. Reprinted with permission from Elsevier and the author.)

4.2.5 Geomagnetic Navigation

Given the exquisite sensitivity of the elasmobranch ampullary receptors, Kalmijn (1974) has proposed an electromagnetic induction theory whereby marine fish moving relative to the earth's magnetic field could detect voltage gradients useful for orientation and navigation (see Section 2).

In fact, there is as yet no direct evidence showing that elasmobranchs use electromagnetic induction to navigate. However, behavioral experiments have established that sharks, skates, and rays *are* sensitive to magnetic fields at geological strengths and can distinguish the direction and polarity of the resulting voltage gradients. A fish need only to swim at 1 cm s^{-1} in a 0.5 Gauss magnetic field to produce the threshold gradient of 5 nV cm^{-1} (Kalmijn 1988). In preliminary observations, leopard sharks (*Triakis semifasciata*) avoided local perturbations (approximately 25%) of the earth's magnetic field imposed by a small induction coil next to the tank, and showed distinct preferences for near-magnetic north locations when resting undisturbed. To test whether elasmobranchs detect the earth's magnetic field under more rigorous conditions the round stingray (*Urolophus*) was trained to choose between compartments located on the east and west sides of the tank to obtain a food reward or avoid punishment. The stingrays were readily conditioned to choose the east compartment, after which they accurately switched to the west compartment (magnetically east) after the magnetic field was reversed with Helmholtz coils that encircled the tank and its enclosure (Kalmijn 1978). A similar conditioning protocol was followed to test for electromagnetic orientation by subjecting stingrays directly to uniform weak electric fields under various magnetic conditions, that is, in a null field, ambient field, or vertical component of the magnetic field (Kalmijn 1982). Stingrays chose the correct compartment, including following reversal of the field polarity, at a threshold potential of 5 nV cm^{-1} under vertical and ambient magnetic field conditions (Fig. 9.5C). In a limited set of experimental observations testing orientation sensitivity in the field, electric fields were imposed on Atlantic stingrays (*Dasyatis americanus*) swimming in the shallow featureless waters of the Bahamas. Electrodes were positioned to simulate a 45° rotation in the local electric field and in five of six encounters the stingrays made appropriate changes in direction.

These behavioral experiments show that elasmobranchs can be trained to respond predictably to currents induced by the earth's magnetic field. Attempts to replicate these results have been less successful with sharks. Pals et al. (1982b) were unsuccessful using food to condition the dogfish, *Scyliorhinus,* to respond to magnetic fields although this species showed orientation responses to dipole fields of much higher field intensity (1.5 μV cm^{-1}). Knudtson and Stimers (1977) found swell and horn sharks to be relatively insensitive to magnetic fields but demonstrated orientation and an avoidance response (burial in the sand) in the round stingray in laboratory experiments. Among elasmobranchs, rays and skates are best suited to detect potential gradients according to the passive model owing to the broad lateral extensions of their ampullary canals. Their dorsoventral flattened shape would be less well suited to detect the vertical gradients of the active model (Tricas 2001).

Evidence suggesting that elasmobranchs may use magnetic fields for electronavigation is still limited. The clearest indication is shown in the nocturnal homing migrations of hammerhead sharks tracked using ultrasonic transmitters from their home seamounts to feeding grounds in the offshore oceanic environment and back (Klimley 1993). These highly oriented migrations involved frequent changes in direction that would rule out passive drift potentials. Sharks did not follow bottom topography nor were their orientations correlated with spectral features of the environment. However, tracks were associated significantly with slope changes in the magnetic intensity based on magnetometer surveys encompassing their routes. Sharks showing homing migrations by following ridges and valleys in magnetic field intensities may be exercising a geomagnetic topotaxis.

The mechanism for geomagnetic sensitivity is still in dispute, however. Recent evidence for magnetoreception in rainbow trout has shown that these fish can be conditioned to discriminate the presence or absence of magnetic field anomalies, and putative receptors containing ferromagnetic crystals have been identified (Walker et al. 1997). Similar behavioral results have been obtained for the short-tailed stingray, *Dasyatis brevicaudata* (Hodson 2000; Walker et al. 2003). Although magnetic receptors have not been identified in elasmobranchs, small magnets attached to the nose of trained stingrays effectively blocks magnetic discrimination. Since a magnet would override the putative magnetoreceptors, as it does in sea turtles (Irwin and Lohmann 2003), these results have been interpreted as supporting magnetoreception as the primary sense—an attached magnet would not interfere with electromagnetic induction. Further study will be necessary to resolve the competing hypotheses for geomagnetic navigation, most likely incorporating the use of magnets in the behavioral paradigms for electric induction.

5. Passive Electroreception in Non-teleost Bony Fishes

The majority of the bony fishes are teleosts that have lost electroreception and reinvented it several times. The relatively few groups of non-teleost bony fishes, excluding only the Holostei (Fig. 9.1), have all retained the ancestral electrosense. These include the brachiopterygians (bichirs and reed fish) and chondrosteans (sturgeons and paddlefish), members of the ray-finned lineage, and lungfish and coelacanths, the lobe-finned fish (Fig. 9.1).

5.1 Lungfish and Coelacanths

Although the electrosensory organs of lungfish were discovered a long time ago (Fahrenholz 1929; Pfeiffer 1968), their function was unknown until Roth (1973) showed that lungfish could be trained to avoid a plastic tube carrying electrodes that produce a weak electric field as low as 0.3 μA cm^{-1}. Prey-catching behavior has been studied by Watt et al. (1999) in the Australian lungfish (*Neoceratodus forsteri*). A small crayfish located in a test chamber in the substrate evoked foraging and prey capture attempts even in the presence of an olfactory distractor. The distractor consisted of a tube leading from the test chamber to a remote site in the experimental arena. Even under these conditions, where the olfactory cues were spatially separated from the electric fields, the lungfish still relied more on its electric sense and was not distracted by the remote olfactory cue.

Evidence for electroreception in the coelacanth *Latimeria* stems from an anatomical study that revealed a well-developed dorsal nucleus in the brain stem (Northcutt 1980). This nucleus is the primary target of electrosensory fibers in all non-teleost electrosensory fish and its presence is strong evidence for electroreception. There are no behavioral studies in *Latimeria* except observations made by Fricke (personal communication, 2003) that *Latimeria* does react to weak electric fields presented through an antenna mounted on a submersible. However, it was not possible to control for the electric and magnetic influence of the submersible.

5.2 Brachiopterygians

The only behavioral investigation within the Brachiopterygii was done by Roth (1973) using the reedfish, *Calamoichthys calabaricus*. The study was done in parallel with the experiments on lungfish described in the preceding section and revealed the same results regarding the avoidance behavior to artificial electrical fields.

5.3 Chondrostean Fishes

The chondrosteans are comprised of the sturgeons and the paddlefishes. There is only one report on the behavior of sturgeons in response to electric fields (Basov 1999). At low field intensities, foraging behavior was observed whereas higher intensities triggered escape. Most electroreceptors are located ventrally around the mouth and the sturgeon feeding habit of searching for food in the riverbed suggests a primary role in prey detection.

Paddlefish (*Polyodon spathula*) are different in their feeding habit as they search for planktonic food in three-dimensional space. Accordingly, their electroreceptors are distributed more evenly between ventral and dorsal surfaces of the head. Another distinct feature is their elongated rostrum, which is covered with an extensive array of Lorenzinian electroreceptors (Jørgensen et al. 1972). The paddlefish has by far the largest number of electroreceptors of any passive (or active) electrosensory animal. Up to 75,000 were found by Nachtrieb (1910), whereas only 2824 were counted in *Sphyrna lewini* (Bodznick and Boord 1986), probably the shark endowed with the greatest number of ampullae. Behavioral data on the function of the electrosensory system in paddlefish are restricted so far to avoidance behavior and prey-catching behavior. No data are available linking the electric sense to orientation or navigation.

5.3.1 Avoidance Behavior

As in other electrosensory animals, larger electrical stimuli reliably trigger an avoidance response in the paddlefish (Jørgensen et al. 1972). Metal rods (17 to 78 cm long and 2.54 cm in diameter), avoided by sharks and catfish because of their electrochemical potential in the water, are very effective as well for the paddlefish (Gurgens et al. 2000). Avoidance typically occurred at distances up to 20 to 30 cm during continuous ram-ventilatory swimming. Rarely did the fish approach to less than 10 cm (Gurgens et al. 2000). Plastic rods or plastic covered metal rods are ineffective and presumably undetectable since, unlike metal, paddlefish frequently collide with these obstacles.

5.3.2 Prey-Catching Behavior

The long rostrum with its enormous number of receptors functions as an electrosensory antenna that enables the paddlefish to locate their minute planktonic prey (Wilkens et al. 1997). Juvenile paddlefish hunt individual plankton particles and can locate them with great accuracy using their electric sense alone. This was shown by Wilkens et al. (2001) in a series of feeding experiments. Young paddlefish were videotaped swimming in place in a flow tank to which plankton were added. The location and distance of each captured *Daphnia* was determined in a reference plane perpendicular to the rostral axis as they approached the fish (Fig. 9.7). All feeding experiments were conducted under infrared illumination, which excluded visual cues. Mechanosensory and chemosensory cues could be excluded by comparing the feeding performances of fish under high background concentrations of plankton extract, under turbulent flow conditions, and when presented with live, agar-coated *Daphnia.* With equal numbers of agar-coated *Daphnia* and "empty" agar pellets, the former accounted for 94% of all captures. Under each of these conditions plankton capture distributions and feeding efficacy were equivalent leading to the conclusion that plankton were detected and captured solely on the basis of their electrical field potentials (see Section 2.1). As added proof, Wojtenek et al. (2001a) showed that artificial electric fields mimicking those of the plankton and presented through silver dipole electrodes are sufficient to evoke prey-catching strikes. However, responses habituated with time, probably because the efforts to feed were not rewarded in this study.

6. Teleosts

Only a few groups of teleost fishes are electroreceptive. Many of them (the weakly electric gymnotiforms and mormyrids) possess electric organs and employ active electrolocation. Two groups, however, have evolved only low-

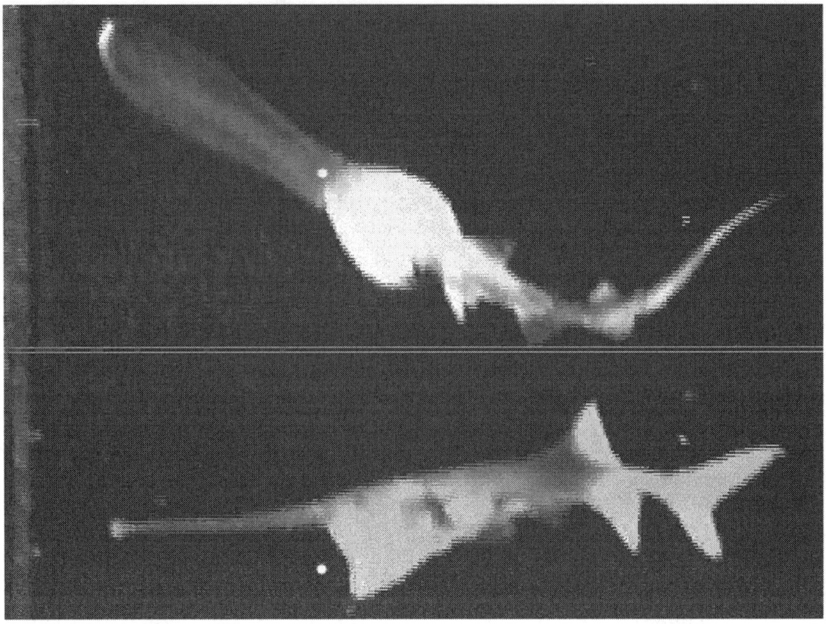

A

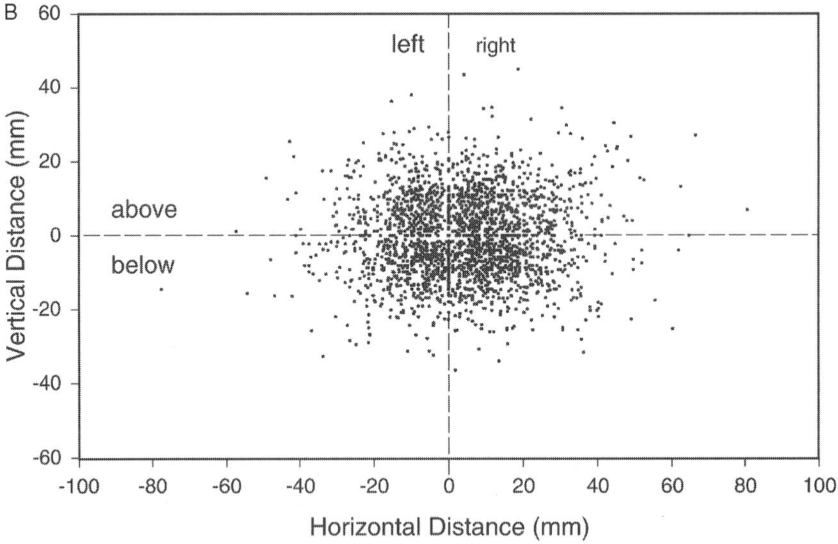

FIGURE 9.7. Plankton catching response of paddlefish in a flow tank. (**A**) Ventral (*above*) and lateral view (*below*) of a paddlefish catching a *Daphnia* (*white dot*). (**B**) Scatterplot of the position of *Daphnia* captures in a reference plane perpendicular to the tip of the rostrum. Most *Daphnia* were detected at a distance of less than 2 cm, but some were as far as 8 cm away.

frequency ampullary organs. These are the xenomystids, a group of fish closely related to the mormyrids, and the large and diverse order Siluriformes (catfishes). There are no behavioral data for the Xenomystidae, but for a few species of catfishes the behavior to both local and uniform electric fields is well investigated.

Catfish are long known for their sensitivity to metallic objects (Parker and van Heusen 1917). Depending on the size and kind of metal used, catfish either show avoidance reactions or are attracted and show various feeding behaviors. Since nonmetallic objects are ineffective, it was suggested that catfish respond to the electric fields caused by the electrochemical potential of metals (Parker and van Heusen 1917). This was confirmed by using batteries as stimuli, which give better control over the intensity of the electric field (Uzuka 1934). The responses obtained in that way range from barbel movements, approach and nibbling to avoidance and escape. In general, weaker stimuli resulted in approach and stronger stimuli in avoidance. However, Uzuka (1934) noted a great variability of the behavioral response, both between individuals and between different presentations of the same stimulus in one and the same individual.

The high sensitivity in catfish to electric fields was also thought to be a possible explanation for the anecdotal observations of earthquake predictions by these fish. The unusual behavior of catfish preceding earthquakes has been noted on several occasions. Although this applies also to nonelectrosensory animals, electrical anomalies preceding an earthquake theoretically could be detected by electrosensory animals (see Ikeya et al. 1998).

After initial speculation that the catfish gustatory system is involved in electroreception (Parker and van Heusen 1917), Lissmann and Machin (1963) and Dijkgraaf (1968) showed that the receptors involved are the small pit organs (ampullary organs) found over most of the head and part of the trunk. However, the behavioral relevance of electroreception in catfish remains elusive, as detection of metallic objects and earthquake prediction alone are not convincing reasons for the presence of electroreception.

Peters and Bretschneider (1972) measured the electric fields in the natural habitat of *Ictalurus* along with the bioelectric fields of some of its animals. Based on the sensitivity of catfish to electric fields, they concluded that bioelectric fields might be used to detect other animals (prey, predators) at close range. Furthermore, they found large-scale hydroelectric fields that could be used by catfish for orientation and navigation. Experimental confirmation for both was subsequently published.

Evidence for electrosensory guided feeding behavior was presented by Peters and Meek (1973). They fed two groups of catfish with either live *Xenopus* larvae or with pieces of meat for some time. Thereafter, the animals could select between a plastic tadpole dummy and a dummy with an artificial electric field simulating live prey. Only the group that had previously been fed live tadpoles preferred the dummies with an electric field over the others. This clearly showed that catfish use their electric sense for prey detection, but it depends on the previous feeding habits. Peters et al. (1999) investigated the discrimination thresholds between artificial dipoles of different sizes. They showed that catfish can discriminate between a 1 cm and a 1.5 cm DC dipole and successfully choose the smaller one in order to obtain a food reward.

Roth (1969) was the first to show that catfish use electrical fields to find hiding places. Catfish like to hide in crevices or, more conveniently for laboratory experiments, in tubes. Roth (1969) presented catfish two tubes, one with an artificial electric field and one without (Fig. 9.8). The fish could easily be trained to choose the tube without an electric field. In a similar experiment, catfish could even discriminate between fields of different polarity. The thresholds were said to be 0.0006 μA mm^{-1}. In addition, Peters and Meek (1973) mentioned preliminary studies on the spontaneous behavior of catfish to uniform and structured electrical fields. Catfish tend to orient themselves parallel to the

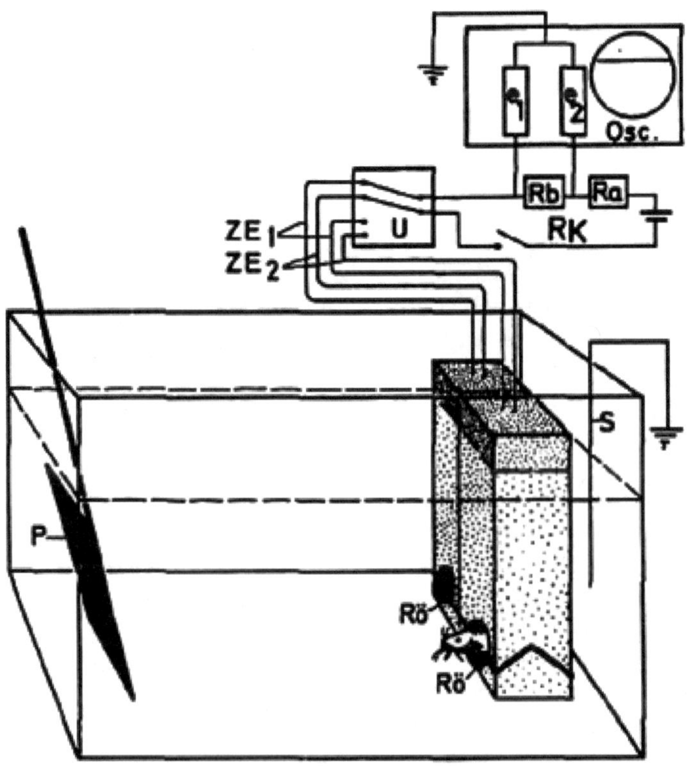

FIGURE 9.8. Experimental setup used by Roth (1969) to test whether catfish can use electric field information to find its home shelter. The fish could choose from two tubes (Rö) with different electric field properties produced by electrodes within the tubes. A black plate (P) was used as a starting point. When forced out of a tube, the catfish hides under the plate. When the plate is removed, the catfish had to choose one of the tubes. Electric fields were controlled by resistors (Ra, Rb) and a switch (U). An oscilloscope (Osc.) monitored the current flow. (From Roth 1969.)

field lines and prefer locations with the lowest current density. These observations were subsequently confirmed and extended (Peters and van Wijland 1974) by training catfish to find food either at the negative pole of an electric field (electrotaxis) or at a position that corresponds to a fixed angle relative to the field lines (an electrocompass reaction). In both cases, catfish were able to find the target. Peters and van Wijland (1974) also found that catfish show some degree of spontaneous compass reaction without conditioning.

Based on their need to seek shelter during daytime, Kalmijn et al. (1976a) trained catfish to find the correct hiding place within a uniform electric field. He could train catfish to find shelter not only at a certain polarity, but also at a certain angle relative to the field lines, thus confirming Peters and van Wijland (1974) results on prey detection. Kalmijn et al. (1976b) went further and demonstrated that catfish can find the correct hiding tube even if up to 12 are arranged radially in a circle at 30° intervals. Later, Baranyuk (1979) confirmed Kalmijn's results in an eight-armed maze.

Asano and Hanyu (1987) investigated the respiratory reflex, an interruption in the normal respiratory rhythm, to determine threshold values for the Japanese catfish *Parasilurus asotus*. This very sensitive method revealed that the animal could detect electric field amplitudes that are equivalent to a 1.5 nV voltage drop between head and tail of 30-cm fish. This brings the sensitivity of catfish into the range of elasmobranchs. Comparative data on the respiratory reflex in other passive electroreceptive animals is lacking.

7. Amphibians

Despite of a long history of research on the nervous system and sensory organs, the electric sense in amphibians was discovered relatively late. Anatomical studies have pointed out similarities between specialized organs in the skin and the Lorenzinian ampullae of elasmobranchs (Coggi 1905; Hetherington and Wake 1979), but it was not until the 1980s that the presence of electroreception was confirmed with anatomical (Fritzsch 1981a,b; Fritzsch and Wahnschaffe 1983; Istenic and Bulog 1984), electrophysiological (Claas et al. 1982; Münz et al. 1982, 1984), and behavioral techniques (see Section 7.1). Among the three groups of amphibians, only urodeles and gymnophians are electroreceptive; there are no reports of any electroreceptive frogs or toads (anurans), including their larvae.

7.1 Urodeles

Avoidance behavior was studied by Roth and Schlegel (1988) in the blind cave salamander, *Proteus anguinus*. If approached by a tube containing a silver wire electrode with a DC electric field, *Proteus* showed spontaneous avoidance or fleeing behavior. Habituation was prevented in successive trials by touching the flank of an animal that did not respond to the electrical stimulus alone. The threshold for avoidance behavior was 0.15 μA cm^{-2}. Later, Schlegel (1997) and Schlegel and Bulog (1997) tested avoidance using electrical stimuli at different frequencies, along with DC pulses, clicks, and noise bursts. They found similar minimum thresholds and best frequencies of 20 to 30 Hz. By investigating the threshold values in different populations of *Proteus*, Schlegel and Bulog (1997) found significant variations. Differences were thought to be due to the degree of adaptation to life in caves because obligatory cave dwelling populations show the greatest sensitivity.

Himstedt et al. 1982 showed that the axolotl *Ambystoma mexicanum* reacts to electric fields of 25 μV cm^{-1} with an arousal reaction involving increased frequency of gill movements. Local fields at somewhat higher intensities elicited prey-catching behavior toward the electrodes (Fig. 9.9). However, the animals had to be aroused and attracted with a visual stimulus prior to the onset of electrical stimulation. Himstedt et al. (1982) also mentioned preliminary studies on larvae of *Salamandra salamandra* and *Triturus alpestris* and suggested a more widespread distribution of electroreception in aquatic urodeles. Schlegel (1997), who studied mainly avoidance reactions, also observed approach in the blind cave salamander *Proteus*, a response that resembles the early stages of prey-catching behavior. Sometimes even biting or snapping was observed. These behaviors occurred only rarely and not in all individuals.

Prey-catching behavior is obviously not dependent on electrosensory stimuli alone, being observed only in combination with other stimuli. Elasmobranchs are usually aroused with chemical stimuli (see Section 4.2.2) and the axolotl

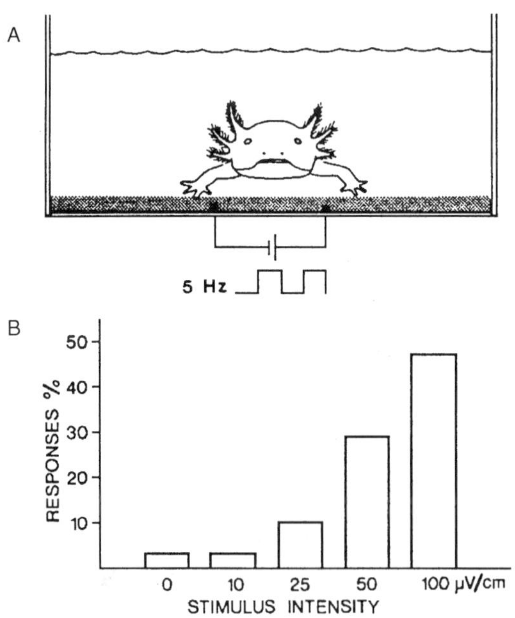

FIGURE 9.9. (**A**) Prey-catching response of the axolotl (*Ambystoma mexicanum*) to a dipole electrode buried in the substrate. (**B**) Histogram of the number of responses at different electric field amplitudes. (From Himstedt et al. 1982.)

with visual cues (Himstedt et al. 1982). The response to electrical stimuli also depends on the previous feeding history. Neither catfish (Peters and Meek 1973) nor amphibians (Himstedt et al. 1982) exhibit a preference to objects with an electric field if not previously fed with electrically active, live prey.

The possibility of orientation or navigation by means of electrical cues has been discussed in amphibians, but so far experimental evidence is lacking. Some urodeles migrate over long distances and it has been shown that they use magnetic cues to adjust their headings (Fischer et al. 2001). Others live in caves where visual landmarks cannot be used. They may also use magnetic cues for orientation (Phillips 1986). Whether magnetoreception is mediated by the electrosensory system, as proposed for elasmobranchs, is not known.

7.2 Gymnophians

In the limbless *Ichthyophis kohtaoensis*, no distinct prey-catching or avoidance behavior was observed. However, electric fields produced by electrodes centered in the behavioral arena increased locomotor activity and attracted the animal (Himstedt and Fritzsch 1990).

8. The Passive Electrosense in Mammals (Monotremata)

The electrosense is represented in only a single group of higher vertebrates, the monotremes of Australia and New Guinea. The semiaquatic platypus (*Ornithorhynchus anatinus*) forages exclusively in streams and ponds, where its nightly intake of invertebrates, mostly crustaceans, worms, and insect larvae, is equal to half its own weight. The platypus bill has been regarded as its primary *underwater* sense organ since the eyes, nares, and auditory canals remain closed as it swims. As a tactile organ with an expanded neocortical representation in the brain the bill undoubtedly performs sensory functions similar to those of the lateral line in fish and amphibians, although innervated by the trigeminal as opposed to lateral line nerve roots. However, the ability of the platypus to supply its prodigious appetite by detecting and capturing fast-swimming crustaceans is predicted to rely in large part on the electrosense (Manger and Pettigrew 1995). Electroreceptors derived from mucous glands in the skin number around 40,000 and are arranged in parasagittal stripes on the bill along with 60,000 more uniformly distributed push-rod mechanoreceptors (Pettigrew 1999). There is as yet no evidence that the platypus uses its electrosense for navigation, although sensitivity to galvanic field potentials has been demonstrated. Rather, the bill functions as an antenna used primarily in detecting aquatic prey in a remarkable example of parallel evolution with the rostral paddle/antenna of the paddlefish (Pettigrew and Wilkens 2003).

Electrosensory behavior has been studied in a laboratory platypussary where the animal is free to move between a dry nest box and a tank where it can swim, forage, or rest out of the water or in underwater structures (Manger and

Pettigrew 1995). In the water the platypus spends a substantial portion of time (41.3%) actively patrolling the bottom of the tank in search of food, swimming with a sweeping undulation of the head at 2 to 3 Hz that triples the sampling area relative to prepatrol swimming. Platypuses abruptly swing their head toward an electrical target simulating its shrimp prey and explore the electrodes with their mouth. To analyze the electrolocation behavior, controlled stimuli were presented while animals rested at one of their underwater stations, a period accompanied by pronounced bradycardia. The behavioral response to a brief electrical pulse (50 to 75 μV cm^{-1}) in a nonpatrolling platypus involves a head saccade toward the stimulus followed by exploratory movements of the head. For a train of stimulus pulses the platypus responds initially with a head saccade and begins to mouth the electrodes. But, after repeated stimulation only the head saccade persists as an involuntary reflex movement of the bill toward the stimulus, with a maximum repetition rate of 12.5 saccades s^{-1}. At a threshold intensity of approximately 50 μV cm^{-1} head saccades are barely visible, but they increase to a 3 cm lateral excursion at 1 mV cm^{-1}. A similar stimulus paradigm was used to measure directionality, for example, by advancing a fixed amplitude stimulus toward the bill from different angles and elevations and noting the distance and field potential at which phase locking of the head saccade occurred. The saccade itself is directional with the bill moving toward the electrodes along either lateral or vertical trajectories. Platypuses are most sensitive to lateroventral directions where the stimulus axis is 80° lateral and approximately 20° forward and ventral relative to the tip of the bill (Fig. 9.10).

The directionality of the saccade is an indication of how the platypus targets and captures its prey. To do so it must reconstruct the pattern of field strengths across the bill by associating the decay of isofield potentials in order to estimate the source direction. The proposed mechanism for this is based anatomically on the parasagittal stripes of electroreceptors on the bill and their topographic

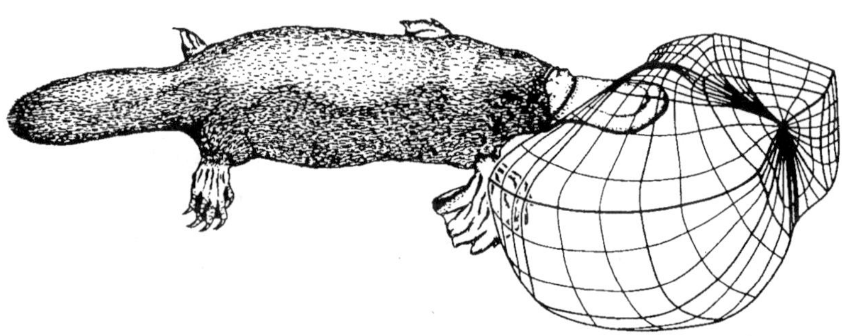

FIGURE 9.10. Iso-threshold distances from the platypus bill at which time-locked head saccades are elicited at 2 Hz for a constant square-wave stimulus. Maximum distance is 25 cm at an angle 80° from the front and 20° down, and corresponds to a field intensity of 0.1 mV cm^{-1}. (After Manger and Pettigrew 1995, with permission from the Royal Society of London.)

projection into the cortex, a "built-in tendency for more neural associations to be carried out along this parasagittal axis" (Pettigrew et al. 1998). Maximum sensitivity is orthogonal to the parasagittal stripes and cortical receptive fields are biased for parasagittal orientations (Krubitzer et al. 1995). Cortical projections from the bill also provide a mechanism for measuring distance to the source, for example, to a shrimp or crayfish (yabbie) attempting to escape predation by a tail-flip response. Such an event would produce an instantaneous electrical signal followed by a hydrodynamic pulse of variable delay, depending on the distance from the prey to the bill. The push-rod mechanoreceptors on the bill project in parallel with the electroreceptors to the S1 cortex, where there is a highly organized array of mechanoreceptive and electroreceptive inputs (Krubitzer et al. 1995). Cortical cells in this area receive bimodal input from both bill modalities, and show marked facilitation depending on the temporal input features, thus providing an attractive model for analyzing distance perception (Pettigrew et al. 1998).

Echidnas, two highly specialized species derived relatively recently (20 mya) from the more aquatic platypus lineage, are the only true terrestrial mammals with electroreceptors (Pettigrew 1999). The short-nose echidna (spiny anteater, *Tachyglossus aculeatus*) occupies a sizeable arid habitat in southeastern Australia where it burrows in sandy terrain to feed on ants, termites, and insect larvae. Its mucus-gland electroreceptors have been reduced to an estimated 100 located at the tip of its snout. Threshold sensitivity is relatively high at 1.8 mV cm^{-1} based on recordings from the infraorbital nerve, but this is equivalent to the behavioral thresholds (1.8 to 73 mV cm^{-1}) obtained from echidnas trained to detect currents in a water bath (Gregory et al. 1989). No electrosensory behavior has been documented, although the only conceivable function for the electrosense in a terrestrial animal would be in feeding. Although a degenerate remnant of the aquatic electrosensory system, echidna electroreception may facilitate feeding as a result of the mucous secretions that keep the tip of the snout moist. Thus, a low-resistance "liquid junction" has been proposed whereby the electroreceptors might sense the field potentials of ants and termites as the echidna probes their humid chambers (Gregory et al. 1989), or where beetle larvae consumed during the spring and summer are detected only when they become active in the warmer months of the year (Andres et al. 1991). Corresponding anatomical and lifestyle features also support a functional role in feeding for the terrestrial electrosense in the long-beaked echidna, *Zaglossus bruijnii*, although direct evidence is still lacking. *Zaglossus* has upwards of 3000 electroreceptors with anatomical features closer to those of the platypus and its range is restricted to the moist montane forests of New Guinea where detection of its favorite diet of earthworms is predicted (Manger et al. 1997). Earthworm giant fiber and muscle potentials are available at the surface of the worm ranging from 50 to approximately 200 μV (Drewes et al. 1983). If the sensitivity of *Zaglossus* electroreceptors is closer to that of platypuses, as the anatomy might suggest, a startled earthworm would provide ample stimulus intensity for the hungry echidna.

9. Summary

Passive electroreception refers to the ability of many aquatic animals to perceive weak electric fields generated by other animals, or by geoelectric or geomagnetic phenomena. In contrast to active electrosensory fish, which monitor the distortions of their own electric discharges by other objects, passive electroreception functions without a "probing" test signal. In contrast to the two groups of active electrosensory fish, passive electroreception is present in lampreys, elasmobranchs, bichirs and reed fishes, sturgeons and paddlefish, catfish, amphibians, and in three species of monotremes. Electric fields detectable by passive electrosensory animals may be divided into (1) electric fields of animate origin, (2) fields caused by physicochemical effects, and (3) electric fields induced by magnetic induction. Fields of animate origin are primarily used to detect prey or to avoid predators. Prey-catching behavior and escape and avoidance are well investigated in almost all groups of electrosensory animals. Geoelectric fields are not well understood but they could be used for orientation or navigation in an aquatic environment with limited visibility. Evidence for geoelectric orientation stems primarily from studies on catfish. Electric fields induced by motion through the earth's magnetic field are proposed as a mechanism by which electroreceptive animals can indirectly perceive the horizontal and vertical components of the magnetic field. Although studies on rays confirmed behavioral responses to magnetic fields, the involvement of the electrosensory system is not proven. There is also evidence for a role of electroreception in communication, social interactions, and courtship, mainly in skates and rays. Many aspects of passive electroreception, however, remain to be explored. So far, we have only scratched the surface of the electrosensory world and we expect many more surprising facts to emerge concerning the function of a sensory system completely escaping the imagination of the human observer lacking this remarkably sensitive sensory mode.

References

Andres KH, von Düring M, Iggo A, Proske U (1991) The anatomy and fine structure of the echidna *Tachyglossus aculeatus* snout with respect to its different trigeminal sensory receptors including the electroreceptors. Anat Embryol 184:371–393.

Asano M, Hanyu I (1987) Sensitivity to electricity in the catfish, *Parasilurus asotus*. Comp Biochem Physiol 86A:485–489.

Baranyuk GV (1979) Differential sensitivity of the brown bullhead *Ictalurus nebulosus* to intensity and frequency of an electric field. Fiziologicheskii Zhurnal Sssr Imeni I M Sechenova 65:826–829.

Basov BM (1999) Behavior of sterlet *Acipenser ruthenus* and Russion sturgeon *A. gueldenstaedtii* in low-frequency electric fields. J Ichthyol 39:782–787.

Bass AH (1986) Electric organs revisited. In: Bullock TH, Heiligenberg W (eds), Electroreception. New York: John Wiley & Sons, pp. 13–70.

Bennett MVL (1970) Comparative physiology: electric organs. Annu Rev Physiol 32: 471–528.

Bodznick D, Boord RL (1986) Electroreception in chondrichthyes. In: Bullock TH, Heiligenberg W (eds), Electroreception. New York: John Wiley & Sons, pp. 225–256.

Bodznick D, Preston DG (1983) Physiological characterization of electroreceptors in the lampreys *Ichthyomyzon unicupsis* and *Petromyzon marinus*. J Comp Physiol 152:209–217.

Bogorov VG, Demenitskaya RM, Gorodnitskiy AM, Kazanskiy MM et al (1969) Character and causes of the vertical variation of the natural electric field in the ocean. Oceanology 9:622–626.

Bratton BO, Ayers JL (1987) Observations on the electric organ discharge of two skate species (Chondrichthyes: Rajidae) and its relationship to behavior. Environm Biol Fishes 20:241–254.

Bullock TH, Bodznick DA, Northcutt RG (1983) The phylogenetic distribution of electroreception: evidence for convergent evolution of a primitive vertebrate sense modality. Brain Res Rev 6:25–46.

Burr HS, Northrop FSC (1939) Evidence for the existence of an electro-dynamic field in living organisms. Proc Natl Acad Sci USA 25:284–288.

Claas B, Fritzsch B, Münz H (1982) Electrosensitivity in the lateral line system of urodele amphibians. Neurosci Lett (Suppl) 10:115–116.

Coggi A (1905) Le ampolle di Lorenzini nei Gimnofioni. Monit Zool Ital 16:49–56.

Dijkgraaf S (1968) Electroreception in the catfish, *Amiurus nebulosus*. Experientia 24: 187.

Dijkgraaf S, Kalmijn AJ (1962) Verhaltensversuche zur Funktion der Lorenzinischen Ampullen. Naturwissenschaften 49:400.

Drewes CD, Callahan CA, Fender WM (1983) Species specificity of giant nerve fiber conduction velocity in oligochaetes. Can J Zool 12:2688–2694.

Fahrenholz C (1929) Über die "Drüsen" und die Sinnesorgane in der Haut des Lungenfisches. Z Mikr-Anat Forsch 16:55–74.

Fields RD, Bullock TH, Lange, GD (1993) Ampullary sense organs, peripheral, central and behavioral electroreception in chimeras (*Hydrolagus*, Holocephali, Chondrichthyes). Brain Behav Evol 41:269–289.

Fischer JH, Freake MJ, Borland SC, Phillips JC (2001) Evidence for the use of magnetic map information by an amphibian. Anim Behav 62:1–10.

Fritzsch B (1981a) The pattern of lateral-line afferents in Urodeles. A horseradish-peroxidase study. Cell Tissue Res 218:581–594.

Fritzsch B (1981b) Electroreceptors and direction specific arrangements in the lateral-line system of salamanders? Z Naturforsch 36C:493–495.

Fritzsch B, Wahnschaffe U (1983) The electroreceptive ampullary organs of urodeles. Cell Tissue Res 229:483–503.

Gregory JE, Iggo A, McIntyre AK, Proske U (1989) Responses of electroreceptors in the snout of the echidna. J Physiol 414:521–538.

Gurgens C, Russell DF, Wilkens LA (2000) Electrosensory avoidance of metal obstacles by the paddlefish. J Fish Biol 57:277–290.

Haine OS, Ridd PV, Rowe RJ (2001) Range of electrosensory detection of prey by *Carcharhinus melanopterus* and *Himantura granulata*. Mar Freshwater Res 52:291–296.

Hetherington TE, Wake MH (1979) The lateral-line system in larval *Ichthyophis* (Amphibia: Gymnophiona). Zoomorphology 93:209–225.

Himstedt W, Fritzsch B (1990) Behavioural evidence for electroreception in larvae of the caecilian *Ichthyophis kohtaoensis* (Amphibia, Gymnophiona). Zool Jb Physiol 94: 486–492.

Himstedt W, Kopp J, Schmidt W (1982) Electroreception guides feeding behavior in amphibians. Naturwissenschaften 69:552–553.

Hodson RB (2000) Magnetoreception in the short-tailed stingray, *Dasyatis brevicaudata*. MSc thesis, University of Aukland, New Zealand.

Ikeya M, Matsuda T, Yamanaka C (1998) Reproduction of mimosa and clock anomalies before earthquakes: are they "Alice in Wonderland" syndrome? Proc Jpn Acad Ser B 74:60–64.

Irwin WP, Lohmann, KJ (2003) Magnet-induced disorientation in hatchling loggerhead sea turtles. J Exp Biol 206:497–501.

Istenic L, Bulog B (1984) Some evidence for the ampullary organs in the European cave salamander *Proteus anguineus* (Urodela Amphibia). Cell Tissue Res 235:393–402.

Johnson CS, Scronce BL, McManus MW (1984) Detection of DC electric dipoles in background fields by the nurse shark. J Comp Physiol A 155:681–687.

Jørgensen JM, Flock A, Wersall J (1972) The Lorenzinian ampullae of *Polyodon spathula*. Z Zellforsch 130:362–377.

Kajiura SM, Holland KN (2002) Electroreception in juvenile scalloped hammerhead and sandbar sharks. J Exp Biol 205:3609–3621.

Kalmijn AJ (1966) Electro-perception in sharks and rays. Nature 212:1232–1233.

Kalmijn AJ (1971) The electric sense of sharks and rays. J Exp Biol 55:371–383.

Kalmijn AJ (1972) Bioelectric fields in the sea water and the function of the ampullae of Lorenzini in elasmobranch fishes. Scripps Inst Oceanogr Ref Ser 72–83:1–21.

Kalmijn AJ (1974) The detection of electric fields from inanimate and animate sources other than electric organs. In: Fessard A (ed), Handbook of Sensory Physiology, Vol. III/3. New York: Springer-Verlag, pp. 147–200.

Kalmijn AJ (1978) Experimental evidence of geomagnetic orientation in elasmobranch fishes. In: Schmidt-Koenig K, Keeton WT (eds), Animal Migration, Navigation, and Homing. Berlin: Springer-Verlag, pp. 347–353.

Kalmijn AJ (1982) Electric and magnetic field detection in elasmobranch fishes. Science 218:916–918.

Kalmijn AJ (1988) Detection of weak electric fields. In: Atema J, Fay RR, Popper AN, Tavolga WN (eds), Sensory Biology of Aquatic Animals. New York: Springer-Verlag, pp. 151–186.

Kalmijn AJ, Kolba CA, Kalmijn V (1976a) Orientation of catfish (*Ictalurus nebulosus*) in strictly uniform electric fields: I. Sensitivity of response. Biol Bull 151:415.

Kalmijn V, Kolba CA, Kalmijn AJ (1976b) Orientation of catfish (*Ictalurus nebulosus*) in strictly uniform electric fields: II. Spatial discrimination. Biol Bull 151:415.

Kleerekoper H, Sibakin K. (1956a) Spike potentials produced by the sea lamprey (*Petromyzon marinus*) in the water surrounding the head region. Nature (Lond) 178:490–491.

Kleerekoper H, Sibakin K (1956b) An investigation of the electrical "spike" potentials produced by the sea lamprey (*Petromyzon marinus*) in the water surrounding the head region. J Fish Res Bd Can 13:375–383.

Kleerekoper H, Sibakin K (1957) An investigation of the electrical "spike" potentials produced by the sea lamprey (*Petromyzon marinus*) in the water surrounding the head region. II. J Fish Res Bd Can 14:145–151.

Klimley AP (1993) Highly directional swimming by scalloped hammerhead sharks,

Sphyrna lewini, and subsurface irradiance, temperature, bathymetry, and geological field. Mar Biol 117:1–22.

Knudtson BK, Stimers JR (1977) Notes on the behavior of elasmobranch fishes exposed to magnetic fields. Bull S Calif Acad Sci 76:202–204.

Krajew AP (1957) Grundlagen der Geoelektrik. Berlin: VEB Verlag Technik.

Krubitzer L, Manger PR, Pettigrew JD, Calford MB (1995) Organization of somatosensory cortex in monotremes: in search of the prototypical plan. J Comp Neurol 352: 261–306.

Ladich F, Popper AN (2004) Parallel evolution in fish hearing organs. In: Manley GA, Popper AN, Fay RR (eds), Evolution of the Vertebrate Auditory System. New York: Springer-Verlag, pp. 95–127.

Lissmann HW, Machin KE (1963) Electric receptors in a non-electric fish (*Clarias*). Nature 199:88–89.

Manger PR, Pettigrew JD (1995) Electroreception and the feeding behaviour of the platypus (*Ornithorhyncus anatinus*: Monotremata: Mammalia). Philos Trans R Soc Lond B 347:359–381.

Manger PR, Collins R, Pettigrew JD (1997) Histological observations on presumed electroreceptors and mechanoreceptors in the beak skin of the long-beaked echidna, *Zaglossus bruijnii*. Proc R Soc Lond B 264:165–172.

Münz H, Claas B, Fritzsch B (1982) Electrophysiological evidence of electroreception in the axolotl *Siredon mexicanum*. Neurosci Lett 28:107–111.

Münz H, Claas B, Fritzsch B (1984) Electroreceptive and mechanoreceptive units in the lateral line of the axolotl *Ambystoma mexicanum*. J Comp Physiol 154:33–44.

Murray RW (1962) The response of the ampullae of Lorenzini of elasmobranchs to electrical stimulation. J Exp Biol 39:119–128.

Nachtrieb HF (1910) The primitive pores of *Polyodon spathula* (Walbaum). J Exp Zool 9:455–468.

New JG (1994) Electric organ discharge and electrosensory reafference in skates. Biol Bull 187:64–75.

Northcutt RG (1980) Anatomical evidence of electroreception in the coelacanth (*Latimeria chalumnae*). Zbl Vet Med C Anat Histol Embryol 9:289–295.

Northcutt RG (1986) Electroreception in nonteleost bony fishes. In: Bullock TH, Heiligenberg W (eds), Electroreception. New York: John Wiley & Sons, pp. 257–285.

Pals N, Schoenhage AAC (1979) Marine electric fields and fish orientation. J Physiol (Paris) 75:349–353.

Pals N, Peters RC, Schoenhage AAC (1982a) Local geoelectric fields at the bottom of the sea and their relevance for electrosensitive fish. Neth J Zool 32:479–494.

Pals N, Valentijn P, Verwey D (1982b) Orientation reactions of the dogfish, *Scyliorhinus canicula*, to local electric fields. Neth J Zool 32:495–512.

Parker GH, van Heusen AP (1917) The responses of the catfish, *Amiurus nebulosus*, to metallic and non-metallic rods. Am J Physiol 44:405–420.

Paulin MG (1995) Electroreception and the compass sense of sharks. J Theor Biol 174: 325–339.

Peters RC, Bretschneider F (1972) Electric phenomena in the habitat of the catfish *Ictalurus nebulosus* LeS. J Comp Physiol 81:345–362.

Peters RC, Meek J (1973) Catfish and electric fields. Experientia 29:299–300.

Peters RC, van Wijland F (1974) Electro-orientation in the passive electric catfish, *Ictalurus nebulosus* LeS. J Comp Physiol 92:273–280.

Peters RC, Loos WJG, Bretschneider F, Baretta AB (1999) Electroreception in catfish: patterns from motion. Belg J Zool 129:263–268.

Pettigrew JD (1999) Electroreception in monotremes. J Exp Biol 202:1447–1454.

Pettigrew JD, Wilkens L (2003) Paddlefish and platypus: parallel evolution of passive electroreception in a rostral bill organ. In: Collin SP, Marshall NJ (eds), Sensory Processing in Aquatic Environments. New York: Springer-Verlag, pp. 420–433.

Pettigrew JD, Manger PR, Fine SLB (1998) The sensory world of the platypus. Philos Trans R Soc Lond B 353:1199–1210.

Pfeiffer W (1968) Die Fahrenholzschen Organe der Dipnoi und Brachiopterygii. Z Zellforsch 90:127–147.

Phillips JB (1986) Magnetic compass orientation in the eastern red-spotted newt (*Notophthalmus viridescens*). J Comp Physiol A 158:103–109.

Ronan M (1986) Electroreception in cyclostomes. In: Bullock TH, Heiligenberg W (eds), Electroreception. New York: John Wiley & Sons, pp. 209–224.

Ronan M (1988) Anatomical and physiological evidence for electroreception in larval lampreys. Brain Res 448:173–177.

Ronan M, Northcutt RG (1987) Primary lateral line projections in adult lampreys. Brain Behav Evol 30:62–81.

Roth A (1969) Elektrische Sinnesorgane beim Zwergwels *Ictalurus nebulosus* (*Amiurus nebulosus*). Z Vergl Physiol 65:368–388.

Roth A (1972) Wozu dienen die Elektrorezeptoren der Welse? J Comp Physiol 79:113–135.

Roth A (1973) Electroreceptors in Brachiopterygii and Dipnoi. Naturwissenschaften 60: 106.

Roth A, Schlegel P (1988) Behavioral evidence and supporting electrophysiological observations for electroreception in the blind cave salamander, *Proteus anguinus* (Urodela). Brain Behav Evol 32:277–280.

Schlegel P (1997) Behavioral sensitivity of the European blind cave salamander, *Proteus anguinus,* and a Pyrenean newt, *Euproctus asper,* to electrical fields in water. Brain Behav Evol 49:121–131

Schlegel P, Bulog B (1997) Population-specific behavioral electrosensitivity of the European blind cave salamander, *Proteus anguinus*. J Physiol (Paris) 91:75–79.

Sisneros JA, Tricas TC, Luer CA (1998) Response properties and biological function of the skate electrosensory system during ontogeny. J Comp Physiol A 183:87–99.

Taylor NG, Manger PR, Pettigrew JD, Hall LS (1992) Electromyographic potentials of a variety of Platypus prey items: an amplitude and frequency analysis. In: Augee ML (ed), Platypus and Echidna. Sydney: The Royal Zoological Society of NSW. pp. 216–224.

Tricas TC (1982) Bioelectric-mediated predation by swell sharks, *Cephaloscyllium ventriosum*. Copeia 1982:948–952.

Tricas TC (2001) The neuroecology of the elasmobranch electrosensory world: why peripheral morphology shapes behavior. Environ Biol Fish 60:77–92

Tricas TC, McCosker JE (1984) Predatory behavior of the white shark (*Carcharodon carcharias*), with notes on its biology. Proc Calif Acad Sci 43:221–238.

Tricas TC, New JG (1998) Sensitivity and response dynamics of elasmobranch electrosensory primary afferent neurons to near threshold fields. J Comp Physiol A 182:89–101.

Tricas TC, Michael SW, Sisneros JA (1995) Electrosensory optimization to conspecifics phasic signals for mating. Neurosci Lett 202:129–132.

Uzuka K (1934) Some notes of the behavior of the catfish, *Parasilurus asotus*, as seen through the responses to weak electric current. Sci Rep Tohoku Imp Univ Ser 4 8: 369–381.

Walker MM, Diebel CE, Haugh CV, Pankhurst PM, Montgomery JC (1997) Structure and function of the vertebrate magnetic sense. Nature 390:371–376.

Walker MM, Diebel CE, Kirschvink JL (2003) Detection and use of the earth's magnetic field by aquatic vertebrates. In: Collin SP, Marshall NJ (eds) Sensory Processing in Aquatic Environments. New York: Springer-Verlag, pp. 53–74.

Watt M, Evans CS, Joss JMP (1999) Use of electroreception during foraging by the Australian lungfish. Anim Behav 58:1039–1045.

Westby GWM (1988) The ecology discharge diversity and predatory behavior of gymnotiform electric fish in the coastal streams of French Guiana. Behav Ecol Sociobiol 22:341–354.

Wilkens LA, Russell DF, Pei X, Gurgens C (1997) The paddlefish rostrum functions as an electrosensory antenna in plankton feeding. Proc R Soc Lond B 264:1723–1729.

Wilkens LA, Wettring BA, Wagner E, Wojtenek W, Russel DF (2001) Prey detection in selective plankton feeding by the paddlefish: is the electric sense sufficient? J Exp Biol 204:1381–1389.

Wojtenek W, Pei X, Wilkens LA (2001a) Paddlefish strike at artificial dipoles simulating the weak electric fields of planktonic prey. J Exp Biol 204:1391–1399.

Wojtenek W, Hofmann, MH, Wilkens LA (2001b) Primary afferent electrosensory neurons represent paddlefish natural prey. Neurocomputing 38:451–458.

Zakon HH (1986) The electroreceptive periphery. In: Bullock TH, Heiligenberg W (eds), Electroreception. New York: John Wiley & Sons, pp. 103–156.

10

Passive Electrolocation and the Sensory Guidance of Oriented Behavior

Carl D. Hopkins

1. Introduction

A central concern of neuroethology is to explain how the sense organs guide the orientation behavior of animals. Whether the guidance serves to turn the organism toward a stimulus or away, to facilitate an approach or a retreat, or to track a target or to avoid it, each of the sensory systems guides motor behavior according to its own set of stimulus parameters and physical cues. For many of the more commonly explored sensory modalities such as vision, hearing, vibration sensing, wind or water flow detection, or surface wave monitoring, we already know a good deal about the physical cues that are available for locating stimuli in space, and we know how the sensory processing of these cues leads to motor output (Fraenkel and Gunn 1940; Schöne 1984). In contrast, much less is known about the mechanisms of sensory guidance for the electrosensory modality although galvanotaxis of unicellular and other nonelectroreceptive organisms was known to early students of orientation by kineses and taxes. This review focuses on electrosensory systems used in passive sensing of external electrical sources of stimuli, particularly in the context of electric communication.

Passive electrolocation occurs when an electroreceptive animal senses the electric signals arising from some external electrogenic source—an electric fish, a prey item, a predator, or an inanimate object. Lissmann recognized that the evolution of passive electrolocation of prey must have preceded the evolution of electric organs (Lissmann 1958; Lissmann and Machin 1958), and that both systems must have been functional before active electrolocation was possible. Active electrolocation depends on sensing reafferent electrosensory information from the animal's own electric discharge. It is discussed by Nelson (Chapter 11) and in studies and reviews by Bastian (1986; 1987a,b) and by Heiligenberg (1977). One of the first convincing behavioral demonstrations of passive electrolocation was the elegant study of A. Kalmijn (1971), who demonstrated that sharks were exquisitely sensitive to weak electric fields from prey fish that were buried in the substrate. Passive electrolocation is more comparable to passive listening, passive lateral line mechanoreceptor sensing, or passive vision which sometimes involves detection of external and unpredictable stimuli from the environment, unrelated to the self-generated stimuli from the animal's own actions. In spite of the conceptual similarity of passive electrolocation with these other passive sensory processes, there are fundamental differences in physical cues available to electroreceptive receivers, and these differences have important consequences for the passive electroreceptive guidance of behavior. These are especially apparent when considering the detection and localization of electric communication signals.

From observations of behavior we know that electroreceptive fishes perform well at electrolocation tasks such as approaching electrogenic fish or dipole electrodes. For example, weakly electric fish that are living in aquaria often show rapid approaches toward electric fish intruders that trespass on their territories (Lissmann 1958; Lissmann and Machin 1958; Black-Cleworth 1970; Westby 1974). They also approach electrical dipoles that are playing conspecific discharges (Schluger and Hopkins 1987; Davis and Hopkins 1988; Crawford 1991; McGregor and Westby 1992, 1993; Shieh et al. 1996). Elasmobranchs that produce electric signals locate their mates using electrosensory cues (Tricas et al. 1995; Sisneros et al. 1998; Sisneros and Tricas 2002a,b).

Electroreceptive fishes also locate prey by using passive electroreception. Sharks approach prey organisms that emit weak electric fields (Kalmijn 1971, 1974; (Kajiura and Holland 2002). The paddlefish (*Polyodon spathula*) orients to planktonic prey using passive electroreception as discussed in Wilkens (Wilkens et al. 2002; Chapter 9). All these are examples of passive electrolocation.

2. Theoretical Difficulties of Locating Electric Sources

Many of the locatable physical cues that we associate with sensory modalities other than electroreception arise because stimuli travel through the environment as propagated waves. This certainly applies to vision, audition, and the vibratory senses. For each of these modalities, a readily available localization cue arises from the velocity vector associated with the propagation of the stimulus through the environment. For sounds, light, and vibrations, the velocity vector usually points to the source of the signal in space.

For the electrosensory system, by contrast, signals of biological importance do not propagate as waves but rather exist as electrostatic fields that are non-propagating. This is because biological signals have most of their signal energy at frequencies ranging from a fraction of a Hertz to several kiloHertz. Propagating electromagnetic waves would have a wavelength measured in kilometers at these spectral frequencies, but the electric fields generated by biological sources attenuate so rapidly in aquatic environments that the field magnitude would be effectively zero at signal ranges of several meters: much less than a wavelength of any electromagnetic wave. For this reason passive electrolocation must depend on physical cues arising from a nonpropagating signal (Hopkins 1986; Kalmijn 1988a; Schluger and Hopkins 1987; Hopkins et al. 1997). With electrostatic fields, there is no velocity vector pointing to the source, no temporal delay associated with the propagation that might be used to locate the direction of the source, and no shadow cast by the receiver's body that could be used for orientation. The electrostatic field vector describes a curved path that points not the source, but in the direction of the current flow through the water. As such it provides an ambiguous cue about source location. These physical characteristics place constraints on the mechanisms of passive electrolocation. What are the electrosensory cues available to these fish, and how do they modify their search algorithms compared to other sensory modalities in order to passively locate signal sources in the environment?

To develop a better basic understanding of the physical properties of electric fields in aquatic environments, Kalmijn (1988a) reviewed and carefully described a theory for the geometry of electric fields from biological sources. He demonstrated that many electric fields from highly localized biological sources such as weakly electric fish, prey organisms, and predators could be considered as a sum of dipole, quadrapole, or higher-order multipole sources. Further he showed that the dipole source contribution dominates the far field, while higher-order multipole fields dominant the near field. This is because in three dimensions the dipole contribution to the electric potential gradient decreases according to the inverse third power of distance while the contributions from quadrapole sources decrease according to the inverse fourth power of distance. Contributions from higher-order multipole source decrease at a correspondingly higher-order rate so that they become vanishingly small at greater and greater distances. The dipole contribution to the field is the dominant contribution at greater distances from the source. The focus in this chapter is on the electrolocation at relatively large distances where the dipole field dominates.

Behavioral studies have shown that electroreceptive organisms are very good at locating natural and artificial sources of electric signals in their environment and make rapid and well-directed approaches to the electrodes generating these signals (Kalmijn 1971, 1974, 1988a; Schluger and Hopkins 1987; Davis and Hopkins 1988; Hopkins 1993; Shieh et al. 1996). Given the relative ease with which fish and other organisms can be induced to approach electric sources, it is not surprising that experimental studies of passive electrolocation provide insights into the sensory and perceptual processes behind signal localization in this unusual sensory modality. Experimental approaches to studies of passive electrolocation have invigorated renewed interest in theories of electrolocation and stimulated renewed estimates of the nature of electric field sources and the geometry of electroreceptor organs and structures that carry them.

Little is known about the integration of spatial electrosensory information and the generation of motor responses, but the behavioral data provide a strong basis for future physiological work on this question. We may only speculate on the complexities of sensory–motor integration and where in the brain spatial information is used to guide behavior.

2.1 Functions of Passive Electrolocation

Passive electrolocation is used for three different behavioral functions. The first is passive electrolocation of predators or prey. This involves sensing the uncontrolled, extremely weak, usually low-frequency electric fields surrounding predator or prey organisms—fields that surround most living organisms in aquatic environments (Kalmijn 1974). The second is passive electrolocation of communication signals: the passive sensing of electric communication signals of one electric fish by another in the context of aggression, reproduction, alarm, parental care, or a variety of other behavioral functions (Schluger and Hopkins 1987; Davis and Hopkins 1988). The third is navigation: the passive sensing of directional or navigational cues in the environment arising from inanimate objects (Kalmijn 1974, 1988a).

Passive electrolocation of DC fields from predators, prey, or inanimate objects tends to operate only at very short ranges owing to the very weak signals produced by muscle activity, gill membranes, skin wounds, or other uncontrolled sources of electric fields around these sources. For example, sharks respond to their prey when passing within tens of centimeters above them (Kajiura and Holland 2002) and they react to electrodes buried in the substrate in a similar way. Mormyrids that have been electrically silenced and are using only their passive electric sense respond to invertebrate prey at distances of several centimeters (von der Emde et al. 1998). By contrast, passive electrolocation of electric communication signals occurs at greater distances, due to the magnitude of the electric organ discharge (EOD) signal compared to that from a predator or prey. A mormyrid or gymnotiform fish may electrolocate a conspecific fish at a distance of one or more meters—many times their own body length (Hopkins et al. 1997). Long-range navigation by passive electrolocation, while theoretically possible (Kalmijn 1974), has not been demonstrated in a natural setting. When passive electrolocation operates at long distances, the electric fields sensed by the receiver tend to be nearly homogeneous in amplitude and direction over the recipient's entire body surface. At short ranges, the signals are far more intense at some parts of the body surface than others, and the field changes markedly in direction over the body surface.

Consider an electric fish such as the gymnotiform, *Gymnotus carapo,* guarding a territory against intruders (Fig. 10.1). The territory owner responds to an EOD of a passing conspecific by approaching the intruder's signal while giving threat displays. These consist of postures and threatening bursts of EODs. Not only does the territory owner need to identify the signal as originating from a conspecific, but it also must locate it in space, and guide its subsequent path in the correct direction for approach. This scenario contrasts with the response of a fish that detects an electric signal from a predator or dominant animal. Detection and identification must be accompanied by withdrawal from the stimulus and removal from danger. For most responses the selection of the appropriate response from the repertoire will be spatially guided by an approach or a withdrawal strategy. Such considerations should apply to electroreceptive organisms

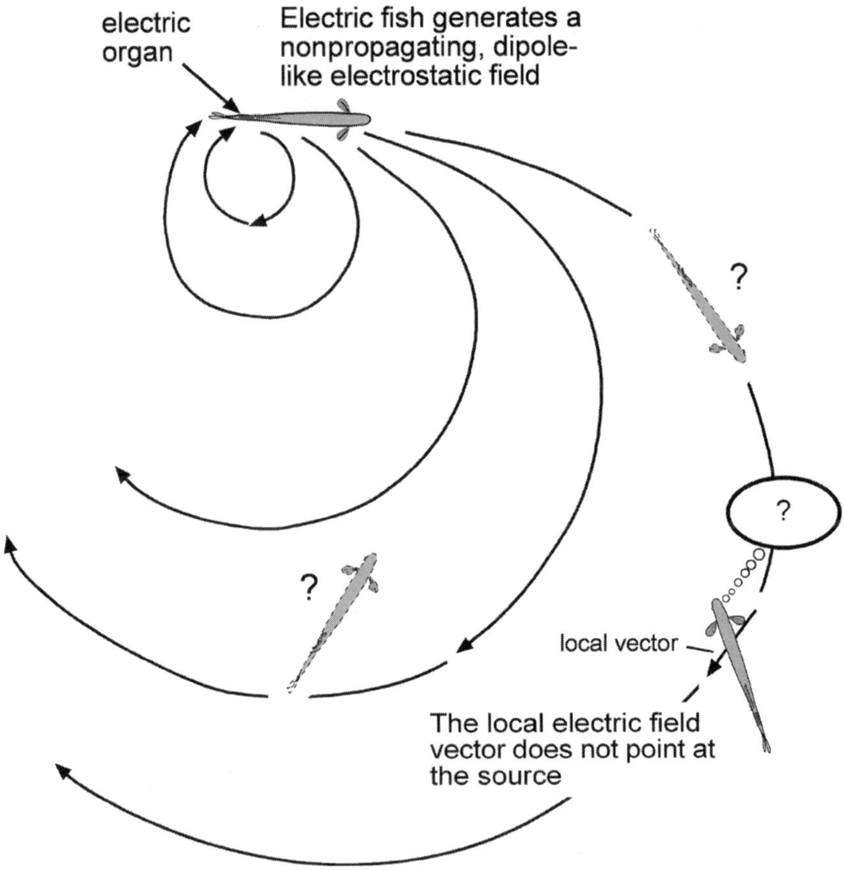

FIGURE 10.1. An electric fish intrudes on a second fish's territory. The electric organ produces a nonpropagating electrostatic field that resembles the field from a dipole. The recipient fish senses the discharge, but does not know where the intruder is located in space. The local electric field vector (*arrow*) does not point toward the source, but follows a curved path. The intruder could be at any number of locations and orientations (*dashed outlines*) that produce the same local field vector. Since the signal is electrostatic, not a propagated wave, there are no time-of-arrival cues to point to the source. Amplitude cues are also ambiguous since the receiver knows neither the size nor the orientation of the signaler in space. Electroreceptive fish must solve the electrolocation problem using alternative behavioral strategies.

as they do for other sensory modalities including vision, hearing, smell, and other external senses.

2.2 Spatial Properties of Electric Signals

Of all the possible sources of electric fields that are of significance to electroreceptive organisms (Kalmijn 1974), we know the most about the discharges of weakly electric fish. The fields have been carefully mapped for several species of electric fish (Knudsen 1975; Heiligenberg 1977; Hopkins et al. 1990; Rasnow et al. 1993; Assad et al., 1998; Stoddard et al. 1999). The DC fields around prey organisms tend to be weak and highly variable which makes mapping the field difficult. By contrast, electric discharges are stable, repeated, and of high amplitude allowing sufficient reliability to make detailed maps. Several early studies of fields from EODs (i.e., Knudsen 1975) made estimates of the electric field properties by measuring the average peak-to-peak amplitude of the EOD in a two-dimensional plane through an electric fish's midline. More recent studies have digitally mapped EOD fields in both three dimensions and as a function of time in order to describe both spatial and temporal characteristics of the signal (Rasnow et al. 1993; Assad et al. 1998, 1999; Stoddard et al. 1999). Knudsen (1975) concluded that electric fish approximate the theoretical field from a dipole source for distances greater than 20 cm from the fish. At these distances, the EOD field or current density decreases in proportion to the one over third power of distance as expected from electrostatic theory. Distances were measured from the electrical reversal point within the electric organ, close to the end of the tail. Knudsen also estimated a dipole moment of the fish's current capacity in an infinite volume conductor, and was able to compare the signal magnitude for different individuals, fish of different sizes and species, and fish in water of different conductivity.

More recently it has been possible to describe the electric field around a fish both in three dimensions, and with microsecond precision over the time period of one EOD cycle. Figure 10.2 shows the shape of the electric field around the gymnotiform fishes, *Eigenmannia* and *Brachyhypopomus*, varying over space and time. The measures were taken by Assad et al. (1998) and by Stoddard et al. (1999), respectively. *Eigenmannia* fields increase and then decrease with little change in vector direction over the time course of one EOD cycle so the small plots in Fig. 10.2 A trace out a straight line. In *Brachyhypopomus* the electric field changes direction during the EOD cycle, so the small plot of the electric field vector traces an open loop rather than a line, especially for recordings made close to the fish. This indicates that the dipole source is shifting position throughout the EOD cycle. In many gymnotiforms, the EOD appears to rapidly travel from anterior to posterior within the electric organ and within the time of one EOD period. In addition, the EOD waveform may change slightly from one recording site to another around a fish. At a greater distance from the fish, where the field is dipole like, the open loop field trajectories tend to become collapsed to a line as the distance increases from the source, so that at several body lengths from an electric field, the field will once again appear unidirectional. No high-resolution electric field maps exist for mormyrid fishes from Africa, but it is probable that the EOD is more synchronized in mormyrids compared to gymnotiforms. In addition, the electric organ occupies a very small part of the caudal peduncle, so that the field it generates probably more precisely resembles a dipole source.

As a fish approaches the source of an electric field, the magnitude and direction of the current can be estimated from basic electrostatic theory. The spatial

A

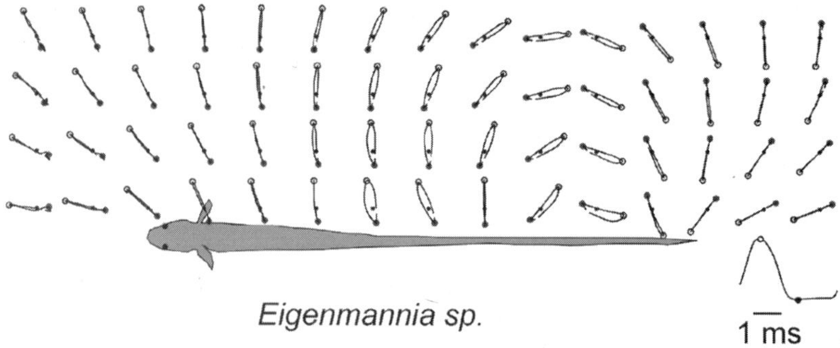

B

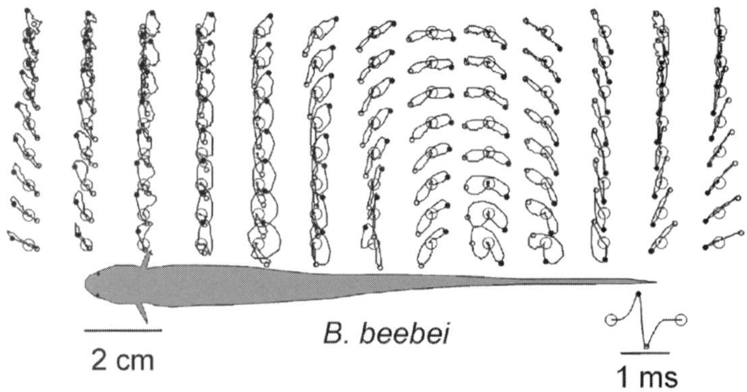

FIGURE 10.2 Spatial and temporal map of the electric field vectors around the gymnotiform fishes, *Eigenmannia* sp. (**A**) and *Brachyhypopomus beebei* (**B**) measured at all phases of a single EOD cycle. (**A**) Top view of an *Eigenmannia* showing the electric field measured in the midbody plane. The EOD of this wave-discharging species is shown in the *lower right inset.* Each small plot shows a Lissajou figure for the trajectory of the electric field vector in the *x–y* plane for the point in space indicated by the *small dot.* The peaks of each EOD cycle are marked by *open circles* (head positive) and *closed circles* (head negative) on the Lissajou figures. The eliptical traces show places where the electric field direction wobbles slightly over the course of the EOD cycle. Linear traces indicate points where the electric field vector does not change direction during the EOD cycle. (**B**) For comparison, the pulse-discharging *Brachyhypopomus beebei* produces EODs that change direction over the course of one EOD cycle, as shown by the *squiggles* representing open Lissajou figures. *Open circles* mark the baseline; *closed circles* show the positive and negative peaks. (**A** and **B** modified from Assad et al. 1999 and from Stoddard 1999, respectively.)

profile of electric field amplitude depends entirely on the geometry of the stimulus, as shown by Kalmijn (1988a). Of course, for artificial sources that are large and extended such as widely separated electrode plates that generate homogeneous field geometries, the amplitude can be constant everywhere. Such fields probably do not occur in nature (Kalmijn 1988a, 1997).

2.3 Algorithms for Sensory Guidance of Passive Electrolocation

Kalmijn (1988a) proposed a theory for how a fish might use passive electrolocation cues to approach a dipole source in a conducting volume—an algorithm for sensory–motor integration that would work to guide a shark's approach to the dipole electric field from a prey organism buried in the substrate. The model is summarized in Figure 10.3. Sharks cruise over the bottom searching for prey. Once they detect something, they turn and make a well-directed approach along a curved path that ends at the dipole center. There is little searching, error, or hesitation. Sharks are able to do this when approaching parallel to the dipole axis, perpendicular to the axis, or at any arbitrary angle. The model proposes a theory for how they might be able to rapidly approach the dipole center when the field lines are curved and never point to the source. According to Kalmijn's hypothesis, the shark swims in a straight line until it first detects the electric field from the prey. At this point the shark fixes the direction of the local field on its body surface and makes turns in the appropriate direction to prevent any apparent perceived rotation of the local electric field vector. Figures 10.3B–D shows the predicted approach trajectories for a fish using this algorithm. Approaches are from three different directions relative to the dipole axis. Kalmijn asserts that the model is an accurate fit to observational data on shark passive electrolocation. In a later article, Kalmijn (1989) applied the same theory to approaches to hydrodynamic fields where, in the near field, particle motion flow takes on a field geometry identical to that of an electric dipole. Coombs et al. (2002) and Nelson et al. (Coombs et al. 2002) provide recent reviews comparing electrosensory location with hydrodynamic fields.

Schluger and Hopkins (1987) and Davis and Hopkins (1988) proposed a nearly identical approach algorithm based on observed data. According to the Schluger/Davis/Hopkins model, one electric fish approaches another discharging fish by turning its body axis parallel to the local electric field vector. Then, by swimming forward it tracks the field lines until they lead it to the source of the dipole field. This model fits the behavioral data well, as illustrated in the examples in the following section.

3. Sensory Guidance: Behavioral Responses to Simulated Electric Fields.

Because weakly electric fish will often vigorously defend their territories against intruding conspecifics, it is possible to provoke an electric fish to approach and attack an electrode (Fig. 10.4). This can be done under controlled laboratory conditions in which the fish's behavior can be accurately monitored. Schluger and Hopkins (1987) first asked if the African mormyrid fish, *Brienomyrus brachyistius* is able to determine the distance and direction of an electrical stimulus

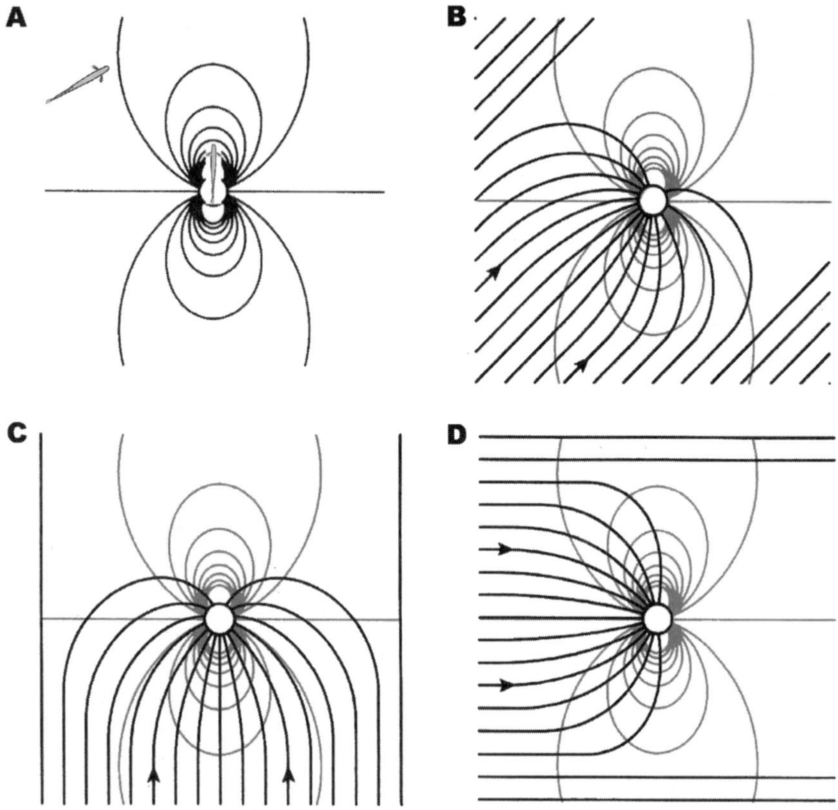

FIGURE 10.3. An algorithm for prey detection and localization proposed by Kalmijn (1988) suggests a simple model for how electroreception might be used to find an electric dipole source in space. (**A**) Top view of the electric field around a dipole source such as an electric fish or a prey organism. According to the model, the receiver should swim until it detects the electric field and then it should fix the direction of local electric field by turning its body in the appropriate direction so as to prevent apparent rotation of the field vector on the skin surface. Doing this, the receiver should eventually get to the source. (**B–D**) Simulations of approaches to the same dipole from different directions. *Black lines* and *arrows* show the predicted approach trajectory toward the electric field lines shown in *gray*. Note how the simulations always get to the dipole center regardless of the angle at which it first detects it. These trajectories are apparently a good fit to the behavior of sharks locating electric dipoles buried in the substrate. (Modified from Kalmijn 1988.)

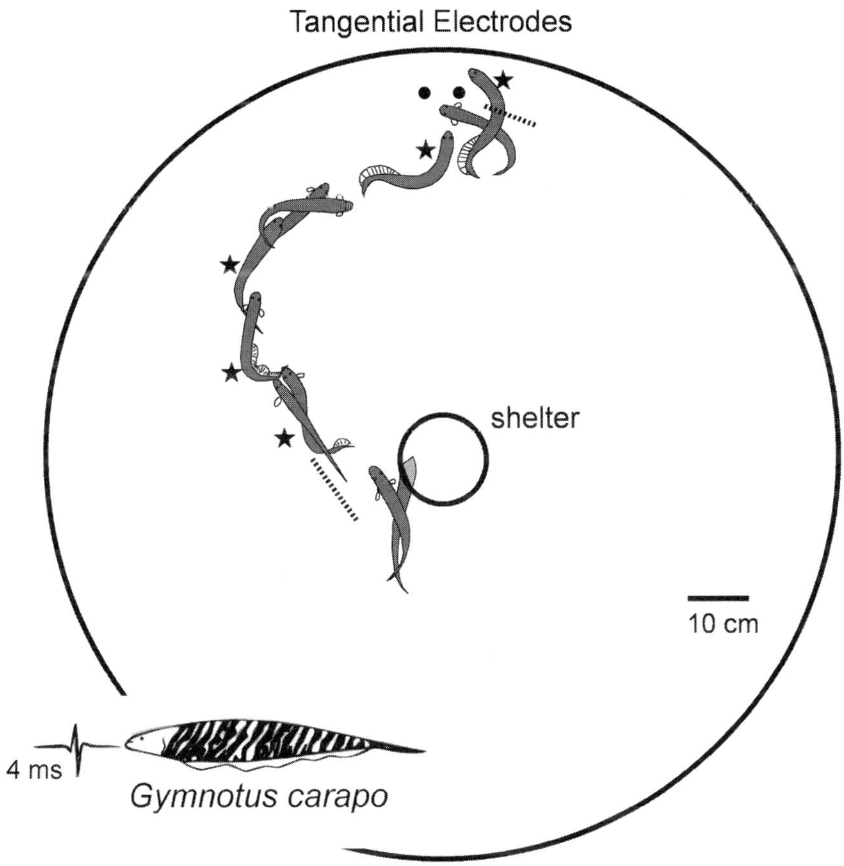

FIGURE 10.4. *Gymnotus carapo* is an aggressive fish that readily defends its territory. Here, an individual is kept in a large circular tank in the dark with a shelter in the center and viewed from above using infrared video. Electrodes are oriented tangential to the tank boundary near the edge and pulse signals resembling the EOD are played into the tank at the amplitude appropriate for a conspecific intruder. The resident emerges from the shelter, turns its electric discharge on and off in an aggressive threat display consisting of short bursts of discharges (marked by *dashed lines*), adopts intense "S" postures in preparation for a lunging attack (*stars*), and swims toward the electrodes, eventually biting them. The species and its EOD are shown in the *inset* (Davis and Hopkins, 1988).

source. Using a large shallow circular arena, they made video observations of *Brienomyrus* approaching a variety of stimulus electrodes in different configurations. Small dipole or bipolar electrodes were placed parallel to the edge of the tank, perpendicular to the edge, or separated on opposite sides of the circle. Schluger and Hopkins found that these fish tended to approach the electrodes with their body axis aligned parallel to the local current lines, which they then followed to the electrode source (Fig. 10.5A–C). As the electrodes were configured into a different geometry, the trajectory was altered to match the new electric field geometry.

A similar study was done with similar results with the gymnotiform, *Gymnotus carapo* (Davis and Hopkins 1988). Like *Brienomyrus*, *Gymnotus* aligns its body parallel to the local electric field vector while it approaches the electrode along the electric field lines. Davis and Hopkins used five different stimulus geometries: radial, tangential, or 45 (and 135) degree orientations relative to the tank radius, bipolar electrodes on opposite sides of the circle, or diverging electrodes on opposite sides (i.e., a point electrode on one side three points on the other side) (Fig. 10.5 D–I). From the traces of the fish's path it is apparent that the fish's trajectory is strongly influenced by the geometry of the stimulus electric field; in every case, the fish aligned itself parallel with the local field lines throughout the entire trajectory even when the track was indirect and longer than the straight path from shelter to intruder. In all of these experiments, stimuli were single-period sine waves adjusted to match the duration of the resident fish's own EOD. For many gymnotiform fish tested there was often a slight preference to approach the electrode that goes positive first (the "head" end) even when the pathway to reach the positive pole was longer than that to the negative pole (as in the case of +135° geometry. Head-end preference has been noted in other behavioral studies as well (Westby 1974, 1981; McKibben et al. 1993; Hopkins 1997), although individuals show varying degrees of positive polarity preference.

3.1 Quantitative Assessment of Passive Electrolocation Behavior

Quantitative measures of the alignment error between the fish's body axis and local field vector demonstrated a strong preference for either 0° or 180° error angles relative to the local electric field, no matter what the geometry of the stimulus (Fig. 10.6) (Davis and Hopkins 1988; Hopkins et al. 1997). To determine the direction of the local electric field, Hopkins et al. (1997) used the method of images, deriving an analytic solution to the Laplace equations for electrostatics to compute the electric field direction from any number of point electrodes positioned within a cylindrical nonconducting boundary. This method permitted a detailed frame-by-frame analysis of the approach behavior in addition to detailed analysis of the dynamics of the responses to the stimulus field, as discussed below.

The behavioral studies have suggested that these fish are using the zero-error algorithm proposed by Schluger and Hopkins (1987) rather than the constant-error algorithm proposed by Kalmijn. These fish are simply aligning the body axis parallel with the local electric field while swimming forward, tracking the field to the source. The response to the hemicircle bipole stimulus geometry shown in Figure 10.5C, F is noteworthy. Although the fish begins the trajectory at some arbitrary angle relative to the local electric field, it quickly turns to

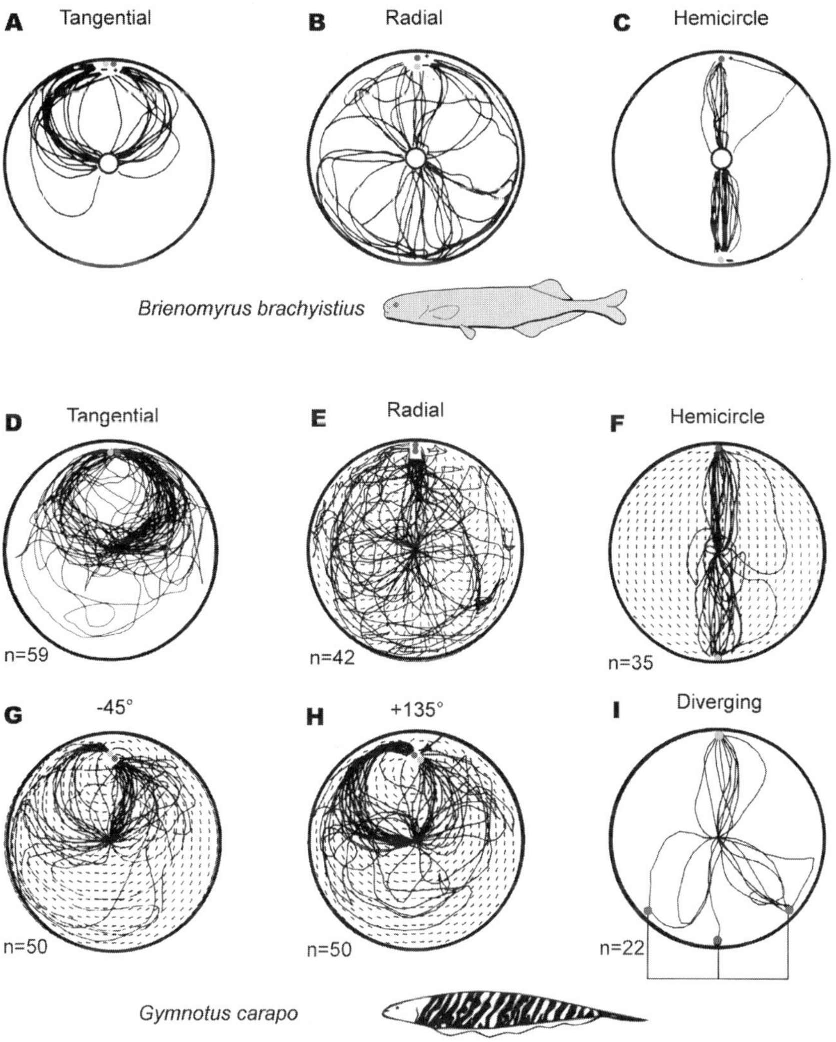

FIGURE 10.5. Superimposed tracks taken by *Brienomyrus brachyistius* (**A–C**) and *Gymnotus carapo* (**D–I**) when approaching electrodes positioned near the tank boundary in varying geometries. The fish starts a trial in a shelter near the middle of the tank and all trials are run in darkness and recorded with infrared video. *Short lines* in **E–I** show the local electric field direction within the tank boundary. The number of tracks is indicated to the left of each set of experiments. (**A**, **D**) Electrodes oriented tangential to the tank boundary, positive electrode ("head") faces right. (**B**, **E**) Electrodes oriented radially, or $+90°$ or $-90°$ with respect to x-axis. (**C**, **F**) Electrodes in hemicircle geometry, positioned on opposite sides of the tank, positive electrode up. (**G**) Electrodes oriented at $-45°$ relative to the x-axis, positive electrode to right. (**H**) Electrodes oriented at 135° relative to the x-axis, positive electrode to left. (**I**) Electrodes arranged in diverging geometry, negative up. (**A–C** modified from Schluger and Hopkins, 1987; **D–I** modified from Davis and Hopkins 1988.)

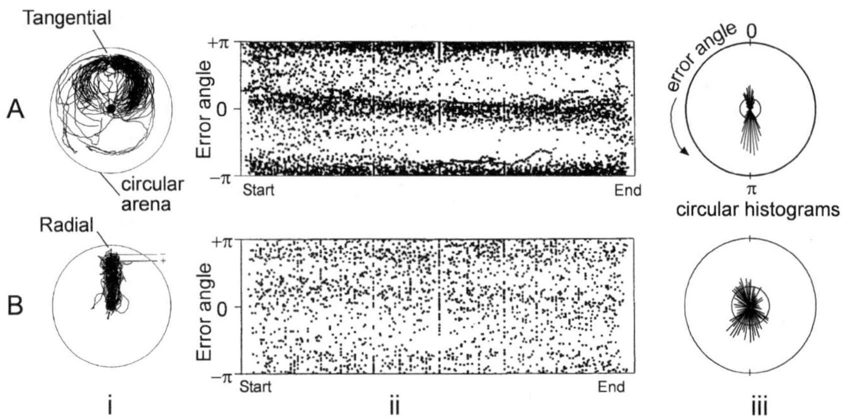

FIGURE 10.6. *Gymnotus carapo* approaching dipole electrodes in the tangential geometry (**A**) compared to radial geometry (**B**). Data are from 113 tracks (tangential) and 117 tracks (radial) using three fish. The tracks are superimposed on diagrams of the circular arena viewed from above (i) with the shelter in the center and the electrodes positioned at the top of each diagram. Each trial begins with the fish in the shelter. (ii) The error angle, that is, the angle between the fish's body axis and the direction of the local electric field is measured for each video frame and for each track and plotted as a function of the track length. All tracks are normalized to the same length in this plot. (iii) Circular histograms of cumulative error angles for all tracks. The *inner circle* shows the total number of tracks divided by the number of bins in the circular histogram; the *outer circle* is proportional to the total number of tracks. The tracks taken in these two electrode geometries differ markedly, but note how the error angles tend to cluster around either 0 or 180° in each case. These fish were usually aligned parallel with the electric field axis, and most often with their head facing opposite to the flow of current. During approaches to the tangential dipoles these fish are closely aligned with the electric field while they appear less well aligned when approaching the radial dipole. (Modified from Hopkins et al. 1997.)

align with the field lines, following them closely until reaching the source. If these fish had used the constant-error algorithm they would have continued on the arbitrary direction until they reached the edge of the tank.

The zero-error algorithm appears to work well for a variety of stationary stimulus geometries discussed earlier, even though the path taken may be longer than the straight-line trajectory from start to finish. One might well ask whether a fish would learn a more efficient approach pathway if it were given opportunity to repeatedly perform from the same distance and direction. Further, one might ask whether the approach algorithm could work for cases when the stimulus is moving or intermittently changing direction. Do these electric fish get more efficient with experience with a given stimulus geometry, and how do they cope with dynamically changing stimulus conditions when electrolocating?

Hopkins et al. (1997) repeatedly tested single specimens of *Gymnotus carapo* using a single-stimulus geometry to see if these fish would show improvement in approach path length with repetition. But even with 40 trials per day for eight days, the fish never improved at the task by taking a shorter and shorter route to the electrodes that were in the tangential orientation. Track lengths were unchanged even after 160 and 80 trials on two fish, respectively. These fish continued to follow the field lines toward the source rather than taking the direct path, even thought the path of the current lines was 1.6 times longer, on average. One fish tested in this way improved in performance only by swimming more rapidly, not by taking a more direct approach.

In a related study, Shieh et al. (1996) asked whether these fish could ever find an electrical dipole without receiving constant feedback or updates on the electric field vector. In other words, could they formulate some sort of cognitive image or cognitive map of a dipole source position based on brief exposure to electric field information monitored at a distance during the early part of an approach? Shieh et al. (1996) presented EOD-like stimuli to *Gymnotus* in the circular area from a variety of stimulus geometries but then turned the stimulus off when the fish had covered half of the distance to the dipole source. The fish became disoriented when the electrode went silent. The forward swimming velocity, which had been steadily increasing as the fish approached the target, immediately slowed or reversed. The distance to dipole, which had been steadily decreasing, immediately started to increase, and the fish hardly ever reached the target electrode once the stimulus went silent. Apparently these fish are unable to predict where the stimulus is and continue on their approach without the constant feedback from the stimulus.

Hopkins et al. (1997) also followed the approaches of these fish to rotating electrodes by attaching a dipole electrode to the end of a slowly rotating rod. When the electrode turned at 10 to 20 revolutions per minute, the fish's approach was often in a highly convoluted path with many twists and turns. One might expect that a fish would abandon its field-following strategy under these conditions since the rotating field would constantly lead it off in the wrong direction. Nevertheless the data clearly show that the fish tracks the rotating field using the same algorithm, maintaining a preferred alignment parallel to the electric field axis as it swims (Fig. 10.7). Some of the approach paths contain closed loops as the fish turns to follow the rotating electric field. Interestingly, loops run more frequently in the direction opposite to that of the rotation of the electrode, showing the fish is indeed tracking the current lines throughout their approach. A simulation shows that a point in space that moves at a constant speed but follows the local electric field vector will trace out loops in the direction opposite to that of a rotating dipole (Hopkins et al. 1997).

As these gymnotiform fish approach a dipole source under these controlled laboratory conditions, they tend to align their body axes relatively precisely with the local electric field vector. Hopkins et al. (1997) demonstrated that the accuracy improves (error dispersion is reduced) for the first 75% of the approach, but then declines during the final 25%. One interpretation is that the fish switches to an alternative strategy when it gets to within a body length of the

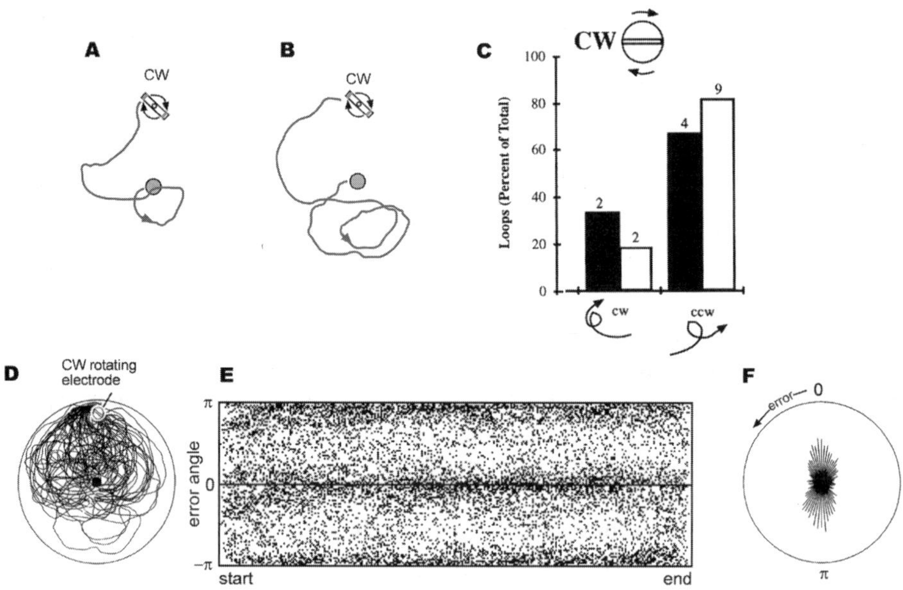

FIGURE 10.7. Even when the stimulus electrodes are rotating, *Gymnotus carapo* appears to follow the electric field lines. This sometimes results in trajectories with closed loops. Two example tracks are shown in **A** and **B**. The electrode is rotating clockwise; note that the loops run counterclockwise. (**C**) In response to clockwise rotation of the electrode, the fish make more counterclockwise loops. *Bars* show percentages of loops that are clockwise (cw) or counter clockwise (ccw) in response to clockwise rotation. *Black bars* are one fish, *white* a second. (**D**) Superposition of 66 tracks of approaches to clockwise rotating electrodes. (**E**) Error angles between the fish's body and the local electric field, determined for each video frame. The abscissa is frame number plotted as a percentage of the total number of frames in the track. These data include 190 trials using two fish. The *dots* clearly cluster around 0° and 180°. (**F**) Circular histogram of error angles for all tracks shown in **E**. (Modified from Hopkins et al. 1997.)

stimulus source. The alternative strategy is probably due to a sharp, highly focused stimulus "hot spot" on the fish's skin near the dipole center. As the electric field changes from globally homogeneous to a localized hot spot, the fish may simply turn toward the part on the body most strongly stimulated. This behavior might well lead to a trajectory similar to that predicted by Kalmijn.

3.2 Dynamics of Passive Electrolocation Behavior

A complete description of passive electrolocation behavior includes not simply the static posture of a fish relative to the local stimulus conditions at a given moment, but also the dynamics of the response: how the animal's behavior changes over time. This is a complex problem that requires analysis of stimulus conditions and behavioral responses at regular time intervals throughout an approach trajectory. This problem is suited to frame-by-frame analysis of behavior and of the electric stimuli that are present.

Hopkins et al. (1997) studied dynamic aspects of passive electrolocation using pulse gymnotiform fishes. This study explored in depth the time dependency of tracking behavior, both for moving and nonmoving stimuli. Figure 10.8 shows a frame-by-frame analysis of an approach to a tangential dipole. The track of the fish emerging from the shelter is represented in Figure 10.8C. The fish backs out of the shelter, reverses direction, and swims in a curved path toward the electrodes. Measurements of the alignment "error" are illustrated in Figure 10.8A, B. In the first few frames the error between the fish's body and the axis of the electric field (ignoring the polarity of the electric field, which appears not to be important) is nearly $+\pi/2$, that is, the axis is pointing nearly 90° to the fish's right. The fish is bent slightly to its left (positive bend angle, β) and it backs up into its V-turn (indicated by 1 in Fig. 10.8C). Then the fish swims forward and bends to the right (negative β). As it progresses, the error angle, ξ, is reduced while the fish picks up speed. At point 2, as the error angle starts to increase, the fish makes a corrective turn at point 3. This reduces the error angle to zero again. The fish continues until at point 4, where the error angle increases sharply in the negative direction. At this point the fish makes a sudden left turn that reduces the error again and also brings it in contact with the dipole. Although this is just one example of many such trajectories, it illustrates a number of characteristics of the dynamics of passive electrolocation: constant error correction, changing velocity, and reversing direction when abrupt changes are called for.

Dynamic analysis of individual tracks shows that they can be quite variable. Nevertheless a number of simple rules appear to apply broadly when a large number of trials are analyzed. Hopkins et al. (1997) analyzed a large number of tracks such as the one illustrated, including different electrode geometries as well as dynamically changing electrodes (rotating). A detailed cross correlation analysis between the fish's bend angle and the stimulus error angle indicated that the fish's bend angle is positively and significantly correlated with the error angle especially at a cross-correlation delay of 0.3 s. It thus appears that the fish bends its body in response to an electric field error that is nonzero. The time delay for the response is approximately 300 ms. When the electric field is to the fish's left (positive error), the fish bends to the left (positive bend). A negative error (right side) predicts a negative bend (right bend). Of course when the fish bends to the left in response to a positive error as it is swimming forward, it results in a turn that reduces the error angle. The electric vector rotates around toward the fish's nose. Interestingly, the cross correlation analysis shows that the sign of the relationship between error angle and bend angle is reversed when the fish is swimming backwards. Since gymnotiform fish can swim both forward and backward with apparent equal ease simply by reversing the direction of the power stroke in the undulating anal fin, the fish has to reverse the relationship between error and bend so that it turns in the correct direction

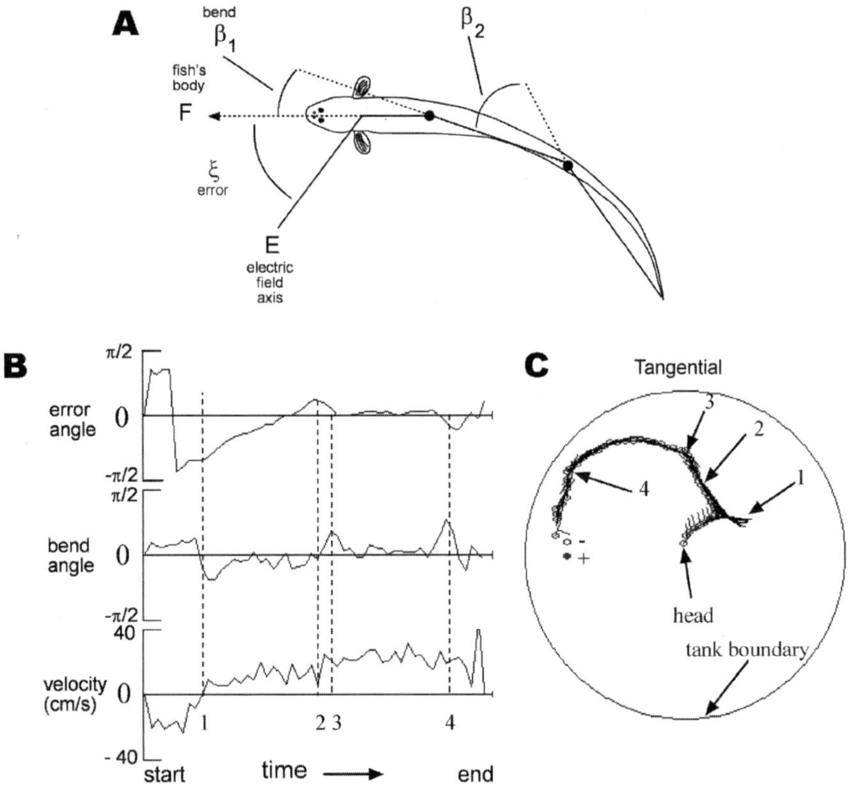

FIGURE 10.8. Dynamic analysis of *Gymnotus carapo* approaching a tangentially oriented dipole. (**A**) Drawing of the fish taken from above shows the **F**-vector defined by the orientation of the head segment of the body, the **E**-vector defined by the direction of the local electric field, and the error angle, ξ, between them. Note that ξ, which varies between −90° and +90°, is the angle between the *axis* of the fish and the *axis* of the electric field, ignoring polarity. Angles β_1 and β_2 are the bend angles for the fish's body while the total bend angle is the sum of the two. (**B**) Frame by frame analysis of error angle, bend angle, and velocity as the fish makes the approach shown in **C**. (**C**) Trajectory of the fish as it backs out of the shelter and approaches the tangentially oriented dipole. Note in **B** how the negative error angle at position 1 precedes a negative bend a few frames later. Later, a positive error angle at position 2 precedes a positive bend at position 3. (Modified from Hopkins et al. 1997.) This single track is highly representative as shown by cross-correlation between error angle and bend angle (see text).

to reduce error while backing up. This was observed in two species of gymnotiforms (Hopkins et al. 1997). Dynamic analysis also reveals a significant correlation between error angle and swimming velocity. The velocity of swimming is inversely correlated with the absolute value of the error angle. In other words, the fish slows down when the error angle gets larger and larger. This is consistent with the observation that velocity reversals (V-turn) occur when the error axis approaches the maximum value, 90°.

Convincing evidence for a causal relationship between electric field direction and dynamic control of turning and velocity comes from an experimental study in which electric fields were manipulated during the course of normal approach (Shieh et al. 1996). To gain experimental control, the dipole orientation was electronically "jumped" from one orientation to another as the fish moved to attack a dipole electrode.

To jump the angle of the dipole electronically, Shieh et al. (1996) produced a quadrapole electrode: two dipoles set mutually perpendicular to each other. A computer was used to determine the x-component and the y-component of a given signal voltage as a function of a preferred dipole axis angle. The signal to the x-axis dipole was equal to the EOD voltage multiplied by the sine of the desired dipole angle, θ, while the signal to the y-axis was equal to the EOD times the cosine of θ. Tests showed that the field could be instantly jumped from one dipole direction to another and that the field maps were dipole shaped. The results of jumping the electric field direction are shown in Figure 10.9.

Figure 10.9A, B shows the results of a typical jump experiment in which the electric field was twice jumped through a vector angle of 45°. The fish was already aligned with the electric field before the jump, so the jump produced an immediate error in alignment. Following each jump in dipole direction, the fish turned to correct the error and realign itself with the field axis. Figure 10.9C shows the averages of 40 different trials. Dots show error, bend, and velocity for each frame; the lines show mean and standard error of the mean of the error angle, bend angle, and velocity. Arrows point to the time when there was either a sudden increase or sudden decrease in error angle, and also to the increase bend probability. Clearly, positive jumps (right column) resulted in an increased bend probability at a delay of 0.3 s. Negative jumps (left column) produce a weak response (arrow in bend response, left column). Velocity is unaffected by the jump in dipole direction. When error angle is cross correlated with bend angle for all experiments involving jumps of dipole angles, there is a strong positive correlation between error and bend at a delay of 0.33 s, as seen in previous observations.

3.3 Approach Algorithm Summary

Passive electrolocation is so important for electroreceptive organisms attempting to find prey, predators, or electric communication signals that one might imagine that the fish would be able to integrate all the information about the direction, curvature, and intensity of the electric field in order to come up with some sort of cognitive or spatial map of the electric field. It would seem reasonable, given certain assumptions about the dipole nature of many electric fields of biological origin and the potential importance, that the fish might be able to then extrapolate and predict the position of a dipole source, especially given the relatively

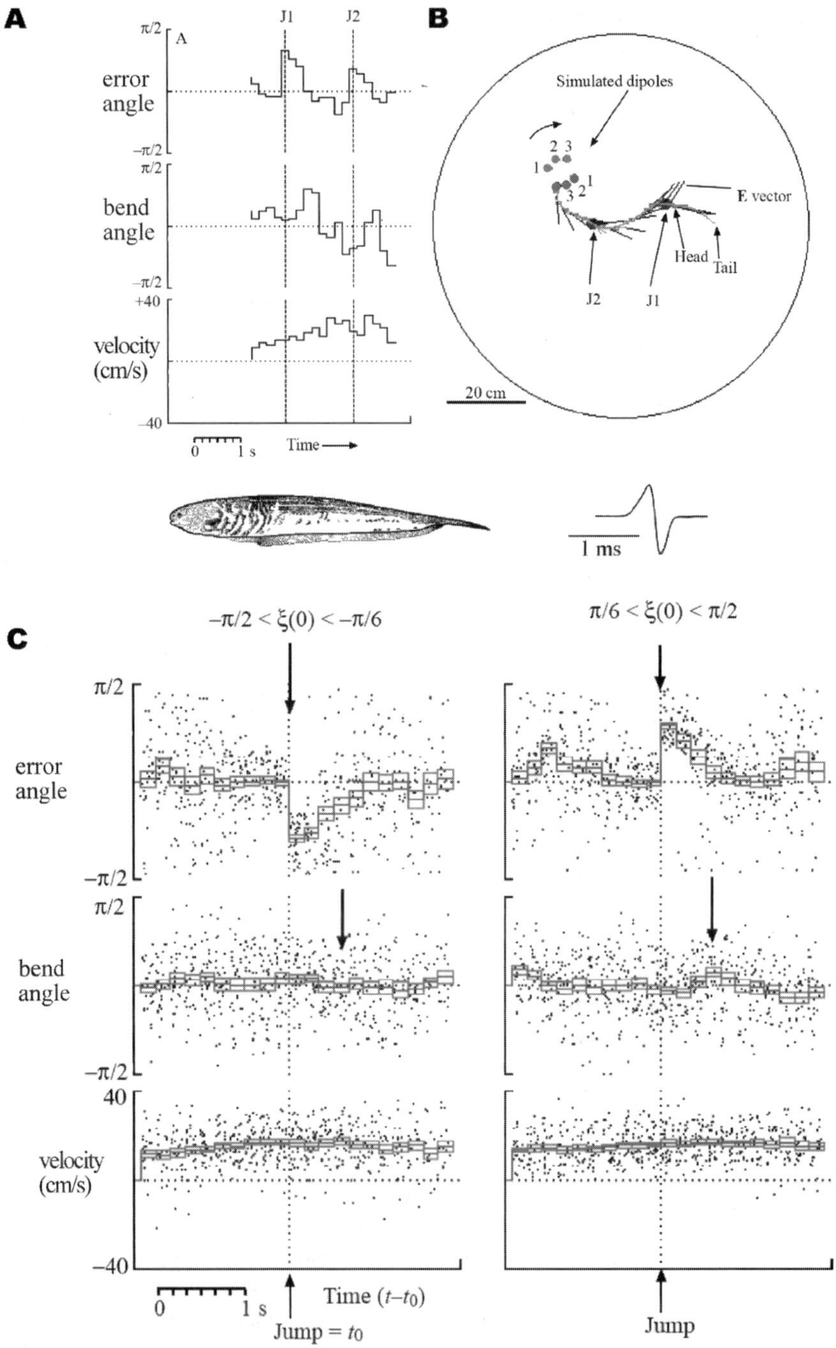
A
J1
J2
π/2
error
angle
−π/2
bend
angle
+40
velocity
(cm/s)
−40
0
1 s
Time
B
Simulated dipoles
E vector
Head
Tail
J2
J1
20 cm
1 ms
C
−π/2 < ξ(0) < −π/6
π/6 < ξ(0) < π/2
error
angle
bend
angle
40
velocity
(cm/s)
−40
Time (t−t0)
Jump = t0
Jump

enormous brain volumes devoted to electrosensory processing ((Erdl 1846; Bullock and Heiligenberg 1986). However, the experimental evidence from simple electrode approaches, dynamic analysis of electric field following, the silencing experiment, the repeated trial learning experiment, and the jump experiments is decidedly against this hypothesis.

Instead, the approach behavior appears to be mediated by a simple set of behavioral rules that guide the fish toward the source. First the fish needs to be able to detect the direction of the local electric field and align its body along that vector direction. Then, by swimming forward, it follows the field lines until it reaches the source. Although the alignment plus movement superficially resembles galvanotaxis described for many unicellular organisms (Loeb 1918; Adler and Shi 1988), the behavior of electric fish is more complex. When the fish are swimming, errors in the electric field direction cause the fish to bend its body to the left or right leading to a subsequent turn. If the electric field axis lies to the fish's left, it bends left and turns left. If the field is to the right, it bends right and turns right. When swimming backwards, the fish reverses the bend angle in response to a given error angle. Also, if the error angle gets large, the fish slows down; and sometimes executes a V-turn. These behavioral strategies will keep the fish aligned with the electric field under virtually any conditions, and allow it to approach the source in a reliable way even under conditions filled with objects that distort and alter the electric field locally. Kalmijn's approach algorithm (Kalmijn 1988a, 1989) appears not to apply to the

FIGURE 10.9. Jump experiments allow the experimenter to jump the direction of an electric dipole from one angular direction to another while a fish is approaching the dipole source. The experimental fish is *Brachyhypopomus diazi* (*inset*), which has a short pulse-like EOD. When the fish approaches the dipole source it aligns its body axis parallel to the axis of the electric field sensed locally. Jumping the direction of the dipole source to a different angular direction produces a sudden increase in the error angle, which then causes the fish to change directions in response. (**A**, **B**) results of a single jump experiment. The track of the fish is shown in **B**, the analysis in **A**. When the trial begins, the electrodes are in position 1 in **B**. The dipole is then jumped to position 2 and then position 3, each time incrementing the dipole axis by 45°. Analysis of the "error angle" in **A** shows that in this trial the jump produces a sudden increase in error angle simultaneous with the jump. Following this, the fish bends to the left to reduce the error back to 0°. The next jump causes a second increase in error, which is similarly corrected. (**C**) Plots of error angles, bend angles, and velocity of the fish in response to jumps in the dipole direction. The jumps that result in positive error angles are shown in the *right column*; those that result in negative errors are shown in the *left column*. The *dots* show data from individual frames from a total of 40 jump experiments. *Horizontal gray lines* show the mean and standard error of the values for each video frame. The *horizontal axis* is time relative to the jump. Note that positive jumps that produce positive error angles result in positive bends 0.2 to 0.5 later (*arrow* in right column, second row of **C**) and that this corrects the error. Negative jumps produce weak bends (*arrow* in left column, second row of **C**), and the error is reduced to zero within a fraction of a second. The velocity is unchanged by the jump. (Modified from Shieh et al. 1996.)

electric fish approaching electric discharges, although the theory appears to describe accurately the behavior of sharks approaching prey, and may also describe fish following hydrodynamic near fields toward a mechanically vibrating source (Kalmijn 1989), although experimental data are yet available. The critical difference between of prey localization and EOD localization is the distance at which oriented responses occur.

4. Electroreception for Passive Electrolocation

Given that electric fish are able to guide their approach to dipole sources, we may well ask about the sensitivity of individual electroreceptors to stimuli from different directions. This has been explored in some detail in the gymnotiform electric fish, *Brachyhypopomus* (McKibben et al. 1993; Yager and Hopkins 1993), by placing a single fish in the center of a large circular tank and delivering homogeneous electric fields to the fish through pairs of electrodes mounted on opposite sides of the tank. With electrode pairs spaced apart at 15° intervals, it is possible to map the directionality of individual receptors in the horizontal plane. Figure 10.10 shows plots of receptor directionality as a function of position on the body surface. Two types of receptors are included: burst duration coders, which are the tuberous electroreceptors that code for amplitude, and pulse markers, the tuberous receptors that code for time. Both types show the same pattern of directionality.

Typically the sensitivity of these electroreceptors follows a cosine function of stimulus angle, that is, maximally sensitive at 0° and 180° relative to the best direction, and minimally sensitive at ± 90° relative to the best direction. On the polar plots of sensitivity or spike number in Figure 10.10, the cosine-like sensitivity profiles are figure-8 shaped. The directionality appears to derive from the receptor's position on the skin surface, not from some inherent directionality of the receptor itself. Figure 10.10 shows how best direction varies systematically over the body surface by arrows pointing in the best direction. On the nose the receptors are most sensitive to fields aligned with the body axis; on the mid-body, they are most sensitive to fields tangential to the skin surface, and on the tail, they are again parallel to the body axis.

From these measurements one should expect that the receptors most strongly stimulated will shift from the nose to the cheek to the flank as the homogeneous field move around from 0° to 45° to 90°. Thus, for the fish to align itself to the electric field, it simply needs to bend its body on the side where the head is most strongly stimulated. If the nose is strongly stimulated and both sides are stimulated equally, the fish should be perfectly aligned with the field.

It is still unknown how electric-field errors are translated into bending or turning movements in these fish, but it is known that electric-field vector directions are represented as a surface map in the optic tectum of catfish, *Ictalurus* spp. (Knudsen 1976). Since the tectum is involved in multimodal sensory integration in a large number of vertebrates (Hartline et al. 1978; Bastian 1982;

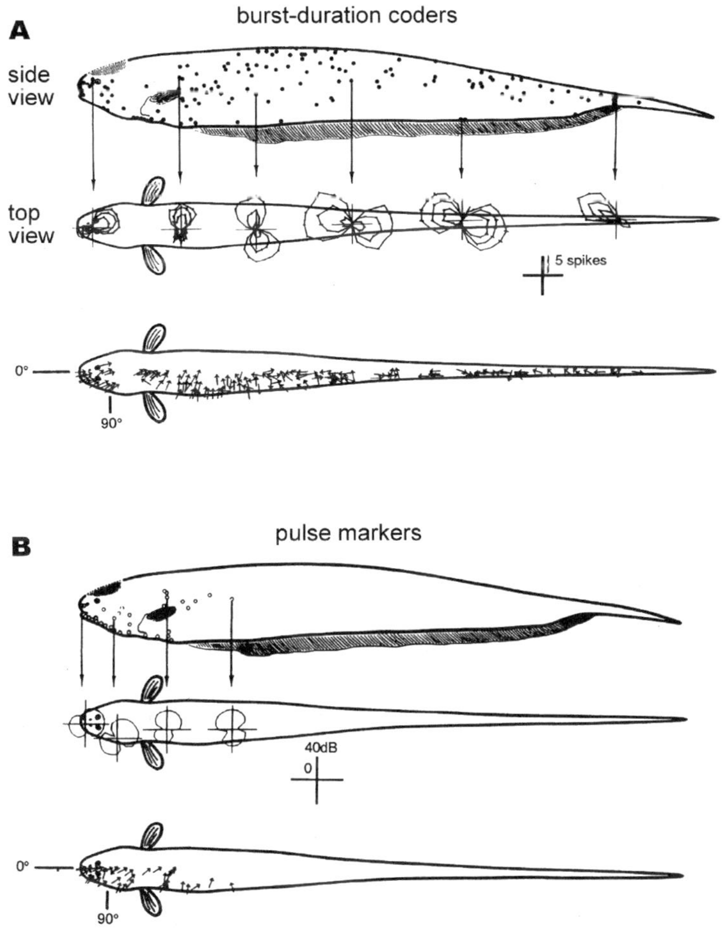

FIGURE 10.10. The directionality of individual electroreceptors was explored in the gymnotiform, *Brachyhypopomus* by recording from individual afferents in the lateral line nerve. The fish was stimulated with homogeneous electric fields originating from electrodes positioned around the edge of a circular tank. Electrodes were spaced apart every 15° around the edge of the tank. Directionality is shown for two types of tuberous electroreceptors: (**A**) burst duration coders, which encode stimulus amplitude, and (**B**) pulse markers, receptors that encode stimulus time. In each set of drawings, the *top picture* shows the location of a single electroreceptor on the skin surface. The *middle drawing* shows the directionality of selected receptors located under the center of the drawing. For burst duration coders, directionality is shown as a polar plot of numbers of spikes as a function of direction. For pulse markers, the polar plot shows sensitivity versus direction. The third drawing shows the directionality of all of the receptors shown in the top drawing, with the *arrow* pointing in the most sensitive direction. Note that the most sensitive direction is usually perpendicular to the skin surface, and that the most sensitive direction of a receptor varies systematically from the head region to the tail. The skin surface is thus a map of the local electric field direction. (Modified from Yager and Hopkins 1993.)

Knudsen 1982; Sparks 1986), it is likely that sensory–motor integration takes place in this structure. Much more work needs to be done on passive electrolocation to understand fully the complexities of sensory guidance in this unique system where vector cues do not point to the source of the stimulus.

5. Summary

Although sensory guidance is of critical importance to nearly all sensory systems, we know comparatively little about sensory guidance in the electrosensory modality, for which the physical cues are limited to amplitude cues and vector cues that do not point to the source of the signal. This creates special problems for fish using electroreception to locate predators or prey or conspecifics with electric discharges. This chapter has discussed a number of studies in which it has been possible to observe and quantify the behavioral strategies used in passive electrolocation by fish that have tuberous or high-frequency electroreceptors designed to detect the EODs of conspecifics. Passive electrolocation of EODs is extremely reliable and repeatable as a behavior, so it can be studied experimentally and quantitatively. From this it has been possible to derive a behavioral algorithm for an orientation behavior. The algorithms used by these fish are simple and reliable, although the resulting behavior can be complex.

To date little is know about the central physiology of passive electrolocation except that the receptors are highly directional in the periphery and that they map somatotopically onto the electrosensory lateral line lobe (Knudsen 1976; Bastian 1982). Although these fish appear not to be able to tell where the stimulus is located, they are fully capable of finding their way to the source of a stimulus simply by following electric field lines. To do this task, they must be able to tell the angle of the electric field vector relative to their body axis and, given this, turn their body in the correct direction to align it with the field axis.

Much remains to be learned about the interactions between electroreceptor maps, known for catfish (Knudsen 1976) and gymnotiform fish (Bastian 1982), and motor output. To fully understand sensory guidance, future work should focus on the details of motor control: the control bending and turning, the control of velocity, and the control of swimming direction. Combined with a more detailed understanding of how the peripheral electroreceptors encode passive electrolocation signals, future studies should extend our knowledge of spatial aspects of field following behavior for this unusual sensory modality. Understanding gained from the study of passive electrolocation may well apply to other systems of orientation, including orientation to near-field sounds and hydrodynamic fields (Kalmijn 1988b; Coombs et al. 2002).

Acknowledgments. The author thanks T.H. Bullock, Arthur Popper, and Richard Fay for helpful comments on the manuscript. The author's own research has been supported by grants from the National Science Foundation (BNS-8810080), the New York State Hatch Act (191426), and the National Institutes of Health (MH37972, DC006206). The studies done here were carried out by Elizabeth Davis, James Schluger, Garry Harned, Willard Wilson, Kwang Shieh, Donald McBride, David Yager, and Michael Winslow.

References

Adler J, Shi W (1988) Galvanotaxis in bacteria. Cold Spring Harbor Symp Quant Biol 53:23–25.

Assad C, Rasnow B, Stoddard PK, Bower JM (1998) The electric organ discharges of the gymnotiform fishes: II. *Eigenmannia*. J Comp Physiol A 183:419–432.

Assad C, Rasnow B, Stoddard PK (1999) Electric organ discharges and electric images during electrolocation. J Exp Biol 202:1185–1193.

Bastian J (1982) Vision and electroreception: integration of sensory information in the optic tectum of the weakly electric fish *Apteronotus albifrons*. J Comp Physiol A 147: 287–297.

Bastian J (1986) Electrolocation: behavior, anatomy and physiology. In: Bullock TH, Heiligenberg W (eds), Electroreception. New York: John Wiley & Sons, pp. 577–612.

Bastian J (1987a) Electrolocation in the presence of jamming signals: behavior. J Comp Physiol A 161:811–824.

Bastian J (1987b) Electrolocation in the presence of jamming signals: electroreceptor physiology. J Comp Physiol A 161:825–836.

Black-Cleworth P (1970) The role of electric discharges in the non-reproductive social behaviour of *Gymnotus carapo*. Anim Behav Monogr 3:1–77.

Bullock TH, Heiligenberg W (1986) Electroreception. New York: John Wiley & Sons.

Coombs S, New JG, Nelson M (2002) Information-processing demands in electrosensory and mechanosensory lateral line systems. J Physiol Paris 96:341–354.

Crawford J (1991) Sex recognition by electric cues in a sound-producing mormyrid fish, *Pollimyrus isidori*. Brain Behav Evol 38:20–38.

Davis EA, Hopkins CD (1988) Behavioural analysis of electric signal localization in the electric fish, *Gymnotus carapo*, Gymnotiformes. Anim Behav 36:1658–1671.

Erdl MP (1846) Über das Gehirn der Fischgattung *Mormyrus*. Gelehrte Anzeig Bayer Akad Wiss 22/23:403–407.

Fraenkel GS, Gunn DL (1940) The Orientation of Animals. Oxford: Oxford University Press.

Hartline PH, Kass L, Loop MS (1978) Merging of modalities in the optic tectum: infrared and visual integration in rattlesnakes. Science 199:56–59.

Heiligenberg W (1977) Principles of electrolocation and jamming avoidance in electric fish. A neuroethological approach. In: Braitenberg V (ed), Studies in Brain Function. New York: Springer-Verlag, pp. 1–85.

Hopkins CD (1986) Temporal structure of non-propagated electric communication signals. Brain Behav Evol 28:43–59.

Hopkins CD (1993) Behavioral analysis of sensory function: active and passive electrolocation. J Comp Physiol A 173:688.

Hopkins CD, Comfort NC, Bastian J, Bass AH (1990) Functional analysis of sexual dimorphism in an electric fish, *Hypopomus pinnicaudatus*, order Gymnotiformes. Brain Behav Evol 35:350–367.

Hopkins CD, Shieh K-T, McBride DW Jr, Winslow M (1997) A quantitative analysis of passive electrolocation behavior in electric fish. Brain Behav Evol 50 (Suppl 1):32–59.

Kajiura SM, Holland KN (2002) Electroreception in juvenile scalloped hammerhead and sandbar sharks. J Exp Biol 205:3609–3621.

Kalmijn AJ (1971) The electric sense of sharks and rays. J Exp Biol 55:371–383.

Kalmijn AJ (1974) The detection of electric fields from inanimate and animate sources other than electric organs. In: Fessard A (ed), Handbook of Sensory Physiology, vol. III/3: Electroreceptors and Other Specialized Receptors in Lower Vertebrates. Berlin: Springer-Verlag, pp. 147–200.

Kalmijn AJ (1988a) Detection of weak electric fields. In: Atema J, Fay RR, Popper AN, Tavolga WN (eds), Sensory Biology of Aquatic Animals. New York: Springer-Verlag, pp. 151–186.

Kalmijn AJ (1988b) Hydrodynamic and acoustic field detection. In: Atema J, Fay RR, Popper AN, Tavolga WN (eds), Sensory Biology of Aquatic Animals. New York: Springer-Verlag, pp. 83–130.

Kalmijn AJ (1989) Functional evolution of lateral line and inner ear sensory systems. In: Coombs S, Görner P, Müntz H (eds), The Mechanosensory Lateral Line: Neurobiology and Evolution. New York: Springer-Verlag, pp. 187–215.

Kalmijn AJ (1997) Electric and near-field acoustic detection, a comparative study. Acta Physiol Scand 161(Suppl): 25–38.

Knudsen EI (1975) Spatial aspects of electric fields generated by weakly electric fish. J Comp Physiol 99:193–198.

Knudsen EI (1976) Midbrain responses to electroreceptive input in catfish: evidence for orientation preferences and somatotopic organization. J Comp Physiol A 109:51–67.

Knudsen EI (1982) Auditory and visual maps of space in the optic tectum of the owl. J Neurosci 2:1177–1194.

Lissmann HW (1958) On the function and evolution of electric organs in fish. J Exp Biol 35:156–191.

Lissmann HW, Machin KE (1958) The mechanisms of object location in *Gymnarchus niloticus* and similar fish. J Exp Biol 35:457–486.

Loeb J (1918) Forced Movements, Tropisms, and Animal Conduct. Philadelphia: Lippincott.

McGregor PK, Westby GWM (1992) Discrimination of individually characteristic electric organ discharges by a weakly electric fish. Anim Behav 43:977–986.

McGregor PK, Westby GWM (1993) Individually characteristic EOD waveforms and discrimination by *Gymnotus carapo*. J Comp Physiol A 173:741.

McKibben JR, Hopkins CD, Yager DY (1993) Directional sensitivity of tuberous electroreceptors: polarity preferences and frequency tuning. J Comp Physiol A 173:415–424.

Rasnow B, Assad C, Bower JM (1993) Phase and amplitude maps of the electric organ discharge of the weakly electric fish, *Apteronotus albifrons*. J Comp Physiol A 172: 481–491.

Schluger J, Hopkins CD (1987) Electric fish approach stationary signal sources by following electric current lines. J Exp Biol 130:359–367.

Schöne H (1984) Spatial Orientation: the Spatial Control of Behavior in Animals and Man. Princeton: Princeton University Press.

Shieh K-T, Wilson W, Winslow M, McBride DW Jr, Hopkins C (1996) Short-range orientation in electric fish: an experimental study of passive electrolocation. J Exp Biol 199:2383–2393.

Sisneros JA, Tricas TC (2002a) Neuroethology and life history adaptations of the elasmobranch electric sense. J Physiol Paris 96:379–389.

Sisneros JA, Tricas TC (2002b) Ontogenetic changes in the response properties of the peripheral electrosensory system in the Atlantic stingray (Dasyatis sabina). Brain Behav Evol 59:130–140.

Sisneros JA, Tricas TC, Luer CA (1998) Response properties and biological function of the skate electrosensory system during ontogeny. J Comp Physiol A 183:87–99.

Sparks DL (1986) Translation of sensory signals into commands for control of saccadic eye movements: role of primate superior colliculus. Physiol Rev 66:118–171.

Stoddard PK, Rasnow B, Assad C (1999) Electric organ discharges of the gymnotiform fishes: III. *Brachyhypopomus*. J Comp Physiol A 184:609–630.

Tricas TC, Michael SW, Sisneros JA (1995) Electrosensory optimization to conspecific phasic signals for mating. Neurosci Lett 202:129–132.

von der Emde G, Schwartz S, Gomes L, Budelli R, Grant K (1998) Electric fish measure distance in the dark. Nature 395:890–894.

Westby GWM (1974) Assessment of the signal value of certain discharge patterns in the electric fish, *Gymnotus carapo* by means of playback. J Comp Physiol A 92:327–341.

Westby GWM (1981) Communication and jamming avoidance in electric fish. Trends Neurosci 4:205–210.

Wilkens LA, Hofmann MH, Wojtenek W (2002) The electric sense of the paddlefish: a passive system for the detection and capture of zooplankton prey. J Physiol Paris 96: 363–77.

Yager DY, Hopkins CD (1993) Directional characteristics of tuberous electroreceptors in the weakly electric fish, *Hypopomus* (Gymnotiformes). J Comp Physiol A 173:401–414.

11

Target Detection, Image Analysis, and Modeling

Mark E. Nelson

1. Introduction

One of the selective advantages afforded by the electric sense is that it enables animals to detect objects in dark aquatic environments where visual cues are limited. This ability, referred to as electrolocation, has functional similarities to the echolocation ability of bats and opens up a rich ecological niche for hunting at night or in turbid waters. Electrolocation also has functional similarities to visual processing because target objects form electrosensory images on a spatially distributed sensor array. This chapter summarizes recent advances in understanding target detection, localization, and image analysis in the electrosensory system, including contributions of mathematical modeling and computer simulation in helping to advance knowledge in this area.

Conceptually, the electrolocation task can be subdivided into three components—detection, characterization, and localization. The *detection* problem involves a determination of whether or not a potential electrolocation target is in the vicinity. In its simplest form, this can be formulated as a binary hypothesis-testing problem, namely one that can be answered with either a "yes" or "no" response. If a potential target is present, the next conceptual stage is one of *characterization*. This stage requires an assessment of various target properties that can vary along a continuum, such as target size, shape, and electrical impedance. From a signal processing perspective, this is a problem in parameter estimation. If the behavioral response requires a directed movement, such as prey capture, then the final component of the task is one of *localization*. Localization involves estimation of quantities that can vary along a continuum, such as direction and distance to the target. More generally, the localization problem is a target-tracking task that may require estimation of dynamic properties of both the fish and prey, such as relative velocity and acceleration. Although the processes of detection, characterization, and localization have been described sequentially, they are tightly coupled and are likely to take place in parallel in the nervous system.

Electrolocation can be mediated by both low-frequency (Wilkens, Chapter 9) and high-frequency (Hopkins, Chapter 10) electrosensory systems. For animals that lack an electric organ, *passive* electrolocation is based on the ability to detect the intrinsic electric fields associated with target objects, such as bioelectric fields generated by prey. In contrast, animals that can generate their own electric field can detect a larger variety of target objects because *active* electrolocation requires only that the electrical properties of the target differ from those of the surrounding water. Aside from the source of the electric field, the functional principles of electrolocation are similar for both active and passive modes. To streamline the presentation, this chapter focuses on active electrolocation (reviewed in Moller 1995; von der Emde 1999; Bastian 2003). Comparisons between active and passive electrolocation are discussed in Nelson et al. (2002) and comparisons with the mechanosensory lateral line system can be found in this volume (Coombs, Chapter 12).

In the late 1950s, Lissmann and Machin (1958) first demonstrated that weakly electric fish (*Gymnarchus niloticus*) can detect and discriminate objects based solely on their electrical properties. They employed a clever experimental design in which artificial targets were placed inside visually opaque but electrically transparent porous pots. Using this approach they determined, for example, that *Gymnarchus* could detect the presence of a small glass rod (threshold diameter, 2 mm) hidden inside a pot using only electrosensory cues. Generalizing from these early results and subsequent experiments using artificial targets, it has long been presumed that weakly electric fish, which are largely nocturnal and found in turbid waters, use electrolocation to hunt for food. Only recently have behavioral studies demonstrated that the electrosensory system is indeed involved in the detection and localization of small live prey (Wilkens et al. 1997; von der Emde and Bleckmann 1998; Nelson and MacIver 1999). This chapter focuses on prey capture as a prototypical electrolocation task that is linked to the animal's natural behavior and sensory ecology.

We begin, in Section 2, with an overview of the physics of active electrolocation and electrosensory image formation. Section 3 discusses the encoding of information by primary electrosensory afferent nerve fibers. Section 4 reviews key aspects of central neural processing, with an emphasis on the first stage of central processing in the hindbrain electrosensory lateral line lobe (ELL). Finally, Section 5 provides a summary of key insights that have been gained from studies of electrolocation and indicates directions for future research.

2. Physics of Active Electrolocation

This section provides an overview of the physics of electric field generation and electrosensory image formation. Physics-based models and computer simulation techniques can be combined to generate accurate representations of the fish's self-generated electric field and the electrosensory images that are formed on the skin when an object is placed in the field. Combining these modeling techniques with movement trajectories from three-dimensional video analysis allows the reconstruction of dynamic electrosensory images of prey capture events. This type of analysis provides a first glimpse of what the world "looks like" when viewed through an electric sense.

2.1 Temporal Aspects of the Fish's Electric Field

A natural starting point for describing the physics of active electrolocation is a description of the electric field that the fish uses to probe its environment. The electric field generated by a weakly electric fish can be characterized in terms of the *temporal* properties of the electric organ discharge (EOD) waveform and the *spatial* properties of the field surrounding the fish's body. The temporal properties of the EOD can be qualitatively assessed by placing a pair of wires in the water near the fish and listening to the EOD signal on an audio monitor. The EOD patterns of weakly electric fish fall naturally into two categories. Pulse-type fish generate brief discharges separated by significantly longer gaps, such that the audio output sounds like a series of clicks. Wave-type electric fish generate EODs with little or no gap between successive discharges, resulting in a periodic signal that sounds like a tone or hum.

Quantitative comparisons of EOD waveforms are typically made by recording the potential difference between the head and tail of the fish. Figure 11.1 illustrates head–tail waveforms for representative pulse-type and wave-type EODs, along with their corresponding power spectra. Because of inherent tradeoffs in time–frequency representations, pulse fish have EOD waveforms that are narrow in the time domain and broad in the frequency domain, while wave fish EODs are broad in the time domain and narrow in the frequency domain. Commonly studied species with pulse-type EODs include most African mormyrids, such as *Gnathonemus petersii* (elephant nose fish), and certain South American gymnotiform knifefish, including *Gymnotus carapo* (banded knifefish) and *Brachyhypopomus pinnicaudatus*. Commonly studied species with wave-type EODs include many other South American gymnotiform knifefish, such as *Apteronotus albifrons* (black ghost knifefish), *Apteronotus leptorhynchus* (brown ghost knifefish), *Eigenmannia virescens* (glass knifefish), and *Sternopygus macrurus* (longtail knifefish), as well as a single African mormyriform with a wave-type EOD, *Gymnarchus niloticus* (Aba aba) (Bass 1986; Zupanc and Bullock, Chapter 2).

The EOD amplitudes of both pulse- and wave-type fish seem to remain constant during electrolocation behavior. Although there can be circadian, seasonal, and socially mediated changes in EOD amplitude, short-term modulations of EOD amplitude in association with electrolocation have not been reported. This is in contrast to some echolocating bats, which are known to decrease the intensity of their ultrasonic calls as they approach a prey target. This sort of intensity compensation behavior has not been observed in weakly electric fish. However, there are technical difficulties in making accurate measurements of

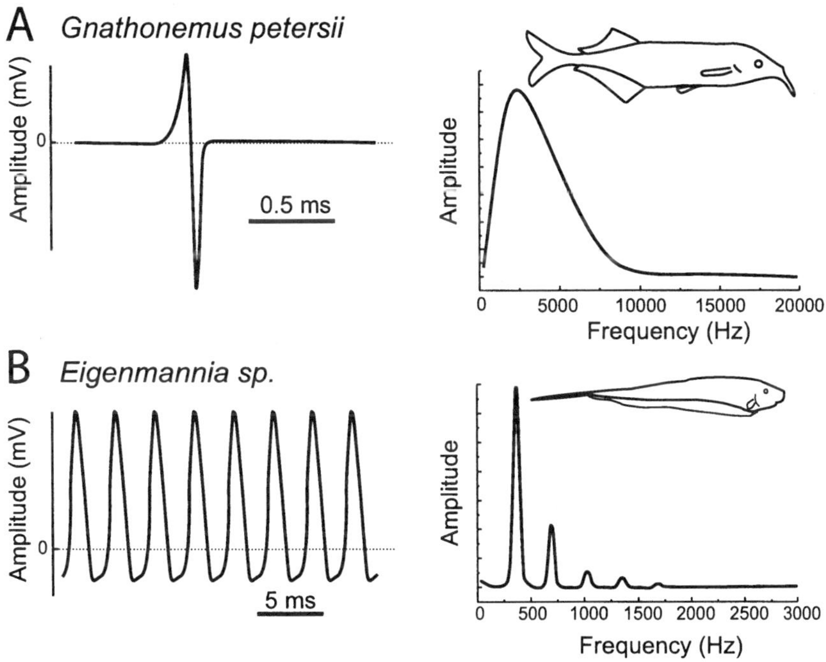

FIGURE 11.1. Electric organ discharge (EOD) waveforms (*left*) and power spectra (*right*) for weakly electric fish with pulse-type (**A**) and wave-type (**B**) discharges. (**A**) The pulse-type mormyrid *Gnathonemus petersii* generates brief pulses that have a broad frequency spectrum. (**B**) The wave-type gymnotiform *Eigenmannia* sp. generates a quasi-sinusoidal waveform with a frequency of a few hundred Hertz with most of the power concentrated at the fundamental frequency. Note that the time and frequency scales are different in (**A**) and (**B**). (From von der Emde 1999, with permission.)

EOD amplitude in free-swimming fish, so stringent tests of the constant-amplitude hypothesis have not been made.

Wave-type fish also seem to maintain a constant EOD frequency during electrolocation behavior, whereas pulse-type fish can alter the EOD repetition rate. In fact, the short-term stability of EOD frequency in wave-type fish places it among the most regular known biological oscillators (Moortgat et al. 1998). Changes in EOD frequency can be induced by environmental factors as well as a variety of social stimuli, but frequency modulation does not appear to play a role in electrolocation behavior. Pulse-type fish, on the other hand, often exhibit increases in EOD rate and regularity in response to detection of a novel electrolocation target. In some ways this is reminiscent of echolocating bats that increase the frequency of their calls as they approach a prey target, culminating in a so-called "feeding buzz." However, changes in EOD rate in pulse fish seem to reflect a more general increase in electrosensory arousal, rather than target-specific modulations (reviewed in Moller 1995). The functional equivalent of a "feeding buzz" has not been reported in pulse fish. For further discussion of factors that can influence EOD properties, see Macadar et al. (Chapter 14).

2.2 Spatial Aspects of the Fish's Electric Field

In contrast to the heterogeneity of EOD temporal properties, the spatial structure of the electric field is similar for all weakly electric fish, including both wave- and pulse-type species. This is a consequence of the physics of electric field generation, which results in a roughly dipolar field pattern surrounding the fish, independent of the temporal structure of the EOD waveform. Figure 11.2 shows a contour map of the empirically measured electric potential in the horizontal midplane of *A. albifrons* (Knudsen 1975). Electric potentials are always reported relative to some reference value. This is analogous to elevations on a topographic map being reported relative to sea level. When visualizing spatial maps of the electric field, the most common convention is to report potentials relative to a remote reference.

The contour levels shown in Figure 11.2 represent the amplitude of the EOD

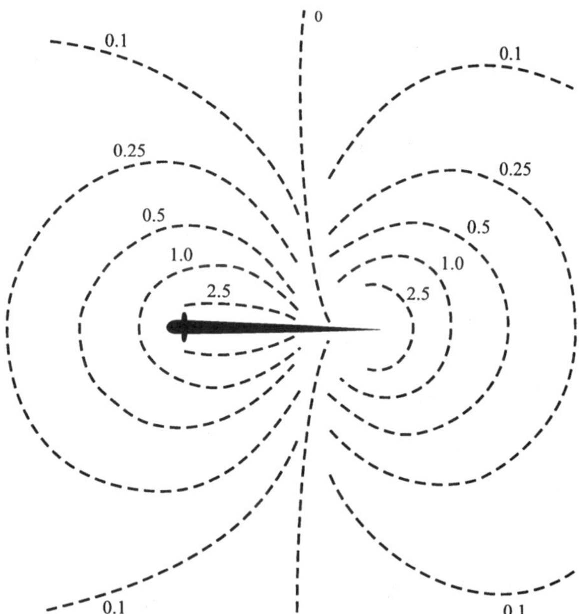

FIGURE 11.2. Equipotential lines (*dashed line*) illustrating the spatial structure of the electric field surrounding a 22 cm long *Apteronotus albifrons*. Potentials are reported as the peak-to-peak amplitude in microvolts relative to a remote reference. (Redrawn from Knudsen 1975, with permission.)

waveform at different points in space. Note that the peak-to-peak values are always positive. If, however, the EOD had been "frozen" at a single instant in time, say when the head region was at the positive peak of the EOD waveform, then the isopotential values in the tail region would have had the opposite polarity from those in the head region.

To a first approximation, the spatial structure of the electric field surrounding an electric fish can be described as a dipole field, with an intensity that varies in proportion to the instantaneous amplitude of the EOD. Figure 11.3 shows an ideal current dipole, consisting of a current source of magnitude $+I$ at location $\mathbf{p}_+$ and a current sink of magnitude $-I$ at $\mathbf{p}_-$ in an infinite volume conductor of conductivity σ. At an arbitrary point $\mathbf{p}$ in three-dimensional space, the electric potential $\phi(\mathbf{p})$ is given by:

$$\phi(\mathbf{p}) = \frac{I}{4\pi\sigma}\left(\frac{1}{|\mathbf{p} - \mathbf{p}_+|} - \frac{1}{|\mathbf{p} - \mathbf{p}_-|}\right) = \frac{I}{4\pi\sigma}\left(\frac{1}{\mathbf{r}_+} - \frac{1}{\mathbf{r}_-}\right) \quad (1)$$

A useful approximation can be derived when the distance r from the point $\mathbf{p}$ to the center of the dipole is much greater than the separation between the two poles d. This case ($r >> d$) is known as the far-field approximation. In the far field, the dipole potential is given by:

$$\phi(r, \theta) = \frac{Id \cos \theta}{4\pi\sigma r^2} \quad (2)$$

The electric field vector $\mathbf{E}$ is given by the spatial gradient of the potential ϕ (Jackson 1999):

$$\mathbf{E} = -\nabla\phi \quad (3)$$

For a dipole field, it is convenient to specify the electric-field vector in spherical coordinates. In spherical coordinates, r and θ are as shown in Figure 11.3, and the third dimension φ (not shown) represents the rotational angle out of the

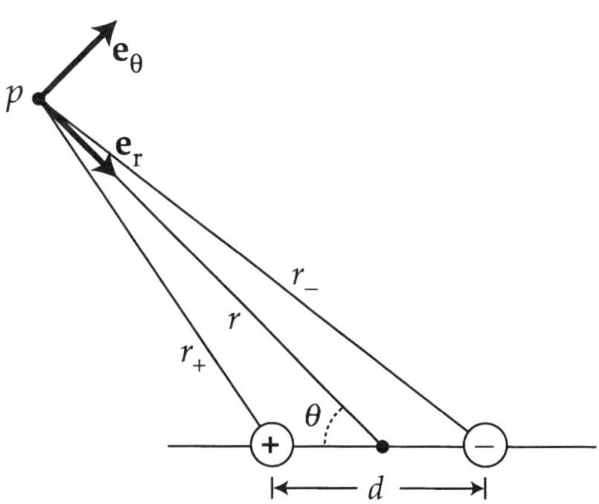

FIGURE 11.3. Coordinate system and definitions for calculating the electric potential at an arbitrary point p in three-dimensional space arising from an ideal current dipole. See text for explanation.

plane of the figure. In this coordinate system, the gradient operator is defined by:

$$\nabla\phi = \frac{\partial\phi}{\partial r}\hat{\mathbf{e}}_r + \frac{1}{r}\frac{\partial\phi}{\partial\theta}\hat{\mathbf{e}}_\theta + \frac{1}{r\sin\varphi}\frac{\partial\phi}{\partial\varphi}\hat{\mathbf{e}}_\varphi \tag{4}$$

Combining Equations (2) to (4), the **E** vector in the far field is given by:

$$\mathbf{E}(r,\ \theta) = \frac{Id}{4\pi\sigma r^3}(2\cos\theta\,\hat{\mathbf{e}}_r + \sin\theta\,\hat{\mathbf{e}}_\theta) \tag{5}$$

where $\hat{\mathbf{e}}_r$ and $\hat{\mathbf{e}}_\theta$ are unit vectors in the radial and tangential directions, as shown in Figure 11.3. Because of the cylindrical symmetry about the axis of the dipole, there is no gradient, and hence no electric field component, in the azimuthal direction $\hat{\mathbf{e}}_\varphi$ (which would be pointing out of the plane of Figure 11.3).

At discharge frequencies relevant to electrolocation (see Fig. 11.1), the electrical properties of the water can be treated as purely resistive. In this case, the current density in the water **J** is related to the electric field **E** by $\mathbf{J} = \sigma\,\mathbf{E}$, which can be viewed as a form of Ohm's law. Thus electric current flows in the direction of the local **E** vector. Because **E** is the gradient of the potential, the **E** vectors are oriented perpendicular to the isopotential lines, as shown in Figure 11.4. Going back to the topographic map analogy, electric current flows downhill along the path of steepest descent.

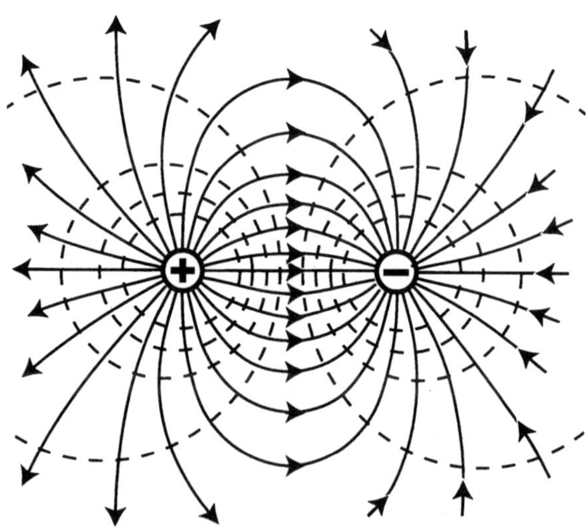

FIGURE 11.4. Illustration of the lines of current flow (*solid lines*) associated with a dipole field. The direction of current flow is orthogonal to the isopotential contours (*dashed lines*), and current flow is from positive to negative.

2.3. Properties of the Near Field

The effective range of electrolocation for detecting small prey is typically less than the fish's body length (MacIver et al. 2001). Hence, the relevant part of the field for understanding target detection is the "near field" rather than the far field. The far field approximation described above is useful for developing a general understanding of the field structure. However, in the near field (less than a few body lengths), the spatiotemporal structure becomes more complex than predicted by a simple dipole model. One indication of this is shown in Figure 11.2, where it can be seen that the isopotentials near the head and trunk are elongated in the direction of the fish's body axis compared to the corresponding isopotentials in the tail region. This arises because the relatively low internal body resistance of the fish effectively extends the poles, making them appear as line sources rather than point sources. Another important effect in the near field is that the power-law scaling is different. In the far field, the magnitude of the electric field falls with distance as r^{-3} (Eq. 5). In the near field, the attenuation with distance cannot be accurately described by a power law with a constant exponent. In general, the field intensity falls off more slowly in the near field than in the far field. Fits to empirical data yield approximate power law attenuation near the body of r^{-1} to r^{-2} depending on location (Rasnow and Bower 1996). Recently, detailed maps of the near field region have been constructed at multiple time points during the EOD cycle for several different species (reviewed in Assad et al. 1999). These dynamic maps reveal interesting spatiotemporal structure that is not evident in the static snapshot shown in Figure 11.2. These empirical maps provide the most detailed and reliable descriptions of the spatiotemporal properties of the near field.

2.4 Target Perturbations and Image Formation

The key principle of active electrolocation is based on the fact that an object that differs in electrical impedance from the surrounding water can perturb the flow of electric current near the fish. These perturbations give rise to changes in the local potential difference across the skin of the fish, referred to as the transdermal voltage. For example, a nonconducting object near the fish would block the flow of electrical current, causing the current lines to flow around the object as shown in Figure 11.5. This decrease in current density causes a localized decrease in voltage drop across the skin. Thus, a nonconducting object casts a sort of electrosensory "shadow" on the surface of the fish. Conversely, a conducting object would increase the local current density, and hence increase the local transdermal voltage, giving rise to an electrosensory "bright spot" on the surface of the fish. Thus, objects near the fish that differ in conductivity from the surrounding water cause local modulations of the transdermal voltage. Electrosensory images correspond to changes in transdermal voltage on the electroreceptor surface of the skin, just as visual images correspond to changes in

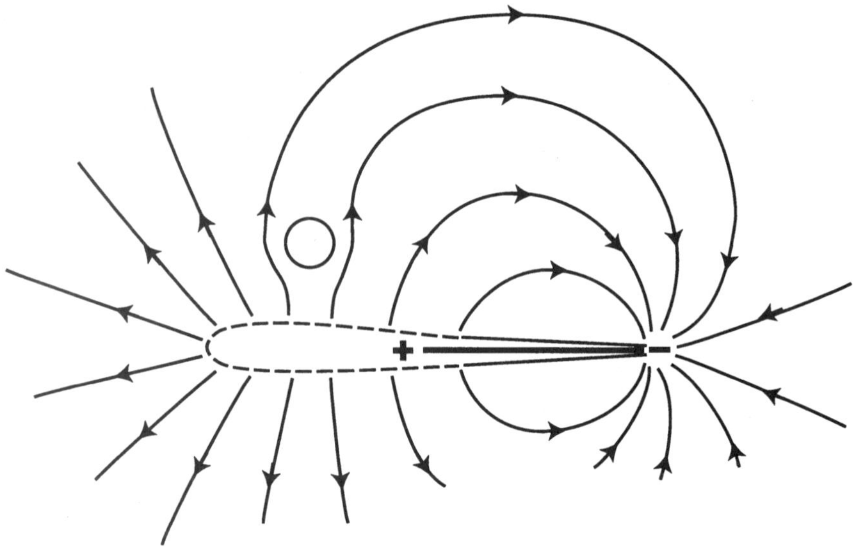

FIGURE 11.5. Schematic representation of the principles of active electrolocation. The electric organ (*solid black bar*) gives rise to a dipolar field pattern around the fish. The electric current follows the local field lines (*solid lines with arrows*). A nonconducting target object (*circle*) perturbs the flow of electric current, causing a local decrease in current density near the object. This decrease in current density translates into a decrease in the transdermal voltage across the skin near the object. The spatial pattern of the transdermal voltage across the sensory surface represents the electric image of the object. (Redrawn from Heiligenberg 1977, with permission.)

light intensity on the photoreceptor surface of the retina. Since there is no functional equivalent to the lens of the eye, electrosensory images are blurred and the degree of blur increases with target distance.

The electrosensory image of a target object depends on multiple factors, including target properties (location, size, shape, and electrical impedance), fish properties (amplitude and spectral characteristics of the fish's electric field at the target location), and environmental properties (water conductivity, background noise characteristics). Some electrosensory targets, such as small prey, can be well approximated as spherical targets. The analytic formula for the voltage perturbation caused by a perfectly conducting sphere of radius a is (Rasnow 1996):

$$\delta\phi\,(\mathbf{r}) = \frac{a^3\mathbf{E}_{\mathrm{f}} \cdot \mathbf{r}}{r^3} = \frac{a^3 E_{\mathrm{f}} \cos\,\theta}{r^2} \tag{6}$$

This equation, like Equation (2), is also the equation of a dipole potential. When a small spherical target is placed in an electric field, a dipole perturbation is induced at the location of the target (Jackson 1999). In this equation, $\delta\phi(r)$

represents the change in potential at position r relative to the center of the *target object*. The a^3 term is proportional to the volume of the target. This is an important result for predicting the effects of differently sized targets. Doubling the radius of a target will produce an eightfold increase in the voltage perturbation. When computing electrosensory images, r will be a vector from the target object to a point on the fish's sensory surface. The term $\mathbf{E}_f$ represents the fish's electric field vector at the location of the target object. The voltage perturbation $\delta\phi$ scales linearly with the strength of the fish's field at the target. The perturbation is strongest when the direction of the fish's electric field vector $\mathbf{E}_f$ is aligned with the direction of the $\mathbf{r}$ vector (i.e., $\cos\theta = 1$). Over much of the head and trunk region of the fish, the **E**-vector is approximately normal to the body surface. Hence, the peak perturbation tends to occur at a point on the surface directly "underneath" the target object. Away from this peak, the decrease in voltage perturbation gives rise to a roughly Gaussian shaped electrosensory image on the sensory surface. Representative electrosensory images are shown in Figure 11.6A.

Equation six specifies the voltage perturbation for a perfect conductor. The more general case takes into account the resistivity and dielectric properties of the target and the surrounding water (Rasnow 1996):

$$\delta\phi(\mathbf{r}) = \frac{\mathbf{a}^3\mathbf{E}_f \cdot \mathbf{r}}{r^3}\left[\frac{(\rho_w - \rho_t) - i\omega\rho_w\rho_t(\varepsilon_w - \varepsilon_t)}{\rho_w + 2\rho_t + i\omega\rho_w\rho_t(2\varepsilon_w + \varepsilon_t)}\right] \tag{7}$$

where ρ_w is the water resistivity, ρ_t is the target resistivity, ε_w is the dielectric constant of water, ε_t is the dielectric constant of the target, $\omega = 2\pi f$ is the angular frequency of the fish's electric field and $i = \sqrt{-1}$. There are several interesting features to note about this relationship. The term in square brackets on the right hand side of Equation (7) represents the electrical contrast χ of the target:

$$\chi = \frac{(\rho_w - \rho_t) - i\omega\rho_w\rho_t(\varepsilon_w - \varepsilon_t)}{\rho_w + 2\rho_t + i\omega\rho_w\rho_t(2\varepsilon_w + \varepsilon_t)} \tag{8}$$

If the resistivity and dielectric constant of the target match that of the water, then the numerator is zero, and the electrical contrast $\chi = 0$, which is to say that the target becomes electrically invisible. For a perfect conductor ($\rho_t = 0$), the electrical contrast is unity $\chi = 1$. Metal spheres closely approximate a perfect conductor and are often used as targets in electrolocation experiments. For many insulating materials, such as rocks, air bubbles, or plastic spheres, the resistivity of the target is much higher than that of the surrounding water ($\rho_t >> \rho_w$) and the dielectric constant is significantly less than water ($\varepsilon_r << \varepsilon_w$). In this case, Equation 8 simplifies to $\chi = -\frac{1}{2}$.

The electrical contrast is a complex number, with both real and imaginary parts. The real part is related to the resistive component of the target impedance, while the imaginary part is related to the target capacitance. Note that the imaginary parts of Equation 8 are proportional to the frequency of the fish's electric field. In general, fish with higher frequency components in their EOD

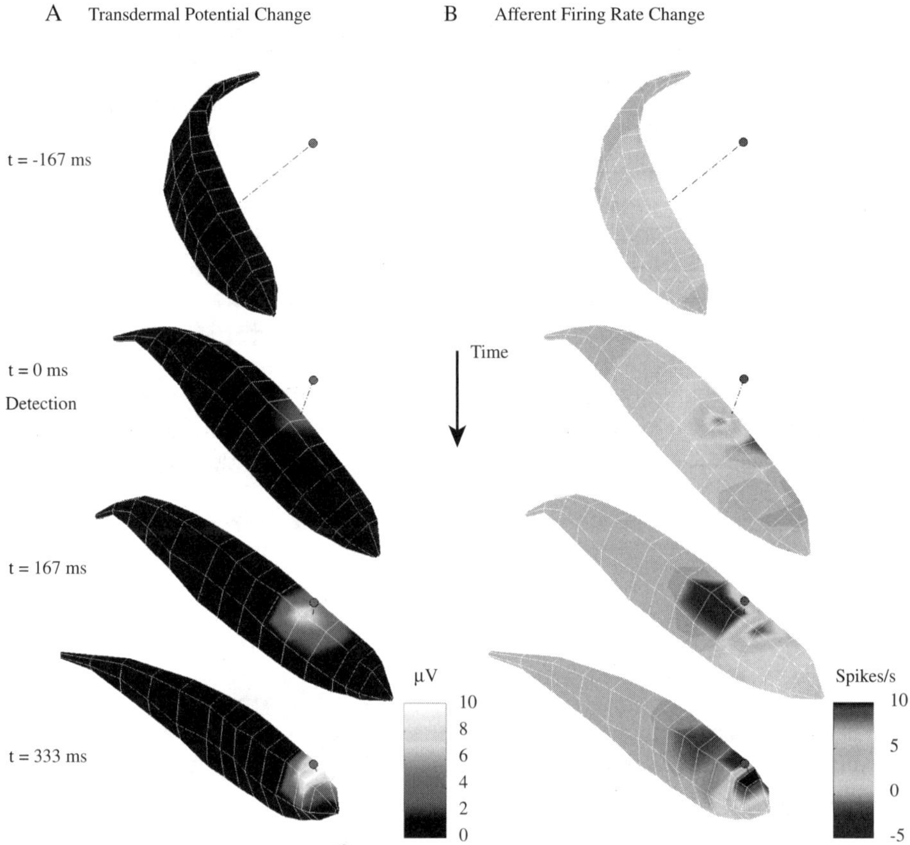

FIGURE 11.6. Reconstructed electrosensory images from a prey capture study. Each column shows snapshots of the model fish and prey at four different times in the prey capture sequence. The *left-hand column* (**A**) shows the change in transdermal voltage induced by the prey (*small sphere*). The *right-hand column* (**B**) shows the predicted change in spike activity on P-type primary electrosensory afferent nerve fibers. The dashed line indicates the shortest distance between the fish and prey. (From MacIver and Nelson 2000, with permission.)

waveform are better able to probe the capacitive components of object impedance. Both pulse-type and wave-type fish have been shown to be able to detect capacitive components of target impedance (reviewed in von der Emde 1999). Pulse-type fish appear to accomplish this by analyzing changes in the EOD waveform, while wave fish are thought to analyze timing shifts between parts of the body at different relative distances from the target (von der Emde 1999).

2.5 Modeling the Electric Field and Electrosensory Images

Several approaches have been used for modeling the spatial pattern of the electric field surrounding the fish and the image properties of electrolocation targets.

Seminal work by Heiligenberg (1975) used a two-dimensional finite-difference technique to numerically solve for the electric potential in the horizontal mid-plane of the fish. This work led to new insights regarding the importance of fish body geometry and impedance characteristics in electrolocation performance. Hoshimiya et al. (1980) extended this approach using a higher resolution finite-element method. Bacher (1983) introduced an analytic technique that treated the electric organ as an extended line charge. Caputi et al. (1998) used a finite difference technique constrained by empirical measurements of skin resistance to accurately describe electrosensory images of resistive targets in pulse fish and extended these results to include objects with complex impedance (Budelli and Caputi 2000). Assad and colleagues have used a three-dimensional boundary element technique to accurately model both spatial and temporal aspects of the field (Assad 1997). Over the years, increasingly sophisticated simulation techniques have led to increasingly accurate models of the electric field. These electric field models provide a foundation for experimental and theoretical studies of electrosensory image formation and neural processing.

2.6 Spatiotemporal Properties of Dynamic Targets

Relative motion between the fish and prey converts a static spatial image into a dynamic spatiotemporal pattern. For example, if an electrosensory image with a spatial width of 1 cm on the skin sweeps across the sensory array at a velocity of 10 cm/s (a typical swimming velocity), then an individual receptor situated along the projected path of the target would experience an activation profile with a temporal width of 0.1 s. Spatiotemporal relationships are more complex during prey capture behavior because target distance and relative velocity are constantly changing. Figure 11.6A shows snapshots of computer-reconstructed electrosensory images that occur during prey capture behavior of *A. albifrons* (Nelson and MacIver 1999). Initially, when the prey is several centimeters from the fish, the electrosensory image is weak and diffuse (Fig. 11.6A, top). The image becomes progressively stronger and narrower as the relative distance between the fish and prey decreases. Just prior to capture, the image sweeps across the head at a distance of just a few millimeters, giving rise to intense, focal images (Fig. 11.6A, bottom). These dynamic changes in spatiotemporal image properties occur over a time course of approximately 500 ms, from the time of initial prey detection to the final capture.

2.7 Motor Strategies Associated with Electrolocation

Motor control of body position and swimming velocity can actively influence the spatiotemporal properties of electrosensory images. Weakly electric fish have been observed to use a variety of motor strategies while exploring their environment, including tail-probing, body bends, back-and-forth scanning, reverse swimming, and body rolls (see Hopkins, Chapter 10). These motor behaviors affect the quality and content of incoming sensory signals, in much the same way that head and eye movements influence visual input. Knifefish are particularly remarkable in this regard because of the high degree of maneuverability afforded by a ribbon fin propulsion system. Knifefish can swim both forwards and backwards with a broad range of velocities and body orientations.

When weakly electric knifefish capture prey, they usually do so by swimming backwards (Lannoo and Lannoo 1993; MacIver et al. 2001). During the search phase, prior to detection, the fish is often swimming forwards. When a target is detected, typically at a distance of a few centimeters for small prey (MacIver et al. 2001), the fish executes a rapid reversal in swimming direction and backs up to capture the target, as shown in Figure 11.6. In addition to backing up, the fish executes body movements that bring the sensory surface closer to the target. In *A. albifrons,* the receptor organ density in the head region is about 5 to 10 times higher than on the trunk (Carr et al. 1982). Functionally, the head region can be considered as an "electrosensory fovea." As the fish executes movements to bring the target toward this high-acuity area, the relative distance between the target and the sensor array decreases with time, thereby providing the fish with a progressively stronger and sharper electrosensory image of the target.

The backward swimming capability of gymnotiform knifefish has been a subject of much discussion and speculation in the literature. Ribbon fin propulsion effectively decouples control of locomotion from trunk movements. Lissman (1958) and others speculated that the ability to locomote while maintaining a rigid body posture might help electric fish avoid electrosensory reafference that would be caused by tail movements during swimming. However, video analysis of feeding behavior in *A. albifrons* has shown that body bends are a common feature of prey-capture behavior (MacIver et al. 2001), suggesting that maintaining a rigid trunk is not a key constraint on the animal's motor strategy. It is now known that neural circuitry in the hindbrain electrosensory nucleus is able to adaptively cancel out most of the reafferent components of the input signal at the first stage of central processing (see Bastian and Zakon, Chapter 8).

2.8 Signal Strength and Signal-to-Noise Considerations

There are two metrics to consider when assessing the strength of a sensory signal. One measures the signal strength relative to some absolute reference level, such as the animal's behavioral threshold in the absence of noise. The other metric measures signal strength relative to the context-dependent background noise, and is typically characterized by a signal-to-noise ratio (SNR). It is possible for a signal to be strong by one metric, but weak by the other. For example, the sound intensity of a soft whisper is about two orders of magnitude above the human auditory threshold, but a whisper can be impossible to detect in a noisy environment, such as a rock concert.

For the high-frequency electrosensory system, behavioral detection thresholds have been estimated to be on the order of 0.1 μV, when expressed as the root mean square (RMS) change in transdermal potential, or about 0.1 μV/cm when expressed as a field gradient in the water outside the skin (Rasnow 1996). This represents about a 0.1% change in the baseline transdermal potential or external field gradient. Based on reconstructed prey-capture events, MacIver et al. (2001) estimated the peak change in transdermal potential at the time of prey detection to be on the order of 1.0 μV RMS, assuming the electrical contrast of the prey to be unity (i.e., a perfect conductor). Factoring in the actual electrical contrast of the prey drops this estimate to about 0.4 μV RMS (Nelson et al. 2002). Thus, prey-related electrosensory signals are estimated to be only modestly above behavioral threshold at the time of detection. This implies that in the absence of noise the fish should be able to detect the prey using information from the high-frequency electric sense. If, however, the fish encounters significant noise in its natural environment, then this weak signal may be difficult to detect.

The SNR provides a measure of the strength of a signal relative to background noise. One significant source of background noise in the electrosensory system is associated with the fish's own movements. During swimming and exploratory movements, changes in the position of the trunk and tail will alter the position of the electric organ relative to the electroreceptor array. Empirically measured tail bend modulations in *A. leptorhynchus* are on the order of 20 to 50 μV RMS for tail bends in the range of 20 to 50°, or approximately 1 μV RMS per degree (Bastian 1995). Based on the RMS value of tail bend angle during prey capture behavior (MacIver et al. 2001), the associated background noise is on the order of 30 μV (RMS), resulting in a SNR $<<$ 1. Fortunately, tail bends are not purely random events and the effects can be predicted. Tail bends are initiated by outgoing motor commands and are monitored by proprioceptors. The nervous system therefore has access to efference copy and sensory feedback signals that are correlated with tail position (Bastian 1995). Neural circuitry in the electrosensory lateral line lobe (ELL) makes use of such signals to adaptively suppress the predictable component of the electrosensory background associated with tail bending (Bastian and Zakon, Chapter 8), thus greatly improving the effective SNR.

In addition to tail-bend modulations, several other sources of background noise can influence target detectability. Depending on the context, the target signal might be obscured by other objects in the electrosensory scene. This sort of background is often referred to as clutter. There is very little empirical information on natural electrosensory clutter or the properties of natural electrosensory scenes. In considering the problem of electrosensory clutter, there are several important points to keep in mind. (1) The electrosensory system is a short-range sense, so the electrosensory scene is dominated by objects that are within about one body length of the fish. This differs from the background clutter problem in long-range modalities such as vision and audition, in which sensory energy from distant objects constitutes a significant component of the background clutter. (2) There is likely to be significant spatial overlap in the images formed by different objects in the scene because electrosensory images are intrinsically blurred. (3) Nonconducting boundaries, such as the water surface or a large rock, can cause large-scale modulations over the entire body surface. The spatial filtering properties of neurons in the ELL may help the animal distinguish local modulations arising from small prey from the global modulations caused by nonconducting boundaries and other large objects in the electrosensory scene.

One additional source of background noise is associated with the sensory transduction process. Transdermal voltage changes are converted into membrane voltage changes in electroreceptor cells, which modulate the release of neurotransmitter and influence spike generation in primary afferent nerve fibers. At the biophysical level, each of these steps involves stochastic processes that potentially introduce noise into the measurement process. The next section discusses electroreceptor and primary afferent properties in more detail. It turns out that intrinsic variability in primary afferent spike trains can also contribute to the background noise that potentially limits the detectability of weak sensory signals.

3. Neural Coding in Primary Electrosensory Afferents

For the nervous system to carry out the computations necessary to support electrolocation, information about the transdermal voltage patterns on the skin must be converted into a neural representation. This conversion is performed by an array of electroreceptor organs that are embedded in the skin of the fish. Functionally, each electroreceptor organ acts as a sort of digital voltmeter, converting analog changes in transdermal voltage into trains of action potentials. The information needed for subsequent neural processing is encoded in the spike train data. The coding strategies implemented by these biological voltmeters are rather sophisticated, involving various forms of input filtering and noise suppression during the encoding process. Different classes of electroreceptor organs filter and encode the voltage data in different ways, providing the central nervous system (CNS) with an efficient representation for performing its computations. Electroreceptor afferents enter the CNS via the eighth cranial nerve and share similarities with afferents from other octavolateralis systems, including auditory, vestibular, and mechanosensory lateral line systems. This section describes some of the basic neural coding properties of primary electrosensory afferents, with an emphasis on amplitude coding in wave-type gymnotiforms. Anatomical, physiological, and biophysical properties of electroreceptors and primary afferents are discussed elsewhere in this volume (Northcutt, Chapter 5; Bodznick and Montgomery, Chapter 6, Kawasaki, Chapter 7).

3.1 Types of Electroreceptor Organs and Primary Afferents

Electroreceptor organs can be grouped into two broad categories based on their frequency filtering characteristics. *Ampullary* organs, which are found in all electroreceptive organisms, are sensitive to low-frequency electric fields (typically less than 50 Hz). Ampullary organs are generally associated with *passive* electrolocation of extrinsic electric field sources. They also respond to low-frequency components of the EOD associated with certain types of electrocommunication signals. *Tuberous* organs are found in all weakly electric fish and in the South American electric eel *Electrophorus* (which has both strong and weak EODs). Tuberous organs are sensitive to high-frequency (typically 0.1 to 10 kHz) electric fields, and typically tuned to frequencies near the peak of the EOD spectrum. Tuberous organs are generally associated with *active* electrolocation involving perturbations of the fish's own electric field. Tuberous organs also respond to electric fields generated by other electric fish, and thus play a role in electrocommunication and social interactions (Hopkins, Chapter 10; Jørgensen, Chapter 3). In general, tuberous receptor organs are broadly distributed over the entire body surface, often with a higher density in the head region. The total number of tuberous organs varies by fish size and species, but is typically in the range of several thousand.

Coding properties of electroreceptor organs are deduced from the spike activity of the electrosensory afferent nerve fibers that innervate them. In some cases, a single afferent fiber carries information from a single receptor organ, while in other cases, a single afferent pools information from multiple organs. It is also possible for a single electroreceptor organ to give rise to multiple afferent fibers. These relationships vary by species and by receptor subtype. The descriptions of coding properties given below generally apply to a "receptor unit" consisting of a single primary afferent nerve fiber and its corresponding set of one or more electroreceptor organs.

Fish with wave-type EODs have two subtypes of tuberous receptor units (Scheich et al. 1973). One type conveys information about stimulus amplitude, while the other type conveys information about stimulus timing. The ongoing quasi-sinusoidal oscillation of the fish's EOD provides a natural "clock" for the system. In wave-type fish, tuberous afferents fire at most one spike per clock cycle. Amplitude information is conveyed by a spike probability code. When no target object is present, probability-coding (P-type) units fire a spike with some baseline probability per EOD cycle, typically around 0.3. When a conducting object approaches the receptor organ, it causes an increase in the local transdermal voltage and an increase in the per-cycle firing probability of the P unit. Similarly, when a nonconducting object approaches the receptor, the local transdermal potential and the P-unit firing probability will decrease. The inter-spike interval distribution is irregular, so P-type units are sometimes described as "sputtering." In contrast, time-coding (T-type) units fire exactly one spike per EOD cycle and are thus extremely regular. In a neurophysiological recording, the spike activity of a T-type unit sounds like a constant tone on an audio monitor. For T-type units, the spike time is tightly phase-locked to the EOD cycle. Local phase shifts in transdermal potential, as might be induced by the capacitive component of target impedance, will cause corresponding shifts in spike timing in T-type units. It is interesting to note that the only African mormyriform with a wave-type EOD, *Gymnarchus*, independently evolved the same electrosensory coding strategy as the South American wave gymnotiforms.

Although the functional properties are very similar, the nomenclature is different in *Gymnarchus*; the functional equivalents of P-type and T-type units are called O type and S type, respectively. The segregation of amplitude and timing information is also a common feature of auditory processing, and many of the coding principles are similar for electrosensory and auditory systems (Carr 1986).

Fish with pulse-type EODs also extract information on amplitude and timing, but use a different coding strategy. Gymnotiform pulse species have pulse markers (M units) that fire a single, short-latency spike following each EOD and are functionally similar to time-coding units in wave species. Amplitude information is encoded by burst-duration coders (B units), which convert stimulus intensity into bursts that range from one spike for a weak stimulus, to more than 10 spikes for a strong stimulus. Information about stimulus intensity is contained both in the burst count and in the latency to the first spike (Yager and Hopkins 1993). Mormyrids with a pulse-type EOD have two main classes of tuberous receptor units. One class (knollenorgan) is specialized for electrocommunication and is not considered further; the other class (mormyromast) is specialized for active electrolocation. Mormyromast afferents respond to each EOD with a short burst of one to three spikes. The intensity of the stimulus is encoded in both the latency to the first spike as well as the number of spikes in the burst. Mormyromast afferents typically operate somewhere in the middle of their dynamic range, such that their response can be modulated in both directions by targets of differing conductivity. Each mormyromast organ contains two types of separately innervated receptor cells. Both types encode stimulus amplitude, but one type is sensitive to small phase shifts that distort the EOD waveform, whereas the other is not (Bell 1990). Comparisons between responses of these two fiber types may allow the fish to assess the resistive and capacitive components of target impedance (reviewed in von der Emde, 1999).

3.2 Information-Coding Properties of P-Type Units

P-type (probability coding) afferents of wave-type gymnotiforms have received considerable attention in terms of understanding their response dynamics and information-coding capabilities and are therefore treated in more detail here. Under baseline conditions, with no target present, the input stimulus for a P unit is the local oscillatory transdermal potential created by the fish's own EOD. The amplitude of these oscillations will increase when a conducting object approaches and will decrease when a nonconducting object approaches. The baseline EOD oscillation serves as a carrier signal and a target object induces an amplitude modulation (AM) of this carrier signal. P-type units are tuned to the carrier frequency of the fish's own EOD, with a V-shaped tuning curve (Hopkins 1976). This filtering helps improve the SNR by filtering out background electrical noise in frequency bands other than those close to the fish's own EOD that are functionally important for electrolocation and electrocommunication.

Another type of filtering takes place in P units that is related to the AM frequency tuning rather than the carrier tuning. The frequency content of an AM signal induced by a target is related to its speed and distance. Nearby, fast-moving targets will cause higher-frequency AMs compared to distant, slow-moving targets. P-type units act as high-pass filters in the AM frequency domain, giving the strongest response to fast components of the AM signals (Bastian 1981a; Nelson et al. 1997). If a target remains stationary relative to the fish, P-unit activity gradually adapts back to baseline firing with a multiscale relaxation time course (Xu et al. 1996). The high-pass AM filtering properties and adaptation characteristics of P units reflect the fact that *changes* in the local transdermal potential carry more behaviorally relevant information than the absolute magnitude of the transdermal potential. The high-pass filter characteristics can be seen in Figure 11.6B.

Another important aspect of P-unit coding is related to the statistical properties of the afferent spike trains. P-unit spike activity is relatively irregular on short time scales, which explains why they are often described as "sputtering." This short-term variability can be quantified by the coefficient of variation of the interspike interval (ISI) distribution, which is typically around 0.5. However, this measure characterizes variability only on short time scales, comparable to the mean ISI, which is on the order of a few milliseconds in *A. albifrons*. Although P-unit spike trains are irregular on short time scales, they show a remarkable degree of regularity on intermediate time scales of 100 to 200 ms (30 to 60 ISIs). Spike train regularization on this time scale significantly enhances the detectability of weak signals by decreasing the effective background noise from spike train fluctuations (Ratnam and Nelson 2000).

Information-theoretic measures have been used to demonstrate that P-type afferents accurately encode information about the temporal waveform of AM signals within the behaviorally relevant bandwidth. Under optimal conditions, the information that can be transmitted about a band-limited random AM signal is within a factor of two of the theoretical maximum. Under these conditions, the information content in the spike train is on the order of one bit per spike, resulting in overall information rates of several hundred bits per second (Gabbiani et al. 1996; Wessel et al. 1996; Gabbiani and Metzner 1999).

3.3 Models of P-Unit Response Dynamics and Spike Train Statistics

Over the last several years a number of P-type afferent models have been developed. Kashimori et al. (1996) presented detailed biophysical models of P- and T-type electroreceptor organs, but did not explicitly address issues of neural coding or information transmission. Nelson et al. (1997) developed a computational model of afferent dynamics that reproduced the AM filtering characteristics and short-term variability of P-unit spike trains. Kreiman et al. (2000) developed a similar model to explain their experimental data on cross-trial variability when P units are given repeated presentations of an identical "frozen noise" stimulus. Under baseline conditions, both the Nelson et al. (1997) and the Kreiman et al. (2000) models produce uncorrelated ISI sequences and hence do not accurately describe longer-term spike train regularization, which is important for improving weak signal detectability.

Chacron et al. (2000) developed a P-unit model that included a dynamic threshold that was transiently elevated following each spike to give rise to a relative refractory period. This model yields an ISI distribution that is consistent with experimental data and also reproduces correlations between neighboring intervals in the ISI sequence. An extended and more detailed version of this model is able to reproduce the frequency response characteristics of P-type afferents as well as the key features of long-term spike train regularization (Chacron et al. 2001). Brandman and Nelson (2002) developed a reduced version of the Chacron model that has been used to simulate afferent activity over the entire population of 15,000 P units in *A. albifrons* during reconstructed prey capture sequences, producing dynamic spatiotemporal images similar to those shown in Figure 11.6B.

4. Central Information Processing

The first stage of information processing in the CNS takes place in the hindbrain electrosensory lateral line lobe (ELL). The ELL is the sole recipient of electrosensory primary afferent input and thus all electrosensory information needed to support electrolocation passes through this structure. Detailed anatomical and physiological descriptions can be found elsewhere in this volume (Bell and Maler, Chapter 4; Kawasaki, Chapter 7). The ELL has structural and functional similarities to the dorsal cochlear nucleus in the mammalian auditory system (Montgomery et al. 1995). The ELL contains multiple topographically organized maps of the fish's body surface (Shumway 1989a). Electrosensory images that appear as spatially localized voltage perturbations on the fish's skin are transformed into spatially localized patterns of neural activity in these somatotopic maps. In general, ELL principal neurons have phasic response properties and center-surround spatial receptive field properties, with both on-center (E-type) and off-center (I-type) varieties (Bastian 1981b). They filter the incoming electrosensory image data with spatiotemporal filtering properties that vary across the different maps (Shumway 1989b). The ELL projects to the midbrain torus semicircularis, which is analogous to the mammalian inferior colliculus, as well as to the dorsal preeminential nucleus, which is part of an important electrosensory feedback loop. The ELL and associated feedback circuits are thought to play an important role in electrosensory image analysis (Rasnow 1996; Berman and Maler 1999; Lewis and Maler 2001). The torus projects to the optic tectum, which has been shown to be a site of convergence for visual and electrosensory representations of electrolocation targets (Bastian 1982), which would be useful under low-light conditions.

4.1 Multiple Maps in the Gymnotiform ELL

As shown in Figure 11.7A, the gymnotiform ELL contains four subdivisions, each with its own somatotopic map of the electroreceptive periphery (Heiligenberg and Dye, 1982; Shumway 1989a, b; Metzner and Juranek 1997). The medial map is devoted to processing information from the ampullary system, which detects low-frequency electric fields of extrinsic origin. The three lateral maps are devoted to processing information from the tuberous system which is tuned to high frequencies near the fish's own EOD frequency. Each tuberous afferent trifurcates as it reaches the ELL, providing each of the three tuberous maps with essentially identical input. Amplitude and phase information is processed in parallel by separate components of the ELL circuitry in each map. In the amplitude pathway, there is a tradeoff between sensitivity and spatial resolution across the three maps, with the lateral map having the highest degree of convergence from electroreceptors and exhibiting the best sensitivity to weak stimuli, but having poor spatial resolution. In contrast, the centromedial map has good spatial resolution but low sensitivity. The lateral map has been shown to play a key role in electrocommunication behavior and the centromedial map

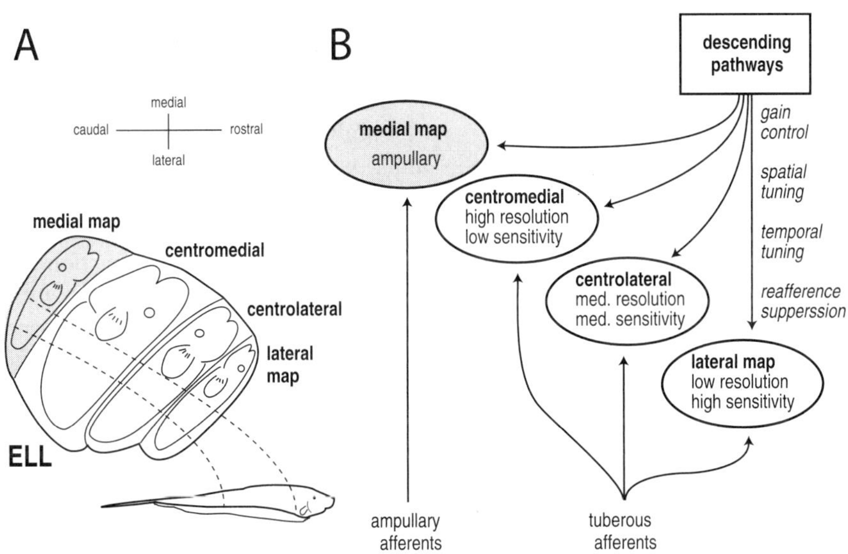

FIGURE 11.7. Schematic of the gymnotiform ELL as a multiresolution adaptive filter array. (A) Dorsoventral projection of the right ELL showing the four somatotopic maps of ampullary (*gray*) and tuberous (*unshaded*) electrosensory input. (Modified from Heiligenberg and Dye 1982.) (**B**) Each map processes the primary afferent input with differing resolution and sensitivity characteristics. Descending signals provide gain control, spatiotemporal tuning, and suppression of reafferent background noise.

is necessary and sufficient for eliciting the jamming avoidance response (Metzner and Juranek 1997; Hopkins, Chapter 10).

4.2 Distance Estimation Using Multiple Maps

Behavioral studies have shown that mormyrid weakly electric fish can measure target distance irrespective of size, shape, or electrical conductivity of the target (von der Emde et al. 1998; von der Emde 1999). In principle, the distance to an electrolocation target can be estimated by combining a measurement of the peak amplitude of an electric image with a measurement of image width (Rasnow 1996) or maximum slope (von der Emde et al. 1998). Modeling studies have shown that topographic maps with broad spatial tuning are better for estimating image amplitude and maps with narrow tuning are better for estimating image width (Lewis and Maler 2001). Distance estimation can therefore be carried out more accurately using two maps with differing spatial resolution than with a single map (Lewis and Maler 2001). Multiple maps in the ELL may thus allow the encoding of additional sensory features, such as object distance, more efficiently than could be accomplished using a single map.

4.3 Stimulus Coding and Feature Extraction

Primary electrosensory afferents that project to the gymnotiform ELL fire at high rates (several hundred Hertz) and faithfully encode temporal modulations of the input signal. In contrast, the output neurons of the ELL fire at lower rates (tens of Hertz) and behave more like feature detectors under many circumstances (Gabbiani et al. 1996; reviewed in Gabbiani and Metzner 1999). There are many factors influencing the coding efficiency and feature extraction performance of ELL neurons. These performance measures depend on the functional properties of the neuron under investigation (e.g., E type versus I type) and the ELL segment in which it is found (Metzner et al. 1998), as well as on the spatial structure (e.g., local versus global) of the stimulus (Bastian et al. 2002; Chacron et al. 2003; Doiron et al. 2003). In many cases the presence of temporal features is best signaled by short bursts of spikes rather than by isolated spikes (Metzner et al. 1998). Target-related coincident activity across multiple neurons can further improve the extraction of behaviorally relevant stimulus features. Simultaneous recordings from pairs of ELL neurons have demonstrated that coincident spikes within a time window of a few milliseconds performed even better as feature detectors than bursts from single neurons. This result suggests that coincident spikes from multiple neurons can be considered as "distributed bursts" (Krahe et al. 2002). Electrophysiological studies coupled with computational models have helped elucidate the key biophysical mechanisms underlying oscillatory and burst discharge mechanisms in the ELL (Turner and Maler 1999; Doiron et al. 2001a,b, 2002, 2003).

4.4 Descending Control of Gain and Spatiotemporal Filtering Characteristics

The gymnotiform ELL receives extensive descending input, which far surpasses the primary afferent input in terms of total number of synapses. These descending pathways carry various types of contextual information that could be useful in adaptive signal processing, including electrosensory feedback signals, sensory signals from other modalities, proprioceptive signals related to body posture, and efference copy signals related to outgoing motor commands (Kawasaki, Chapter 9). Lesioning the descending pathway has been shown to influence both the gain and spatiotemporal response properties of pyramidal cells in the ELL. Descending pathways interact with several classes of inhibitory interneurons in the ELL which are thought to selectively affect specific filtering functions, including gain, temporal adaptation, size of receptive field center, and size of receptive field surround. As shown in Figure 11.7B, the ELL can thus be visualized as a bank of adaptive spatiotemporal filters. It has been suggested that the different types of sensory processing carried out in the ELL—adaptive gain control, spatial filtering, temporal filtering, reafference suppression, and common mode rejection—are likely to be representative of a class of canonical signal processing strategies employed in sensory systems in general (reviewed in Berman and Maler 1999).

4.5 Generation and Subtraction of Sensory Expectation

A number of studies have demonstrated that the neural circuitry of the ELL plays an important role in suppressing certain "expected" components of incoming electrosensory information, particularly those associated with sensory reafference in which an animal's motor actions have predictable sensory consequences (Bastian and Zakon, Chapter 8). Adaptive reafference suppression in the ELL was first discovered in mormyrid pulse-type fish, in which the neural circuitry adaptively constructs a "negative image" of the expected pattern of sensory input over a period of several minutes (reviewed in Bell 2001). Over the past several years the neural mechanisms underlying this phenomenon in the ELL have been worked out in considerable detail, leading to the present view that the negative image is formed by anti-Hebbian mechanisms operating at descending pathway synapses (Bastian and Zakon, Chapter 8). Modeling studies have helped explain the functional benefits of temporal asymmetry in the synaptic modification rule (Roberts 1999; Roberts and Bell 2000). This negative image mechanism is responsible adaptively suppressing reafferent signals associated with tail bending in gymnotiform weakly electric fish (Bastian 1995; Bastian and Zakon, Chapter 10).

5. Directions for Future Research

Although much progress has been made, many gaps remain in our understanding of electrolocation. One fruitful area for future investigation involves a better characterization of the statistical properties of natural electrosensory scenes. Compared to other sensory modalities, such as vision and audition, we have much less knowledge of sensory signal and background characteristics under naturalistic conditions. There is also a gap in our knowledge related to a lack of behavioral data from the field. Almost all electrolocation studies to date have been carried out in laboratory settings. Because weakly electric fish are nocturnal, live in turbid waters, and are found in relatively remote locations, the technical challenges of recording and analyzing electrolocation behavior in the field are significant. Consequently, we know very little about the natural foraging and feeding strategies that these animals use in their natural habitats.

On the electrophysiology front, we need to learn more about neural activity patterns while the animal is actively performing an electrolocation task. Almost all in vivo electrophysiology related to electrolocation has been carried out on restrained animals that are not engaged in any sort of behavioral task. (An exception is the jamming avoidance response, in which a complete behavioral loop, from sensory input to motor output, can be studied in a restrained animal.) Because the gain and spatiotemporal filtering properties of sensory neurons can be modulated at early levels of sensory processing, it is likely that neurons at the level of the ELL and higher will show task-dependent response properties. Investigating adaptive aspects of the neural algorithms underlying target detection, characterization, and localization may require that the fish be actively engaged in an electrolocation task while neural responses are being probed. Ideally, one would like to be able to record and monitor neural activity from multiple neurons in a freely swimming fish. Again, the technical challenges are significant, but developing multineuron chronic recording and telemetry capabilities could dramatically advance our understanding of the adaptive aspects electrolocation.

Another major area for future investigation lies on the motor side of the electrolocation problem. The sensory processing side of the task has been the primary focus of attention for several decades, whereas little work has been done on motor aspects related to active repositioning of the sensor array and target interception. The control of the ribbon fin in knifefish is a particularly appealing problem from a number of perspectives. Although there has been some work on the biomechanical and hydrodynamic aspects of gymnotiform locomotion, little is known about the neural control of the traveling waves on the ribbon fin. Also, little is known about the central representation of the motor space and target location in the nervous system. It would also be interesting to explore what sorts of motor planning and control strategies the fish might use to optimize the acquisition of sensory information from the environment.

6. Summary

Weakly electric fish can detect, localize, and characterize sensory targets in their environment using electrosensory cues. Passive electrolocation relies on detection and analysis of the intrinsic electric fields generated by the target object, while active electrolocation relies on target-induced perturbations in the fish's self-generated electric field. The active mode allows the fish to detect a broader spectrum of target objects, because it requires only that the electrical impedance of the target differ from the surrounding water. The active electric sense has a limited working range of a few centimeters for detecting small prey and about one body length for larger objects. Fish have been shown to be capable of discriminating differences in size, shape, distance, and impedance of target objects.

Some species of weakly electric fish emit brief pulse-type electric organ discharges with a broad frequency spectrum, while others have continuous wave-type discharges with the energy concentrated at the fundamental frequency and the first few harmonics. In contrast to certain echolocating bats, weakly electric fish do not appear to make adaptive adjustments to the amplitude or frequency of their discharge as they approach a target.

The spatial structure of the fish's electric field is approximately dipolar, although the near field is considerably more complex. The presence of a target object induces a local perturbation that is proportional to the strength of the fish's electric field at the target location. For small spherical targets, the perturbation can be accurately modeled as an induced dipole. When mapped onto the receptor surface, the voltage perturbation gives rise to an electrosensory image with a Gaussian-like intensity distribution. Peak image intensity falls rapidly with increasing target distance (approximately as r^{-3} in the near field), and image width is roughly equal to target distance. Computer modeling has been an important tool for understanding the spatial structure of the fish's electric field, the effects of object perturbations, and the spatiotemporal properties of electrosensory images at the receptor surface.

Image intensity scales linearly with the target's volume and electrical contrast—a measure of the relative difference in impedance between the target and the water. Weakly electric fish can assess complex impedance properties by analyzing amplitude and phase information, which is typically encoded by separate classes of primary afferent nerve fibers. Modeling studies have been pivotal in characterizing the response dynamics and coding properties of primary afferents, providing insights into how transdermal voltage patterns are converted into spatiotemporal patterns of spike activity in the CNS.

Primary electrosensory afferents project to the hindbrain ELL. In gymnotiforms, primary afferents faithfully encode the temporal structure of the input signal, while ELL pyramidal neurons seem to behave more like feature detectors. The ELL is divided into multiple somatotopic maps with differing tradeoffs between sensitivity and spatial resolution. Descending pathways are involved in modulating the gain and spatiotemporal filtering properties of ELL pyramidal cells and in suppressing reafferent backgrounds. The functional role of multiple maps in electrolocation is not well understood. One hypothesis is that different maps may subserve different phases of electrolocation behavior, with one map playing a key role during detection and others during localization and characterization. Modeling studies suggest that target distance could be estimated more accurately using multiple maps than with a single map. Electrophysiological models are helping illuminate the biophysical basis of neural information processing mechanisms in the ELL, such as gain control and burst coding.

Exploring principles of electrolocation at a deeper level will require an integrative approach that combines aspects of anatomy, physiology, behavior, sensory ecology, and neural information processing principles. A solid foundation has been laid, as evidenced by the wealth of information presented in this volume, but much more remains to be discovered before we fully understand this fascinating sensory capability.

Acknowledgments. This work was supported by grants from NSF (IBN-0078206) and NIMH (R01-MH49242).

References

Assad C (1997) Electric field maps and boundary element simulations of electrolocation in weakly electric fish. PhD thesis, California Institute of Technology, Pasadena, CA.

Assad C, Rasnow B, Stoddard PK (1999) Electric organ discharges and electric images during electrolocation. J Exp Biol 202:1185–1193.

Bacher M (1983) A new method for the simulation of electric fields generated by electric fish and their distortions by objects. Biol Cybern 47:51–58.

Bass AH (1986) Electric organs revisited: evolution of a vertebrate communication and orientation organ. In: Bullock TH, Heiligenberg W (eds), Electroreception. New York: John Wiley & Sons, pp. 13–70.

Bastian J (1981a) Electrolocation I. How the electroreceptors of *Apteronotus albifrons* code for moving objects and other electrical stimuli. J Comp Physiol 144:465–479.

Bastian J (1981b) Electrolocation II. The effects of moving objects and other electrical stimuli on the activities of two categories of posterior lateral line lobe cells in *Apteronotus albifrons*. J Comp Physiol 144:481–494.

Bastian J (1982) Vision and electroreception integration of sensory information in the optic tectum of the weakly electric fish *Apteronotus albifrons*. J Comp Physiol 147: 287–298.

Bastian J (1995) Pyramidal-cell plasticity in weakly electric fish: a mechanism for attenuating responses to reafferent electrosensory inputs. J Comp Physiol A 176:63–78.

Bastian J (2003) Electrolocation. In: Arbib M (ed), The Handbook of Brain Theory and Neural Networks. Cambridge, MA: MIT Press, pp. 391–398.

Bastian J, Chacron MJ, Maler L (2002) Receptive field organization determines pyramidal cell stimulus-encoding capability and spatial stimulus selectivity. J Neurosci 22:4577–4590.

Bell CC (1990) Mormyromast electroreceptor organs and their afferent fibers in Mormyrid fish III. Physiological differences between two morphological types of fibers. J Neurophysiol 63:319–332.

Bell CC (2001) Memory-based expectations in electrosensory systems. Curr Opin Neurobiol 11:481–487.

Berman NJ, Maler L (1999) Neural architecture of the electrosensory lateral line lobe: adaptations for coincidence detection, a sensory searchlight and frequency-dependent adaptive filtering. J Exp Biol 202:1243–1253.

Brandman R, Nelson ME (2002) A simple model of long-term spike train regularization. Neural Comput 14:1575–1597.

Budelli R, Caputi AA (2000) The electric image in weakly electric fish: perception of objects of complex impedance. J Exp Biol 203:481–492.

Caputi AA, Budelli R, Grant K, Bell CC (1998) The electric image in weakly electric fish—physical images of resistive objects in *Gnathonemus petersii*. J Exp Biol 201: 2115–2128.

Carr CE (1986) Time coding in electric fish and barn owls. Brain Behav Evol 28:122–133.

Carr CE, Maler L, Sas E (1982) Peripheral organization and central projections of the electrosensory nerves in gymnotiform fish. J Comp Neurol 211:139–153.

Chacron MJ, Longtin A, St-Hilaire M, Maler L (2000) Suprathreshold stochastic firing dynamics with memory in P-type electroreceptors. Phys Rev Lett 85:1576–1579.

Chacron MJ, Longtin A, Maler L (2001) Negative interspike interval correlations increase the neuronal capacity for encoding time-dependent stimuli. J Neurosci 21:5328–5343.

Chacron MJ, Doiron B, Maler L, Longtin A, Bastian J (2003) Non-classical receptive field mediates switch in a sensory neuron's frequency tuning. Nature 423:77–81.

Doiron B, Longtin A, Turner RW, Maler L (2001a) Model of gamma frequency burst discharge generated by conditional backpropagation. J Neurophysiol 86:1523–1545.

Doiron B, Longtin AAE, Berman N, Maler L (2001b) Subtractive and divisive inhibition: effect of voltage-dependent inhibitory conductances and noise. Neural Comput 13: 227–248.

Doiron B, Laing C, Longtin A, Maler L (2002) Ghostbursting: a novel neuronal burst mechanism. J Comput Neurosci 12:5–25.

Doiron B, Chacron MJ, Maler L, Longtin A, Bastian J (2003) Inhibitory feedback required for network oscillatory responses to communication but not prey stimuli. Nature 421:539–543.

Gabbiani F, Metzner W (1999) Encoding and processing of sensory information in neuronal spike trains. J Exp Biol 202:1267–1279.

Gabbiani F, Metzner W, Wessel R, Koch C (1996) From stimulus encoding to feature extraction in weakly electric fish. Nature 384:564–567.

Heiligenberg W (1975) Theoretical and experimental approaches to spatial aspects of electrolocation. J Comp Physiol 103:247–272.

Heiligenberg W (1977) Principles of electrolocation and jamming avoidance studies of brain function. Vol 1. Berlin: Springer-Verlag.

Heiligenberg W, Dye J (1982) Labeling of electroreceptive afferents in a gymnotoid fish by intracellular injection of horseradish peroxidase: the mystery of multiple maps. J Comp Physiol 148:287–296.

Hopkins CD (1976) Stimulus filtering and electroreception: tuberous receptors in three species of gymnotoid fish. J Comp Physiol A 111:171–207.

Hoshimiya N, Shogen K, Matsuo T, Chichibu S (1980) The *Apteronotus* electric organ discharge field waveform and electric organ discharge field simulation. J Comp Physiol 135:283–290.

Jackson JD (1999) Classical Electrodynamics, 3rd ed. New York: John Wiley & Sons.

Kashimori Y, Goto M, Kambara T (1996) Model of P- and T-electroreceptors of weakly electric fish. Biophys J 70:2513–2526.

Knudsen EI (1975) Spatial aspects of the electric fields generated by weakly electric fish. J Comp Physiol 99:103–118.

Krahe R, Kreiman G, Gabbiani F, Koch C, Metzner W (2002) Stimulus encoding and feature extraction by multiple sensory neurons. J Neurosci 22:2374–2382.

Kreiman G, Krahe R, Metzner W, Koch C, Gabbiani F (2000) Robustness and variability of neuronal coding by amplitude-sensitive afferents in the weakly electric fish *Eigenmannia*. J Neurophysiol 84:189–204.

Lannoo MJ, Lannoo SJ (1993) Why do electric fishes swim backwards? An hypothesis based on Gymnotiform foraging behavior interpreted through sensory constraints. En. Bio. Fishes 36:157–165.

Lewis JE, Maler L (2001) Neuronal population codes and the perception of object distance in weakly electric fish. J Neurosci 21:2842–2850.

Lissmann HW (1958) On the function and evolution of electric organs in fish. J Exp Biol 35:156–191.

Lissman HW, Machin KE (1958) The mechanism of object location in *Gymnarchus niloticus* and similar fish. J Exp Biol 35:451–486.

MacIver MA, Nelson ME (2000) Body modeling and model-based tracking for neuroethology. J Neurosci Methods 95:133–143.

MacIver MA, Sharabash NM, Nelson ME (2001) Prey-capture behavior in gymnotid electric fish: motion analysis and effects of water conductivity. J Exp Biol 204:543–557.

Metzner W, Juranek J (1997) A sensory brain map for each behavior. Proc Natl Acad Sci USA 94:14798–14803.

Metzner W, Koch C, Wessel R, Gabbiani F (1998) Feature extraction by burst-like spike patterns in multiple sensory maps. J Neurosci 18:2283–2300.

Moller P (1995) Electric Fishes: History and Behavior. London: Chapman and Hall.

Montgomery J, Coombs S, Conley RA, Bodznick D (1995) Hindbrain sensory processing in lateral line, electrosensory and auditory systems: a comparative overview of anatomical and functional similarities. Audit Neurosci 1:207–231.

Moortgat KT, Keller CH, Bullock TH, Sejnowski TJ (1998) Submicrosecond pacemaker precision is behaviorally modulated: the gymnotiform electromotor pathway Proc Natl Acad Sci USA 95:4684–4689

Nelson ME, MacIver MA (1999) Prey capture in the weakly electric fish *Apteronotus albifrons:* sensory acquisition strategies and electrosensory consequences. J Exp Biol 202:1195–1203.

Nelson ME, Xu Z, Payne JR (1997) Characterization and modeling of P-type electrosensory afferent responses to amplitude modulations in a wave-type electric fish. J Comp Physiol A 181:532–544.

Nelson ME, MacIver MA, Coombs S (2002) Modeling electrosensory and mechanosensory images during the predatory behavior of weakly electric fish. Brain Behav Evol 59:199–210.

Rasnow B (1996) The effects of simple objects on the electric field of *Apteronotus*. J Comp Physiol A 178:397–411.

Rasnow B, Bower JM (1996) The electric organ discharges of the Gymnotiform fishes: I. Apteronotus leptorhynchus. J Comp Physiol A 178:383–396.

Ratnam R, Nelson ME (2000) Nonrenewal statistics of electrosensory afferent spike trains: implications for the detection of weak sensory signals. J Neurosci 20:6672–6683.

Roberts PD (1999) Computational consequences of temporally asymmetric learning rules: I. Differential Hebbian learning. J Comput Neurosci 7:235–246.

Roberts PD, Bell CC (2000) Computational consequences of temporally asymmetric learning rules: II. Sensory image cancellation. J Comput Neurosci 9:67–83.

Scheich H, Bullock TH, Hamstra RH (1973) Coding properties of two classes of afferent nerve fibers: high-frequency electroreceptors in the electric fish, *Eigenmannia*. J Neurophysiol 36:39–60.

Shumway CA (1989a) Multiple electrosensory maps in the medulla of weakly electric Gymnotiform fish I. Physiological differences. J Neurosci 9:4388–4399.

Shumway CA (1989b) Multiple electrosensory maps in the medulla of weakly electric Gymnotiform fish II. Anatomical differences. J Neurosci 9:4400–4415.

Turner RW, Maler L (1999) Oscillatory and burst discharge in the apteronotid electrosensory lateral line lobe. J Exp Biol 202:1255–1265.

von der Emde G (1999) Active electrolocation of objects in weakly electric fish. J Exp Biol 202:1205–1215.

von der Emde G, Bleckmann H (1998) Finding food: senses involved in foraging for insect larvae in the electric fish *Gnathonemus petersii*. J Exp Biol 201:969–980.

von der Emde G, Schwarz S, Gomez L, Budelli R, Grant K (1998) Electric fish measure distance in the dark. Nature 395:890–894.

Wessel R, Koch C, Gabbiani F (1996) Coding of time-varying electric field amplitude modulations in a wave-type electric fish. J Neurophysiol 75:2280–2293.

Wilkens LA, Russell DF, Pei X,Gurgens C (1997) The paddlefish rostrum functions as an electrosensory antenna in plankton feeding. Proc Soc Lond B 264:1723–1729.

Xu Z, Payne JR, Nelson ME (1996) Logarithmic time course of sensory adaptation in electrosensory afferent nerve fibers in a weakly electric fish. J Neurophysiol 76:2020–2032.

Yager DD, Hopkins CD (1993) Directional characteristics of tuberous electroreceptors in the weakly electric fish, *Hypopomus* (Gymnotiformes). J Comp Physiol A 143:401–414.

12

Comparing Octavolateralis Sensory Systems: What Can We Learn?

SHERYL COOMBS AND JOHN C. MONTGOMERY

1. Introduction

For a variety of historical and anatomical reasons, the electrosensory systems of primitive aquatic vertebrates are inextricably linked to their mechanosensory counterpart, the "ordinary" lateral line. In fact, before the discovery of electrosensory function in the late 1950s and early 1960s (e.g., Bullock et al. 1961; Fessard and Szabo 1961), both were regarded as a single lateral line system. Their singular classification is certainly understandable, given the state of knowledge at the time and striking anatomical similarities in (1) the spatial distribution of superficial sense organs on the head and body; (2) surrounding structures such as pits, pores, and subcutaneous canals; (3) peripheral innervation patterns; and (4) nerve termination sites in the medulla (see Bullock 1981 and Bodznick 1989 for review).

Not only did the electrosense escape notice as a distinct sensory system in its own right, but so, in some respects, did the lateral line system. Indeed, the lateral line was viewed by many as an accessory auditory system under the umbrella of the "acousticolateralis" or "octavolateralis" system. This all-encompassing view was originally based on mistaken ideas of how the two systems developed and an inability to discern separate, but adjacent projection areas in the hindbrain (reviewed in Popper et al. 1992). Biophysical theories of how the two systems "should" work (van Bergeijk 1964) and electrophysiological demonstrations of lateral line nerve responses to low frequency "sound" (e.g., underwater loud speaker) vibrations (e.g., Harris and van Bergeijk, 1962) gave further credence to this idea, despite cogent arguments to the contrary (e.g., Dijkgraaf 1963; reviewed in Sand, 1981).

Although it is often assumed that many of the anatomical characteristics shared by octavolateralis systems are due to a common evolutionary origin, there is no definitive cladistic evidence to indicate that the earliest auditory (reviewed by Popper et al. 1992) or electrosensory systems evolved from the mechanosensory lateral line system, as is often claimed in the older literature. Rather, the otolithic organs of the inner ear, the mechanosensory lateral line, and electrosensory ampullary organs can all be traced back to the earliest craniates (Wever 1976; Bullock et al. 1983; Northcutt 1989).

In contrast to the lateral line and inner ear, which persist in all but perhaps a few (e.g., myxiniform hagfish) aquatic anamniotic vertebrates, the electrosense was lost with the origin of neopterygian bony fishes and apparently reinvented again in a few teleost lineages (Bullock et al.1983; New 1997). Peripheral structures enabling the otolithic end organs of the inner ear to respond to acoustic pressure (e.g., Weberian ossicles, swimbladder connections to the ear) also arose for the first time in several different teleost groups (Popper and Coombs 1982; Braun and Grande 2002). It is perhaps significant that these auditory specializations are present in all electroreceptive taxa (i.e., Siluriformes, Gymnotiformes, Mormyridae, and *Xenomystus*).

Although electrosensory, mechanosensory lateral line (hereafter referred to as lateral line), and auditory systems have many characteristics in common and may be stimulated by some of the same sources (e.g., a nearby prey or conspecific), there is now little doubt that they are nevertheless distinct sensory systems. Distinctions are based on the fact that each system: (1) is activated by different physical aspects of the same (or different) source (see Section 4), (2) is innervated by separate cranial nerves, and (3) has ascending central nervous system (CNS) pathways that are largely segregated at least at the level of the hindbrain (e.g., McCormick and Braford 1988). To a somewhat lesser extent, submodalities within each system also share these distinctions. Moreover, we now know from recent work on salamanders that electrosensory (ampullary) and lateral line (neuromast) organs arise from a series of common dorsolateral placodes that are distinct from the single dorsolateral placode that gives rise to the two inner ears (e.g., Northcutt, Chapter 5). Despite clear anatomical and functional distinctions, interest in comparative questions surrounding the octavolateralis suite of sensory systems persists (e.g., Montgomery et al. 1995; Bell et al. 1997; Coombs et al. 2002).

Why should a volume on electrosensory systems include other octavolateralis systems? The simple answer is that there is still a lot to be learned by comparing the similarities and differences between the systems—not only from an evolutionary and developmental perspective (see Northcutt, Chapter 5) but also from a functional point of view. Why is it, for example, that lateral line and auditory systems share a common efferent system, whereas electrosensory systems lack one altogether? What is the functional significance of cerebellar-like structures and similar ascending and descending connections in octavolateralis medullary nuclei? In this chapter, we try to address questions such as these in an attempt to identify and understand common principles of operation and organization. Although both phylogenetic and developmental constraints have undoubtedly played a role in shaping the functional organization of these systems, we focus instead on the overall behavioral use of these systems and the interplay between information-processing demands and the physical constraints of stimulus transmission and transduction.

A conceptual outline of this chapter is presented in Figure 12.1. We are

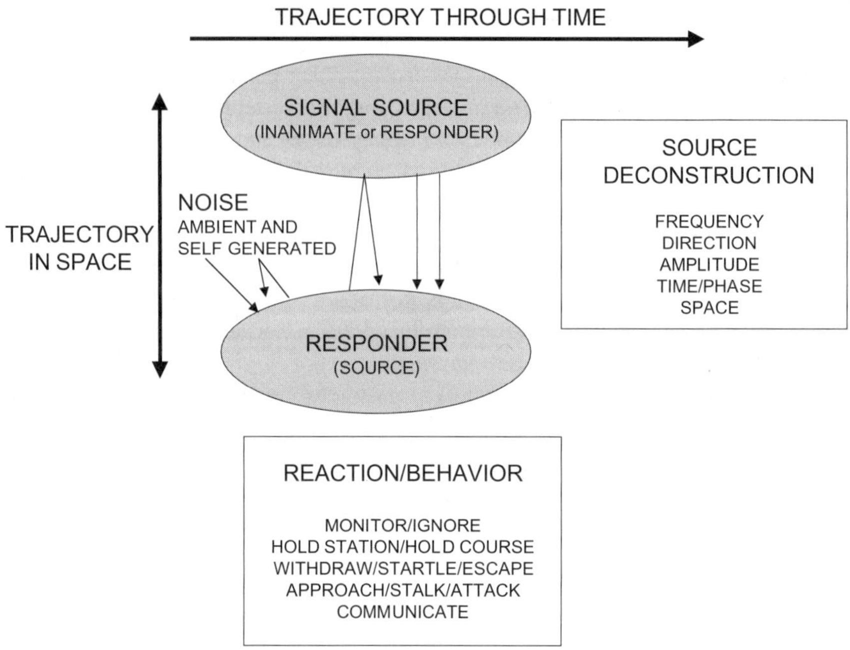

FIGURE 12.1. Schematic diagram illustrating the conceptual framework of this chapter. Fish are viewed as receivers of information, responders to external sources of stimulation, and sources of both signals and noise. Source information is partitioned by the peripheral nervous system along different physical dimensions before being combined in the central nervous system to effect the appropriate behavioral response.

interested in comparing the hydrodynamic, acoustic, and electrosensory capabilities of fish, as receivers of information and as responders to external sources of stimulation, whether they be animate or inanimate. We begin with a general overview of different behavioral contexts and their likely requirements for peripheral and CNS processing. Next, we contrast the physical nature of mechanical and electrical stimuli produced by sources of interest and the way in which these stimuli are transduced and encoded by peripheral sense organs. In particular, we examine how information from multiple sources and different features of the same source are partitioned by the different modalities and submodalities.

Because animate sources may themselves receive reciprocal information from the responding fish, they may seek to limit the "footprint" of their presence (e.g., to escape predation), or to actively engage in an interaction (e.g., courtship and spawning). As Figure 12.1 indicates, any "responder" is also a source of stimulus energy that can be used in an active sensory mode. However, self-generated motions and electrical fields also have the potential to compromise the detection of external sources, and this is discussed in a section on noise rejection mechanisms. Next, we examine central processing and in particular, the integration of multisensory information in the midbrain in relation to orienting behaviors.

Throughout the chapter, we try to place the detailed consideration of electrosensory systems covered elsewhere in the book within the broader context of octavolateralis systems. We also try to identify areas where our understanding of mechanosensory modalities might inform progress in electrosensory systems and vice versa. In the concluding sections of the chapter, we summarize the major similarities and differences among the octavolateralis suite of senses and some of the common evolutionary themes and principles of organization that have emerged from the comparison.

2. Behavioral Overview

Sensory inputs operate on, and with, other CNS activity to generate behavior. At a very basic level, animals need to monitor their own movements and orientation in space and to make correct decisions about whether to approach, avoid, interact with, or otherwise ignore different biotic and abiotic entities in the environment. Decisions such as these depend on the animal's ability to detect, segregate, identify, and locate multiple sources that are often simultaneously present.

In the case of reflexive behavior to external sources, the level of CNS processing can be quite minimal. For example, behavioral responses to acoustic and lateral line stimuli associated with the rapid advance of a predator neither require nor allow sophisticated CNS processing. Time is of the essence and escape responses such as the Mauthner mediated C-starts are initiated with very short latencies. Additional contextual information or sensory processing may then shape the escape response to take account of adjacent obstacles or the precise direction and nature of the response.

For more sophisticated behaviors, the nature of multiple sources in the environment, for example, their size, location, and movement, may need to be reconstructed from the information available to the sensory periphery. This task is made more difficult by signal sources (e.g., prey) that have camouflaged or masked their presence, such that the "signal" is now easily confused with or barely distinguishable from the surrounding "noise." As illustrated in Section 4 sensory information is broken down into recognizably separate sensory channels, both within and across different modalities.

This fundamental feature of specificity in sensory processing may stem from the need to decompose complex stimulus waveforms into their respective component parts so that (1) biologically significant "features" (e.g., direction of movement) of single sources can be extracted; (2) biologically relevant (signal) sources can be discriminated from background (noise) sources; and finally, in a more general sense, so that (3) multiple noise and signal sources can be segregated and identified from within a complex environment. At higher levels of the CNS, these processed streams of information are presumably reintegrated to yield the appropriate, context-dependent information and motor output required for an appropriate behavioral response. Understanding more sophisticated behaviors, such as predation, requires an understanding of how information is extracted and represented by the peripheral (sensory) nervous system and how that information is decoded by the CNS—that is, the neural algorithms that transform the sensory streams and reintegrate them to generate the appropriate, context-specific behavior.

Feeding behavior has been extensively studied in both mechanosensory and electrosensory contexts and provides some nice examples of how the basic requirements for feature extraction and CNS processing can vary with behavioral context. Mottled sculpin (*Cottus bairdi*) show an unconditioned orienting and strike response to a vibrating bead (Hoekstra and Janssen 1985) and a similar behavior has been observed in the common bully (*Gobiomorphus cotidanus*) (Bassett et al. unpublished). In these cases, small, chemically inert vibrating targets are equated with food and there is no evidence for, and perhaps no requirement for, discrimination between small active targets. The same appears to be the case with the passive electrosensory systems of elasmobranchs. Small electrically active targets, including simple dipole electrodes, produce an orientation and strike response. Similarly, weakly electric black ghost knifefish (*Apteronotus albifrons*) track and attack *Daphnia* (e.g., Nelson and MacIver 1999). Yet it is not known if artificial prey with electrical properties that are different from those of *Daphnia* would be treated differently. So the evidence to date is that in both the hydrodynamic and electrosensory contexts, simply equating small active targets with prey is a reasonable rule to follow.

Having detected such a target, the issue them becomes orchestrating the approach and strike behavior. As predicted, the level of sophistication depends on the nature of the interaction and the escape capabilities of the prey. Passive electroreception has a relatively short range and prey may already be in the strike zone at the point of detection. In bioelectrically mediated predation by swell shark, small teleost prey are swept into the mouth and the requirement of the predator is simply to open and close its mouth in the appropriate sequence (Tricas 1982). In contrast to this suction-only-mode of engulfing prey, predators may initiate a final strike behavior that also involves a rapid acceleration of the body (ram) toward the prey.

The search behavior and relative motion of the predator and prey can be very important determinants of predator–prey interaction and rival the complex interactions of interspecific communication. Marauding white tip sharks flush prey out of their night time shelter sites on the reef (as documented in "Blue Planet—A Natural History of the Oceans," BBC Wildlife Programs). The disoriented prey are attacked when they blunder into the strike zone of the sharks. Lateral line search behavior may involve passive detection of small targets in the stream drift (Torrent fish), or active search movements, pausing on the substrate at regular intervals to "listen" for prey (Antarctic notothenioids). In some cases, detection of a target prompts a repositioning of the fish to bring the target closer to the mouth. The approach trajectory of the little skate (*Raja erinacea*) to a small electrical dipole, and likewise that of the mottled sculpin to a small mechanical dipole, involves a series of repositioning steps as the animal approaches the prey. Saltatory approach algorithms also suggest that some fish have adopted behavioral strategies (stop and listen) for reducing their own, self-generated noise interference.

In contrast, the approach trajectory of the weakly electric black ghost knifefish is smooth reversal of the forward motion to bring the prey into the strike zone in front of the mouth (Nelson and MacIver 1999). The predatory sequence between *Electrophorus* and *Gymnotus* observed by Westby (1988) provides interesting insight into the elements of predatory behavior in which active electroreception is being used by both parties to the interaction. The intermittent suppression of electric organ discharges (EODs) by both predator and prey during the course of this interaction illustrates one of the disadvantages of being so dependent on an active sensory system. During the periods of suppression each of the fish presumably loses the ability to determine the whereabouts of the other party. However, as soon as the fish goes active again its own presence is revealed.

3. Stimulus Generation and Transmission

A generalized version of Ohm's law provides a convenient framework for comparing some of the physical properties of electric and mechanosensory stimuli (Table 12.1). Ohm's law states that in an ideal conductor, the magnitude of the current (I) is directly proportional to the applied electromotive force, E, and inversely proportional to the impedance, Z. The hydrodynamic equivalent of the electric current (flowing electrons/charged ions) is incompressible flow (flowing water molecules) and the equivalent of the electromotive force is a pressure drop per unit length. Similarly, the acoustic (propagated sound wave) equivalent of current is particle velocity and pressure gradients are the applied force.

TABLE 12.1. Electric, hydrodynamic, and acoustic transmission.

	Electric current	Incompressible flow	Sound wave particle velocity
Magnitude (I):	Coulombs/s (amps)	Cubic meters/s	m/s
Applied force (E):	Voltage gradient (V/m)	Pressure gradient (Pa/m)	Pressure gradient (Pa/m)
Impedance (Z):	Impedance (ohms)	Flow impedance (p/v)	Acoustic impedance (p/v)
Signal transmission rate:	Nearly instantaneous (speed of light)	Variable: from < mm/s to 1500 m/s	Fast (approximately 1500 m/s in water)

As a ratio of pressure/particle velocity (p/v), the specific acoustic impedance is somewhat analogous to electrical impedance ($Z = E/I$) in that both represent a force/velocity ratio. This should be distinguished from the pressure/volume ratio associated with flow in a pipe, which is a measure of incompressible flow resistance. In the case of mechanical disturbances around accelerating sources, the acoustic impedance increases as a function of distance from the source because the flow velocity falls off at a faster rate than the pressure. As a result, incompressible flow dominates the region close to the source, which produces a complex mixture of both hydrodynamic (incompressible flow) and purely acoustic (propagated pressure wave) signals that are detectable by both lateral line and auditory systems of fish (see Section 4).

The rapid rates at which information is transmitted by electromagnetic (near the speed of light) and sound (approximately 1500 m/s) waves is due to the fact that energy, rather than mass, is being transmitted. This energy transfer rate should not be confused with the current magnitude, which is proportional to the speed at which mass (electrons, charged ions, or water molecules) is transported.

The nature and speed of hydrodynamic signal transmission can vary widely, depending on the forces governing the signal. For example, the near-field flow associated with an accelerating body is transmitted (set up) at the same rate as the propagated pressure wave—at the speed of sound (Kalmijn 1988a). In contrast, surface waves, such as those created by a struggling insect at the water surface, propagate at extremely slow rates (in the centimeter per second range) that are frequency dependent (Bleckmann et al. 1989). In addition, there is a wide range of potential hydrodynamic stimuli for which transmission speeds are governed by the transport of water mass. Many of these (e.g., tidal currents and waves in the ocean and gravity-fed currents in streams or rivers) represent wide-scale, ambient flows, rather than hydrodynamic signals from discrete sources. However, the vortices shed behind a rock in a stream or a swimming fish represent a source-generated signal that is transported by bulk flow at rates that depend on the stream velocity or the swimming velocity of the fish.

The transport of water mass in bulk flows is governed by two other important physical principles, viscosity and inertia (the resistance of mass to acceleration), which have dramatic consequences for the very nature and structure of the flow. Unlike that of solids, the shape of fluid masses can be readily distorted by shear stress (force per unit area), and the measure of how much a fluid resists the rate of distortion is called viscosity (see Vogel 1994 for an excellent and more comprehensive treatment of this subject as it pertains to biological systems). If a liquid flows over a solid surface, the fluid sticks to the surface, resulting in what is called the "no-slip" condition, such that the velocity of fluid flow at the fluid–solid interface is always zero. The region over which the velocity goes from zero to its free-stream velocity is called the boundary layer and this is a velocity-gradient region in which incompressible flow is impeded relative to its maximum value. Highly viscous fluids such as honey resist distortion more than fluids of low viscosity such as water or air, and hence their boundary layers are thicker and the viscosity contribution to flow resistance is greater.

The relative importance of inertial and viscous forces is conveniently described by a dimensionless term known as the Reynolds number (Vogel 1994). At low Reynolds numbers (approximately 10), viscous forces dominate, boundary layers are thick and flow is laminar; at high Reynolds numbers (>>200,000), inertial forces dominate, boundary layers are thin, and flow is turbulent. Size and speed (or frequency of movement) also matter. The Reynolds number associated with the water motions of a small copepod swimming at 0.2 m/s is around 300 whereas that associated with the water motions created by a large whale swimming at 10 m/s is approximately 300 million (Vogel 1994). Finally, as viscous forces give way to inertial forces, flow regimes change from laminar, uniform, and predictable to highly turbulent, nonuniform, and unpredictable. Intermediate regimes often include structures such as vortices, yet another consequence of viscosity.

The different flow regimes represent a whole level of complexity in both the temporal and spatial structure of hydrodynamic stimuli that have few, if any, counterparts in the electric or acoustic world of fish. Similarly, the sound reflections, refractions, and scatter associated with propagated pressure waves (e.g., Rogers and Cox 1988) have few counterparts in the hydrodynamic and electrosensory worlds. The nearly instantaneous transmission of electric signals in the absence of any significant temporal corruption (e.g., reflections) means that the temporal fidelity of electric signals will be unsurpassed among octavolateralis systems (Hopkins 1999). On the other hand, the stimulus complexities associated with hydrodynamic and propagated acoustic signals, although limiting in some respects (e.g., corruption of temporal fidelity), may be enriching in others (e.g., use of reflections in echolocation; see Section 3.1).

3.1 Inanimate Stimulus Sources

Water movements, such as large-scale ocean currents, tidal currents, or stream flows, are of themselves significant stimuli for the lateral line system and perhaps also the vestibular, if not auditory portion of the inner ear. In addition, these flows also generate electric fields in the marine environment (Kalmijn 1984, 1988b, 1989; Paulin 1995). Without an external frame of reference, there are significant limits to the information available to the animal from these flow and electric fields, but they may provide some clues to guide orientation and movement (Table 12.2). With either tactile or visual contact with the substrate, fish can use hydrodynamic information to orient with respect to water currents or to move upstream (positive rheotaxis) or downstream (negative rheotaxis), as is the case with selective tidal stream transport (Harden Jones et al. 1979). The ambient sounds associated with coral reef formations, which may be both animate and inanimate in origin, may also be used as an orientation cue for the presettlement of pelagic larval fish (Tolimieri et al. 2000).

In the presence of flow, inanimate objects in the environment distort the flow, leaving a hydrodynamic signature of their presence (Table 12.3). The nature of the distortion depends on object geometry (including size) and flow velocity,

Table 12.2 Behavioral examples of passive orientation/navigation.

Modality	Submodality	Behavior	Animal	Reference
Electrosense	A A	Orientation to uniform electric fields	Stingray Catfish	Kalmijn (1974); Paulin (1995) Peters and vanWijland (1974)
Lateral line	SN	Rheotaxis Olfactory released rheotaxis	Blind cavefish Trout	Montgomery et al. (1997) Montgomery et al. (2003)
Acoustic	OT	Rheotaxis Orientation to coral reefs	Dace Larval fish	Pavlov and Tjurjukov (1995) Tolimieri et al. (2000)

Submodality key: A, ampullary organs; CN canal neuromasts; OT, otolithic end organ; SB, swimbladder or other air cavity; SN, superficial neuromasts; T, tuberous organs.

but higher flow velocities will generally produce a trailing wake of vortices. Although inanimate objects in a uniform electric current may also cause current distortions, depending on conductivity differences between the object and surrounding water, there is nothing comparable to the wake behind a stone in a stream.

In a special case of this effect, animals can also detect object-created distortions in their own self-generated stimulus fields (Table 12.4). The flow field generated by a swimming fish, for example, is distorted by objects in the near vicinity, and fish such as blind cavefish are capable of detecting these distortions to generate "hydrodynamic images" of their immediate surroundings (e.g., Hassan 1985). Electrosensory systems also lend themselves to active exploration of the environment. The EOD from weakly electric fish is distorted by these objects, which are then "imaged" by the electrosensory system (see Nelson, Chapter 11). An obvious acoustic parallel is echolocation, even though the underlying physical principals (reflection versus distortion) are somewhat different. Although underwater echolocation by fish may be limited by their low-frequency hearing capabilities, there is some evidence to suggest that sea catfish (*Arius felis*) are able to locate and avoid close-range obstacles by listening to the returned echos from their own short, broad-band (from less than 100 Hz to approximately 1500 Hz) sound pulses (Tavolga 1975).

3.2 Animate Stimulus Sources

Animals respire, feed, move, and pump ions to maintain internal homeostasis. These activities generate characteristic hydrodynamic, electrical, or acoustic signatures that can be passively detected by other animals (Table 12.2). With appropriate flow visualization techniques, even the feeding currents of animals as small as copepods can be seen (Fields and Yen 2002), and respiratory flows generated by crabs are visible out to tens of centimeters (Montgomery and Ham-

TABLE 12.3. Behavioral examples of passive orientation to discrete sources.

Modality	Submodality	Behavior	Animal	Reference
Electrosense	A	Prey attraction	Dogfish,	Kalmijn (1971)
	A		Paddlefish	Wilkens et al. (2001)
	A		Swell shark	Tricas (1982)
	A	Mate attraction	Stingray	Tricas et al. (1995)
Lateral line	CN	Attraction to stationary prey in stagnant water	Mottled sculpin	Hoekstra and Janssen (1985); Coombs et al. (2001)
		Attraction to stationary or swimming prey in stagnant water	Many species	Reviewed in Coombs and Montgomery (1999)
	SN + CN	Prey attraction in flowing water	Torrentfish	Montgomery and Milton (1993)
		Attraction to respiratory flow of prey	Scorpionfish	Montgomery and Hamilton (1997)
		Hydrodynamic trail following	Nocturnal catfish	Pohlmann et al. (2001)
	SN + CN	Obstacle entrainment	Trout	Sutterlin and Waddy (1975); Montgomery et al. (2003)
		Predator avoidance	Cichlid	Canfield and Rose (1996)
Acoustic		Prey attraction	Shark	Myrberg et al. (1972)
		Mate attraction	Midshipman	Bass et al. 1999; McKibben and Bass (1998)
	OT + SB	Predator avoidance	Goldfish	Canfield and Rose (1996)

ilton 1997). The whole-body swimming motions of fish and other aquatic animals will also generate flow fields.

In general, small slow-moving bodies will have low Reynolds numbers and will thus give rise to predictable flow patterns that should be relatively easy for a receiving system to analyze and interpret. In contrast, large fast-moving objects will have high Reynolds numbers and complex turbulent flow patterns. This may make it more difficult for the nervous system to extract information about the source properties, but it will also produce additional information in the form of vortices that persist for a considerable time after the animal has passed. Hence frequency, size, and speed can affect the fundamental structure of hydrodynamic flows associated with a moving body.

Furthermore, the locomotor activity of swimming animals creates spatially complex flow fields (reviewed in Bleckmann et al. 2003). For example, forward motion of the head is likely to result in a dipole-like flow field, whereas the undulatory motions of the trunk and tail will leave behind a distinctive hydrodynamic trail that is quite different from that left by an object passively moved through the water (Hanke et al. 2000). Movements of the pectoral fins, especially those in highly agile fish (e.g., Tetrodontiformes), will result in localized flows near the fins and/or local alterations in the flow fields generated by whole body movements or ambient water conditions.

TABLE 12.4. Behavioral examples of active localization or "imaging" of discrete sources.

Modality	Submodality	Behavior	Animal	Reference
Electrosense	T	Prey detection	Ghost knifefish	Nelson and MacIver (1999)
	T	Object distance determination	Elephantnose fish	von der Emde et al. (1998)
Lateral line	CN	Hydrodynamic imaging of objects	Blind cavefish	Weissert and von Campenhausen (1981) Abdel-Latif et al. (1990)
Acoustic		Echolocation of objects	Sea catfish	Tavolga (1975)

The whole notion of the lateral line as a vortex and turbulence detector is one that has only recently been explored, but we now know from the recent studies of Pohlman et al. (2001) and Dehnhardt et al. (2001) that both seals and fish are able to follow vortex trails at relatively far distances from the source using their respective hydrodynamic sensors (see also Bleckmann et al. 2003). In contrast, the hydrodynamic information available in advance of a fast swimming streamlined animal is quite restricted in extent. The passive hydrodynamic detection of fast swimming prey can be achieved, but may require the prey to traverse a section of the receptor array, as in the piper (Saunders and Montgomery 1985), or require an extremely rapid strike reflex, as in the stargazer (Montgomery and Coombs 1998). The damming effect or bow-wave in front of a swimming fish may also be important to the fish itself in providing information to detect and avoid obstacles (Dijkgraaf 1963; Weissert and von Campenhausen 1981).

Although the swimming motions of animals are seldom discussed as a source of acoustic stimulation, they can, with sufficient acceleration, provide a low-frequency stimulus to the inner ears of fish (see Section 4) and may be a more important and generalized source of acoustic stimulation than previously recognized (Kalmijn 1988a, 1989). More often cited sources of "purposeful" fish sounds include moving body parts (e.g., stridulation of bony elements) and muscle drumming against swimbladder walls (reviewed in Fine et al. 1975). Sound production by fish has been associated with a myriad of behaviors, including mate attraction, courtship and spawning, territorial defense, and schooling (Tables 12.3 to 12.5) (reviewed in Fine et al. 1975; Myrberg, 1981).

Living animals also provide quite strong but localized electric targets. Ion pumping by membranes such as gill epithelia or the specialized rectal gland of elasmobranchs provides an electronmotive force with return current pathways through the animals body and the surrounding water. These electrical fields are often modulated by respiratory movements generating low-frequency oscillations of the field (Bodznick et al. 1992). In some instances, muscle action potentials generate detectable external fields, and indeed modified muscle forms the basis

TABLE 12.5. Behavioral examples of interactive communication with conspecifics.

Modality	Submodality	Behavior	Animal	Reference
Electrosense	T	Jamming avoidance response	Glass knifefish	Heiligenberg (1986)
	T-knollenorgans	Echo response	Elephantnose fish	Bell et al. (1974)
	T-knollenorgans	Courtship	Many species	Reviewed in Hopkins (1988, 1999); Hagedorn (1986)
Lateral line		Courtship/spawning	Salmon	Satou et al. (1991)
		Schooling	Saithe	Partridge and Pitcher (1980)
Acoustic		Schooling?	Herring?	Gray and Denton (1991); Moulton (1960)
		Courtship	Many species	Reviewed in Myrberg (1981); Crawford (1997a)
		Territorial defense	Many species	Reviewed in Myrberg (1981)

of most electric organs which provide not only the source for active electrolocation, but also a strong signal for other nearby receivers.

4. Peripheral Extraction and Encoding of Information

4.1 Stimulus Transduction and Partitioning by Mechanosensory Modalities

Although there are obvious and clear physical differences between electro- and mechanosensory stimuli, the lines between what constitutes an effective lateral line, auditory or vestibular stimulus can often be blurred for primitive aquatic vertebrates, especially since all three systems share the same receptor cells (hair cells) and can be stimulated by the same external source by virtue of the good mechanical coupling between the receiving fish and the surrounding water. Moreover, unlike the mammalian inner ear, which has a dedicated organ of hearing (the cochlea), fish ears do not always have a single, easily identified organ of hearing. A prime difficulty is that we do not yet have a clear, universally accepted set of criteria for defining "underwater" hearing. Possible criteria include (1) the nature of the stimulus source, (2) the nature of the transduced (or proximal) stimulus to the ear, (3) the particular sense organ that is activated, (4) the CNS pathways that are recruited, and (5) the overall behavioral response/context. Everyone might agree that the response of a damselfish to the purposeful sounds of a conspecific in the context of territorial defense constitutes hearing and sound communication. But would everyone agree that a Mauthner-mediated escape response to the rapid advance of a predator's body also involves hearing? The presumed auditory (saccular) nerve fiber may be activated in both cases, but the behavioral output in the first case is likely to involve the ascending auditory pathway, whereas the second involves reflexive brain stem and spinal circuits. Furthermore, our own terrestrial biases of what constitutes an effective sound stimulus (high-frequency, propagated pressure waves rather than low-frequency, near-field disturbances) also clouds the issue. Regardless of how one defines underwater sound and hearing, the biophysics of how the inner ear and lateral line respond to mechanical disturbances in water are fairly well understood.

Stimulus transduction mechanisms vary according to how the hair cells are packaged relative to various peripheral structures and the biomechanical consequences of the "packaging" (Fig. 12.2). Hair cells in the lateral line system form discrete patches (neuromasts) that are overlain by a gelatinous cupula. These are distributed on the head and body either as superficial neuromasts on the skin surface and/or as canal neuromasts just under the skin in fluid-filled canals that are open to the exterior through a series of pores. Inner ear hair cells also form discrete patches (maculae) that are either overlain by a gelatinous cupula (crista ampullaris in semicircular canals) or a single dense stone (otolithic end organs). The inner ear of fish can be divided into the semicircular canals

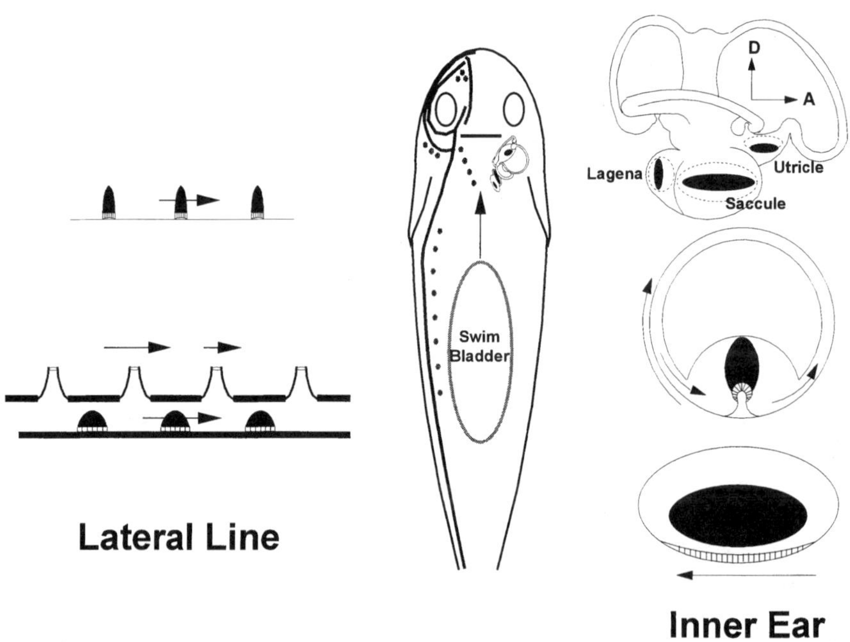

FIGURE 12.2. Schematic diagram of the spatial distribution, number and types of mechanosensory endorgans in the lateral line and auditory systems of fish. (Adapted from Platt, Popper, and Fay 1989.)

and three otolithic end organs (saccule, lagena, and utricle), each oriented at different angles in the head. In most species, the saccule is believed to be the primary acoustic organ, whereas the lagena and utricle are thought to have either dual vestibular and acoustic functions or vestibular functions alone. A glaring exception to this general rule is the utricle in clupeiform species, which is mechanically linked to a swimbladder-derived air bubble in the head (e.g., Blaxter et al. 1981). In this case, the utricle is believed to have the primary acoustic function and may underlie a unique ability (among fishes) for ultrasound detection (Mann et al. 2001).

Whereas the semicircular canals of the inner ear respond to the angular accelerations of the fish's body, the otolithic end organs respond to the differential motion between the animal's body, including the sensory epithelium, and the overlying otolith—or, in other words, to the linear accelerations of the fish's body (de Vries 1950; Kalmijn 1988a, 1989) (Fig. 12.2). This is the so–called "direct" and most primitive mode of inner ear stimulation (Popper and Fay 1999). In terms of the transduction mechanism and proximal stimulus (i.e., body acceleration), the direct route in response to an external source is no different from so-called vestibular stimulation caused by gravity, ambient water motions, or self-generated body movements. Yet the nervous system is somehow able to sort this all out.

Detection of the pressure component of an underwater stimulus field, the so-called "indirect" channel, can occur when a compressible air cavity, such as a gas-filled swimbladder, converts pressure variations into displacements that stimulate the auditory portion of the inner ear (the saccule in ostariophysine fish) (Popper and Fay 1999) (Fig. 12.2), or, in rare cases, adjacent areas of the lateral line (Blaxter et al. 1981; Webb and Smith 2000). Thus, even a single otolithic end organ can be stimulated by two distinct pathways (direct acceleration-sensitive pathway and an indirect, pressure-sensitive pathway). Finally, differential movement between the animal and the surrounding water, or, in other words, water flowing over the skin surface, results in stimulation of the lateral line system (e.g., Denton and Gray 1983) (Fig. 12.2). Propagated sound waves may be important for acoustic sensing by a minority of species that possess pressure transducers (Popper and Coombs 1982; Braun and Grande 2002), but the nearfield, incompressible region of any accelerating body (e.g., a swimming fish) is likely to be important for both lateral line and acoustic sensing by the majority of fish species (Kalmijn 1988a, 1989).

What is perhaps most significant about the aforementioned description is that many moving sources, as long as they are close enough and low enough in frequency, are capable of simultaneously stimulating a variety of octavolateralis systems. However, each of these systems (as well as submodalities within each system) will respond in different ways, thus conveying different types of information about the source. Of particular interest are the dimensions along which information is partitioned (Fig. 12.1) and how the different bits of information are then integrated in the CNS to effect adaptive behavioral responses.

4.2 Frequency Partitioning

Superficial and canal neuromasts have different response properties owing to the intrinsic properties of the receptors and the biomechanical filtering properties of the structural interface (e.g., cupula and canal) between the hair cells and the surrounding water. Lateral line canals essentially operate as high-pass filters—the smaller the canal, the more effective the filter and the higher the low-frequency cutoff (Denton and Gray 1983, 1988). In contrast, the superficial neuromast effectively acts as a low-pass filter. The filtering action of these peripheral structures in concert with the intrinsic frequency limitations of the receptors themselves, produces a response dichotomy that loosely parallels that between ampullary and tuberous electroreceptive organs. Thus, afferent fibers from both ampullary organs of electrosensory systems and superficial neuromasts of the mechanosensory lateral line show sustained or tonic responses to low-frequency signals (less than 30 Hz). In contrast, afferent fibers from canal neuromasts of the lateral line system and the so-called tuberous class of electroreceptors, which includes the knollenorgans and mormyromasts of weakly electric mormyrids, respond best to frequencies above approximately 30 Hz (Hopkins 1976; Zakon 1986; Münz 1989).

In response to direct stimulation of the inner ear, auditory (saccular) afferents in species with and without adaptations for pressure reception can be similarly classified into low- and high-frequency groups (Fay and Edds-Walton 1997b). However, there is no obvious dichotomy of end organs analogous to the ampullary/tuberous or superficial/canal neuromast dichotomies. Rather, frequency selectivity arises from within a single otolithic end organ, apparently from regional differences in micromechanics (e.g., in the stiffness of ciliary bundles and/or their friction coupling to the overlying otolith) and also possibly to the electrical resonance of hair cell membranes.

Low- and high-frequency subsystems in both lateral line and electrosense also appear to differ in their innervation patterns (Coombs et al. 2002), although the data in support of this notion are limited to a few species. Whereas single afferent fibers may innervate up to 10 or so superficial neuromasts in *Tilapia*, rarely does a single afferent innervate more than one canal neuromast (Münz 1979, 1985). A similar dichotomy may apply to electrosensory systems as well, with ampullary organs having relatively high sense organ-to-afferent fiber ratios and tuberous organs having much lower ratios (Zakon et al. 1998). Most importantly, there is at present no evidence that organs in the two subsystems are innervated by the same fiber (Münz 1979, 1985; Zakon et al. 1998). Thus, it would appear that information from the two subsystems is relayed to the brain along independent channels with no cross-coupling at the level of primary afferent fibers. Afferent nerve fibers appear to integrate information from multiple end organs in one subsystem (superficial neuromasts and ampullary organs), but to segregate information in the other (canal neuromasts and tuberous organs).

4.3 Directional Partitioning

One of the defining features of octavolateralis hair cells is their directional response properties. At the apical surface of each hair cell is a bundle of stereovilli and one eccentrically placed and elongated kinocilium. The location of the kinocilium determines the directional response properties of the cell such that bending of the kinocilium away from the stereovilli results in an excitatory response and bending in the opposite direction results in an inhibitory response (Flock 1965).

In the lateral line system, each canal or superficial neuromast contains two populations of oppositely oriented hair cells that are spatially intermingled, but that have separate innervation (Fig. 12.3A, B). In canal neuromasts, the axis of best sensitivity is parallel to the canal such that water motion in one direction along the canal will excite roughly half of the hair cells while simultaneously inhibiting the other half (Fig. 12.3B). In contrast, water motion past superficial neuromasts is not similarly constrained and the axis of best sensitivity depends on neuromast orientation and location on the body surface; both rostrocaudal

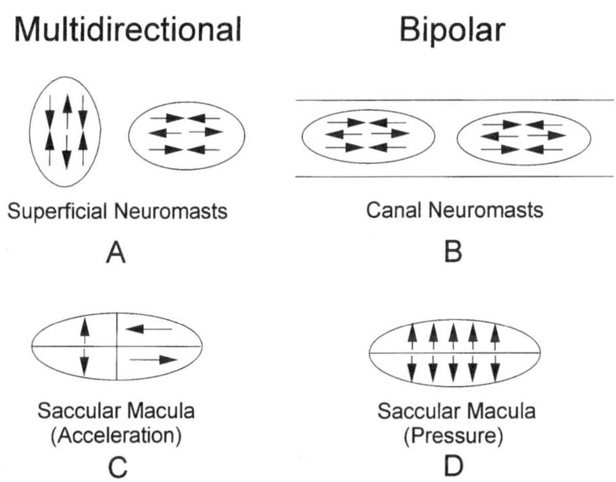

FIGURE 12.3. Illustration showing the directional organization of mechanosensory hair cells in different submodalities of the lateral line (**A**, **B**) and auditory (**C**, **D**) systems of fish. *Arrows* show the best excitatory direction of hair cells in different regions of the sensory epithelium. In the lateral line system, adjacent hair cells have opposing directions which are aligned along a single axis within a given neuromast. The alignment of adjacent canal neuromasts is always confined to the long axis of the canal, whereas that of adjacent superficial neuromasts can vary. Hair cells associated with pressure-sensitive auditory systems (**D**) are generally organized into two opposing groups with adjacent hair cells in each group having the same orientation. Hair cells associated with inertial auditory systems (**C**) are generally organized into at least four groups, two that are opposed along the rostrocaudal axis and two along the dorsoventral axis.

and dorsoventral axes are commonly present in adjacent neuromasts (Fig. 12.3A).

Hair cells in the otolith maculae of the inner ear also exhibit a similar organization of opposing hair cell directions. In this case, however, hair cells in a given region of the macula are all oriented in the same general direction, whereas those in adjacent regions are in opposing directions (Fig. 12.3C, D). The typical saccule in the majority of species is divided into four regions, two of them with opposing hair cells along a rostrocaudal axis and two along a dorsoventral axis (Fig. 12.3C) (Popper and Coombs 1982). In many pressure-sensitive species, the pressure-sensitive end organ (e.g., the sacculus in goldfish and the utricle in herrings) appears to be specialized for the reception of polarity (phase) information, having only two populations of oppositely oriented hair cells (Fig. 12.3D). Hair cells in one half of the macula are maximally stimulated when the air in the swimbladder is compressed, whereas those in the other respond maximally when the volume of air in the swimbladder expands.

To summarize, both the acceleration-sensitive component of the auditory system and the superficial neuromasts of the lateral line system are capable of responding to stimuli from all directions. In contrast, the pressure-sensitive auditory system and lateral line canal organs are constrained to bidirectional stimulus directions along the canal axis or the axis of swimbladder—inner ear motion. This means that stimulation patterns along an array of canal neuromasts might carry instantaneous information about the movement polarity of a nearby source (e.g., whether it is moving up or down, left or right, or forwards or backwards) (Coombs et al. 2002), but that those along a group of superficial neuromasts are more likely to convey information about the general direction of, for example, an ambient current. Likewise, whereas the pressure-sensitive mode of audition is likely to provide basic information on the phase of source movement relative to the fish (e.g., toward or away), the acceleration-sensitive mode will provide information about the angle or precise axis of source movement.

By comparison, electroreceptive organs do not have the same type of bipolar organization as the lateral line and ear, as all of the receptor cells within a given end organ display an identical excitability to either inward or outward current (Zakon 1986). Furthermore, even though tuberous electroreceptors in *Hypopomus* show polarity preferences and directional sensitivities as a function of body location, these preferences are shaped largely by skin resistance and best directions for transepidermal current flow, rather than any intrinsic properties of the receptor cells themselves (McKibben et al. 1993; Yager and Hopkins 1993). Although electroreceptive end organs themselves do not exhibit a bipolar organization, bipolar sensitivity is nevertheless achieved in many electroreceptive teleosts at the level of the first-order brain stem via two different types of principal cells (e.g., basilar versus non-basilar pyramidal cells)—one that receives direct excitatory input on its basilar dendrites from primary afferent terminals and the other that receives indirect input via an inhibitory interneuron (Bastian 1986).

4.4 Amplitude and Phase Partitioning

Conspicuously absent in the lateral line is a parallel for the two subclasses of tuberous organs specialized for encoding time (phase) and amplitude information in the electrosensory system (Carr 1986). Canal neuromast fibers exhibit tonic to slowly adapting responses that might loosely be regarded as the equivalent of the amplitude-encoding fibers in tuberous electrosensory systems. The same can be said of both acceleration- and pressure-sensitive fibers in fish auditory systems, although adaptation rates tend to be steeper for the high-frequency fibers (Coombs and Fay 1985). Furthermore, few, if any, of the features commonly associated with phase-sensitive pathways (e.g., large afferent terminals or calyceal synapses, round adendritic postsynaptic cells in the brain stem, calcium-binding proteins, separate ventral pathways in the CNS, etc.) (Carr 1986, 1993) have been identified in either the lateral line or auditory system of fish. One explanation for their apparent absence is that both of these systems operate at relatively low frequencies (less than 200 Hz for the majority of species), where both amplitude and phase can be adequately represented in a single channel and where demands for temporal precision are fewer. Nevertheless, phase information is likely to be very important for analyzing acoustic communication signals (see Section 5) and perhaps other behaviors, such as the rapid and highly coordinated schooling maneuvers of clupeid fishes.

4.5 Spatial Partitioning and Patterns of Activation

As a spatial array of sensors, both lateral line and electrosensory receptors are capable of gathering, in fine spatial resolution and at a single instant in time, information about spatial discontinuities in the near field of a stimulus source. This kind of instantaneous information is unavailable to auditory sensors, which are few in number and more or less clustered at a single location in the cranial cavity. Recent computational models of how the stimulus field around either a mechanical (vibrating sphere) or electric dipole is transduced by sensor arrays have revealed how different source characteristics can be extracted from the spatial array. Although perfect mechanical and electrical dipoles rarely exist in nature and may not represent all of the complexities and nuances of more biologically realistic sources, they nevertheless capture many essential features of real sources (Kalmijn 1988a,b, 1989, 2000), and in some cases are even capable of eliciting unconditioned behaviors (e.g., prey capture behaviors; Hoekstra and Janssen 1985). Moreover, dipole fields can be generated in one of two different ways: (1) when the source itself produces the hydrodynamic or electric current (e.g., water currents generated by a moving prey or the currents arising from the prey's bioelectric field) or (2) when a noncurrent generating source (e.g., an inanimate object) is placed in a current field (e.g., a rock in a stream or a rock in the EOD field of an electric fish).

Both lateral line and electrosensory models have converged on several common principles of how different dipole source features are represented in stimulation patterns along the animal's body surface (reviewed in Coombs et al. 2002) (Fig. 12.4). For example, the spatial location of the peak excitatory region carries information about the location of the source relative to the animal's body (Fig. 12.4A), whereas the width and/or slopes of the peak excitatory region carry information about source distance (Fig. 12.4B). The sign (polarity) of the peak excitatory region and the overall shape of the stimulation pattern correspond to the polarity (e.g., fore versus aft) of movement and the axis of vibration for a mechanical dipole (Fig. 12.4C, D). Similarly, the sign and shape of electroimages of live *Daphnia* correspond to the orientation of the *Daphnia*'s body (Wojtenek et al. 2001). Finally, for actively generated electric images of stationary objects, the sign (in this case, either an increase or decrease in received EOD amplitude) corresponds to the conductivity of the object relative to the surrounding water (von der Emde 1999).

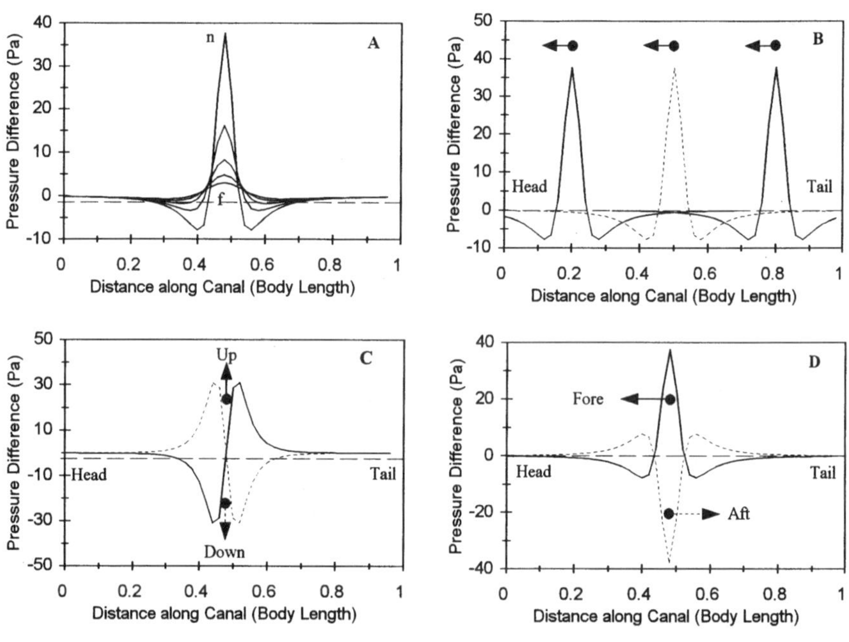

FIGURE 12.4. Examples of how spatial patterns of activation along a two-dimensional array of sensors (e.g., canal neuromasts along the trunk canal of fish) can convey different types of information about a vibrating source, including *distance*, from near (n) to far (*f*) (**A**); location, from head to tail (**B**); polarity of movement (dashed versus solid lines in **C**, **D**); and orientation (axis of vibration) (head/tail axis in **C** versus up/down axis in **D**). Level of activation is expressed as the pressure difference across the two canal pores surrounding each neuromast. (See Coombs et al. 1996 for details.)

5. Organization and Interconnections of First-Order Brain Stem Nuclei

The organization and interconnections of the first-order brain stem nuclei are briefly sketched here as a prelude to the discussion of noise reduction mechanisms (see Section 6) and as an example of striking parallels in the organization of the CNS in different octavolateralis systems. The medial octavolateralis nucleus (MON) of the lateral line and electrosensory regions of the brain stem [the dorsal octavolateralis nucleus (DON) of electroreceptive non-teleosts and the electrosensory lateral line lobe (ELL) of teleost fishes] all share several fundamental features that have been reviewed previously in great detail (Montgomery et al.1995). In brief, these include (1) a superficial molecular layer of parallel fibers derived from granule cells; (2) a principal cell layer of large, multipolar cells with apical dendrites that extend into the molecular layer and ventral dendrites that extend into deeper layers; and (3) deeper layers, in which the terminals from primary afferent fibers contact the ventral dendrites of principal cells either directly or indirectly through inhibitory interneurons. Similarities in the interconnections of these nuclei and the descending (acoustic) nucleus include (1) ascending projections from principal cells to secondary brain stem nuclei and the midbrain; (2) indirect inputs from motor command, somatosensory, and primary afferents via granule cells in the cerebellar eminentia granularis (teleosts) and dorsal granular ridge (elasmobranchs); and (3) descending projections from secondary brain stem nuclei (Fig. 12.5) Many of these shared features, including the parallel fiber inputs from the granule cell regions, function as part of an adaptive filter mechanism (see Section 7) for suppressing sensory reafference (Montgomery et al. 1995; Bell et al. 1997).

Despite striking anatomical similarities, there remain large differences in the level of complexity and number of cell types between tuberous electrosensory regions of the brain stem and mechanosensory regions. Based on Golgi preparations in the goldfish, for example, there are at most 4 layers and 5 main cell types in the MON (New et al. 1996). The number of cell types and layers in first-order acoustic nuclei appears to be equally restricted. In contrast, the mormyromast zone of the ELL has 5 or 6 identifiable layers and approximately 14 main cell types (Meek et al. 1999) and likewise, the tuberous region of the ELL in weakly electric gymnotids has at least 8 layers and 11 main cell types (Maler and Mugnaini 1994). One factor that potentially contributes to this overall level of complexity is the fact that information from tuberous electrosensory afferents are processed in multiple somatotopic maps within the ELL (Carr and Maler 1986). Each map is known to have different spatial and temporal tuning properties (Shumway 1989a,b). Although somatotopic mapping occurs in ampullary brain stem regions (Bodznick and Boord 1986; New and Singh 1994) and in the MON of the lateral line (New and Singh 1994), there is no evidence of multiple maps in either system. Similarly, ampullary brain stem regions of both elasmobranchs (Bodznick and Boord 1986) and teleosts (Finger 1986) are more

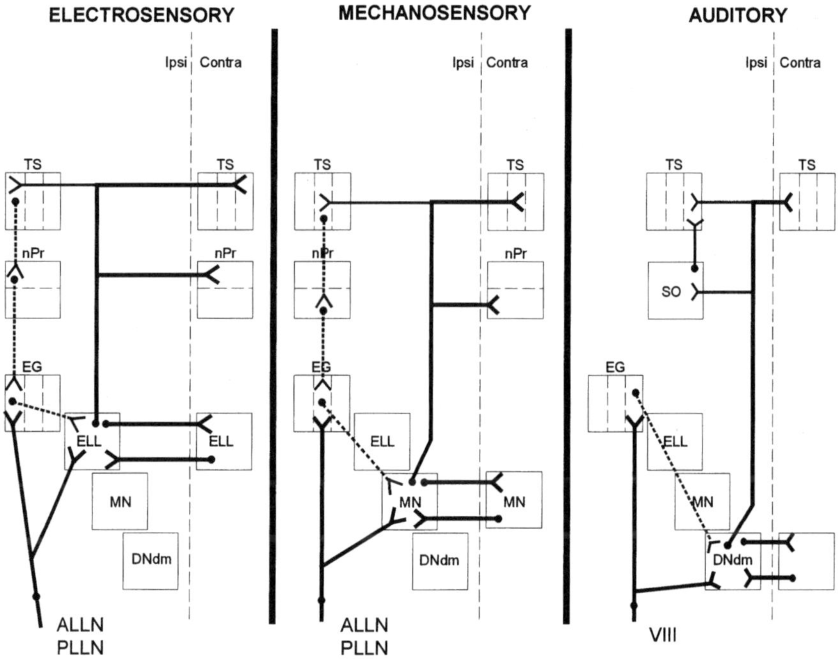

FIGURE 12.5. Schematic diagram of ascending pathways and commissural connections in the electrosensory (*left*), lateral line (*middle*), and acoustic (*right*) regions of the brain (based on compiled data from various teleost taxa). Primary afferents: anterior (ALLN) and posterior lateral line nerves (PLLN), saccular branch of the VIIIth nerve. First-order hindbrain nuclei: eminentia granularis (EG), electrosensory lateral line lobe (ELL), lateral line medial nucleus (MN) and the dorsomedial portion of the descending (auditory) nucleus (DNdm). Second-order hindbrain nuclei: nucleus praeminentialis (nPr), superior olive (SO). Midbrain nuclei: torus semicircularis (TS). All three systems show second and third-order projections to distinct regions of nPr and TS. (See Montgomery et al. 1995 for details.)

comparable to the MON than to the tuberous ELL, having 3 or 4 layers with no more than 5 or 6 main cell types.

6. Noise Rejection and Sensory Expectation Mechanisms

One of the most interesting comparative stories in recent years is the range of strategies and mechanisms for improving signal-to-noise ratios. Given that the same source (e.g., ventilatory movements of the operculum, EODs) may serve as either a signal or noise source, depending on the context and the receiver, it is not surprising that there are a number of potential strategies and mechanisms that fish might use to reduce or filter out different kinds of noises in different behavioral contexts. These range from simple static filters at the periphery to dynamic mechanisms involving complex neural circuits and overt behaviors (e.g., the jamming avoidance response). Many, but not all, of these strategies and mechanisms are shared by both mechano- and electrosensory systems.

Hair cells systems are well known for their exquisite mechanical sensitivity, with calculated threshold responses corresponding to movements measured in diameters of a hydrogen atom. Ampullary electroreceptors are also impressively sensitive with threshold responses measured in the nanovolt per centimeter range. Arguably then, both mechanosensory and electrosensory systems are not limited by sensitivity per se, but rather by their ability to isolate the biologically useful signals against a background of ambient and self-generated "noise." Under these circumstances, five principal strategies exist for signal/noise discrimination: the separation of signal and noise energies along frequency, intensity, spatial, or temporal dimensions and prior knowledge. The first four operate predominantly at the level of the peripheral sense organ, whereas "prior knowledge" is at the heart of the adaptive filters that operate in the hindbrain cerebellar-like structures associated with these sensory modalities.

6.1 Frequency Separation

Frequency separation is one of the standard methods of improving signal/noise ratios. It relies on the signal having different spectral qualities than the noise. Frequency tuning of tuberous organs to the animal's own EOD frequency in weakly electric, waveform species is a perfect example of this phenomenon. In its simplest form both mechanical and electrical systems may have significant and changing "DC" levels that disrupt the detection of the smaller "AC" signals of interest. The common solution for electrophysiologists or electrical engineers is to "AC" couple the signal input. In effect, "AC" coupling is a "high-pass" filter that nulls changing DC levels at the input. Low-pass filters, on the other hand, attenuate high frequencies and at a more sophisticated level, band-pass filters are used to acquire the signals of interest while suppressing spectral energy outside the bandwidth of the filter. As bandwidth decreases, it then becomes appropriate to talk about tuning the receiver to particular signals of interest.

Frequency/response analysis reveals frequency separation as an important component of signal processing in both mechanosensory and electrosensory systems, as evidenced by submodality tuning to low and high frequencies (see Section 4). Relatively few biological receptors respond in a tonic mode to a sustained stimulus, so most receptors are in effect AC coupled. However, hair cell mechanoreceptors can provide a sustained response to a sustained displacement of their cilia when adaptation is undesirable. For example, hair cells in some vestibular organs (e.g., utriculus) show tonic sensitivity to gravity (Baird and Lewis 1986) and hair cells in superficial neuromasts of the lateral line system show sustained responses to DC flows (Voigt et al. 2000). Despite this, most hair cell mechanoreceptors and all electroreceptors adapt to DC stimuli in a way that restores their sensitivity to transient stimuli. In the case of hair cells, response adaptation is due to a unique, active mechanical process involving sliding tip links between adjacent stereocilia and a shift in the sensitive range of the transduction process (e.g., Hudspeth et al. 2000). In the case of electroreceptors, accommodation to standing DC potentials is an intrinsic electrical property of the receptors, and changing background DC levels does not affect receptor sensitivity (Bodznick et al. 1993).

Care must be taken when describing receptor tuning curves to specify the appropriate stimulus parameters. This is generally not a problem in electrosensory systems where the applied stimulus translates to a voltage drop across the receptor, and the electrostatic nature of the fields eliminates propagation delays and response dynamics (see Section 3). However, in mechanosensory systems, response dynamics become very important owing to the interacting elements of mass, elasticity, and viscosity. The effective stimulus is thus very dependent on the architecture of structures surrounding the receptor cells (see Section 4).

Taking the example of lateral line receptors, superficial neuromast responses are largely proportional to flow velocity with respect to the skin surface, whereas canal neuromast responses are more nearly proportional to flow acceleration. Similar distinctions can be made for low- and high-frequency auditory (saccular) fibers of fishes. The responses of low-frequency saccular (auditory) nerve fibers may be nearly proportional to body displacement, whereas high-frequency fibers may be nearly proportional to body acceleration (Fay 1997). In each of these cases, the frequency response curves will appear differently, depending on whether the response is plotted in a displacement, velocity, or acceleration frame of reference. Nevertheless, the response curves are effectively equivalent and for the most part, simply linear transformations across different coordinate systems. Unfortunately, auditory response curves are more often expressed in terms of sound pressure, regardless of whether the transmission pathway to the ear is direct (driven by whole-body motions), indirect (driven by pressure), or a frequency-dependent combination of both. In this case, response curves in a pressure frame of reference can be quite misleading, especially if the ear is responding to body motion in a small experimental test tank, where the relationship between pressure and body motion is typically unpredictable.

All of these examples illustrate that the chosen frame of reference influences the way we describe and think about the response characteristics of the system under study. In the lateral line system, there is a developing consensus to plot response curves for both superficial and canal neuromasts with respect to an isovelocity stimulus. Holding the coordinate system constant for velocity provides a better comparison of the different lateral line submodalities. In these coordinates, a typical superficial neuromast has a low-pass characteristic, and a typical canal neuromast has a band-pass characteristic. The upper roll-off in both cases appears to be due to the receptor cell properties (Montgomery and Coombs 1992), whereas the low-frequency roll-off of the canal receptors is due to the mechanical filtering properties of the canal itself (Denton and Gray 1983) (see Section 4).

Viewed in the velocity frame of reference, lateral line canals can be seen as having a major function in the suppression of low-frequency noise. An elegant demonstration of the effectiveness of this filter is the observation that both superficial and canal neuromast fibers respond to an oscillating stimulus in still water. In the presence of background flow, however, canal neuromasts retain their responsiveness to the oscillating source, whereas superficial receptors, saturated by the DC flow, cease to respond to the oscillating stimulus (Engelmann et al. 2002). Narrow canals, such as those found in the mottled sculpin, are more effective than wide canals at reducing flow noise, and thus it is not surprising that the ability of mottled sculpin to detect an oscillating source is little changed over a four fold increase in ambient flow velocity (Kanter and Coombs 2003). In contrast, species with wider canals are more sensitive to hydrodynamic stimuli in stationary water, but lose that advantage in the presence of background flow (Bassett et al., personal communication). Typically, superficial neuromasts are friction coupled to water flow over the surface of the receptor, which is where they get their velocity-sensitive characteristics. In very low noise environments such as the deep sea, the superficial neuromasts on some fish have a papillate structure and hair cell orientations that imply they are projected above the boundary layer and are likely to be sensitive to water displacement.

When low- and high-frequency response curves of saccular (auditory) fibers to whole-body motions are viewed in an acceleration frame of reference (Fay and Edds-Walton 1997b), the high-frequency fibers may be similarly seen as having a major function in the suppression of low-frequency accelerations. Likewise, the function of high-frequency (tuberous) systems of weakly electric fish may be interpreted in a similar light. It is interesting to note that in all of these examples, the ability to detect exogenous signal sources may be compromised by some of the same low-frequency sources of noise interference—namely, the animal's own incidental movements (e.g., respiratory movements) and ambient motions of the surrounding water. In contrast, high-frequency signal sources are considerably more diverse (see Section 7).

6.2 Efferent Innervation and Intensity Separation

The octavolateralis efferent system is also capable of suppressing noise, in this case self-generated noise. This mechanism is completely absent from electrosensory systems and has several features that distinguish it from other noise-suppression mechanisms. First, it operates directly and rapidly on the peripheral nervous system by inhibiting lateral line receptor cells and their afferent fibers before and during movements of the animal (Roberts and Meredith 1989). Second, it can be activated by unexpected or highly arousing visual stimuli (e.g., prey) (Tricas and Highstein 1991) or by vigorous and rapid, self-body movements (Roberts and Meredith 1989). Efferent modulation is appropriate as a form of active gain control only when there is intensity separation between signal and noise. Noise levels can be variable, but the signal of interest must always be detectable above the background noise level. In this case, reducing the sensitivity of the afferent system, such that it is not saturated by the noise but is still responsive to the more intense stimulus, is an effective noise-suppression strategy.

As Bodznick (1989) points out, the striking absence of efferent innervation in independently evolved electrosensory systems in both primitive and more recently derived species argues for a functional rather than phylogenetic explanation. That is, efferent systems are absent in the electrosense because they are largely ineffective. For example, intensity separation via efferent suppression may be not workable for the electrosense, if self-generated noise is always greater in amplitude than signals from more distant sources. Alternatively, whereas vigorous, self-movements are likely to create potent hydrodynamic stimuli that compromise the sensitivity of the lateral line system to exogenous stimuli, the electric potentials generated by such movements may not be so intense as to severely compromise the sensitivity of the electrosensory system (Bodznick 1989). Another possibility is that the spatially complex and often turbulent and unpredictable nature of mechanosensory reafference may require a more direct, and in some respects simpler, "all or none" mechanism for suppressing reafference. Such a mechanism may be unnecessary or even undesirable in the electrosense, given the availability of both adaptive filter and common-mode rejection (see Section 6.3) mechanisms that are more specific and controllable in their actions.

6.3 Common-Mode Rejection and Spatial Separation

Common-mode rejection is a third mechanism by which sensory reafference may be reduced. This mechanism appears to play an important role in electrosensory systems, but a less significant role for the lateral line system. It takes advantage of the fact that reafference due to ventilatory movement of the gills is common mode among ampullary receptors at different locations and on different sides of the body (Kalmijn 1974; Montgomery 1984). In the little skate, *Raja erinacea*, commissural cells in the DON have inhibitory connections for subtracting out the common mode signal (New and Bodznick 1990; Bodznick et al. 1992). Inhibitory commissural cells such as these are a common feature in deep layers of both ampullary and tuberous regions of the brain stem nucleus in many different taxa. Cells with commissural connections are also present in lateral line (e.g., Zorn et al. 1998) and acoustic hindbrain nuclei (e.g., Edds-Walton 1998), but are not usually distinguishable as a separate ventral layer.

Hydrodynamic reafference due to ventilatory motion is unlikely to be common mode between receptors organs at different locations on the body. That is, both the direction and amplitude of the self-induced flow will vary as a function of location along the fish's head and body. Thus, common-mode rejection mechanisms for suppressing lateral line reafference, if they exist at all, may not be as well developed or play as significant a role as they do in electrosensory systems. Nevertheless, ventilatory flows are likely to vary along each side of the body in a bilaterally symmetrical way and inhibitory, commissural connections could provide the neural substrate for a point-by-point common-mode subtraction of symmetrical flows.

In general, exogenous sources are unlikely to produce common-mode or bilaterally symmetrical fields except when they are in particular orientations or locations (e.g., the median vertical plane of the fish). In this context, it is possible that the same kind of common-mode network that functions in the suppression of self-generated noise could also function in binaural computations of source location. When the inputs to commissural cells on the left and right side of the brain are equal, the output of the network will be zero, signifying that the source is somewhere in the median vertical plane. When the output of the network is nonzero, however, the information is far less ambiguous, with each different output signifying a unique location in space. Owing to the increased speed of sound in water and minimal density differences between the fish and the surrounding water, binaural time and intensity cues have long been thought to be unavailable to the ears for underwater sound localization. However, Fay and Edds-Walton (1997a) point out that the ears' output may contain response time and magnitude differences useful in binaural processing owing to the directionality inherent in the ears' response to body acceleration. In addition, spatial nonuniformities in the flow field and differential activation of lateral line neuromasts on different sides and locations along the body surface could very well provide either unilateral or bilateral cues for localization purposes.

6.4 Temporal Segregation and Pattern Recognition

For weakly electric fish, especially those that produce pulse-like EODs and for the majority of fish that produce communication sounds, signals consist of a series of short pulses that are rich in temporal structure. EOD patterns may include (1) the tonic, unmodulated activity of fish when they are resting, which can be either periodic (regular) or aperiodic (irregular); (2) increases or decreases in the EOD rate (i.e., frequency modulations); (3) silent interruptions in the ongoing pattern; and (4) interactive timing of EODs among two or more individuals (e.g., the echo response of mormyrids) (Bell et al. 1974). Acoustic pulses can be regularly spaced in a pulse train to produce "hum-like" qualities. The envelope of the pulse-train waveform can also be amplitude modulated, resulting in (1) short bursts of pulse trains that sound like "grunts" or (2) more gradual and continuous changes in the overall amplitude that sound like "moans." Temporal patterns of signals such as these are likely to be salient features for communication purposes (e.g., Myrberg 1981). To the extent that temporal patterns of signals are distinguishable from those of background noises, temporal mechanisms of segregation will undoubtedly play a role in the signal-to-noise processing capabilities of these systems. A neural basis for temporal selectivity has been hypothesized for the acoustic midbrain of a weakly electric and sonic species of fish (Crawford 1997b).

6.5 Adaptive Filters

Among noise-reduction mechanisms, perhaps the most complex and interesting is the adaptive filter in the first-order brain stem nuclei of both electrosensory and lateral line systems for suppressing unwanted, self-generated noises (see Montgomery et al. 1995 and Bell et al. 1997 for review). The principal cell types and cerebellar-like structure of brain stem nuclei in several different taxa (see Section 5) have now been shown by a number of investigators to support an adaptive filter or modifiable efference copy mechanism such as that first discovered in the ELL of weakly electric mormyrids (Bell 1982). This adaptive filter uses in effect "prior knowledge" to construct a negative image of the expected temporal pattern of reafferent input. Essentially, hindbrain principal cells in the adaptive network compare inputs to their apical dendrites with those to their ventral dendrites. Whereas the apical dendrites receive information from the parallel fiber system about the animal's own movements, the ventral dendrites receive information from sensory afferents, which includes both self- and exogenously generated stimuli. Correlated activities between apical and ventral dendrites are used to construct a negative image, which then cancels or suppresses the components of the sensory input associated with the animal's own expected movements (e.g., the animal's own breathing movements). This negative image can be slowly modified, typically over a time course of several minutes, to adapt to changes in the reafferent signal.

7. Multimodal Integration

One of the most primitive and highly conserved sites of multisensory and sensory/motor integration is the optic tectum or its mammalian homolog, the superior colliculus. Somatosensory, visual, and auditory maps of sensory space converge in register at this sensory–motor interface to orchestrate various orienting behaviors of the eyes, head, or entire body in a wide variety of vertebrates. In primitive aquatic vertebrates, lateral line and electrosensory maps also converge at these sites (e.g., Carr and Maler 1986; Schellart and Kroese 1989). The general conclusion from these and more extensive studies in other animals across a range of senses (e.g., Stein and Meredith 1993) is that midbrain maps provide a representation of sensory space surrounding the animal.

In only a few selected instances, however, do we have a precise understanding of the computational algorithms that transform the peripheral sensory inputs into specified directional information. Octavolateralis midbrain cells clearly show evidence of directional selectivity, whether it be to the direction of acoustic particle motion (e.g., Ma and Fay 2002), surface wave propagation (Claas et al. 1989), or object motion (Bleckmann et al. 2003). However, we know very little about how directional information from different submodalities is combined within or across modalities. Do submodalities carrying similar types of information (e.g., direction of source movement) from different senses combine in the midbrain to enhance the overall sense of direction? Alternatively (or in addition) do submodalities carrying different, but complementary pieces of information, (either within or across senses) combine to form a more integrated picture of the "whole"? For example, is information about the *direction* of source movement (theoretically carried by superficial neuromasts and otolithic end organs of the inner ear) combined with information about the *location* of the source relative to the receiver, as encoded by spatial excitation patterns along electrosensory or lateral line sensor arrays? As another example, information from both pressure- and acceleration-sensitive auditory submodalities must, in theory, be combined in order to resolve 180° ambiguities inherent to each channel (Schuijf 1975). That is, the acceleration-sensitive channel alone is insufficient for discriminating between a prey advancing from the left and one fleeing to the right (both moving in same direction), whereas the pressure-sensitive channel alone is insufficient for discriminating a prey advancing from the right and one advancing from the left (both causing a compression). Are these two different, but complementary pieces of information integrated in the CNS to resolve these ambiguities, and if so, how? Finally, how is information from spatially distributed sensor arrays (electrosense and lateral line) combined with that from paired sensors (e.g., vision and audition)?

Recent behavioral experiments to determine the relative contributions of different lateral line submodalities to various orienting behaviors have shed some light on these questions, revealing that information from different submodalities may be combined in different ways under different behavioral contexts (reviewed in Montgomery et al. 2002). Whereas a single submodality [either superficial (SN) or canal (CN) neuromasts] is sufficient for relatively simple tasks [e.g., rheotaxis (SN) or prey orientation in stagnant water (CN)], both submodalities are required for other, more demanding tasks (e.g., prey orientation in flowing water) (Table 12.2). Recent physiological experiments to measure the responses of CNS cells to vibrating and moving sources suggest that information from the two lateral line channels may be largely segregated—at least up to the midbrain (reviewed in Bleckmann et al. 2003). Both behavioral and physiological lines of evidence suggest that superficial neuromast channels function in the spatiotemporal *integration* of information across multiple receptors (e.g., to compute the general direction of a uniform or large-scale current), and conversely, canal neuromasts function in the spatiotemporal *segregation* of information (e.g., to determine the exact location of a current-generating or current-distorting source) (reviewed in Kanter and Coombs 2003). This behavioral dichotomy loosely parallels the ampullary-based abilities of sharks and catfish to orient to uniform electric fields (Table 12.2) and the tuberous-based, active electrolocation abilities of weakly electric fish (Table 12.4). Experiments to determine if there is a similar division of labor between low- and high-frequency auditory channels have not been done, primarily because submodalities are not easily localized to separate and identifiable endorgans or populations of neurons.

8. Common Emergent Principles

Despite major and obvious differences between the three major groups of octavolateralis sensory systems, a number of shared characteristics can be identified. All three octavolateralis systems support passive (Tables 12.2 and 12.3) and active (Table 12.4) sensing of animate and inanimate features of the environment. Likewise, all of these systems support interactive communication with conspecifics (Table 12.4). All three systems also share fundamental patterns of organization, including parallel, ascending pathways in the CNS (Fig. 12.5), submodality partitioning of information along different stimulus dimensions, cerebellar-like structures in the hindbrain, and multisensory convergence at tectal and subtectal levels of the midbrain.

In addition, a number of shared features can be attributed to two but not all three modalities. For example, *lateral line* and *electrosensory systems* both have anywhere from hundreds to thousands of receptor organs that are nearly overlapping in their spatial distributions. As might be expected, this organization is largely preserved in the CNS with somatotopic maps, which are likely to encode various stimulus features (e.g., location and distance) in similar ways (via spatial excitation patterns). For dipole-like sources, both also have very short ranges that are limited by the regions of steep attenuation with distance of their respective stimuli. In contrast, the acoustic sense is limited to a few receptor organs that are clustered in the cranial cavity of the fish (Fig. 12.2). Owing to the extended transmission range of propagated sound waves, acoustic ranges in fish are likely to extend beyond the ranges of either the electrosensory or lateral line system.

By comparison, *lateral line* and *acoustic systems* share a common efferent system and the same directionally sensitive mechanoreceptors, both absent in the electrosense. Furthermore, hair cells in both mechanosensory systems are organized according to best directional sensitivity, being grouped into submodalities that can be distinguished by the presence or absence of peripheral structures that restrict the proximal stimulus to two opposing directions (Fig. 12.3B, D). Bidirectional submodalities (canal neuromasts and pressure-sensitive auditory system) (Fig. 12.3B, D), provide instantaneous information about the polarity of a moving source (e.g., whether it is moving toward or away from the receiving animal), whereas multidirectional submodalities (superficial neuromasts and acceleration-sensitive auditory system) (Fig. 12.3A, C) presumably provide continuously graded information about the angular direction of approach or withdrawal.

Although *acoustic and electrosensory systems* differ dramatically in their peripheral organization and the form of physical energy to which they respond, they both nevertheless respond to rapidly transmitted signals that function in a variety of behavioral contexts involving interspecific interactions (e.g., courtship, aggression, appeasement, coordinated movements, etc.). The representation of information in the temporal domain is likely to be very important for both systems, where convergence on common principles and organizations for preserving temporal precision is apparent (e.g., Carr 1986; Kawasaki 2000). A striking example of parallel functions for these two senses can be found within a single species, the weakly electric mormyrid fish, *Pollimyrus isidori*, which uses both electric and acoustic signals in a gender-specific and interactive way during courtship behavior (Crawford 1997a).

Insofar as we know, the repertoire of known acoustic and electric communication signals is relatively rich in comparison to hydrodynamic signals. Although lateral-line mediated schooling and courtship behaviors may qualify as communication behaviors (Table 12.4), other lateral-line mediated behaviors known to function in social communication are few and understudied relative to communication behaviors in weakly electric and sound-producing fish. Furthermore, there is at present no evidence for a specialized class of communication effectors or receptors for sending and receiving hydrodymamic signals. In contrast, weakly electric fish have evolved specialized effectors for generating electric fields and in some cases, classes of sensors dedicated to the sole purpose of communication (e.g., the knollenorgans of mormyrids). Likewise, many sonic species have specialized drumming muscles (with some of the fastest contraction rates known) for producing communication sounds. Nevertheless, the extent to which communication via the lateral line is limited by physical constraints, rather than our own inability to recognize and categorize hydrodynamic communication signals, remains to be determined.

8.1 The Functional Evolution of High-Frequency Channels

From an evolutionary perspective, the most primitive submodalties in both electrosensory (ampullary) and acoustic (acceleration-sensitive) systems are low-frequency channels. High-frequency tuberous organs in the active electrosense and pressure-sensitive channels in the acoustic sense are clearly more recently derived features among teleost fishes. Although a major adaptive function of high-frequency (band-pass) channels is likely the suppression of low-frequency noises common to all octavolateralis systems (see Section 6), the flip side of this coin is that the origin and function of high-frequency signals passed by these channels may vary considerably. High-frequency channels in active electrosensory systems (tuberous organs) appear to have largely coevolved with species-specific EOD signals. Indeed, one is struck by the two independently evolved groups of teleost fishes with active electrolocation. The reinvention of passive (low-frequency) electroreception appears to have been closely followed by the evolution of active (high-frequency) electrolocation in both mormyrid and gymnotid fishes. Active signal production may qualify as a "key innovation" from which the subsequent hypertrophy of the brain and radiation of these groups occurs. Active electrolocation, although restricted in phylogenetic distribution, then becomes a rich source of information about both animate and inanimate sources, and opens a potent communication channel for intraspecific interaction. With its rich potentiality, active electrosense tends to dominate the sensory world and apparatus of these fish in a way that rivals vision (at least in murky water) for the degree of detail and information it can provide. Given that most, if not all, weakly electric fish have small eyes and live in murky waters, a diminished visual world may very well have been a primary selection pressure in the evolution of an active, high-frequency electrosense.

In all cases that have been examined, high-frequency (greater than approximately 200 Hz) channels in fish auditory systems have coevolved with specializations for sound-pressure sensitivity (e.g., Popper and Fay 1999). Although the selection pressures are unknown, pressure sensitivity among extant species is clearly not universally correlated with active sound production (Ladich 1999) or limited visual capabilities. Indeed, at least some ostariophysines (e.g., goldfish) are mute and many are diurnally active in relatively clear, freshwater habitats. Moreover, unlike the hypertrophy of hindbrain nuclei associated with active signal production in the electrosense, that associated with the addition of high-frequency, pressure-sensitive channels is modest at best (McCormick 1999). Finally, many sonic species (e.g., damselfish, toadfish) appear to lack peripheral specializations for pressure sensitivity.

The acquisition of sound pressure sensitivity would appear to have many benefits, including (1) early warning capabilities through extended distance ranges, (2) improved detection capabilities over a wider dynamic and frequency range, and (3) the ability to encode the initial compression or rarefaction (polarity) of a moving source. The teleostean sister groups Ostariophysi and Clupeomorpha represent the two largest radiations of pressure-sensitive species, with each group having evolved different and nonhomologous structural adaptations for pressure sensitivity. Although the current evidence is limited and largely circumstantial, predator avoidance, rather than active sound communication, may have played an initial role in the evolution of pressure-sensitive systems in these two groups. In ostariophysines such as the goldfish, pressure-sensitive (saccular) inputs to the Mauthner cell have been shown to dominate the initial phase of C-start escape responses, which in non-ostariophysine fish are dominated by visual and/or lateral line inputs (Canfield and Eaton 1990). Given that many ostariophysine fish inhabit shallow, freshwater environments, it is possible that pressure sensitivity might confer an adaptive advantage to species that can respond rapidly to an initial compression of the water surface caused by aerial or terrestrial predators. In clupeomorphs, the transient pressure pulses produced by the rapid maneuvers of schooling fish suggest that pressure cues may likewise play a vital role in the ability of these fish to respond to predator attacks in a rapid and coordinated fashion (Moulton 1960; Gray and Denton 1991). It has also been proposed that the ultrasonic hearing capabilities of some clupeid species has evolved in response to predation by echolocating dolphins (Mann et al. 2001).

8.2 Adaptive Filter Mechanisms

Probably one of the most exciting discoveries of emergent principles across octavolateralis systems is the use of cerebellar-like structures for adaptive cancellation of predictable noises. Many, if not most, of these similarities also extend to the dorsal cochlear nucleus of mammals, but adaptive filtering of self-generated or expected noise has not yet been demonstrated for the mammalian auditory system. In this regard, it is interesting to note that the parallel fiber system of the molecular layer (the cerebellar crest in teleosts) is also intimately associated with the dorsomedial portion of the descending (acoustic) nucleus (DNdm) in several groups of fish (McCormick 1999). This portion of the first-order acoustic nucleus is particularly prominent in species for which the inner ear is coupled to a pressure-transducing air bladder. In these groups, the DMd receives inputs from the otolithic end organ that is coupled to the air bladder (the saccule in otophysans and mormyrids and the utricle in clupeids). In contrast, the regions of the hindbrain that receive input from acceleration-sensitive end organs (usually the lagena and utricle), as well as the semicircular canals, are quite ventral to the molecular layer. Thus, these ventrally located regions are unlikely to receive parallel fiber inputs, which provide the necessary information for identifying stimuli caused by self-motion. Given that vestibular systems are geared toward detecting rather than rejecting self-motion, this arrangement would seem to make perfect sense. The appearance of parallel fiber inputs to regions of the hindbrain that receive otolithic inputs may therefore prove to be an important clue for distinguishing between auditory and vestibular function in many groups of fish for which there is often no easily identified auditory end organ. In any event, a physiological analysis of parallel fiber input and function in auditory regions of the brain in both primitive aquatic and more recently derived terrestrial vertebrates remains to be done.

In addition to adaptive filter functions, a number of other functions have been proposed for the recurrent feedback pathways common to many, if not all, octavolateralis systems (Fig. 12.5). These include adaptive control of sensitivity or spatiotemporal filter properties, a sensory searchlight mechanism for attentional control, coincidence detection for enhanced sensitivity to weak signals, and spike burst generation for feature extraction (see Berman and Maler 1999; Turner and Maler 1999; Gabbiani and Metzner 1999 for reviews).

8.3 Neural Algorithms for Extracting Information from Two-Dimensional Arrays

It is likely that spatially distributed electrosensory and lateral line systems will share similar neural algorithms for extracting various stimulus features. Computational models of stimulation patterns along sensor arrays in both systems certainly predict that peripheral representations of different source features, including azimuth, distance, and orientation will be similar (Fig. 12.4) (see Section 4). In particular, the mechanisms underlying the more rudimentary orientation behaviors of passive electrosense and lateral line (Tables 12.2 and 12.3) should prove instructive for understanding the intrinsically more complex behaviors associated with active electrolocation (Tables 12.3 and 12.4). In this case, the reduced complexity of both lateral line and passive (ampullary) electrosensory regions of the hindbrain relative to active (tuberous) electrosensory regions (see Section 5) may simplify the task of identifying fundamental principles of how, for example, source locations are reconstructed by the brain and how orienting behaviors are orchestrated.

9. Summary

Passive electrosense is a primitive sense that is restricted to relatively few basal lineages, perhaps amounting to hundreds of species, diverse phylogenetically, but not so diverse from a functional perspective. In contrast, both the lateral line and the inner ear are effectively ubiquitous across fish species, meaning that there are tens of thousands of species distributed across every imaginable habitat utilizing these modalities to a multiplicity of purposes (Tables 12.2 to 12.5). Some of these species may have particular lateral line or auditory specializations, but many more will utilize vision as their primary sense, so that the specific contribution of hydrodynamic and acoustic senses will vary enormously across different species, and even within a species across different behaviors.

In this regard, it is interesting to reflect that passive electrosense provides a sparse, comparatively simple and slow "picture" of the world at short range. Information on nearby animate sources and uniform electrical fields can be obtained through a single group of relatively homogeneous, low-frequency receptor (ampullary) organs that are utilized for behaviors such as predation, mate detection, and general orientation (Tables 12.2 and 12.3). General orientation to large-scale, ambient fields appears to be a very primitive function of low-frequency submodalities in all octavolateralis systems (Table 12.2).

By comparison with the ampullary electrosense, the hydrodynamic world is arguably both richer and more difficult to interpret. The mechanical nature and diversity of the stimuli introduce additional physical complexities (e.g., viscosity), which, when coupled with the availability (and perhaps necessity) of both low- and high-frequency receptor classes, extends the amount of available information and the variety of behaviors that can be supported (Tables 12.2 to 12.5). The low-frequency acoustic world of fish with primitive (inertial) ears is intricately linked to the hydrodynamic world of moving sources, and thus extends even further the available mechanosensory information and the number and variety of potential behaviors (Tables 12.2 to 12.5). Finally, although the signal sources and selection pressures leading to the evolution of high-frequency channels may have varied considerably across octavolateralis systems and taxa, these channels nevertheless can be viewed as providing two basic functions: (1) the suppression of low-frequency noise, often from similar sources (e.g., the animal's own respiratory movements) and (2) the preservation of information in the spatial and/or temporal nonuniformities of the stimulus field.

In the subsequent water-to-land transition of vertebrates, the lateral line and electrosensory components of the octavolateralis suite of senses were lost. The somatotopic and/or directotopic organizations dominating these primitive senses largely gave way to a tonotopic organization of the tetrapod ear and a dramatic expansion in the number of high-frequency auditory channels among mammals. Whereas primitive aquatic vertebrates presumably rely on directotopic partitioning of information for determining the direction of a sound source, terrestrial vertebrates rely instead on binaural time and intensity cues. This fundamental difference in available cues is a direct, biophysical consequence of density differences between the sound-conducting medium (air or water) and the body of the receiving animal. In water, for which the density differences are minimal, there is a good mechanical coupling between the fish and the surrounding water and a direct transfer of directional information from the motions of the water to the body motions of the fish. In addition, the spatial discontinuities in the hydrodynamic near field of nearby sources provides spatial information through the lateral line. In air, the density differences are appreciable and sound travels more slowly, resulting in usable binaural time and intensity cues. The aerodynamic near field, however, has ceased to be a viable source of information for most land vertebrates. In essence, an important source of mechanosensory spatial information has been lost and replaced by different cues and mechanisms for extracting spatial information. In this regard, the ability to analyze a complex auditory waveform into its component sinusoids for the purpose of identifying and segregating simultaneously present sound sources may be one of the major adaptive advantages of multiple, high-frequency auditory channels (Cortopassi and Lewis 1998; Fay and Popper 2000). If this is true, it seems reasonable to ask whether or not this auditory capability arose for the first time with land vertebrates or whether, instead, it is a primitive feature found in all vertebrate auditory systems (Fay and Edds-Walton 1997b).

In the final analysis, it is impossible to do justice to all of the possible comparisons that can be made across different octavolateralis systems and the lessons that can be learned. Although we have limited most of our comparisons to primitive aquatic vertebrates, comparisons between aquatic and terrestrial vertebrates have likewise revealed striking parallels in the way that temporal information is processed by systems and animals as divergent as the electrosensory system of fish and the auditory systems of barn owls (Carr 1986; Carr and Soares 2002) and echolocating bats (Covey 2000). It is undoubtedly the case that comparisons such as these will continue to reveal converging principles of information processing by the nervous system (Eisthen and Nishikawa 2002).

Acknowledgments. We wish to acknowledge discussions with and the published contributions of colleagues whose ideas and perspectives on comparative aspects of octavolateralis systems made our job so much easier. In particular, we thank Ted Bullock, Dave Bodznick, Catherine Carr, Curt Bell, Dick Fay, Carl Hopkins, Ad Kalmijn, Catherine McCormick, John New, Chris Platt, and Art Popper. A special thanks to Peter Rogers for helping us sort out the appropriate acoustic equivalents to Ohm's law (don't blame him if we got it wrong!). We also thank Weihai Liu for assistance with references and Dick Fay for a careful editing of the chapter and helpful suggestions for improvement.

References

Abdel-Latif H, Hassan ES, von Campenhausen C (1990) Sensory performance of blind mexican cavefish after destruction of the canal neuromast. Naturwissenschaften 77: 237–239.

Baird RA, Lewis ER (1986) Correspondences between afferent innervation patterns and response dynamics in the bullfrog utricle and lagena. Brain Res 369:48–64.

Bass, AH, Bodnar D, Marchaterre M (1999) Complementary explanations for existing phenotypes in an acoustic communication system. In: Hauser MD, Konishi N (eds), The Design of Animal communication. Cambridge, MA: MIT Press, pp. 493–514.

Bastian J (1986) Gain control in the electrosensory system mediated by descending inputs to the electrosensory lateral line lobe. J Neurol 6:553–562.

Bell CC (1982) Properties of a modifiable efference copy in an electric fish. J Neurophysiol 47:1043–1056.

Bell CC, Meyers JP, Russel CJ (1974) Electric organ discharge patterns during dominance-related behavioral displays in *Gnathonemus petersii*. J Comp Physiol 92: 201–228.

Bell CC, Bodznick D, Montgomery JC, Bastian J (1997) The generation and subtraction of sensory expectations within cerebellum-like structures. Brain Behav Evol 50:17–31.

Berman NJ, Maler L (1999) Neural architecture of the electrosensory lateral line lobe: adaptations for coincidence detection, a sensory searchlight and frequency dependent adaptive filtering. J Exp Biol 202:1243–1253.

Blaxter JHS, Denton EJ, Gray JAB (1981) Acousticolateralis system in clupeid fishes. In: Tavolga WN, Popper AN, Fay RR (eds), Hearing and Sound Communication in Fishes. New York: Springer-Verlag.

Bleckmann H, Gottfried T, Blübaum-Gronau E (1989) Lateral line system of surface-feeding fish: anatomy, physiology and behavior. In: Coombs S, Görner P, Münz H (eds), The Mechanosensory Lateral Line: Neurobiology and Evolution. New York: Springer-Verlag, pp. 501–526.

Bleckmann H, Mogdans J, Dehnhardt G (2003) Processing of dipole and more complex hydrodynamic stimuli under still- and running-water conditions. In: Collin SP, Marshall NJ (eds), Sensory Processing in Aquatic Environments. New York: Springer-Verlag, pp. 108–121.

Bodznick D (1989) Comparisons between electrosensory and mechanosensory lateral line systems. In: Coombs S, Görner P, Münz H (eds), The Mechanosensory Lateral Line: Neurobiology and Evolution. New York: Springer-Verlag, pp. 653–678.

Bodznick D, Boord RL (1986) Electroreception in chondricthyes: central anatomy and Physiology. In: Bullock TH, Heilegenberg W (eds), Electroreception. New York: John Wiley & Sons, pp. 225–256.

Bodznick D, Montgomery J, Bradley DJ (1992) Suppression of common mode signals within the electrosensory system of the little skate. J Exp Biol 171:127–138.

Bodznick D, Hjelmstad G, Bennett MV (1993) Accommodation to maintained stimuli in the ampullae of Lorenzini: how an electoreceptive fish achieves sensitivity in a noisy world. Jpn J Physiol 43:S231–237.

Braun CB, Grande T (2002) Evolution of the octavolateralis system: a phylogenetic assessment. Bioacoustics 12:118–120.

Bullock TH (1981) Comparisons of the electric and acoustic senses and their central processing. In: Tavolga WN, Popper AN, Fay RR (eds), Hearing and Sound Communication in Fishes. New York: Springer-Verlag, pp. 527–571.

Bullock TH, Hagiwara S, Kusano K, Negishi K (1961) Evidence for a category of electroreceptors in the lateral line of gymnotid fishes. Science 134:1426–1427.

Bullock TH, Bodznick DA, Northcutt RA (1983) The phylogenetic distribution of electroreception: evidence for convergent evolution of a primitive vertebrate sense modality. Brain Res Rev 6:25–46.

Canfield JG, Eaton RC (1990) Swimbladder acoustic pressure transduction initiaties Mauthner-mediated escape. Nature 347:760–762.

Canfield JG, Rose GJ (1996) Hierarchical sensory guidance of Mauthner-mediated escape responses in goldfish (*Carassius auratus*) and cichlids (*Haplochromis burtoni*). Brain Behav Evol 48:137–156.

Carr CE (1986) Time coding in electric fish and barn owls. Brain Behav Evol 28:122–133.

Carr CE (1993) Processing of temporal information in the brain. Annu Rev Neurosci 16:223–243.

Carr CE, Maler L (1986) Electroreception in gymnotiform fish: central anatomy and physiology. In: Bullock TH, Heilegenberg W (eds), Electroreception. New York: John Wiley & Sons, pp. 225–256.

Carr CE, Soares D (2002) Evolutionary convergence and shared computational principles in the auditory system. Brain Behav Evol 59:294–311.

Claas B, Münz H, Zittlau KE (1989) Direction coding in central parts of the lateral line system. In: Coombs S, Görner P, Münz, H (eds), The Mechanosensory Lateral Line: Neurobiology and Evolution. New York: Springer-Verlag, pp. 409–419.

Coombs S, Fay R (1985) Adaptation effects on the detection of amplitude modulation: neurophysiological and behavioral assessment in the goldfish auditory system. Hear Res 19:57–71.

Coombs S, Montgomery JC (1999) The enigmatic lateral line system. In: Popper AN, Fay RR (eds), Comparative Hearing: Fishes and Amphibians. New York: Springer-Verlag, pp. 319–362.

Coombs S, Hastings M, Finneran J (1996) Measuring and modeling lateral line excitation patterns to changing dipole source locations. J Comp Physiol 178:359–371.

Coombs S, Braun CB, Donovan B (2001) Orienting response of Lake Michigan mottled sculpin is mediated by canal neuromasts. J Exp Biol 204:337–348.

Coombs S, New JG, Nelson M (2002) Information-processing demands in electrosensory and mechanosensory lateral line systems. J. Physiol 96:341–354.

Cortopassi, KA, Lewis ER (1998) A comparison of the linear tuning properties of two classes of axons in the bullfrog lagena. Brain Behav Evol 51:331–348.

Covey E (2000) Neural population coding and auditory temporal pattern analysis. Physiol Behav 69:211–220.

Crawford JD (1997a) Hearing and acoustic communication in mormyrid electric fishes Mar Fresh Behav Physiol 29:65–86.

Crawford JD (1997b) Feature-detecting auditory neurons in the brain of a sound-producing fish. J Comp Physiol A 180:439–450.

Denhardt G, Mauck B, Hanke W, Bleckmann H (2001) Hydrodynamic trail-following in harbor seals (*Phoca vitulina*). Science 293:102–104.

Denton EJ, Gray JAB (1983) Mechanical factors in the excitation of clupeid lateral lines. Proc Soc Lond B 218:1–26.

Denton EJ, Gray JAB (1988) Mechanical factors in the excitation of the lateral line of fishes. In: Atema J, Fay RR, Popper AN, Tavolga WN (eds), Sensory Biology of Aquatic Animals. New York: Springer-Verlag, pp. 595–618.

de Vries HI (1950) The mechanics of the labyrinth otoliths. Acta Oto-Laryngol 38:262–273.

Dijkgraaf S (1963) The functioning and significance of the lateral line organs. Biol Rev 38:51–105.

Edds-Walton PL (1998) Anatomical evidence for binaural processing in the descending octaval nucleus of toadfish (*Opsanus tau*). Hear Res 123:41–54.

Eisthen HL, Nishikawa KC (2002) Evolutionary convergence as a tool in neuroscience. Brain Behav Evol 59 (entire issue).

Engelmann J, Hanke W, Bleckmann H (2002) Lateral line reception in still- and running water. J Comp Physiol A 188:513–526.

Fay RR (1997) Frequency selectivity of saccular afferents of the goldfish revealed by revcor analysis. In: Lewis ER, Long GR, Lyon RF, Narins PM, Steele CR, Hecht-Poinar, E (eds), Diversity in Auditory Mechanics. Singapore: World Scientific pp. 69–75.

Fay RR, Edds-Walton P (1997a) Directional response properties of saccular afferents of the toadfish, *Opsanus tau*. Hear Res 111:1–21.

Fay RR, Edds-Walton PL (1997b) Diversity in frequency response properties of saccular afferents of the toadfish, *Opsanus tau*. Hear Res 113:235–246.

Fay RR, Popper AN (2000) Evolution of hearing in vertebrates: the inner ears and processing. Hear Res 149:1–10.

Fessard A, Szabo T (1961) Mise en evidence d'un recepteur sensible a l'electricite dans la peau d'un mormyre. CR Acad Sci 243:1859–1860.

Fields DM, Yen J (2002) Fluid mechanosensory stimulation of behaviour from a planktonic marine copepod, *Euchaeta rimana* Bradford. J Plankton Res 24:747–755.

Fine ML, Winn HE, Olla BL (1975) Communication in fishes. In: Sebeok T (ed), How Animals Communicate. Bloomington, IN: Indiana University Press.

Finger TE (1986) Electroreception in catfish: behavior, anatomy and electrophysiology. In: Bullock TH, Heilegenberg W (eds), Electroreception. New York: John Wiley & Sons, pp. 287–317.

Flock A (1965) Electron microscopic and electrophysiological studies on the lateral-line canal organ. Acta Oto Laryngol (Suppl) 199:1–90.

Gabbiani F, Metzner W (1999) Encoding and processing of sensory information in neuronal spike trains. J Exp Biol 202:1267–1279.

Gray JAB, Denton EJ (1991) Fast pressure pulses and communication between fish. J Mar Biol Assoc UK 71:83–106.

Hagedorn M (1986) The ecology, courtship, and mating of gymnotiform electric fish. In: Bullock TH, Heiligenberg W (eds), Electroreception. New York: John Wiley & Sons, pp. 497–525

Hanke W, Bruecker C, Bleckmann H (2000) The ageing of low-frequency water disturbances caused by swimming goldfish and its possible relevance to prey detection. J Exp Biol 203:1193–1200.

Harden Jones F, Arnold, GP, Greer-Walker M, Scholes P (1979) Selective tidal stream transport and the migration of plaice (*Pleuronectes platessa* L.) in the southern North Sea. J Cons Int Explor Mer 38:331–337.

Harris GG, van Bergeijk WA (1962) Evidence that the lateral line organ responds to nearfield displacments of sound sources in water. J Acous Soc Am 34:1831–1841.

Hassan El-S (1985) Mathematical analysis of the stimulus for the lateral line organ. Biol Cybern 52:23–36.

Heiligenberg W (1986) Jamming avoidance responses: model systems for neuroethology. In: Bullock T, Heiligenberg W (eds), Electroreception. New York: John Wiley & Sons, pp. 613–649.

Hoekstra D, Janssen J (1985) Non-visual feeding behavior of the mottled sculpin, *Cottus bairdi*, in Lake Michigan. Environ Biol of Fishes 12:111–117.

Hopkins CD (1976) Stimulus filtering and electroreception: tuberous electroreceptors in three species of gymnotoid fish. J Comp Physiol 111:171–207.

Hopkins CD (1988) Neuroethology of electric communication. Annu Rev Neurosci 11: 497–535.

Hopkins CD (1999) Design features for electric communication. J Exp Biol 202:1217–1228.

Hudspeth A.J, Choe Y, Mehta AD, Martin P (2000) Putting ion channels to work: mechanoelectrical transduction, adaptation, and amplification by hair cells. Proc Natl Acad Sci USA 97:11765–11772.

Kalmijn AJ (1971) The electric sense of sharks and rays. J Exp Biol 55:371–383.

Kalmijn AJ (1974) The detection of electric fields from inanimate and animate sources other than electric organs. In: Fessard A (ed), Handbook of Sensory Physiology, Vol. III/3. Berlin: Springer-Verlag, pp. 147–200.

Kalmijn AJ (1984) Theory of electromagnetic orientation: a further analysis. In: Bolis A, Keynes RD, Madrell SHP (eds) Comparative Physiology of Sensory Systems, Cambridge, MA: Cambridge University Press, pp. 525–560.

Kalmijn AJ (1988a) Hydrodynamic and acoustic field detection. In: Atema J, Fay RR, Popper AN, Tavolga WN (eds), Sensory Biology of Aquatic Animals. New York: Springer-Verlag, pp. 83–130.

Kalmijn AJ (1988b) Detection of weak electric fields. In: Atema J, Fay RR, Popper AN, Tavolga WN (eds), Sensory Biology of Aquatic Animals. New York: Springer-Verlag, pp. 151–188.

Kalmijn AJ (1989) Functional evolution of lateral line and inner ear sensory systems. In: Coombs S, Görner P, Münz H (eds), The Mechanosensory Lateral Line: Neurobiology and Evolution. New York: Springer-Verlag, pp. 187–215.

Kalmijn AJ (2000) Detection and processing of electromagnetic and near-field acoustic signals in elasmobranch fish. Philos Trans R Soc Lond B 355:1135–1141.

Kanter M, Coombs S (2003) Rheotaxis and prey detection in uniform currents by Lake Michigan mottled sculpin (*Cottus bairdi*). J Exp Biol 206:59–60.

Kawasaki M (2000) Phylogenetic evolution of computational algorithms: nonparametric approach to knowledge. Discovery 8:77–80.

Ladich F (1999) Did auditory sensitivity and vocalization evolve independently in otophysan fishes? Brain Behav Evol 53:288–304.

Ma WLD and Fay RR (2002) Neural representations of the axis of acoustic particle motion in nucleus centralis of the torus semicircularis of the goldfish, *Carassius auratus*. J Comp Physiol A 188:301–313.

Maler L, Mugnaini E (1994) Correlating gamma-aminobutyric acidergic circuits and sensory function in the electrosensory lateral line lobe of a gymnotiform fish. J Comp Neurol 345:224–252.

Mann DA, Higgs DM, Tavolga WN, Souza MJ, Popper AN (2001). Ultrasound detection by clupeiform fishes. J Acoust Soc Am 109:3048–3054.

McCormick CA (1999) Anatomy of the central auditory pathway of fish and amphibians. In: Fay RR, Popper AN (eds), Comparative Hearing: Fishes and Amphibians. New York: Springer-Verlag, pp. 155–217.

McCormick CA, Braford MR Jr (1988) Central connections of the octavolateralis system: evolutionary considerations. In: Atema J, Fay RR, Popper AN, Tavolga WN (eds), Sensory Biology of Aquatic Animals. New York: Springer-Verlag, pp. 341–364.

McKibben JR, Bass AH (1998) Behavioral assessment of acoustic parameters relevant to signal recognition and preference in a vocal fish. J. Acoust Soc Am 104:3520–3533.

McKibben JR, Hopkins CD, Yager DD (1993) Directional sensitivity of tuberous electroreceptors: polarity preferences and frequency tuning. J Comp Physiol A 173:415–424.

Meek J, Grant K, Bell C (1999) Structural organization of the mormyrid electrosensory lateral line lobe. J Exp Biol 202:1291–1300.

Montgomery JC (1984) Noise cancellation in the electrosensory system of the thornback ray: common mode rejection of input produced by the animal's own ventilatory movement. J Comp Physiol 155A:102–111.

Montgomery J, Coombs S (1992) Physiological characterization of lateral line function in the Antarctic fish, *Trematomus bernacchii*. Brain Behav Evol 40:209–216.

Montgomery JC, Coombs S (1998) Peripheral encoding of moving sources by the lateral line system of a sit-and-wait predator. J Exp Biol 201:91–102.

Montgomery JC, Hamilton AR (1997) Sensory contributions to nocturnal prey capture in the dwarf scorpion fish (*Scorpaena papillosus*). Mar Fresh Behav Physiol 30:209–223.

Montgomery JC, Milton RC (1993) Use of the lateral line for feeding in torrentfish (*Cheimarrichthys fosteri*). New Zeal J Zool 20:121–125.

Montgomery JC, Coombs S, Conley RA, Bodznick D (1995) Hindbrain sensory processing in lateral line, electrosensory and auditory systems: a comparative overview of anatomical and functional similarities. Audit Neurol 1:207–231.

Montgomery JC, Baker CF, Carton AG (1997) The lateral line can mediate rheotaxis in fish. Nature 389:960–963.

Montgomery JC, McDonald F, Baker CF, Carton AG (2002) Hydrodynamic contributions to multimodal guidance of prey capture behavior in fish. Brain Behav Evol 4:190–198.

Montgomery JC, McDonald F, Baker CF, Carton AG, Ling N (2003) Sensory integration in the hydrodynamic world of rainbow trout. Proc R Soc Lond B Biol Sci 270 (suppl 2):195–197.

Moulton JM (1960) Swimming sounds and the schooling of fish. Biol Bull Mar. Biol Lab Woods Hole 119:210–223.

Münz H (1979) Morphology and innervation of the lateral line system in *Sarotherodon niloticus* L. (Cichlidae, Teleostei). Zoomorpholgie 93:73–86.

Münz H (1985) Single unit activity in the peripheral lateral line system of the cichlid fish, *Sarotherodon niloticus* L. J Comp Physiol A 157:555–568.

Münz H (1989) Functional organization of the lateral line periphery. In: Coombs S, Görner P, Münz H (eds), The Mechanosensory Lateral Line: Neurobiology and Evolution. New York: Springer-Verlag, pp. 285–298.

Myrberg AA (1981) Sound communication and interception in fishes. In: Tavolga WN,

Popper AN, Fay RR (eds), Hearing and Sound Communiation in Fishes. New York: Springer-Verlag, pp. 395–426.

Myrberg AA Jr, Ha SJ, Walewski S, Banbury JC (1972) Effectiveness of acoustic signals in attracting epipelagic sharks to an underwater sound source. Bull Mar Sci 22:926–949.

Nelson ME, MacIver MA (1999) Prey capture in the weakly electric fish *Apternonotus albifrons*: sensory acquisition strategies and electrosensory consequences. J Exp Biol 202:1195–1203.

New JG (1997) The evolution of vertebrate electrosensory systems. Brain Behav Evol 50:244–252.

New JG, Bodznick D (1990) Medullary electrosensory processing in the little skate. II. Suppression of electrosensory reafference via a common-mode rejection mechanism. J Comp Physiol A 167:295–307.

New JG, Singh S (1994) Central topography of anterior lateral line nerve projections in the channel catfish, *Ictalurus punctatus*. Brain Behav Evol 43:34–50.

New JG, Coombs S, McCormick CA, Oshel P (1996) The cytoarchitecture and internal organization of the medial octavolateralis nucleus in the goldfish, *Carassius auratus*. J Comp Neurol 364:1–13.

Northcutt RG (1989) The phylogenetic distribution and innervation of craniate mechanoreceptive lateral lines. In: Coombs S, Görner P, Münz H (eds), The Mechanosensory Lateral Line: Neurobiology and Evolution. New York: Springer-Verlag, pp. 17–78.

Partridge BL, Pitcher TJ (1980) The sensory basis of fish schools: relative roles of lateral line and vision. J Comp Physiol A 135:315–325.

Paulin MG (1995) Electroreception and the compass sense of sharks. J Theor Biol 174: 325–339.

Pavolov DS, Tjurjukov SN (1995) Reactions of dace to linear accelerations J Fish Biol 46:768–774.

Peters RC, van Wijland F (1974) Electro-orientation and the passive electric catfish, *Ictalurus nebulosus*. J Comp Physiol 92:273–280.

Platt C, Popper AN, Fay RR (1989) The ear as part of the octavolateralis system. In: Coombs S, Görner P. Münz H (eds), The Mechanosensory Lateral Line: Neurobiology and Evolution. New York: Springer-Verlag, pp. 633–651.

Pohlmann K, Grasso FW, Briethaupt T (2001) Tracking wakes: the nocturnal predatory strategy of piscivorous catfish. Proc Natl Acad Sci USA 98:7371–7374.

Popper AN, Coombs S (1982) The morphology and evolution of the ear in Actinopterygian fishes. Am Zool 22:311–328.

Popper AN, Fay RR (1999) The auditory periphery in fishes. In: Fay RR, Popper AN (eds), Comparative Hearing: Fish and Amphibians. New York: Springer-Verlag, pp. 43–100.

Popper AN, Platt C, Edds PL (1992) Evolution of the vertebrate inner ear: an overview of ideas. In: Webster DB, Fay RR, Popper AN (eds), The Evolutionary Biology of Hearing. New York: Springer-Verlag, pp. 49–61.

Roberts BL, Meredith GE (1989) The efferent system. In: Coombs S, Görner P, Münz H (eds), The Mechanosensory Lateral Line: Neurobiology and Evolution. New York: Springer-Verlag, pp. 445–460.

Rogers PH, Cox M (1988) Underwater sound as a biological stimulus. In: Atema J, Fay RR, Popper AN, Tavolga WN (eds), Sensory Biology of Aquatic Animals. New York: Springer-Verlag, pp. 131–150.

Sand O (1981) The lateral line and sound reception. In: Tavolga WN, Popper AN, Fay

RR (eds), Hearing and Sound Communication in Fishes. New York: Springer-Verlag, pp. 459–480.

Satou M, Shiraishi A, Matsushima T, Okumoto N (1991) Vibrational communication during spawning behavior in the Hime salmon (landlocked red salmon, *Oncorhynchus nerka*). J Comp Physiol A 168:417–428.

Saunders AJ, Montgomery JC (1985) Field and laboratory studies of the feeding behaviour of the piper *Hyporhamphus ihi* with reference to the role of the lateral line in feeding. Proc R Soc Lond B 224:209–221.

Schellart AM, Kroese ABA (1989) Interrelationship of acousticolateral and visual systems in the teleost midbrain. In Coombs S, Görner P, Münz H (eds), The Mechanosensory Lateral Line: Neurobiology and Evolution. New York: Springer-Verlag, pp. 421–443.

Schuijf A (1975) Directional hearing of cod (*Gadus morhua*) under approximate free field conditions. J Comp Physiol 98:307–332.

Shumway C (1989a) Multiple electrosensory maps in the medulla of weakly electric gymnotiform fish. I. Physiological differences. J Neurosci 9:4388–4399.

Shumway C (1989b) Multiple electrosensory maps in the medulla of weakly electric gymnotiform fish. II. Anatomical differences. J Neurosci 9:4400–4415.

Stein BE, Meredith MA (1993) The Merging of the Senses. Cambridge, MA: MIT Press.

Sutterlin AM, Waddy S (1975) Possible role of the posterior lateral line in obstacle entrainment by brook trout (*Salvelinus fontinalis*). J Fish Res Board Can 32:2441–2446.

Tavolga WN (1975) Acoustic obstacle detection in the sea catfish, *Galeichthys felis* In: Schuiff A, Hawkins AD (eds), Sound Reception in Fish. Amsterdam: Elsevier, pp. 185–204.

Tolimieri N, Jeffs A, Montgomery J (2000) Ambient sound as a cue for navigation by the pelagic larvae of reef fish. Mar Ecol Prog 207:219–224.

Tricas TC (1982) Bioelectric-mediated predation by swell sharks, *Cephaloscyllium ventriosum*. Copeia 1982:948–952.

Tricas TC, Highstein SM (1991) Action of the octavolateralis efferent system upon the lateral line of free-swimming toadfish, *Opsanus tau*. J Comp Physiol A 169:25–37.

Tricas TC, Michael SM, Sisneros JA (1995) Electrosensory optimization to conspecific phasic signals for mating. Neurosci Lett 202:129–132.

Turner RW, Maler L (1999) Oscillatory and burst discharge in the apteronotid electrosensory lateral line lobe. J Exp Biol 202:1255–1265.

van Bergeijk WA (1964) Directional and non-directional hearing in fish. In: Tavolga WN: Marine Bioacoustics Oxford: Pergamon Press, pp. 185–204.

Vogel S (1994) Life in Moving Fluids: The Physical Biology of Flow. Princeton: Princeton University Press.

Voigt R, Carton AG, Montgomery JC (2000) Responses of lateral line afferent neurones to water flow. J Exp Biol 203:2495–2502.

von der Emde G (1999) Active electrolocation of objects in weakly electric fish. J Exp Biol 202:1205–1215.

von der Emde G, Schwarz S, Gomez L (1998) Electric fish measure distance in the dark. Nature 395:890–894.

Webb JF, Smith WL (2000) The laterophysic connection in chaetodontid butterflyfish: morphological variation and speculations on sensory function. Philos Trans R Soc Lond B 355:1125–1129.

Weissert R, von Campenhausen C (1981) Discrimination between stationary objectes by the blind cave fish (*Anoptichthys jordani*: Characidae). J Comp Physiol 143:375–381.

Westby GWN (1988) The ecology, discharge diversity and predatory behavior of gymnotiform electric fish in the coastal streams of French Guina. Behav Ecol Sociobiol 22:341–354.

Wever EG (1976) Origin and evolution of the ear of vertebrates. In: Masterton RB, Bitterman ME, Campbell, CBG, Hotton N (eds), Handbook of Sensory Physiology, Vol V/I: Auditory System. Berlin: Springer-Verlag, pp. 423–454.

Wilkens LA, Wettring B, Wagner E, Wojtenek W, Russel D (2001) Prey detection in selective plankton feeding by the paddlefish: is the electric sense sufficient? J Exp Biol. 204:1381–1389.

Wojtenek W, Hofmann MH, Wilkens LA (2001) Primary afferent electrosensory neurons represent paddlefish natural prey. Neurocomputing 38–40:451–458.

Yager DD, Hopkins CD (1993) Directional characteristics of tuberous electroreceptors in the weakly electric fish, *Hypopomus* (Gymnotiformes). J Comp Physiol A 143:401–414.

Zakon H (1986) The electroreceptive periphery. In: Bullock TH, Heiligenberg W (eds), Electroreception. New York: John Wiley & Sons, pp. 103–156.

Zakon H., Lu Y, Weisleder P (1998) Sensory cells determine afferent terminal morphology in cross-innervated electroreceptor organs: implications for hair cells. J Neurosci 18:2581–2591.

Zorn T, Coers S, New JG (1998) Parallel commissural electrosensory pathways in the medulla of the channel catfish, *Ictalurus punctatus*. In: Proceedings of the Fifth International Congress on Neuroethology, p. 34.

13

Diversity and Phylogeny of Neotropical Electric Fishes (Gymnotiformes)

James S. Albert and William G.R. Crampton

1. Introduction to Gymnotiform Diversity

The evolutionary radiations of Neotropical electric fishes (Gymnotiformes) provide unique materials for studies on the evolution of specialized sensory systems and the diversification of animals species in tropical ecosystems (Hopkins and Heiligenberg 1978; Heiligenberg 1980; Heiligenberg and Bastian 1986; Moller 1995a; Crampton 1998a; Stoddard 1999; Albert 2001, 2002). The teleost order Gymnotiformes is a clade of ostariophysan fishes most closely related to catfishes (Siluriformes), with which they share the presence of a passive electrosensory system (Fink and Fink 1981, 1996; Finger 1986). Gymnotiformes also possess a combined electrogenic–electroreceptive system that is employed for both active electrolocation, the detection of nearby objects that distort the self-generated electric field, and also electrocommunication, the signaling of identity or behavioral states and intentions to other fishes (Carr and Maler 1986). Active electroreception allows gymnotiforms to communicate, navigate, forage, and orient themselves relative to the substrate at night and in dark, sediment-laden waters, and contributes to their ecological success in Neotropical aquatic ecosystems (Crampton and Albert 2005). The species-specific electric signals of gymnotiform fishes allow investigations of behavior and ecology that are simply unavailable in other groups. Because these signals are used in both navigation and mate recognition (i.e., prezygotic reproductive isolation) they play central roles in the evolutionary diversification and ecological specialization of species, as well as the accumulation of species into local and regional assemblages.

The derived features of electrosensory and electrogenic structures not withstanding, patterns of diversity in gymnotiform fishes are similar to those of many diverse tropical taxa. Gymnotiform diversity is especially pronounced at the species level, and the group is considerably more diverse than has previously been recognized. About 78%, or 135 of the 173 known gymnotiform species, have been formally described, and perhaps half again as many species remain undiscovered in the wild. Gymnotiforms are widely distributed throughout the humid Neotropics, from southern Mexico (15°N) to northern Argentina (36°S), with representatives in Middle and South America, and in both cis- and trans-Andean basins. The Amazon–Orinoco–Guianas superbasin is the center of diversity for the group, from where the majority of gymnotiform genera (77%) and species (73%) may be found. These geographic distributions allow comparisons of divergence times calculated from independent molecular and geological datasets, and the testing of hypotheses regarding the role of geography in the origin and accumulation of species diversity. Gymnotiform clades are ancient and many genera are distributed in polyphyletic regional species assemblages.

A central theme of this chapter is that an evolutionary understanding of electric signaling in gymnotiform fishes requires knowledge of signal diversity, ecology, biogeography, and phylogenetic interrelationships—all at the species level. As in the evolutionary analysis of any taxon the main features in the phylogeny of electric signaling arise from patterns in cladogenesis (speciation) and anagenesis (adaptive change), processes that operate within and among species. In this chapter we review the current state of knowledge about gymnotiform taxonomic and species diversity and the diversity of phenotypic specializations associated with electrogenesis and active electroreception. These primary taxonomic and phylogenetic observations are used to examine patterns of evolution in body size and shape, of the electric organs and electric organ discharges (EODs), and of habitat use. We conclude this chapter with a review of recent work on the biogeography and historical ecology of gymnotiforms and a summary of recent findings on the origins and maintenance of species-rich gymnotiform faunas.

2. Taxonomic and Species Diversity

The order Gymnotiformes is considerably more diverse than has previously been recognized, with the number of valid species having increased from 94 to 135 in the past 10 years (Fig. 13.1). New methods of sampling, identifying, and collecting electric fishes in the wild have unveiled numerous species in previously unexplored habitats and regions (Crampton 1996a, 1998a; Hagedorn and Keller 1996; Lundberg et al. 1996; Albert and Crampton 2001; Albert et al. 2005a). The use of new techniques for characterizing morphology, and genetic differences have demonstrated that much of what was once regarded as intraspecific variation in fact represents interspecific differences (Albert and Miller 1995; Campos-da-Paz and Costa 1996; Fernandes-Matioli et al. 1998a; b, 2000; Albert et al. 1999; Albert and Crampton 2001, 2003a; Fernandes-Matioli and de Almeida-Toledo 2001; Crampton and Albert 2003; Crampton et al. 2004a, b, 2005). Compilations of the numbers of valid (described) and manuscript (undescribed) gymnotiform genera and species are provided in Tables 13.1 and 13.2. Many of the undescribed species known from museum collections are being described as of this writing. Extrapolating from current rates of discovery

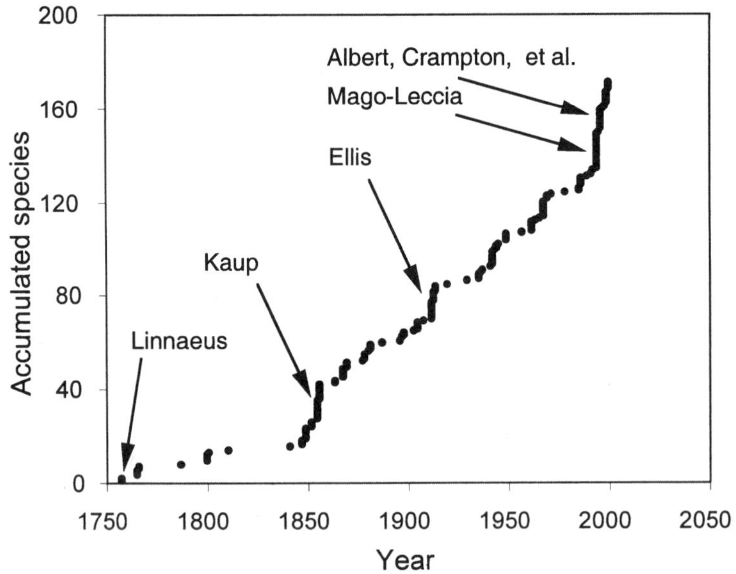

FIGURE 13.1. Accumulation of gymnotiform species from Linnaeus to the present. Note the curve is not approaching an asymptotic value, indicating the relatively incomplete state of knowledge of gymnotiform species diversity.

50 to 100 additional species are anticipated from newly explored field localities in tropical America.

Histories of the classification of gymnotiform fishes are provided in Campos-da-Paz and Albert (1998) and Albert (2001). A key to 28 genera is provided in Albert (2001) and diagnostic features of three genera subsequently recognized as valid are provided in the original citations (*Megadontognathus,* Mago-Leccia 1994; *Stegostenopos,* Triques 1997; *Humboldtichthys,* Gayet and Meunier 2000).

TABLE 13.1. Gymnotiform families with genus and species diversity estimates.

		Species			
Family	Genera	Valid	MS	Total	Spp./Genus
Apteronotidae	14	45	16	61	4.7
Gymnotidae	2	33	3	36	18.0
Hypopomidae	7	16	13	29	4.1
Rhamphichthyidae	3	12	0	12	4.0
Sternopygidae	6	29	6	35	5.8
Total	32	135	38	173	7.3

Valid, published and not junior synonym. MS, manuscript names (undescribed). Taxa are arranged alphabetically.

TABLE 13.2. Gymnotiform genera with original citation and species diversity estimates.

Family	Genus	Author	Date	Species Valid	Species MS	Species Total
Apteronotidae	*Adontosternarchus*	Ellis	1912	4	2	6
	Apteronotus s.s.	Lacépède	1800	13	2	15
	"*Apteronotus*"		—	6	2	8
	Compsaraia	Albert	2001	1	2	3
	Magosternarchus	Lundberg, Cox, and Albert	1996	2	0	2
	Megadontognathus	Mago-Leccia	1994	2	0	2
	Orthosternarchus	Ellis	1912	1	0	1
	Parapteronotus	Albert	2001	1	0	1
	Platyurosternarchus	Mago-Leccia	1994	1	0	1
	Porotergus	Ellis	1912	2	0	2
	Sternarchella	Eigenmann	1905	4	2	6
	Sternarchogiton	Eigenmann and Ward	1905	1	2	3
	Sternarchorhamphus	Eigenmann	1905	1	0	1
	Sternarchorhynchus	Castelnau	1855	6	3	9
	n. Gen.	—	—	0	1	1
Gymnotidae	*Electrophorus*	Gill	1864	1	0	1
	Gymnotus	Linnaeus	1758	32	3	35
Hypopomidae	*Brachyhypopomus*	Mago-Leccia	1994	7	12	19
	Hypopomus	Gill	1864	1	0	1
	Hypopygus	Hoedeman	1962	2	1	3
	Microsternarchus	Fernàndez-Yépez	1968	1	0	1
	Racenisia	Mago-Leccia	1994	1	0	1
	Stegostenopos	Triques	1997	1	0	1
	Steatogenys	Boulenger	1898	3	0	3
Rhamphichthyidae	*Gymnorhamphichthys*	Ellis	1912	4	0	4
	Iracema	Triques	1996b	1	0	1
	Rhamphichthys	Müller and Troschel	1848	7	0	7
Sternopygidae	*Archolaemus*	Korringa	1970	1	0	1
	Distocyclus	Mago-Leccia	1978	2	0	2
	Eigenmannia	Jordan and Evermann	1896	8	3	11
	Humboldtichthys†	Gayet and Meunier	2000	1	0	1
	Rhabdolichops	Eigenmann and Allen	1942	8	2	10
	Sternopygus	Müller and Troschel	1849	9	1	10
Total				135	38	173

MS, manuscript names (undescribed). Taxa are arranged alphabetically by family and genus.

Gymnotiform families are recognized on the basis of the presence or absence of a caudal fin, dorsal organ (i.e., "dorsal thong"), oral teeth, and the shape of the head and snout. The families Gymnotidae and Electrophoridae were combined by Albert and Campos-da-Paz (1998). Salient diagnostic characters for families and genera are illustrated in Figure 13.2.

Many gymnotiform genera are recognized from features of head morphology and the oral jaws, especially the relative shape and proportions of the snout and mouth. As a result, the extent to which the current taxonomy of gymnotiform genera expresses morphological diversity strongly emphasizes trophic speciali-

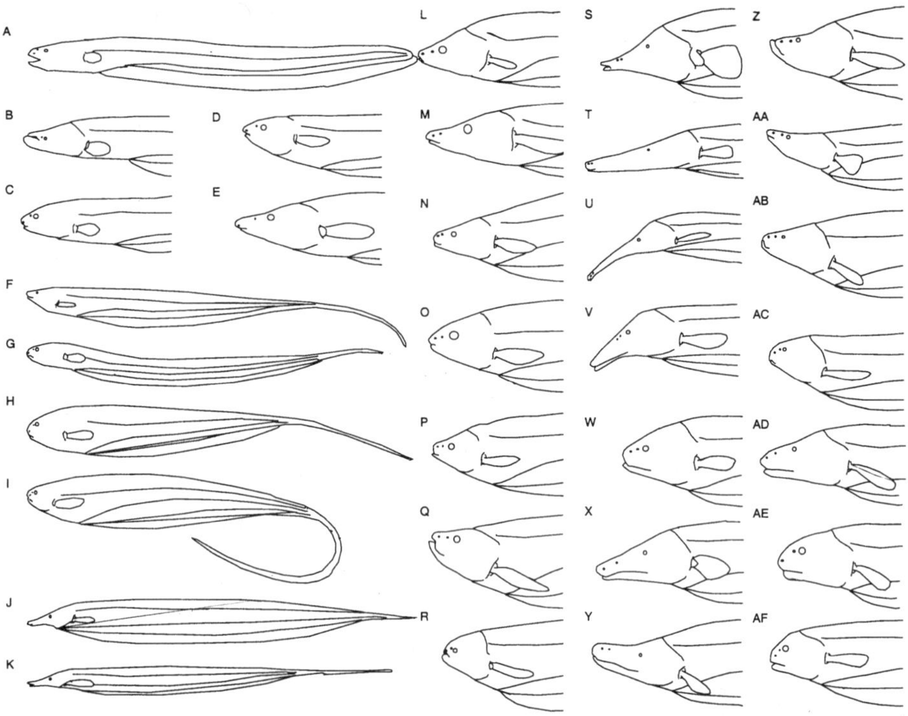

FIGURE 13.2. Line drawings of specimens representing 27 gymnotiform genera in profile, illustrating some important features used in identifications. Drawings not to scale. (**A**) *Electrophorus electricus*, (**B**) *Gymnotus mamiraua*, (**C**) *Microsternarchus bilineatus* (head), (**D**) *Brachyhypopomus occidentalis*, (**E**) *Hypopomus artedi*, (**F**) *M. bilineatus* (body), (**G**) *Racenisia fimbriipinna*, (**H**) *Hypopygus lepturus*, (**I**) *Steatogenys elegans*, (**J**) *Rhamphichthys marmoratus*, (**K**) *Gymnorhamphichthys rondoni*, (**L**) *Sternopygus xingu*, (**M**) *Archolaemus blax*, (**N**) *Distocyclus conirostris*, (**O**) *Eigenmannia humboldtii* (200 mm), (**P**) *E. humboldtii* (350 mm), (**Q**) *Rhabdolichops troscheli*, (**R**) *Adontosternarchus sachsi*, (**S**) *Sternarchorhamphus muelleri*, (**T**) *Orthosternarchus tamandua*, (**U**) *Sternarchorhynchus oxyrhynchus*, (**V**) *Platyurosternarchus macrostomus*, (**W**) *Apteronotus albifrons*, (**X**) *Apteronotus leptorhynchus*, (**Y**) *Parapteronotus hasemani* (mature male), (**Z**) *Magosternarchus duccis*, (**AA**) *Magosternarchus raptor*, (**AB**) *Sternarchella schotti*, (**AC**) *Sternarchella sima*, (**AD**) *Compsaraia compsa*, (**AE**) *Porotergus gimbeli*, (**AF**) *Sternarchogiton nattteri*.

zations. This bias may be observed in the numbers of species per genus among gymnotiform families, in which the Gymnotidae is much less divided into genera than are the other families (Table 13.1). Diversity of head and snout morphology also results from pronounced sexual dimorphism in many species of Apteronotidae in which individuals engage in male–male conflict. Sexual dimorphism in Apterontoidae has historically resulted in an overestimation of the number of certain apteronotid genera and species; for example, the nominal taxon *"Oedemognathus exodon"* is now known to a sexually mature male of *Sternarchogiton nattereri,* and the nominal taxon *"Apteronotus anas"* is a sexually mature male of *Parapteronotus hasemani* (Cox-Fernandes 1998a; Albert 2001; Cox-Fernandes et al. 2002).

2.1 Gymnotidae

Gymnotids may be distinguished from other gymnotiforms by the following unique combination of characters: a cylindrical body with the adult body depth 55% to 90% the body width at the anal-fin origin, the absence of cranial fontanels on the dorsal surface of the head of adults, a very long body cavity with 31 to 51 (more than 100 in *Electrophorus*) precaudal vertebrae, and the absence of displaced hemal spines at the posterior end of the body cavity. The Gymnotidae is represented by two genera. *Gymnotus* is the most diverse gymnotiform genus, with 32 described and at least three additional undescribed species (Campos-da-Paz 2003; Albert and Crampton 2003a; Albert and Crampton 2003a; Albert et al. 2005b). *Gymnotus* species occur in all major river systems in the humid Neotropics and inhabit a wide variety of lowland aquatic habitats. *Gymnotus* is the most geographically widespread of all gymnotiform genera, including the full range of the order (Albert and Crampton 2003a). The type species *Gymnotus carapo* (L) is distributed throughout the Amazon and Orinoco Basins (below 500 m), the Island of Trinidad, the drainages of the Guyanas Shield, and the Atlantic drainages of northeastern Brazil. Adult body size in *Gymnotus* ranges almost an order of magnitude, from 80 to 160 mm in mature specimens of *G. coropinae* from the Amazon Basin, to 1 m in *G. inaequilabiatus* from the Rio Paraná drainage. *Gymnotus* species are all aggressive nocturnal predators of fishes and other small aquatic animals, and most are also territorial (Black-Cleworth 1970; personal observation). Adult males of the small-bodied species *G. coropinae* (90 to 120 mm total length) guard territories in undercut banks of small rainforest streams, spaced at intervals of about 1.0 to 1.5 m (J. Albert and P. Moller, personal observation). The males of at least three *Gymnotus* species form nests and guard larvae (Crampton and Hopkins 2005). *Gymnotus carapo* is reported to mouth brood its eggs and larvae (Kirschbaum and Wieczorek 2002). Most if not all *Gymnotus* species utilize aerial respiration in hypoxic conditions (Liem et al. 1984; Crampton 1998b).

The monotypic *Electrophorus electricus* is unique among gymnotiforms in possessing a strong electric discharge of up to 600 V in mature specimens (Bennett 1971). *Electrophorus electricus* is also unique among gymnotiforms in possessing a vascularized oral respiratory organ, a body cavity extending to the caudal tip of the body (i.e., no postcelomic "tail"), the continuous addition of vertebrae throughout life, and three anatomically distinct hypaxial electric organs; the Main, Hunter's, and Sachs organs. *Electrophorus electricus* grows to the largest body size among gymnotiforms, attaining a total length more than 2

m. A specimen of 7 feet 4 inches (2.24 m) is reported by Ellis (1913) from Guyana. Males achieve larger body size, attaining reproductive maturity at about 120 cm total length, whereas females mature at about 70 cm (Assunção and Schwassman 1995). Specimens of *Electrophorus* have been maintained alive in aquaria for more than 20 years.

2.2 *Hypopomidae*

Hypopomids may be distinguished from other gymnotiforms by the following unique combination of characters: short snout, no oral teeth, tubelike infraorbital bones, anterior nares located outside gape; anal-fin origin below or posterior to pectoral-fin base; no caudal fin or dorsal organ. The family Hypopomidae is represented by seven genera. *Brachyhypopomus* is the most diverse hypopomid genus, with 7 described and at least 12 additional species as yet undescribed from localities throughout tropical South America and Panama (Albert and Crampton 2003b). *Brachyhypopomus* species occur in all major river systems in the humid Neotropics and inhabit a wide variety of lowland aquatic habitats. *Hypopomus* is known from a single species, *H. artedi*, which is endemic to the Guyanas Shield. *Hypopygus* is represented by three species; *H. lepturus* from the Amazon, Orinoco, and Guianas regions; *H. neblinae* from the Guianas, Rio Negro, and Amazon Basins; and a recently discovered new species from the Venezuelan Amazon (Crampton and Albert, personal observation). *Hypopygus* is the smallest-bodied gymnotiform taxon, attaining reproductive maturity at 50 to 90 mm total length (Nijssen and Isbrücker 1972; Crampton and Albert, personal observation). *Microsternarchus* is known from a single described species, *M. bilineatus*, which is distributed in the Amazon, Orinoco, and Guianas basins. Substantial genetic variation has been reported in both *Hypopygus* and *Microsternarchus* and the actual species diversity of these taxa is underappreciated (Aadland et al. 2003). *Racenisia* is represented by a single described species, *R. fimbriipinna*, known only from the Guianas region of Venezuela and Brazil (Mago-Leccia 1994; F. Lima, personal communication). *Stegostenopos* is represented by a single described species, *S. cryptogenys*, which is known from the Rio Negro basin of Brazil (Triques 1997) and Venezuela (Crampton personal observation), and which has also been recently found in blackwater streams in the Western Amazon of Peru (Albert personal observation). All species of *Hypopygus*, *Microsternarchus*, *Racenisia*, and *Stegostenopos* are restricted to the slowly moving waters or pools of terra firme streams. *Steatogenys* is represented by three species, including *S. elegans*, which inhabits large rivers and whitewater and blackwater floodplain systems; *S. ocellatus*, which inhabits blackwater floodplain systems, and *S. duidae*, which is restricted to terra firme forest streams (Crampton et al. 2004a). Whereas *S. elegans* occurs in Amazon and Orinoco basins and parts of the Guianas, and *S. duidae* occurs in the Amazon and Orinoco Basins, *S. ocellatus* is restricted to the Upper Amazon basin (Crampton et al. 2004a).

2.3 *Rhamphichthyidae*

Rhamphichthyids may be distinguished from other gymnotiforms by the following unique combination of characters: a highly elongate snout, a small mouth, no teeth in oral jaws, infraorbital canal present as a single membranous tube, location of anterior nares entirely within gape, anterior narial pore sessile, and absence of a caudal fin or dorsal organ. The family Rhamphichthyidae is represented by three genera. *Rhamphichthys* is known from seven species, *Gymnorhamphichthys* from four described species, and *Iracema* from one described and one undescribed species (Ferraris 2003). The species-level taxonomy of *Rhamphichthys* is perhaps presently the least well understood among Gymnotiformes with confusions as to species boundaries and the nature of intraspecific variation versus interspecific differences (Campos-da-Paz and Paepke 1994; Triques 1999; Albert 2001).

2.4 *Sternopygidae*

Sternopygids may be distinguished from other gymnotiforms by the following unique combination of characters: multiple rows of small, villiform (brushlike) teeth on premaxilla and dentary; relatively large eye (diameter equal to or greater than distance between nares); large baglike infraorbital bones with expanded bony arches; anterior nares located outside the gape; anal-fin origin at isthmus; absence of urogenital papilla; no caudal fin or dorsal organ. The Sternopygidae is represented by five genera (Albert 2003b). *Archolaemus* is known from a single described species (*A. blax*) that inhabits rapids in Amazonian black and clear water rivers. *Distocyclus* is known from two species, *D. conirostris* from Amazonian white water rivers and *D. goajira* from the trans-Andean Maracaibo, Magdalena, and Baudo (Pacific slope) basins. *Eigenmannia* is the most species-rich nominal sternopygid genus, with eight described and at least three undescribed species currently known, ranging from the Pacific Slope and Magdalena basins of Colombia throughout the Orinoco–Amazon basin, to the La Plata basin of Argentina. The monophyly of *Eigenmannia* is uncertain and there is some evidence that the type species of the genus *E. virescens* is phylogenetically closer to *Rhabdolichops* than to *E. humboldtii* (Albert 2001). A species of *Eigenmannia* is the only gymnotiform known to inhabit caves (Triques 1996a). *Rhabdolichops* is known from eight described and at least two undescribed species, from the Amazon, Orinoco and Guyanas regions, all of which inhabit large rivers (Albert and Crampton, personal observation). Several *Rhabdolichops* species are planktivorous with well-developed gill rakers and others feed on small aquatic invertebrates (Lundberg and Mago-Leccia 1986; Crampton 1996b). *Sternopygus* is known from nine described at least one undescribed species. *Sternopygus* exhibits the largest geographical range of the family, extending beyond that of *Eigenmannia* into Panama and into the Rio Salgado basin in the State of Ceará in Northeastern Brazil (Albert 2001). *Sternopygus macrurus* is the most widely distributed and most eurytopic of all gymnotiform species, inhabiting all hydrogeographical regions of tropical South America and most lowland aquatic habitats (Hulen et al. 2005).

2.5 *Apteronotidae*

Apteronotids may be distinguished from other gymnotiforms by the presence of a caudal fin and a fleshy dorsal organ, the latter of which has often been interpreted as a modified adipose fin. These two features led previous workers to the conclusion that apteronotids are primitive gymnotiforms (Kaup 1856; Chardon and de la Hoz 1974; Triques 1993; Gayet et al. 1994). The family Apteronotidae is characterized by much greater morphological diversity than in other gymnotiform families (Fig. 13.2). This is expressed primarily as variation in head structure related to trophic specialization and, in part, to male–male aggression. In at least four separate apteronotid clades (*Parapteronotus hasemani*, *Apteronotus leptorhynchus* + *A. (Ubidia) magdalenensis*, *"Apteronotus" bonapartii*, and *Compsaraia* n. sp. A.) sexually mature males develop elongate jaws (Cox-Fernandes 1998a; Albert 2001; Cox-Fernandes et al. 2002). In *Sternarchogiton nattereri* and in some species of *Sternarchorhynchus*, males develop prominent external teeth on the jaws and snout.

The Apteronotidae is the most speciose family of gymnotiform fishes, with 45 species described at present, allocated to 14 genera, and at least an additional 16 undescribed forms known in museum collections (Albert 2003a). The majority of apteronotid species (47 of 61, 77%) inhabit the deep channels (up to 25 m) of large rivers, a habitat surveyed systematically only in recent years (Lundberg et al. 1987, 1996; Crampton 1996b, 1998a; Albert 2001, 2003a). As a result, many apteronotid species were until recently rare or unknown in museum collections.

3. Phylogenetic Overview of Gymnotiformes

There have been four cladistic studies of interfamily relationships among Gymnotiformes. Based on analyses of morphological data, Triques (1993) and Gayet et al. (1994) place Apteronotidae as basal among gymnotiforms, and regard the presence of the caudal fin in apteronotids as primitive. Alves-Gomes et al. (1995) present the first analysis of molecular data, using 718 aligned base pairs of ribosomal mitochondrial DNA, including 211 informative sites among 19 gymnotiform taxa. In combination with analysis of characters of the electromotor system they conclude that *Sternopygus* is the basal gymnotiform taxon, and that this taxon retains a primitive lack of a jamming avoidance response (JAR). The preferred topology of Alves-Gomes et al. (1995, fig. 6) is a strict consensus of 32 equally most parsimonious trees recovered using three weighting schemes. This topology is used in the analysis of Alves-Gomes (2001).

The phylogeny of Figure 13.3A is the result of a maximum parsimony anal-

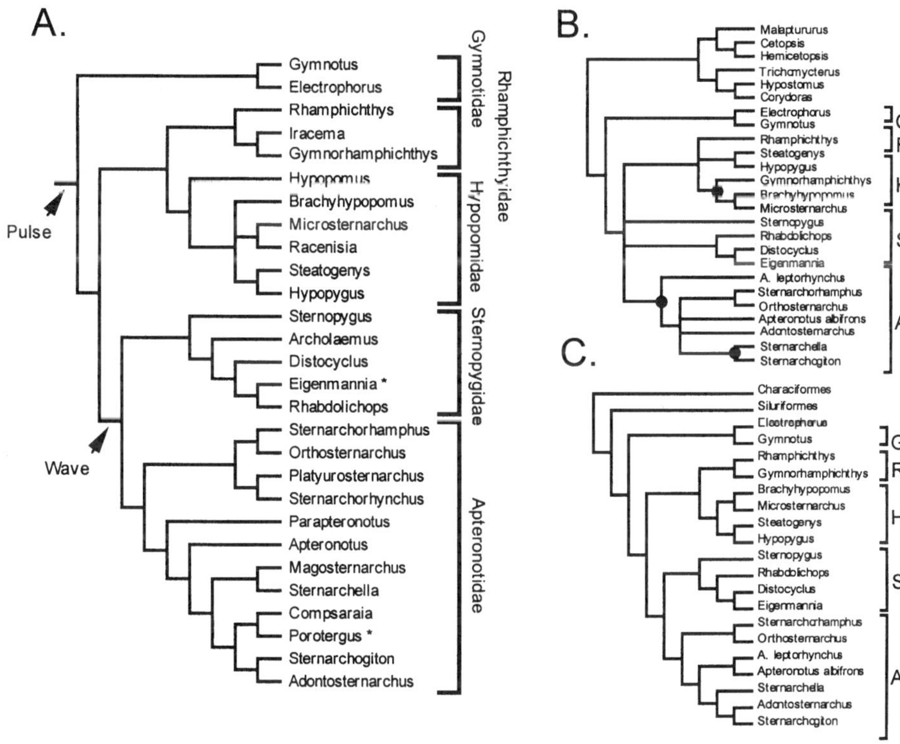

FIGURE 13.3. Alternative hypotheses of gymnotiform interrelationships. (**A**) Tree topology from maximum parsimony analysis of combined morphological and molecular data (Albert 2001). See text for details. *, Generic monophyly uncertain; A, Apteronotidae; G, Gymnotidae; H, Hypopomidae; Pulse, pulse-type EOD; R, Rhamphichthyidae; S, Sternopygidae; Wave, wave-type EOD. Note pulse-type EOD is plesiomorphic and wave-type EOD is derived. (**B**) Tree topology from maximum parsimony analysis of 12S + 16S rRNA (mt DNA). 25 OTUs, 718 bp, 178 informative sites, strict consensus of 9 trees, each l = 733, CI = 0.49, RC = 0.31. (Data from Alves-Gomes et al. 1995, topology from analysis of Albert 2001). *Black circles* indicate clades incongruent with topology of **C**. (**C**) Tree topology from maximum parsimony analysis of morphology + 12S/16S rRNA; 935 characters (CI = 0.57, RC = 0.39) (Albert 2001). Note family-level interrelationships and generic composition are the same in all three analyses.

ysis (Albert 2001) of combined morphological and molecular datasets, including all data then available (Triques 1993; Gayet et al. 1994; Alves-Gomes et al. 1995; Albert and Fink 1996; Sullivan 1997; Albert and Campos-da-Paz 1998). The morphological data in Albert (2001) include 249 characters of squamation, pigmentation, laterosensory canals, morphometrics, fin ray counts, osteology of the oral jaws and dentition, suspensorium, neurocranium, branchial skeleton, pectoral girdle, and axial skeleton, neuromorphology of the whole brain, sensory organs, developmentally defined brain regions, nuclei, tracts, and cranial nerves, and morphology of the musculature and electric organs.

Albert (2001) also reanalyzed the sequence data presented by Alves-Gomes et al. (1995) using maximum parsimony regarding all characters as unordered and weighting transitions equal with transversions. Down-weighting transitions is common in molecular systematic studies when there is evidence that transitional changes are saturated. The rRNA data of Alves-Gomes et al. (1995), however, do not appear to be saturated at these sites (Alves-Gomes et al. 1995, figs. 4 and 5). A strict consensus tree of the nine most parsimonious topologies consistent with the equally weighted mitochondrial sequence data is presented in Figure 13.3B. The single most parsimonious tree resulting from analysis of a combined data matrix (Eernisse and Kluge 1993) of 935 morphological and molecular characters is presented in Figure 13.3C.

These results show strong agreement in the structure of the molecular and morphological datasets. All topologies presented in Figure 13.3A–C support the monophyly of Sternopygoidei, Rhamphichthyoidea, *Steatogenys* + *Hypopygus*, Eigenmannini, Apteronotidae, and Sternarchorhynchinae. The relative positions of *Sternopygus*, Eigenmanninae, and Apteronotidae are unresolved in Figure 13.3B. Nonetheless, the tree provided by the molecular data alone is consistent with the hypothesis that gymnotiform taxa with a wave-type EOD form a monophyletic group, the Sinusoidea (Apteronotidae + Sternopygidae). The trees derived from equally weighted sequence data are inconsistent with the morphological data with respect to the position of *Gymnorhamphichthys* and *Apteronotus leptorhynchus*.

Species diversity and systematics of the Hypopomidae are treated by Sullivan (1997) from a study of molecular sequence and morphological data. These data include 181 parsimony informative sites from 802 aligned base pairs of the 12S and 16S rRNA mitochondrial genes, sequenced from 33 specimens representing 18 rhamphichthyoid and a single gymnotiform outgroup species. Additional data were provided from 1065 aligned base pairs of the cytochrome *b* mitochondrial gene for 12 specimens representing 11 rhamphichthyoid and a single gymnotiform outgroup species. Sullivan (1997) also provides observations on color, external morphology, osteology, meristics, and electric organ morphology for 15 hypopomid species, including 6 undescribed species. The main difference between the topologies of Sullivan (1997) and that of Figure 13.3A is in the position of the Steatogenini (*Steatogenys* + *Hypopygus*); the Steatogenini is included within the Hypopomidae in Figure 13.3A and is the sister taxon to *Rhamphichthys* + *Gymnorhamphichthys* in Sullivan (1997). The monophyly and interrelationships of Hypopomidae is currently unresolved (Albert 2001).

Among gymnotiform family-level taxa there are abundant morphological and molecular data supporting the monophyly of the Rhamphichthyoidea (Rhamphichthyidae + Hypopomidae) and Sinusoidea (Sternopygidae + Apteronotidae). The interrelationships among Gymnotidae, Rhamphichthyoidea, and Sinusoidea are less well supported (Albert 2001). Among the characters used to place Gymnotidae as the sister taxon to other Gymnotiformes are several features that are unique and unreversed in the Sternopygoidei (Rhamphichthyoidea + Sinusoidea); for example, Gymnotidae lacks reduced oral jaws (Albert 2001), fasciculated laterosensory afferents (Albert et al. 1998), and displaced hemal spines at the posterior margin of the body cavity (Albert 2001). However, Sternopygidae is unique among gymnotiforms in possessing the highly developed (plesiomorphic) visual system of other teleost fishes, with large eyes and an accessory optic system (Lazar et al. 1987; Albert et al. 1998). Further, Apteronotidae possesses the plesiomorphic caudal fin of other teleost fishes with a hypural plate and segmented fin rays. Therefore, regardless of the tree topology, the phylogenetic distribution of these features requires a complex history of character evolution involving multiple losses and/or reversals. Maximum parsimony optimization of these features on the tree topology in Figure 13.1 suggests the developed visual system of Sternopygidae and caudal fin of Apteronotidae are atavistic (derived) reversals to the complex (plesiomorphic) teleostean condition. A less parsimonious but possible alternative is that the Sternopygidae retains a primitive (developed) visual system, and the reduced visual system of Gymnotidae, Rhamphichthyoidea, and Apteronotidae was derived independently. By a similar logic, the caudal fin of Apteronotidae could also be a plesiomorphic retention, having been lost independently in Gymnotidae, Rhamphichthyoidea, and Sternopygidae.

Investigations into the species-level diversity and phylogeny of Gymnotiformes are accelerating and the actual dimensions of the fauna are at last becoming clear. Phylogenetic revisions of eight species-rich gymnotiform clades are now being undertaken by the authors to complete the descriptive stage of this research program (Fig. 13.4): 1, *Gymnotus* species—group B (13 spp.); 2, *Rhamphichthys* (8 spp.); 3, *Brachyhypopomus* (22 spp.); 4, "*Eigenmannia*" (11 spp.); 5, *Sternarchorhynchus* (9 spp.); 6, *Apteronotus sensu stricto* (15 spp.); 7, *Sternarchella* (6 spp.); and 8, "*Apteronotus*" *sensu lato* (8 spp.). These eight clades represent 51% of known gymnotiform species and include 39 species with pulse- and 49 species with wave-type EODs. These investigations include: (1) monographic revisions of the alpha taxonomy, species diversity, and geographical distributions based on an exhaustive survey of existing museum materials and from new collections in remote regions; (2) phylogenetic hypotheses of species relationships using morphological and molecular data from all available species; (3) biogeographical analyses testing hypotheses on the role of geomorphology in the origin and accumulation of species diversity; (4) behavioral analyses of intra- and interspecific electric signal differences to test alternative hypotheses on the role of electric signals in the maintenance of species diversity; and (5) investigations into the diversity of electric organ morphology and cellular physiology that underlies EOD diversity.

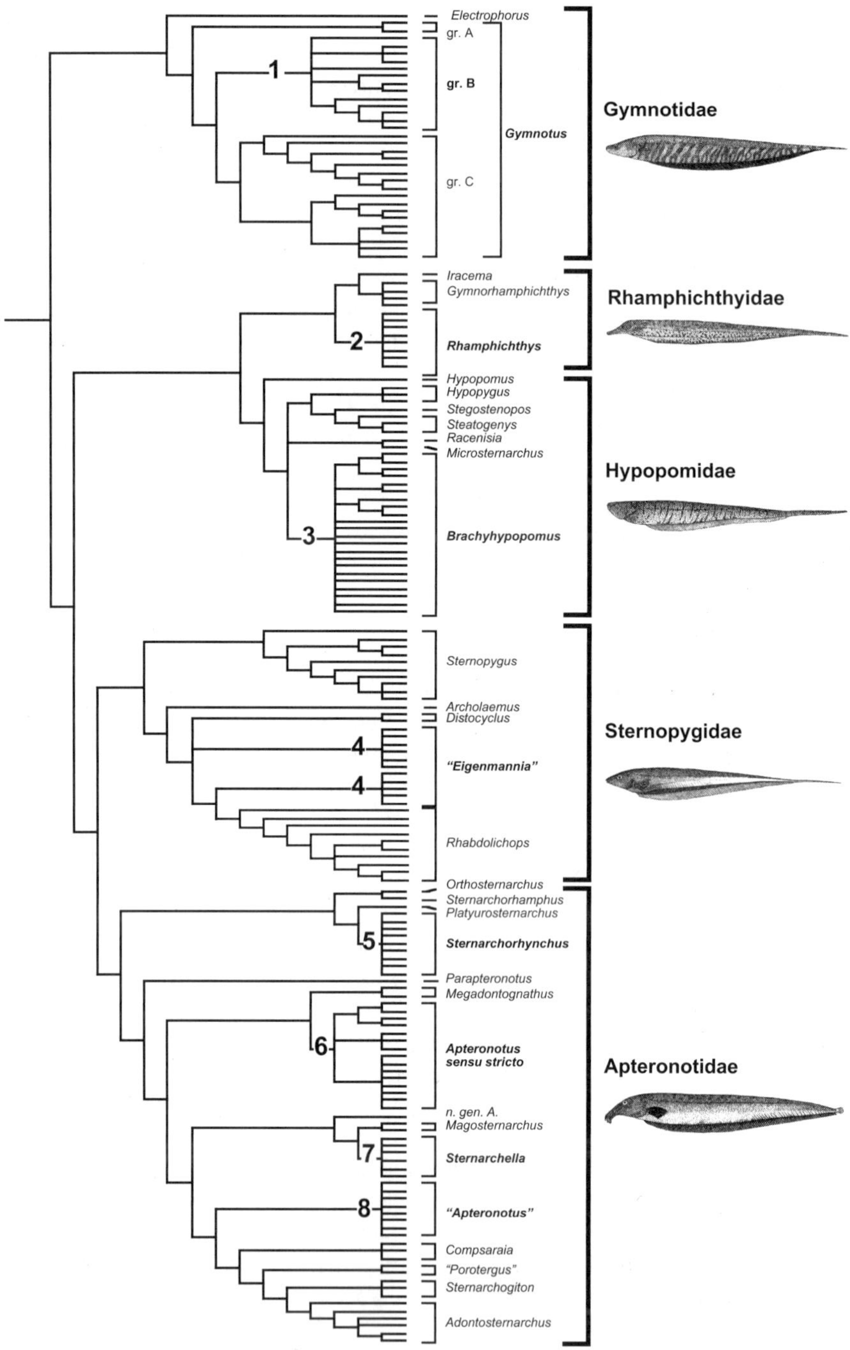
Electrophorus
gr. A
1
gr. B
Gymnotus
gr. C
Gymnotidae
Iracema
Gymnorhamphichthys
2
Rhamphichthys
Rhamphichthyidae
Hypopomus
Hypopygus
Stegostenopos
Steatogenys
Racenisia
Microsternarchus
3
Brachyhypopomus
Hypopomidae
Sternopygus
Archolaemus
Distocyclus
4
4
"Eigenmannia"
Rhabdolichops
Sternopygidae
Orthosternarchus
Sternarchorhamphus
Platyurosternarchus
5
Sternarchorhynchus
Parapteronotus
Megadontognathus
6
Apteronotus sensu stricto
n. gen. A.
Magosternarchus
7
Sternarchella
8
"Apteronotus"
Compsaraia
"Porotergus"
Sternarchogiton
Adontosternarchus
Apteronotidae

3.1 Age of Taxa

Extant gymnotiform taxa are undoubtedly of considerably antiquity. Estimates of the minimum age of taxa may be inferred directly from fossil evidence and indirectly from age estimates of sister tax, molecular datasets, and biogeographic distributions. These sources of evidence suggest a chronology for gymnotiform evolution with origins and early divergence in the Upper Cretaceous and Lower Tertiary, and the acquisition of essentially modern phenotypic and species diversity by the late Middle Miocene. The following discussion combines information from each of these sources and presents information relevant to the timing of gymnotiform divergences.

The only known fossil gymnotiforms are 10 fragments from the Yecua Formation (Upper Miocene, c. 8 to 10 Ma) in Bolivia, ascribed to *Humboldtichthys* (formerly *Ellisella*) *kirschbaumi* (Gayet et al. 1994, Gayet and Meunier 2000). Although fragmentary, these fossil specimens can unambiguously be identified as Gymnotiformes by highly elongate body and anal fin, and specialized ball-and-socket anal-fin articulation with the pterygiophores (bony fin-ray supports). Because of incomplete preservation, many of these specimens cannot be ascribed to a modern family. In one specimen the caudal portion of the body is preserved with a regenerated caudal appendage and other aspect of morphology diagnostic of the extant sternopygid *Distocyclus*. The holotype of *H. kirschbaumi* shares the presence of a deeply striated opercle with some extant species of *Sternopygus* and *Distocyclus* and can also be placed in the Recent gymnotiform family Sternopygidae (Albert and Fink, in review).

Gymnotiformes is the sister taxon to Siluriformes, which is itself first known from fossil specimens ascribed to modern families (Ariidae, Diplomystidae) in the Campanean (83 to 73 Ma; Benton 1993), setting a minimum date for the origin of the line leading to modern Gymnotiformes. Gymnotiform taxa are entirely restricted to the Neotropics and it was once thought they originated after the final separation of South America from Africa in the Upper Cretaceous (c. 100 Ma; Lundberg 1993). Recent studies on the higher level interrelation-

FIGURE 13.4. Composite species-level phylogenetic hypothesis of 173 gymnotiform species. Topology from multiple sources, using both morphological and molecular data; Gymnotidae (Albert et al. 2004; J. Albert, W. Crampton, N. Lovejoy, unpublished observations); Rhamphichthyoidea (Rhamphichthyidae + Hypopomidae) (Albert 2001; Sullivan 1997; J. Albert, W. Crampton, D. Thorsen, unpublished observations); Sternopygidae (Albert 2001; Hulen et al. 2004; J. Albert, W. Crampton, S. Correa, unpublished observations); Apteronotidae (Mago-Leccia et al. 1985; Albert 2001; J. Albert and W. Crampton, unpublished observations). Monophyly of *Eigenmannia* and *Porotergus* not confirmed. Numbered clades are polytomies (areas of phylogenetic uncertainty) prioritized for current and future studies: 1, *Gymnotus* species-group B; 2, *Rhamphichthys*; 3, *Brachyhypopomus*; 4, "*Eigenmannia*"; 5, *Sternarchorhynchus*; 6, *Apteronotus sensu stricto*; 7, *Sternarchella*; 8, "*Apteronotus*" *sensu lato*. Species in these eight clades represent of 88 of 173 (51%) known gymnotiforms, including 39 species with pulse- and 49 species with wave-type EODs.

ships of the first two outgroups (Siluriformes and Characiformes) have revealed numerous trans-Atlantic clades in both groups, suggesting origins predating the final breakup of Gondwana (de Pinna 1998; Lundberg 1998; Vari 1998). Because the line leading to modern Gymnotiformes originated before the final breakup of Gondwana the group may have originated in the western portion of Gondwana, in the area of modern northern South America (Albert 2001). It is also possible that gymnotiforms once exhibited a broader distribution, having since become extinct on the eastern portion of Gondwana, in the area of modern Central Africa, or that they were excluded from this region by the prior presence of electrosensory mormyrids.

Alves-Gomes (1999) calculated absolute divergence times among 22 ostariophysan species representing 17 genera and all 5 orders, and including 13 gymnotiform species representing 8 genera and all 5 families. Sequence divergences among 810 base pairs of mitochondrial 12S + 16S rRNA genes were calibrated using data from the ostariophysan fossil record to estimate minimum times of divergence for four clades. Using a constant mutation rate of 0.13% 10^6 years^{-1} for the conserved stretches (stems), Alves-Gomes (1999) estimated minimal divergence times for Siluriphysi (Siluriformes + Gymnotiformes) of 79.39 to 117.56 Ma, and for the sternopygid *Eigenmannia* of 16.7 Ma. Examples of the use of historical biogeography to estimate minimum divergence times in gymnotiforms are provided in Section 6.3.

As in other Neotropical fish groups, the distribution of taxa with cis (east) and trans (west) Andean distributions have proven tremendously useful in estimating the phylogenetic age of taxa (Albert et al. 2005b). There are at least 12 trans-Andean gymnotiform clades, including examples in 6 genera and 4 of the 5 families. These clades are: 1, *Gymnotus cylindricus* + *G. maculosus*; 2, *G. panamensis*; 3, *G. choco* + *G. ardilai*; 4, *G. esmeraldas* + *G. henni*; 5, *Brachyhypopomus occidentalis* + *B. diazi*; 6, *Sternopygus macrurus* + *S. arenatus*; 7, *S. aequilabiatus* + *S. dariensis*; 8, *Eigenmannia virescens*; 9, *E. humboldtii*; 10, *Distocyclus goajira*; 11, *Apteronotus albifrons* + *A. mariae* group; and 12, *A. cuchillo* + *A. (Ubidia) magdalenensis* + *A. rostratus*.

4. Survey of Phenotypic Diversity

4.1 Culteriform Body Plan and Active Electroreception

The body plan of gymnotiform fishes is highly specialized in association with electrogenesis and active electroreception (Bastian 1986; Fink and Fink 1996; Albert 2001). A combination of derived features give gymnotiform fishes a knifelike or "culteriform" body shape. Gymnotiforms achieve propulsion by undulations of their elongate anal fin rather than swimming with alternating constrictions of the axial muscles, as do most fishes. Gymnotiforms also use their pectoral fin for fine adjustments of body position. All gymnotiform fishes possess a highly elongate body, with at least 50 (*Hypopygus*) and as many as

300 (*Electrophorus*) vertebrae, and an elongate anal fin containing 100 to 350 rays. The anal fin extends along the majority of the ventral portion of the body such that the position of the anus lies anterior to the body midlength in adults, often at the branchial isthmus, and sometimes at a vertical with the eye. Each anal-fin ray has a ball-and-socket articulation with the bony supports of the axial skeleton (pterygiophores) facilitating the use of anal-fin undulation in locomotion. All gymnotiform fishes entirely lack dorsal, adipose, and pectoral fins and their bony supports (i.e., pterygiophores, girdles).

All Gymnotiformes possess a unique capacity to regenerate the entire caudal (postcoelomic) portion of the body (Ellis 1913; Anderson and Waxman 1981; Anderson 1987). The postcranial portion of the body is buttressed with numerous and highly branched intermuscular bones. In Rhamphichthyoidea (Rhamphichthyidae + Hypopomidae) and Sternopygidae an elongate caudal appendage (sometimes called the "caudal filament" or "tail") extends beyond the last anal fin ray. This caudal appendage is a complex structure composed of electrocytes, and depending on taxon and extent of regeneration, may include axial and pterygiophore muscles, vertebral centra, neural and hemal spines, intermuscular bones, blood vessels, spinal cord, and spinal nerves (Kirschbaum and Meunier 1981; Albert 2001). In gymnotiforms the caudal fin is either small (Apteronotidae) or absent (other families).

The culteriform body shape facilitates the use of the external body surface as a sensory array. The integument of gymnotiforms is invested with numerous tuberous and ampullary electroreceptor organs, and in at least one species with a wave-type EOD (*A. leptorhynchus*) individuals learn to associate the proprioceptive feedback from bending the tail with increases in the activity of electroreceptors (Bell et al. 1997). Most of the body surface is covered with high frequency–tuned tuberous electroreceptor organs, which in most taxa are most densely arrayed on the surface of the snout (Zakon 1986; Albert 2001). In Apteronotidae tuberous organs are also densely arrayed on the midsagittal dorsal organ (Franchina and Hopkins 1996). In Gymnotiformes the low-frequency–tuned ampullary organs are less densely arrayed over the head and body, and in adults are organized in clusters called "rosettes" of 15 to 20 ampullary organs per cluster in which each cluster is innervated by a single afferent nerve fiber (Zakon 1986). The primary rhombencephalic electrosensory centers of the brain (i.e., the electrosensory lateral line lobe and corpus cerebellum) constitute approximately 50% of the total brain volume (Albert et al. 2000). This hypertrophy of electroreceptive brain structures is accompanied by a substantial reduction in other sensory systems. The eye is small and subdermal in all gymnotiform taxa except two sternopygids genera (*Sternopygus* and *Archolaemus*), and the accessory optic system is absent in all gymnotiforms except the Sternopygidae (Finger and Karten 1978; Lazar et al. 1987). Gymnotiformes do not possess extraoral taste buds, Schreckstoff, or integumental club cells (Albert 2001).

The culteriform body shape also facilitates production of a stereotyped three-dimensional electric field (Rasnow and Bower 1996; Assad et al. 1998). The hypaxial electric organ of gymnotiform fishes extends along 80% to 90% of the fish's body length and generates spatiotemporally complex electric fields and waveforms (Watson and Bastian 1979; Lorenzo et al. 1988; Caputi 1999; Stoddard et al. 1999). The external electric field approximates an oscillating dipole (a dumbbell-shaped field) that envelops the head and tail of the fish with alternating head-positive and head-negative phases that cycle with the EOD repetition rate. A transverse plane of zero voltage is located in the posterior region of the body, near the base of the caudal appendage (caudal end of the anal fin) in most taxa, and at about the level of the lateral-line rami in *Gymnotus*. The field amplitude decays from the body surface with the inverse cube of distance so that the effective range of sensory perception is 5–10 body lengths. Active electroreception is therefore sensitive to objects in the near field only. The electric field results from the weighted sum of individual electrocytes arranged in series and in parallel, with each electrocyte having a potentially different time-waveform and internal resistance (Caputi 1999; Aguilera et al. 2001; Aguilera and Caputi 2003).

4.2 Body Shape and Size Evolution

A combination of functional and phylogenetic considerations suggests that the active electrosensory system of Gymnotiformes has placed strong constraints on the evolution of adaptive diversity. Despite the substantial diversification of Gymnotiformes into numerous family and generic-level taxa and with significant adaptive changes in body size and trophic morphology, no gymnotiform lineage has deviated from the basic culteriform body plan. Active electroreception in Gymnotiformes involves physiological interactions between electrosensory and electrogenic structures in a stable geometric configuration (Caputi 1999). The culteriform body of Gymnotiformes with anal-fin locomotion and a semirigid body surface is required to form and maintain the electroptopic mapping of electroreceptors on the body surface to the primary electrosensory brain centers (Carr and Maler 1986).

Constraints on the phylogenetic diversification of Gymnotiformes are apparent in comparison with that of its sister taxon, Siluriformes (catfishes) (Fink and Fink 1981). There are approximately 3600 species of catfishes currently described, allocated to 436 genera and 37 families with a cosmopolitan distribution, including freshwaters of all continents except Antarctica, and two families that are found in coastal marine tropical and temperate waters worldwide (Eschmeyer 2003). Gymnotiformes, by contrast, are an order of magnitude less diverse in terms of species numbers, and much more conservative in terms of overall morphological and ecological disparity. Sister taxa are by definition of equal age and share the same unique common ancestor. Thus, the disparity species richness, morphological variety, and ecological breadth between Gymnotiformes and Siluriformes is not simply a result of differing clade ages or initial genetic variation.

Patterns of diversity in Gymnotiformes also suggest that the evolution of body size is influenced by EOD type. Adult body size in Gymnotiformes as a whole ranges over an order of magnitude, from as small as 50 mm total length in mature specimens of *Hypopygus* n. sp from the Venezuelan Amazon to more than 2 m in *Electrophorus electricus*. Gymnotiform species with pulse-type EODs are, on the whole, smaller and yet exhibit more size diversity than species with wave-type EODs (Fig. 13.5). The size distribution of species with pulse-type EODs is significantly skewed to smaller sizes (Kolmogorov-Smirnoff test, $p < 0.05$) whereas the size distribution of species with wave-type EODs does not differ significantly from a normal distribution. Constraints on size diversification in taxa with wave-type EODs may arise directly from the physiology of electrogenesis itself, or indirectly from epistatic factors associated with respiration, locomotion, navigation, and communication in the deep river channel habitats that characterize many species with wave-type EODs (Julian et al. 2003). Alternatively, the relatively higher size diversity of gymnotiforms with a pulse-type EOD may reflect the greater age and broader distribution of these taxa, in terms of both habitat and geographical region.

Nevertheless, there has been less net diversification (speciation minus extinction) in gymnotiforms with pulse-type EODs (77 species) as compared to wave-type EODs (96 species). In many deep-channel gymnotiforms with a wave-type EOD, notably species of Eigenmannini and Navajini, the entire body is modified for foraging and locomotion in a flowing environment. In these species the mouth is often terminal and the body cavity is relatively short and slender. The

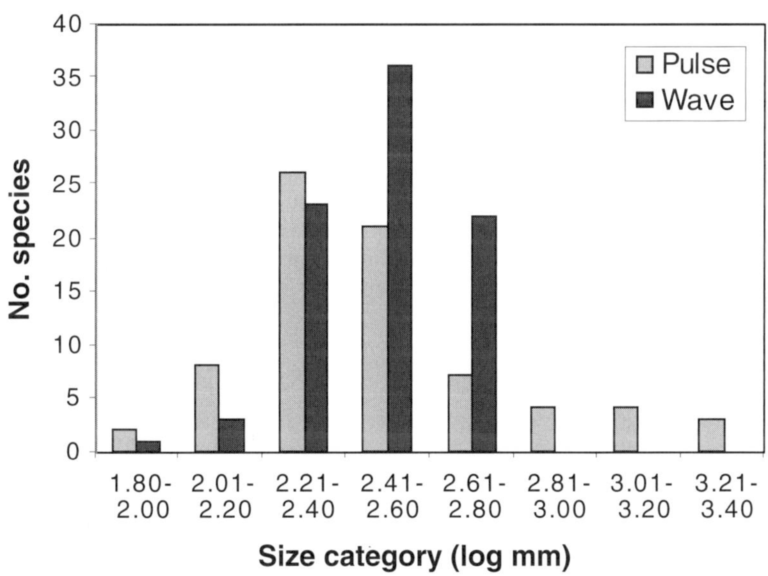

FIGURE 13.5. Size distributions for Gymnotiformes. Maximum total lengths for all described pulse- and wave-type gymnotiform species. Note species with a pulse-type EOD possess a smaller modal and median sizes and a larger body size range.

entire body is strongly compressed laterally, the lateral midbody scales are large, and the squamation is reduced anteriorly and dorsally. In addition, the extent of endosekletal ossification is substantially reduced, through one of several mechanisms including delayed onset of mineralization (i.e., paedomorphosis) and active bone remodeling (Albert 2001).

4.3 Electric Organ and Electric Signal Diversity

Gymnotiform fishes exhibit a wide range of electric organ and EOD specializations (Bennett 1971; Hopkins 1983; Bass 1986; Dye and Meyer 1986). Gymnotiform EODs fall into two main physiologically defined categories: pulse-type EODs comprising a train of relatively short and stereotyped pulses with one to six phases of alternating polarity separated by periods of silence; and wave-type EODs comprising a continuous periodic wave, with one to four phases. The EOD is itself the product of a temporal pattern of electrocyte activation across a regionally differentiated electric organ. Species-specific electric organ discharge patterns may result from differences in several morphological features of the electric organ and its constituent electrocytes: (1) neural innervation patterns and the coordinated activation of electrocyte faces; (2) size and shape of electrocytes at different positions along the body axis; (3) lengths, positions, and number of longitudinally arranged electrocyte tubes; and (4) orientation and position of insulating bands of connective tissue that channel ion flow across the electrocytes membranes within each tube. Of special interest for interspecific comparisons is the axial extent of double-innervated electrocytes (on both anterior and posterior faces) in relation to the magnitude of the first head-negative EOD phase (V1–V2 of Caputi 1999; P0 of Crampton et al. 2003).

The phylogenetic distributions of salient electric organ and EOD characters are summarized in Table 13.3. In two clades (Gymnotidae and Rhamphichthyoidea) electrogenic organs develop from paired bundles of the hypaxial muscle and extend along the entire ventral margin of the body, from below the pectoral fins to the tip of the caudal appendage. Members of these clades retain a plesiomorphic pulse-type EOD and larval hypaxial electric organs into maturity.

TABLE 13.3. Gymnotiform taxa with states of adult electric organ and EOD parameters.

Taxon	Main adult EO	EOD type	Accessory EOs
Apteronotidae	Spinal motoneurons	Wave	None
Gymnotidae	Hypaxial muscles	Pulse	None
Hypopomidae	Hypaxial muscles	Pulse	Several
Gymnorhamphichthys	Hypaxial muscles	Pulse	Mental
Sternopygus	Hypaxial and epaxial muscles	Wave	None
Eigenmanninae	Pterygiophore muscles	Wave	None

See text for descriptions of accessory electric organs.

In Sinusoidea (Apteronotidae + Sternopygidae) the larval hypaxial electric organ degenerates at metamorphosis to be replaced with a derived adult organ and a derived wave-type EOD (Mills et al. 1992; Unguez and Zakon 1998; Albert 2001).

Gymnotids possess hypaxial myogenic electric organs with barrel- or coin-shaped electrocytes. The plesiomorphic condition for gymnotids is a head-positive monophasic pulse type EOD as seen in *E. electricus* (the most basal species) and the two Middle American species of *Gymnotus, G. cylindricus* and *G. maculosus* (Albert and Miller 1995; Kirschbaum 1995; Moller 1995a, fig. 8.8). All South American species of *Gymnotus* possess a derived EOD with two to five phases (Albert 2001; Albert et al. 2003, 2005b). The electric organ of *Gymnotus*, although physically not compartmentalized, exhibits regional specialization in patterns of nervous innervation, size and shape of individual electrocytes, and the rostral extent and number of electrocyte tubes, all of which yield physiological heterogeneities in the electric field. During two decades of laboratory investigations Caputi and colleagues have generated a detailed cellular-level functional model for the physiological and anatomical basis of the complex multiphasic EOD waveform in one species of *Gymnotus*, *G. inaequilabiatus* (Trujillo-Cenóz et al. 1984; Caputi et al. 1989, 1994; Caputi and Aguilera 1996, 2003; Aguilera et al. 2001). This model is summarized by Caputi (1999) and has been used as the basis for understanding the diversity of EOD waveforms observed in other gymnotiform taxa (Caputi et al. 1994; 1998). The electromotor system is repetitively activated by impulselike signals generated by the pacemaker nucleus in the brain stem medulla. This nucleus activates a set of relay cells whose axons descend along the spinal cord and project to electromotor neurons, which, in turn, project to the electrocytes of the electric organ. These relay neurons, electromotor neurons, and electrocytes are layers of a network arranged in a lattice hierarchy, which coordinates the spatiotemporal pattern of postsynaptic and action currents generated by the electrocyte membranes. The size, shape, and configuration of electrocytes within the electric organ vary dorsoventrally and anteroposteriorly, and in some cases are differentiated into morphologically discrete compartments. Neural innervation of the electrocytes may be at either their rostral or caudal faces, or at both faces, depending on the site within the organ and on the species. The depolarization activity of electrocyte faces is synchronized by a synergistic combination of delay lines and coordinated in a precise sequence to produce the orderly recruitment of electromotor neurons according to their position along the spinal cord. The body of the animal filters the electric organ output electrically, and the whole fish is transformed into a distributed electric source.

Electrophorus electricus is unique among gymnotiform fishes in possessing three anatomically distinct electric organs (i.e., Sachs, Hunter's, and Main electric organs), all of which are derived developmentally from a germinal zone located on the ventral margin of the hypaxial musculature (Bennett 1971; Schwassman, personal communication). The specialized electric tissues of *E. electricus* are used as a model system in membrane electrophysiology and biochemistry (Gotter et al. 1998). The Main electric organ and the anterior two thirds of the more ventral Hunter's electric organ generate occasional trains of strong (up to 600 V) discharges. These trains of strong discharges are under voluntary control and are used in aggressive and predatory behaviors. The posterior one third of the Hunter's organ, together with the organ of Sachs lying dorsal and posterior to the other two electric organs, generate a continuous, variable, low-frequency (1 to 5 Hz) weak (1 to 10 V) electric discharge. The continuous weak EOD of *Electrophorus* is a head-positive monophasic pulse that generates the electric field used in navigation and sexual communication (Bass 1986).

Hypopomids possess a single hypaxial myogenic electric organ with barrel-shaped to cigar-shaped electrocytes, and a pulse-type EOD with one to three phases in *Brachyhypopomus*, two phases in *Microsternarchus* and *Racenisia*, and three to five phases in *Hypopygus*, *Steatogenys*, and *Stegostenopos*. Several hypopomids have evolved accessory electric organs that increase the complexity and presumably species specificity of the waveform generated by the summed discharges of the electric organs. Members of the Steatogenini (*Steatogenys*, *Stegostenopos*, *Hypopygus*) are characterized by paired postpectoral organs, which assume different configurations in each genus. *Steatogenys* also has paired villiform mental organs. Both the postpectoral and villiform mental organs are innervated by spinal nerves emerging from the region immediately behind the pectoral fin (Bennett 1971; Moller 1995b, fig. 16.7; D. Thorsen, personal communication). The EOD repetition rate is more stable, and the medullary pacemaker nucleus is larger, in *Steatogenys elegans* which inhabits swiftly flowing large rivers, than in *S. duidae*, which inhabits small, slowly moving streams (Crampton and Albert 2005). At least one (undescribed) species of *Brachyhypopomus* possesses paired accessory electric organs in the opercular region. This appears, however, to be unusual in *Brachyhypopomus*. Accessory electric organs are not known from *Hypopomus*, *Racenisia*, and *Microsternarchus*. Many species of *Brachyhypopomus* display sex differences in the EOD (Hopkins 1972; Stoddard et al. 1999). In *B. pinnicaudatus*, for example, the EOD is different in shape and up to twice as long in mature males as in females or immature males with the peak frequency of the Fourier transform hence being lower. The tail filament also becomes elongated (Hopkins et al. 1990). Hopkins et al. (1990) suggested that the sex difference in EODs has arisen because of female choice for long duration signals. They also observed a proportional decrease in the amplitude of the electric discharge of males of a given body size compared to females and suggested that the production of a long EOD causes a reduction in amplitude. The concluded that males may have evolved longer electric organs with more electrocytes to compensate for the loss of signal amplitude. Recently, it has become clear that some species of *Brachyhypopomus* (Hypopomidae) can rapidly modulate both the duration and amplitude of their EODs. These modulations can occur transiently, in response to a fright stimulus, and, in breeding males, under the influence of the natural circadian rhythm

(Franchina and Stoddard 1998). The presence of males rather than females appeared to stimulate these changes, a pattern observed independently for *B. brevirostris* and *B. beebei* (W.G.R.C., unpublished observation).

Rhamphichthyids possess a hypaxial myogenic electric organ with barrel-shaped to cigar-shaped electrocytes, and a pulse-type EOD with three to five phases. Individual specimens of *Gymnorhamphichthys* also possess a mental accessory electric organ that develops as an anterior extension of the hypaxial musculature along the ventral margin of the head.

Juvenile sternopygids possess the larval hypaxial organ characteristic of other gymnotiforms. *Sternopygus* has in addition to the hypaxial an epaxial organ. Both of these organs extend along the majority of the length of the body and caudal appendage. In the Eigenmanninae the larval hypaxial organ degenerates at metamorphosis and is replaced by an adult electric organ derived from anal-fin pterygiophore muscles. The individual electrocytes of adult sternopygid electric organs are cigar shaped (elongate on the body axis) and innervated on their posterior margin. Sternopygids possess a wave-type EOD characterized by a monophasic hyperpolarization from a negative baseline. Transient and spontaneous modulations of EOD amplitude have been observed in at least two species: *Sternopygus macrurus* (McAnelly et al. 2003) and *S. branco* (Crampton et al. 2004b).

There is considerable variation in EOD repetition rate in the sternopygids. *Sternopygus* generate the lowest rates. One recently described species, *S. branco* (Crampton et al. 2004b), generates EODs at a rate of 24 to 35 Hz, the lowest rate known for any gymnotiform with a wave-type EOD. Other *Sternopygus* species generate EODs from around 40 to 300 Hz (Crampton 1998a). *Sternopygus macrurus* has sexually dimorphic EODs, with females generating EODs at approximately twice the frequency of males (approximately 40 to 80 Hz in males, 80 to 160 Hz in females). EOD sexual dimorphism was not noted in *S. branco* (Crampton et al. 2004b). All Sternopygidae have simple sinusoid-like discharges with no complex inflections or extra phases contained within each discharge cycle. The dominant component of the Fourier power spectra of sternopygids is invariably the fundamental frequency.

In Apteronotidae the larval hypaxial organ also degenerates at metamorphosis, where it is replaced in adults by proliferation and elongation of the spinal electromotor neurons that innervated the larval organ (Kirschbaum 1977). The neurogenic electric organ of adult apteronotids facilitates a high frequency wave-type EOD. Apteronotidae generate the highest EOD repetition rates (up to 2200 Hz), have the largest medullary pacemaker nucleus, and possess many other derived features of the nervous system associated with living in deep river channels and other fast-flowing environments (Crampton 1998a; Albert 2001). Waveform shape varies considerably in the Apteronotidae, from a simple sinusoid-like cycle to complex waveforms with more than one peak in a single cycle. In many but not all species the dominant component of the Fourier power spectrum is the fundamental frequency.

4.4 Gymnotiform Chromosomal Diversity

Karyotypic diversity is most well known in two species-rich genera, the pulse-type *Gymnotus* and wave-type *Eigenmannia*. A systematic survey of chromosomal diversity has not yet been undertaken in Gymnotiformes. Karyotype work shows the following diploid number ($2n$) for selected taxa: *Gymnotus* (*G. carapo*, $2n = 54$, *G. pantherinus* $2n = 52$, *G. inaequilabiatus* $2n = 52$, Fernandes-Matioli et al. 1998; *G. sylvius*, $2n = 40$, Albert et al. 1999; *G.* n. sp. "pan," $2n = 40$, Fernandes et al., unpublished observations), *Electrophorus* ($2n = 52$, Fonteles-Santos et al. 2002), *Eigenmannia* ($2n = 28$, 36, 38, and 31/32, de Almeida-Toledo et al. 2002), and *Brachyhypopomus* ($2n = 42/41$, de Almeida-Toledo et al. 2000a). High intraspecific conservatism in karyotypes was detected in four species of the genus *Gymnotus* from Southeastern Brazil and the karyotypes are species specific in relation to diploid number, NOR-bearing chromosomes, and C-band pattern (Fernandes-Matioli et al. 1998b). There is also a natural case of triploidy in *G. carapo* (Fernandes-Matioli et al. 1998). No geographic chromosome races have yet been reported in *Gymnotus*. A multiple sex chromosome system has been reported in species representing two gymnotiform families, the hypopomid *Brachyhypopomus pinnicaudatus* and the sternopygid *Eigenmannia gr. virescens* (de Almeida-Toledo et al. 1998, 2000a,b). Five distinct karyotypes are known from *E. gr. virescens* living in sympatry in Southeastern Brazil, with $2n = 36$, and $2n = 38$, a $2n = 36–38$ (de Almeida-Toledo et al. 2002). This karyotypic diversity may result in part from the presence of sexual chromosomes, hybridization, or multiple (cryptic) species (de Almeida-Toledo 1998).

5. Gymnotiform Species Assemblages

Gymnotiforms occur in most lowland Neotropical aquatic habitats, including rainforest streams and swamps, xeric coastal streams and estuaries, caves, flooded forests and grasslands, and large river channels and cataracts. There have been significant advances in the understanding of gymnotiform species assemblages over the last two decades, with an emphasis on the study of the Amazon and Orinoco basins (Henderson and Walker 1986; Crampton 1996a,b, 1998a,b, 1999; Cox-Fernandes 1998b; Motta-Bührnheim and Cox-Fernandes 2001; Stewart et al. 2002), and the smaller coastal drainages of Central and South America (Westby 1988; Hagedorn 1985; Castro and Casatti 1997; Giora 2004). In terms of relative abundance and ecological importance, Gymnotiformes are a predominant component of the fauna in flooded forests and floating aquatic macrophytes of Amazonian floodplains (Crampton 1996a,b). Two genera in particular with pulse-type EODs (i.e., *Gymnotus* and *Brachyhypopomus*), are especially diverse and abundant in floodplain aquatic macrophytes (floating meadows). Gymnotiformes with wave-type EODs, notably species of Eigenmannini and Navajini, dominate the benthos of deep river channels with more than 75% of the biomass in trawl samples from both blackwater and whitewater rivers (Lundberg et al. 1987, 1996; Marrero 1987, 1993; Goulding et al. 1988).

Three distinct gymnotiform communities can be delineated, the exact species composition of which may vary from region to region, but which are similar in terms of species numbers, generic composition, and morphological diversity (Crampton 1998a; Crampton and Albert 2005): (1) terra firme streams and small rivers not exposed to seasonal flooding; (2) seasonally inundated floodplains, including lakes and flooded forests; and (3), deep, swiftly flowing river channels.

5.1 Terra Firme Systems

Terra firme forest streams of the Amazon and Orinoco Basins and the Guyanas host specialized communities of miniaturized gymnotiform species that shelter in submerged vegetation, root mats, and leaf litter. Larger species often hide in holes or caverns in undercut banks. One species, *Gymnorhamphichthys rondoni*, burrows into sand during the day—emerging to forage at night. This species forms small communities on submerged sand banks. In the Tefé region of the Brazilian Amazon streams host a fauna of four *Gymnotus*, four *Brachyhypopomus*, two *Hypopygus*, one *Microsternarchus*, one *Gymnorhamphichthys*, two *Sternopygus*, and one *Eigenmannia*. As with deep river channel and varzea floodplains, gymnotiforms represent a dominant component of the numbers and biomass of stream-dwelling fishes. Some stream dwelling fishes are able to move over land to inhabit ephemeral rainwater pools. In addition to the well-known land-walkers such as *Rivulus* spp. *Erythrinus erythrinus* and *Callichthys callichthys*, one species of *Gymnotus* (*G. curupira*) from the Western Amazon basin of Brazil and Peru can move over land on rainy nights from streams to pools. Terra firme forest streams and their flooded lower reaches are typified by very-low-conductivity water. Conductivity, as a metric of the concentration of nutrients (salts), is often correlated to fish distributions owing to its influence on primary productivity and correlation to other important water quality variables such as pH. Conductivity is also a measure of the external resistance through which electric fishes must drive electric current. As such it could also impose direct effects on the biology of electric fishes—either on the physiology of their electric organs and electroreceptors, or on their biogeographical distribution. One genus, *Brachyhypopomus*, exhibits clear correlations between the structure of the electric organ (and therefore the shape of the caudal appendage) and the ambient conductivity. First, the caudal portion of the electric organ tends to be long and slender in fishes that inhabit low conductivity water (less than 20 μs/cm); and short and thick in fish from higher conductivity waters (100 μs/cm or more). Second, there are structural differences in electric organs related to water conductivity, with those species occurring in low-conductivity waters having only two to three parallel columns of electrocytes on each side of the midline, whereas those from higher conductivity waters having up to five columns on each side.

Long, thin tails represent a predominantly serial configuration of the electrocytes, whereas short, thick tails represent a predominantly parallel configuration. Hopkins (1999) used a mathematic model to demonstrate that a deployment of electrocytes in series should theoretically maximise power output where the external resistance is high. In contrast, a deployment of electrocytes in parallel should theoretically maximise power output where the external resistance is low. Therefore, the electric organ structure and associated morphology of the caudal appendage in *Brachyhypopomus* appears to be adapted to the conductivity of the environment.

5.2 Floodplains

The expansive and productive whitewater floodplains or "várzeas" of the Amazon and Orinoco basins, and parts of the Paraguay–Paraná basins, contain diverse and abundant communities of gymnotiforms dominated by taxa that are able to withstand prolonged periods of hypoxia or anoxia. The decomposition of leaf litter in seasonally flooded forests causes seasonal anoxia throughout the water column of Amazonian várzeas and excludes all species that are intolerant of such conditions. Consequently only gymnotiforms with air breathing adaptations are able to take advantage of the exceptional productivity of these ecosystems. *Electrophorus electricus* is an obligate air breather, employing the hypervascularized and convoluted lining of its mouth as an air-breathing organ (Evans 1929). *Gymnotus* species breathe air using the gas bladder, an accessory air-breathing organ (Liem et al. 1984). Some species of *Brachyhypopomus* take up atmospheric oxygen via their gills by inflating the opercular chambers with air and floating at the water surface (Carter and Beadle 1931; Hopkins et al. 1990; Crampton 1998b). Many gymnotiform fishes enter varzea systems via paraná channels that transport well-oxygenated river water from the adjacent parent river(s). Floating rafts of macrophytes along the borders of these channels are important spawning sites for many apteronotid and sternopygid species that are intolerant of hypoxia. As water levels rise and oxygen levels decline at the onset of the high water season, juvenile apteronotids and sternopygids hitch rides out of the varzea in drifting meadows, while the adults retreat to the main river channels, which remain well oxygenated throughout the year.

Floating meadows of macrophytes in várzeas lakes are extremely productive habitats that support diverse communities of two pulse-generating genera: *Gymnotus* and *Brachyhypopomus*. In the Tefé region a syntopic breeding assemblage of eight permanently resident species of *Brachyhypopomus* and seven resident species of *Gymnotus* occurs in whitewater floodplain habitats. Only two other resident species occur, including *Eigenmannia gr. virescens* and *Rhamphichthys marmoratus*. A further three species of *Brachyhypopomus* and three species of *Gymnotus* occur sympatrically in low-conductivity systems of the Tefé region. The electric organ discharge waveforms of syntopic (but not necessarily allotropic) species of *Brachyhypopomus* and *Gymnotus* show clear species differences, which presumably mediate reproductive isolation (Crampton et al. 2003).

Gymnotiform fishes with wave-type EODs are less hypoxia tolerant and are less likely to be found in hypoxic habitats than weakly electric gymnotiforms with pulse-type EOD, suggesting that differences in metabolism resulting from EOD type affects habitat choice. Julian et al. (2002) report O_2 consumption rates during EOD generation for 34 gymnotiforms representing 23 species, all 5 families and 17 (59%) of the 28 genera. Over the size range sampled (0.4 g to 125 g), O_2 consumption of gymnotiform fishes was found to be dependent on body mass, fitting a power function with a scaling exponent of 0.74, but the O_2 consumption rate was generally about 50% of that expected by extrapolation of temperate teleost metabolic rates to a similar ambient temperature (26°C). The O_2 consumption rate was not found to be dependent on EOD type, but rather on maintenance of active scan swimming, suggesting that the locomotor costs of amplitude modulations underlie the exclusion of wave-type gymotiforms from hypoxic floodplain environments. Scan-swimming is a behavior observed only in fishes with wave-type EODs, in which the fish continually moves forwards and backwards near the substrate, thereby creating amplitude-modulated changes in the voltage detected by the electroreceptors (Julian et al. 2003).

Gymnotiform fishes form a dominant component of the biomass and numbers of fishes living in floating meadows (Crampton 1996a; Albert 2002). Likewise, the electric eel, *E. electricus*, is a dominant predator in seasonally flooded forests of varzea habitats (Crampton 1996a). The abundance of gymnotiform fishes has also been noted in seasonally flooded grassland habitats of the Venezuelan llanos (Marrero 1987; Lasso et al. 1997) and savannahs of Roraima, Brazil (Alves-Gomes 1997). In the Central Amazon basin nutrient-poor and relatively sediment-free rivers form smaller areas of seasonally inundated floodplain. In the Tefé region of the Brazilian Amazon, small patches of floating meadows in oxbow lakes support breeding communities of three species of *Brachyhypopomus*, one *Gymnotus* and two *Sternopygus* species. The diversity of *Brachyhypopomus* and *Gymnotus* is substantially lower than in whitewater floodplain systems, a pattern repeated at the generic level in several other fish groups. These discrepancies in species diversity are probably related to productivity, habitat size and availability, the amount of interconnectance, differential immigration/emigration levels, and, perhaps most importantly, differential levels of disturbance.

5.3 Deep River Channels

The benthos of large, swiftly flowing and deep (up to 25 m) river channels of the Amazon basin host diverse and specialized communities of gymnotiforms with species that are rarely found in other habitats. These communities are dominated by apteronotids and sternopygids generating wave-type discharges. Few deep river channel gymnotiform species generate pulse-type discharges but these can be abundant in terms of numbers and biomass. Common pulse-type species include *Steatogenys elegans, Rhamphichthys* spp., and *Gymnorhamphichthys* spp. Gymnotids are entirely absent. Pioneering studies by John Lundberg and colleagues used trawling equipment to investigate the species diversity of deep channel gymnotiform communities in the Orinoco and Amazon basins and revealed a plethora of new species including two species of the new genus *Magosternarchus* (Lundberg et al. 1996), two new species of *Adontosternarchus* (Mago-Leccia et al. 1985), and four new species of *Rhabdolichops* (Lundberg and Mago-Leccia 1986). Although there are some correlations between water quality and species occurrences (Cox-Fernandes 1998b), the deep-channel fauna of turbid, nutrient rich whitewater rivers is similar to those of relatively sediment free and nutrient poor blackwaters in terms of species composition (Garcia 1995; Crampton 1996a, 1998a; personal observation). Three apteronotid genera are restricted to large river channels of the Amazon Basin: *Magosternarchus* with two species, and the monotypic *Orthosternarchus* and *Parapteronotus*.

Intensive studies of deep river channels communities in the Tefé region of the lowland Amazon basin revealed a fauna of 55 species comprising one *Distocyclus*, three *Eigenmannia*, seven *Rhabdolichops*, three *Sternopygus*, two *Gymnorhamphichthys*, three *Rhamphichthys*, one *Steatogenys*, five *Adontosternarchus*, five *Apteronotus*, three *Compsaraia*, two *Magosternarchus*, one *Orthosternarchus*, one *Parapteronotus*, one *Platyurosternarchus*, two *Porotergus*, four *Sternarchella*, three *Sternarchogiton*, one *Sternarchorhamphus*, six *Sternarchorhynchus*, and one undescribed genus and species (Crampton 1996b; 1998a,b; 1999; Crampton and Albert 2005). All species, with the exception of the following, are found in both whitewater and blackwater channels: *Sternopygus branco*, *Adontosternarchus* n. sp. A and n. sp. B, *Orthosternarchus tamandua*, *Platyurosternarchus macrostomus*, *Sternarchella schotti*, and *Sternarchella* sp. A are known from whitewaters only. *Gymnorhamphichthys rosamariae* is known from blackwaters only. In terms of species diversity, gymnotiforms represent the dominant group in deep river channels, catfishes being the other main groups with some 30 species specialized for life on the benthos of deep river channels. More than 75% of the biomass of trawl samples from both blackwater and whitewater rivers during the low-water season in the Tefé region was represented by gymnotiform fishes. These samples did not intercept large pimelodid catfishes, which are also expected to form a large biomass component of these communities. Gymnotiform fishes are a dominant component of the diet of large and commercially important pimelodid catfishes, notably *Pseudoplatystoma* spp. (Reid 1983; Lundberg et al. 1987). It is unlikely that the biomass of large pimelodid catfishes in these systems outweighs that of their principal food source. Some deep channel electric fishes (e.g., *Sternarchella* spp.) are found in the lower estuarine reaches of the Amazon River around Ilha do Marajó and in tidal channels that connect the Amazon and lower Rio Tocantins. During the flood season, these species can occur many miles offshore. During the dry season wedges of brackish water push further inland and displace gymnotiform fishes (Barthem 1985).

Some physical aspects of deep river channel habitats have resulted in the evolution of a number of morphological and physiological adaptations unique to gymnotiforms (Albert et al. 1998; Albert 2001). The absence of light at the bottom of deep river channels has resulted in a number of taxa with highly reduced eyes and pigmentation and a greatly increased dependence on the electric sense. Elongate snouts with small terminal mouths have evolved independently in at least three gymnotiform lineages (Rhamphichthyidae, Sternarchorhynchinae, *Apteronotus sensu stricto* [including *Ubidia*]), in each case involving different sets of bones (Albert 2001; figs. 15 and 16). These elongate and sometimes decurved snouts are used for extracting larvae of Chironomidae, Plecoptera, Ephemeroptera, and other insects from small burrows in the clay substrates of river beds (Marrero 1987; Winemiller and Adite 1997; Crampton 1998a). In many places the clay beds of the Amazon River and its major white-water tributaries are riddled with insect burrows, and the biomass of benthic insect larvae is substantial (Junk and Robertson 1997). Many deep-channel gymnotiforms, notably species of Eigenmannini and Navajini, are highly specialized for life at the bottom of deep Amazonian rivers. Some derived species of *Rhabdolichops* have long, ossified gill rakers used to filter plankton. Many apteronotids (e.g., *Sternarchella*) are aggressive predators with short, strong jaws and robust oral teeth. *Magosternarchus* is specialized to nip the tails off of other gymnotiforms. Other apteronotids have gracile jaws in which the oral teeth are reduced (*Sternarchogiton*) or absent (*Adontosternarchus*). These species forage on small aquatic animals, and one species (*Sternarchogiton nattereri*) eats sponge spicules.

6. Macroevolutionary Patterns

6.1 Descriptive Biogeography

Extant gymnotiform species are distributed in the continental waters of all South and Middle American countries except Chile and Belize, ranging from the Rio Salado of Argentina (36°S) to the Rio San Nicolás of southern Chiapas, Mexico (15°N) (Albert 2001). The nine hydrogeographic regions defined in Figure 13.6A are circumscribed by watershed divides and other geomorphological features (i.e., fall line of the Guyanas Shield, Purus Arch). These boundaries were selected to best represent the species composition of regional gymnotiform faunas (Albert 2001). Numbers of gymnotiform species in each genus for these nine hydrogeographic regions are summarized in Table 13.4.

As with most Neotropical taxa the Amazon Basin is the center of diversity for gymnotiforms, with 99 species (57%) in the Western Amazon (WA), and 79 species (46%) in the Eastern Amazon (EA) regions, and 128 species (74%) in these two regions combined. The Guyanas–Orinoco region (GO) has 62 species (35%). The disproportionately high species richness of this region for its size is due in part to its historical and current hydrological connections to the Amazon basin. The Paraguay–Paraná region (PA) has 23 species (13%), which is disproportionately low for its area owing in part to its relatively high latitude and cool temperatures (Silva et al. 2003). The Northeast region of Brazil (NE) with only 10 known species (5%) is also relatively depauperate for its size, largely as a result of the aridity of this region. The three trans-Andean regions

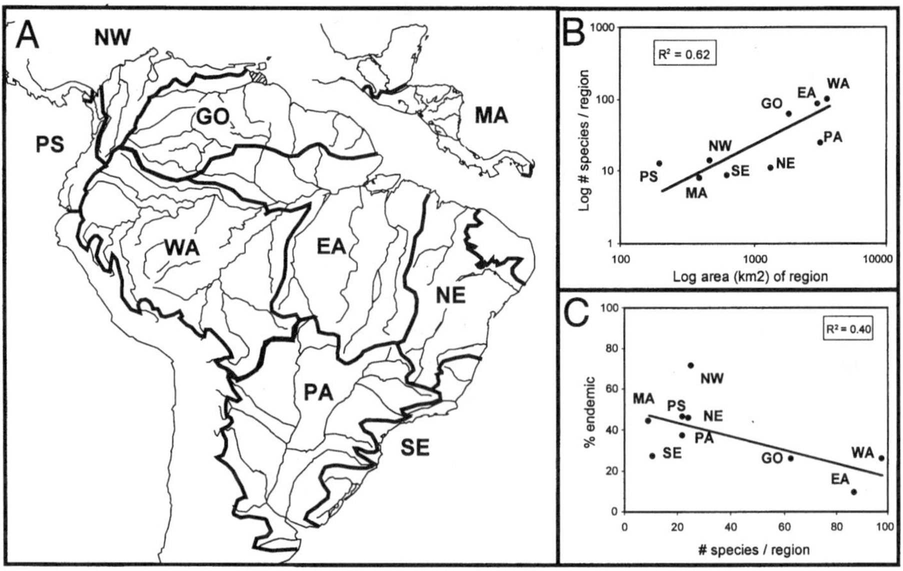

FIGURE 13.6. Descriptive biogeography of Gymnotiformes. (**A**), Neotropical hydrogeographic regions used in biogeographic analysis of Gymnotiformes. (Regions modified from Albert 2001.) EA, Eastern Amazon Basin, east of Purus Arch and all tributaries below fall-line of Guyana Shield (2,985,000 km^2); GO, Guyanas–Orinoco Region, including island of Trinidad and Upper Rio Negro drainages above fall line (1,843,000 km^2); MA, Middle America, Atlantic and Pacific slopes, from the Motagua to Tuyra Basins (393,000 km^2); NE, Northeast Brazil coastal drainages, including Parnaíba, Piaui, São Francisco, and Jequitinhonha Basins (1,357,000 km^2); NW, northwestern South America including the Magdalena and Maracaibo Basins, and the north slope of Venezuela (471,000 km^2); PA, Paraguay–Paraná Basin including Dulce-Salí and Salado Basins of Argentina (3,185,000 km^2); PS, Pacific Slope of Colombia and Ecuador, from Baudó to Guayaquil Basins, including the Atrato (Caribbean) Basin (200,000 km^2); SE, Southeast Brazil and Uruguay, coastal drainages from the Docé to Lagoa Mirim Basins (628,000 km^2); WA, Western Amazon Basin, west of Purus Arch, below about 500 m elevation (3,556,000 km^2). (**B**) Species-area relationship of Gymnotiformes in the hydrogeographic regions of **A**. (**C**) Species-endemism relationships of Gymnotiformes in the hydrogeographic regions of **A**. Note the regions with highest species diversity (EA, GO, WA) have the lowest levels of species endemicity, indicating a history of extensive faunal exchange.

(MA, NW, PS) contribute 32 species (19%) to the gymnotiform total, and include numerous examples of taxa with cis-trans Andean distributions useful for estimating the phylogenetic age of taxa. Examples of dating clades are provided for two taxa in Section 6.3 (*Gymnotus* and *Sternopygus*).

Several gymnotiform species are distributed over remarkably large areas of South America. A single morphospecies (*Eigenmannia virescens*) is known

TABLE 13.4. Gymnotiform species diversity by genus for the nine hydrogeographic regions of Figure 13.6A.

Family	Genus	MA	PS	NW	GO	WA	EA	NE	SE	PA
Apteronotidae	*Adontosternarchus*	0	0	0	4	6	5	0	0	0
	Apteronotus s.s.	1	4	4	3	1	1	2	1	4
	"Apteronotus"	0	0	0	2	4	2	0	0	3
	Compsaraia	0	0	0	1	3	1	0	0	0
	Magosternarchus	0	0	0	0	2	2	0	0	0
	Megadontognathus	0	0	0	1	0	1	0	0	0
	Orthosternarchus	0	0	0	0	1	1	0	0	0
	Parapteronotus	0	0	0	0	1	1	0	0	0
	Platyurosternarchus	0	0	0	1	1	1	0	0	0
	Porotergus	0	0	0	1	2	2	0	0	0
	Sternarchella	0	0	0	1	5	3	0	0	1
	Sternarchogiton	0	0	0	2	3	2	0	0	0
	Sternarchorhamphus	0	0	0	1	1	1	0	0	0
	Sternarchorhynchus	0	0	0	1	8	5	1	0	1
	n. Gen.	0	0	0	0	1	0	0	0	0
Gymnotidae	*Electrophorus*	0	0	0	1	1	1	0	0	0
	Gymnotus	3	3	1	5	14	8	1	3	4
Hypopomidae	*Brachyhypopomus*	1	1	2	8	11	9	2	3	4
	Hypopomus	0	0	0	1	0	0	0	0	0
	Hypopygus	0	0	0	2	2	2	0	0	0
	Microsternarchus	0	0	0	1	1	1	0	0	0
	Racenisia	0	0	0	1	0	0	0	0	0
	Steatogenys	0	0	0	1	3	2	0	0	0
	Stegostenopos	0	0	0	0	1	1	0	0	0
Rhamphichthyid.	*Gymnorhamphichthys*	0	0	0	3	3	2	0	0	1
	Iracema	0	0	0	0	0	2	0	0	0
	Rhamphichthys	0	0	0	3	3	5	1	0	2
Sternopygidae	*Archolaemus*	0	0	0	1	0	1	0	0	0
	Distocyclus	0	0	1	1	1	1	0	0	0
	Eigenmannia	2	2	2	5	6	5	2	2	2
	Humboldtichthys†	0	0	0	0	1	0	0	0	0
	Rhabdolichops	0	0	0	7	9	6	0	0	0
	Sternopygus	1	3	3	4	4	5	1	1	1
No. of species		8	13	13	62	99	79	10	10	23
Percentage of Gymnotiformes		5	8	8	36	57	46	6	6	13
No. of endemics		3	8	9	15	32	8	3	5	11
Percentage endemic		38	62	64	23	32	11	33	50	48

Taxa arranged alphabetically by family and genus.

from populations in all nine hydrogeographic region of Figure 13.6A, although the conspecific nature of these populations remains to be validated with molecular data. *Sternopygus macrurus* is known from all eight regions in South America (Hulen et al. 2005), including populations on both the Pacific and Atlantic slopes of the Colombian Andes. Three gymnotiform species are known from five regions (*Apteronotus albifrons*, *Brachyhypopomus pinnicaudatus*, *Brachyhypopomus* n. sp. E), 32 of 173 (18%) species in three regions, 38 of 173 (22%) species in two regions, and 95 of 173 (55%) species are endemic to a single region.

In general most gymnotiform clades have broad geographical distributions, occurring in two or more of the hydrogeographic regions in Figure 13.6A. Five gymnotiform genera are known from all nine regions (*Gymnotus*, *Brachyhypopomus*, *Sternopygus*, *Eigenmannia*, and *Apteronotus sensu stricto*), and 24 genera are present in two to eight regions. The Amazon–Orinoco–Guianas superbasin (WA + EA + GO) is the center of diversity at the generic level, and the majority of gymnotiform genera (24 of 32, 75%) are known from this region. The pattern of gymnotiform clades having broad geographical distributions is so strong that no gymnotiform clade comprising more than one species is restricted to a single hydrogeographic region. Five genera are restricted to a single biogeographic region, all of which are monotypic, and of which only two (the terra firme stream hypopomids *Hypopomus* and *Racenisia*) are known sufficiently to be able to have confidence in these restricted distributions. Most closely related species are allopatric; in only one genus (*Hypopygus*) are all the extant species both sympatric and syntopic in at least part of their range (Upper Orinoco basin). Rhamphichthyidae is the only gymnotiform family that is entirely restricted to Atlantic (cis-Andean) drainages.

The production of electric fields in freshwaters is highly sensitive to water quality parameters such as temperature and conductivity, and gymnotiforms are strictly excluded from marine or brackish waters. Similarly most gymnotiform species are physiologically intolerant of temperatures below approximately 17°C, and this may be the reason why they are restricted to altitudes below about 500 m elevation and to tropical or subtropical latitudes (Silva et al. 2003; see Section 6.3). A few species have adapted locally to cooler winter temperatures (c. 8°C) in the Brazilian state of Rio Grande do Sul, and in Uruguay and Argentina. One species of *Gymnotus* inhabits whitewater rivers in the Magdalena basin of the Colombian Andes at about 1000 m elevation (Maldonado-Ocampo and Albert 2003), and one (undescribed) species of *Sternarchorhynchus* is known from about 500 m in the Huallaga basin of Peru (H. Ortega, personal communication).

6.2 Polyphyletic Species Assemblages

A prominent feature of gymnotiform evolution and ecology is the absence of monophyletic regional species assemblages. In all genera, regional assemblages are polyphyletic and composed of species with disparate phylogenetic affinities and widespread geographical distributions. This feature is exemplified by the case of the gymnotiform fauna of the Tefé region, in the Western Amazon, from where 25/32 (78%) genera are known, representing all five families and nearly every major gymnotiform clade (Crampton 1998a; Crampton and Albert 2005). Of the species in the Tefé region 66 of 89 (74%) also occur outside the Western Amazon.

Sharing of species between the Amazon–Orinoco–Guianas superbasin (WA + EA + GO) is very extensive. Although these regions have the highest gymnotiform species diversity they exhibit the lowest levels of species endemism (32%, 11%, and 23% respectively). These patterns are illustrated by the positive regression in Figure 13.6B and the negative regression in Figure 13.6C. The Western Amazon, Eastern Amazon, and Guyanas–Orinoco are interlinked by two major modern hydrological connections: the Casiquiare canal (Amazon–Orinoco), and the Rupununi wetlands (Rio Branco–Essequibo), and have probably been subject to more intimate connections in recent geological history (Hoorn et al. 1995; Lovejoy et al. 1998; Lundberg et al. 1998; Lovejoy and deAraújo 2000). The Paraná–Paraguay drainages (PA), Southeast Brazilian drainages (SE), and Northeast Brazilian drainages (NE) all have considerably higher levels of endemicity despite their lower diversity (48%, 50%, and 33%, respectively). This is undoubtedly related to a longer history of isolation and a lack of hydrological interconnections. The upper Paraná is hydrologically linked only intermittently to the Upper Madeira at the Pantanal wetlands. Middle America, and the Pacific slope and North Western regions of South America have comparatively high levels of endemicity (38%, 57%, and 69%, respectively) owing to their complete isolation by high mountain ranges (4000 m +) since the late Middle Miocene (c. 12 Ma) or the narrow Isthmus of Panama since the Pliocene (c. 3 Ma).

The distribution of species among the three structurally defined Neotropical aquatic ecosystems is summarized in Table 13.5A. The two main patterns that emerge from this analysis relate to diversity and ecosystem endemism (specificity). The first observation is that species with wave-type EODs are more diverse in deep river channels, and much less so in streams and floodplains. In contrast, species with pulse-type EODs are most diverse in streams and floodplains, and substantially less so in deep river channels. The second observation is that the great majority (81%) of gymnotiform species occur in just one of these three ecosystems and, remarkably, only two species (*Sternopygus macrurus* and *Apteronotus albifrons* with wave-type EODs), representing only about 1% of gym-

TABLE 13.5. Ecological distribution of pulse- and wave-type gymnotiform species among three structurally defined lowland aquatic ecosystems in the Neotropics.

	Pulse-type EOD	Wave-type EOD
Terra firme streams	49/67 (73%)	18/67 (27%)
Floodplains	35/51 (69%)	16/51 (31%)
River channels	12/90 (13%)	78/90 (87%)

Number of species with a given EOD type in a given ecosystem is followed by the total number of species in that ecosystem (with percentage in parentheses). For example, 49 of 67 (73%) gymnotiforms living in terra firme streams have a pulse-type EOD.

notiforms, occur in all three ecosystems. A more complete analysis of habitat and electric signal diversity is provided by Crampton and Albert (2005).

The polyphyletic nature of regional gymnotiform species assemblages is the result of evolutionary diversification at a continental scale. The gymnotid *Gymnotus* is currently the best understood gymnotiform genus in terms of species-level interrelationships and geographic variation. The most species-rich local assemblage of *Gymnotus* is the area of Tefé in the Western Amazon from where 11 species occur in sympatry (Crampton et al. 1998a, b, 2005; Albert and Crampton 2001; Albert et al. 2003a; Crampton and Albert 2003). Species in this assemblage are members of at least four distinct clades, each with a sister taxon in another part of South America (Albert et al. 2005b). The capacity of this region to permit the coexistence of so many *Gymnotus* species in sympatry involves multiple ecological and historical factors, as well as the species-specific nature of their electric signals (Crampton and Albert 2005). Similarly, the assemblages of *Gymnotus* species in all nine hydrogeographic regions are not monophyletic. In other words, the species present in each of these regions do not represent the result of local or regional radiations, but rather were assembled incrementally over a lengthy history by means of a combination of processes. These include in situ speciation, extinction, immigration, and ecological factors allowing coexistence in sympatry. In this regard Amazonian species richness is not strictly a consequence of local or regional processes. These patterns are representative of other highly diverse groups of Neotropical fishes, and do not resemble those of monophyletic, rapidly generated species flocks in isolated aquatic systems (Greenwood 1984; Kornfield and Smith 2000).

6.3 Historical Biogeography and Ecology

The lineage leading to modern Gymnotiformes originated before the final breakup of Gondwana into South America and Africa some 100 Ma. During the Upper Cretaceous and early Cenozoic the gymnotiform line acquired the suite of derived phenotypes associated with bioelectrogenesis and active electroreception (e.g., electric organs, tuberous electroreceptors, anal-fin locomotion, caudal regeneration). It was during this early period, when South America was an isolated island continent and large river basins were being formed, that Gymnotiformes radiated into the full range of body plans (e.g., family and genera), habitats, and EOD types known today.

Dating by vicariance biogeography and molecular clocks combined with the phylogenetic distribution of modern habitat preferences suggests that the origins of pulse-type electric signals occurred before 80 Ma in terra firme (non-floodplain) forest streams and small rivers (Albert 2001). The transition from pulse to wave-type EODs can be inferred to have occurred in the Sinusoidea (Sternopygidae + Apteronotidae) within the giant rivers of northern South America between 20 and 80 Ma. The early gymnotiform radiations coincided in time and space with the development of the extraordinarily diverse Neotropical biota as a whole, including other ostariophysan fishes, the modern angiosperm-dominated tropical rainforests, and the radiations of insects and other sources of food for modern Gymnotiformes.

With the uplift of the Eastern Cordillera (c. 11.8 to 12.2 Ma) the Amazon–Orinoco, Maracaibo, Magdalena, and Atrato–Pacific Slope regions became isolated (Hoorn et al. 1995; Hoorn 1996; Diaz de Gamero 1996; Guerrero 1997; Gregory-Wodzicki 2000). According to a simple vicariance model this event would have separated sister taxa with existing distributions across northwestern South America, or caused the split. Such Miocene origins of multiple gymnotiform taxa with cis-trans Andean distributions resemble other groups of Neotropical fishes for which species level phylogenies have been proposed (Vari 1988, 1995; Vari and Weitzman 1990; Bermingham et al. 1997; Bermingham and Martin 1998; Martin and Bermingham 2000; Albert et al. 2005a).

The existence of at least 12 trans-Andean gymnotiform clades, including examples within six genera, suggests origins for this level of phenotypic diversity before the most recent geological isolation of the trans-Andean regions. Miocene tectonism in the northeastern Andes resulted in the formation of the modern watersheds of northwestern South America, including the Western Amazon, Orinoco, Maracaibo, and Magdalena Basins (Hoorn et al. 1995; Paxton et al. 1996; Guerrero 1997; Lundberg et al. 1998; Gregory-Wodzicki 2000). Also during the Middle Miocene the Choco Block underlying the modern San Juan and Atrato Basins was accreted to the northwest corner of South America (Duque-Caro 1990). Before the Middle Miocene most of the area of the modern Western Amazon drained northward to a delta located in the area of the modern Maracaibo Basin, and lowland Amazonian ichthyofaunas were present throughout much of northwestern South America (Lundberg 1997, 1998; Lundberg et al. 1998).

Phylogenetic data on at least two gymnotiform genera support the general chronology of generic diversification by the late Middle Miocene (c. 12 Ma). Both the genera *Gymnotus* and *Sternopygus* include multiple sister-taxon pairs on the Pacific (trans-Andean) and Atlantic (cis-Andean) slopes of northwestern South America (Albert et al. 2005b; Hulen et al. 2005). In both of these genera the cis–trans-Andean sister taxa occupy terminal positions within the phylogeny, suggesting minimum dates for the origins of these clades in the late Middle Miocene (Hoorn et al. 1995; Lundberg et al. 1998). In the case of *Gymnotus*, the basal division is between clades endemic to Middle and South America, suggesting an origin before the Pliocene Panamanian landbridge (Fig. 13.7). Under a vicariance model *Gymnotus* clade E dispersed to Middle America before the speciation of *Gymnotus* clades D and E; under a dispersal model *Gymnotus* clade E dispersed to Middle America after speciation of *Gymnotus* clades D and E within South America. In combination these geographic and phylogenetic patterns suggest that *Gymnotus* and *Sternopygus* predate the uplift of the Eastern Cordillera c. 12 Ma, and that *Gymnotus* became emplaced in Middle America long before the Pleistocene landbridge. These patterns of diversity and biogeography are consistent with those of other groups of Neotropical freshwater fishes indicating a long history extending for tens of millions of years (Lundberg

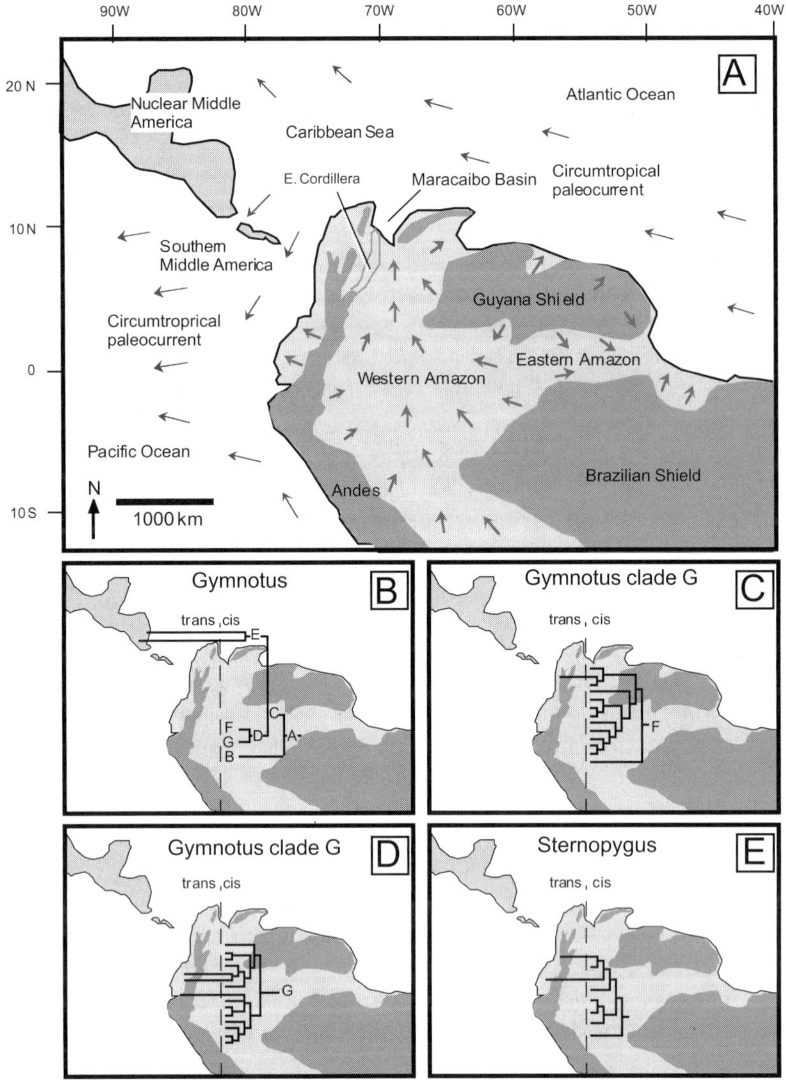

FIGURE 13.7. Historical biogeography of trans-Andean Gymnotiformes. (**A**) Paleogeographic reconstruction of northern South America and southern Middle America immediately before the uplift of the Eastern Cordillera in the late Middle Miocene (c. 12 Ma). (Map redrawn from Hoorn et al. 1995 and Ituralde-Vincent and MacPhee 1999. *Long arrows* indicate prevailing directions of marine paleocurrents; *short arrows* indicate direction of continental sediment paleotransport. *Dark shading* indicates highlands, *light shading*, lowlands. Note mouth of paleo-Amazon in area of modern Maracaibo Basin. (**B**) Basal four nodes of Gymnotidae (clade A) superimposed on paleogeographic recon-

1993, 1997; Bermingham et al. 1997; Bermingham and Martin 1998; Lundberg 1998; Reis 1998; Lovejoy et al. 1998; Lovejoy and deAraújo 2000; Albert 2001).

The last 12 million years of active tectonism in the northern Andes resulted in dramatic geomorphological changes to the region and the closures of the modern watersheds of northwestern South America. Paramount among these changes was the capture of the Western Amazon basin by the Eastern Amazon basin, forming the modern Atlantic draining Amazon (c. 8 to 10 Ma), the largest freshwater basin on earth with by far the largest hydrological discharge (Goulding et al. 2003). This period also witnessed additional multiple changes in the hydrogeographic connections of the Amazon, Orinoco, and Paraná–Paraguay systems and other basins of northern South America. Extensive interbasin exchanges of species during these hydrological upheavals are reflected today by the low levels of geographical endemism in the Atlantic drainages of South America. The time frame represented by these late Tertiary events represents a relatively recent portion (less than 12%) of the whole history of Gymnotiformes.

7. Summary and Prospectus

Recent years have seen tremendous advances in our understanding of the diversity and phylogeny of gymnotiform fishes, opening new avenues of research into ecology and ethology of electric signaling and the evolution of diverse tropical animal species assemblages. The continuous, stereotyped, species-specific EODs of gymnotiform fishes play a role in sexual communication and reproductive isolation. These signals are easy to record and quantify, facilitate the identification of morphologically cryptic species, and provide an unparalleled window on communication and species recognition that is simply unavailable in most other fish groups. New methods of sampling, identifying, and collecting gymnotiform in the wild have unveiled numerous species in previously unexplored habitats and regions. Gymnotiformes is a considerably more diverse order than was previously recognized, with 135 valid species as of this writing, assigned to 32 genera in 5 families, and an additional 38 undescribed species currently known in museum collections. The actual number of gymnotiform species in the wild probably lies between 200 and 300.

FIGURE 13.7. *Continued*
struction (Albert et al. 2004b). Note basal split in *Gymnotus* (clade C) between clades endemic to South America (clade D) and Middle America (clade E). (**C**) Phylogenetic relationships of *Gymnotus* clade F. Note the single cis–trans–Andean species pair nested at a terminal position in the phylogeny. (**D**) Phylogenetic relationships of *Gymnotus* clade G. Note the two cis–trans–Andean taxa nested at terminal positions. (**E**) Phylogenetic relationships of *Sternopygus* (Hulen et al. 2004). Note the two cis–trans–Andean taxa nested at terminal positions in the phylogeny. Physical position of terminals on maps only informative with respect to locations in Middle America and South America, or cis- and trans-Andean watersheds.

Gymnotiform fishes are an important component of the nocturnal ichthyofauna of Neotropical freshwaters, both in terms of relative abundance and ecological importance. Gymnotiforms occur in most lowland Neotropical aquatic habitats and are especially diverse and abundant in floodplain aquatic macrophytes and deep river channels. Gymnotiform species diversity may be roughly divided eqally between those with pulse-type (45%) and wave-type (56%) electric signals. Species with pulse-type EODs are most diverse in streams and floodplains whereas species with wave-type EODs are more diverse in deep river channels. In terms of biomass and numbers of individuals two clades with pulse-type EODs (*Gymnotus* and *Brachyhypopomus*) dominate floodplain aquatic macrophytes, and two clades with wave-type EODs (Eigenmannini and Navajini) dominate deep river channels. The majority (81%) of gymnotiform species are specialized to occur in just one of these three ecosystems, and only two species (both with wave-type EODs) are generalized to occur in all three ecosystems. Adult body size in Gymnotiformes ranges over an order of magnitude (from about 5 to 200 cm total length). Gymnotiform species with pulse-type EODs are smaller on average than species with wave-type EODs, and exhibit more size and habitat diversity.

A combination of functional and phylogenetic considerations suggests that the active electrosensory system of Gymnotiformes, which on the one hand has allowed these fishes to exploit new habitats and dominate whole ecosystems, has also placed strong constraints on the evolution of adaptive diversity. Overall phenotypic diversity in Gymnotiformes is relatively conservative compared to other Neotropical taxa of comparable geological age and geographical range. Active electroreception in Gymnotiformes involves a physiological coupling of electrosensory and electrogenic structures interacting in a fixed geometry.

Many of the diagnostic characters of Gymnotiformes are salient features of body form and are functionally coupled with active electroreception. The elongate, knife-shaped culteriform body plan of Gymnotiformes with anal-fin locomotion and a semirigid body surface facilitates the ontogenetic formation of two-dimensional electroptopic maps on the surface on the body and primary electrosensory brain centers. The culteriform body shape also facilitates production of a stereotyped three-dimensional electric field from the electric organ. Despite the substantial diversification of Gymnotiformes into numerous family and generic-level taxa with more than a hundred species and significant adaptive changes in body size and trophic morphology, no gymnotiform lineage has deviated from the basic culteriform body plan.

Despite or perhaps because of these constraints, the gymnotiform radiations have produced substantial diversity in the organization and development of the electrogenic and electrosensory systems. Two clades, the Gymnotidae and Rhamphichthyoidea (Rhamphichthyidae + Hypopomidae), retain the plesiomorphic pulse-type EOD, and also retain the larval hypaxial electric organ into maturity. Myogenic accessory electric organs have evolved in at least four species-rich clades of rhamphichthyoid fishes with a pulse-type EOD, in each case on or around the head; a neurogenic mental accessory electric organ is known in one species-rich clade with a wave-type EOD (*Adontosternarchus*). Sinusoidea (Sternopygidae + Apteronotidae) possess a derived wave-type EOD, and the larval electric organ of these taxa degenerates at metamorphosis to be replaced by an adult organ derived from anal-fin pterygiophore muscles (Sternopygidae) or the electromotor neurons (Apteronotidae) that innervated the larval organ. Sinusoids also possess numerous derived features related to the utilization of deep rivers channels, including specializations of the central nervous system. Approximately half of total brain volume in gymnotiform fishes is dedicated to primary electrosensory structures. Apteronotids have the largest pacemaker nucleus of all gymnotiforms, the group of medullary neurons that regulates the rhythmic EOD rate. The EOD of some apteronotid species is the fastest and most stable known biological oscillator, achieving a maximum repetition rate of 2179 Hz in *Sternarchella schotti*. In terms of waveform diversity the Sternopygidae with a wave-type EOD is the most conservative gymnotiform taxon. Waveform diversity is highest in two clades with a pulse-type EOD (*Gymnotus* and *Brachyhypopomus*), which together include 54 known species, or about 31% of all gymnotiform species. Species assemblages of these two genera are most diverse in floodplains of the Western Amazon, where up to 11 species of each may occur in sympatry.

Two species-rich clades (i.e., Rhamphichthyidae and Sternarchorhynchinae) have independently evolved elongate tubular snouts with small mouths for foraging on insect larvae in the benthos of deep river channels. Apteronotidae is characterized by much greater morphological diversity than other gymnotiform families, which is expressed primarily as variation in head structure related to trophic specialization and, in part, to male–male aggression. In at least four separate apteronotid clades (*Parapteronotus, Apteronotus sensu stricto*, *"Apteronotus" sensu lato*, and *Compsaraia*) sexually mature males develop elongate jaws. In *Sternarchogiton nattereri* and in some *Sternarchorhynchus*, males develop prominent external teeth.

The lineage leading to modern Gymnotiformes originated before the final breakup of Gondwana into South America and Africa some 100 Ma. During the Upper Cretaceous and early Cenozoic the gymnotiform line acquired the suite of derived phenotypes associated with bioelectrogenesis and active electroreception (e.g., electric organs, tuberous electroreceptors, anal-fin locomotion, caudal regeneration). It was during this early period, when South America was an isolated island continent and large river basins were being formed, that Gymnotiformes radiated into the full range of body plans (e.g., family and genera), habitats, and EOD types known today. Dating by vicariance biogeography and molecular clocks suggest that the origins of pulse-type electric signals occurred before c. 80 Ma. in terra firme (non-floodplain) forest streams and small rivers. Given the phylogenetic distribution of modern habitat preferences, the transition from pulse to wave-type EODs can be inferred to have occurred in the Sinusoidea (Sternopygidae + Apteronotidae) within the giant rivers of northern South America between 20 and 80 Ma. The early gymnotiform radiations coincided in time and space with the development of the extraordinarily diverse

Neotropical biota as a whole, including other ostariophysan fishes, the modern angiosperm-dominated tropical rainforests, and the radiations of insects and other sources of food for modern Gymnotiformes.

Most if not all gymnotiform lineages were essentially modern in terms of phenotypic and species diversity by the late Middle Miocene (c. 12 Ma). There are at least 12 gymnotiform clades, in 6 genera and 4 of the 5 gymnotiform families, with cis–trans-Andean distributions, suggesting minimum dates for the origins of these clades by 12 Ma as a consequence of the final closure of the Andean cordillera during the Magdalenian orogeny. Moreover, in several of these clades the cis–trans-Andean sister taxa occupy terminal positions within the phylogeny, indicating that many modern species-level taxa had evolved by the late Middle Miocene. The only known fossil gymnotiforms are from the Late Miocene of Bolivia. These fossils resemble modern sternopygids and set a minimum date for the origin several derived features associated with active electroreception, including the specialized culteriform body shape, ball-and-socket anal-fin ray articulation, and regeneration of the caudal appendage. Since the Middle Miocene the South American river basins have undergone massive modifications and sharing of faunas resulting in very extensive exchanges of species. This is reflected by low levels of geographical endemism in the area corresponding to the highest electric fish diversity, and by the polyphyletic structure of regional assemblages.

Understanding the forces governing the number and identity of gymnotiform species in a regional assemblage requires phylogenetic, biogeographic, and ecological data as well as information about electric signals. These phylogenetic and biogeographic patterns suggest that diversification in most gymnotiform taxa has been a continental, not basin-wide, phenomenon, occurring over periods of tens of millions of years. In this regard Amazonian species richness is not strictly a consequence of local or regional processes. These patterns are representative of other highly diverse groups of Neotropical fishes, and do not resemble those of monophyletic, rapidly generated species flocks in isolated aquatic systems.

Investigations into phylogenetic, biogeographic, and ecological aspects of gymnotiform diversity are accelerating and the actual dimensions of the fauna now coming to be more fully appreciated. Current projects to complete the descriptive stage of this research program include phylogenetic revisions of eight species-rich gymnotiform clades including 51% of known gymnotiform species. Species-level understanding of these geographically widespread and diverse taxa is advancing the use of Neotropical electric fishes as a model group for studies on the origins and maintenance of species and signal diversity. The results of these systematically based studies, combined with comparative studies of electric organ and electrocyte morphology and electrocyte membrane physiology, are advancing understanding how gymnotiform fishes generate stereotyped, species-specific signals. These results are also contributing to the general area of vertebrate neurophysiology concerned the coordination of complex motor output patterns. The wealth of knowledge accumulated from comparative studies of electric fish phylogeny, biogeography, ecology, and ethology is allowing perhaps for the first time a synthetic understanding of animal communication in the evolutionary context.

Acknowledgments. We thank T. Bullock, C. Hopkins, and A. Popper for the invitation to write this chapter and for critical reviews of the manuscript. We also thank C. Aadland, S. Albert, J. Alves-Gomes, C. Braun, R. Campos-da-Paz, A. Caputi, F. Fernandes, W. Fink, M. Goulding, K. Hulen, D. Julian, F. Kirschbaum, S. Kullander, W. Lamar, M. Lannoo, K. Lester, N. Lovejoy, L. Malabarba, R. Lowe-McConnell, J. Lundberg, J. Maldonado, R. Miller, P. Moller, H. Ortega, L. Page, L. Parenti, R. Reis, R. Royero, D. Santana, H. Schwassman, G. Smith, J. Sullivan, D. Stewart, P. Stoddard, D. Thorsen, M. Triques, R. Vari, S. Wolgemuth, and H. Zakon for their generosity of data and ideas. We acknowledge the Neodat project (NSF/AID DEB Grant 90-24797) for collection information. Aspects of this research were supported by grants to J.S.A. from the U.S. National Science Foundation (NSF-DEB 0215388, 0317278, 0138633) and to W.G.R.C. from the Fisheries Society (UK) and CNPq (Brazil). This chapter is dedicated to the late F. Mago-Leccia for his enormous contributions to our understanding of gymnotiform diversity.

References

Aadland CR, Braun CB, Stölting KN, Alves-Gomes JA (2003) Intraspecific differences in the electric organ discharge of two species of sympatric gymnotiforms. In: Abstracts of the 2003 meeting of the American Society of Ichthyologists and Herpetologists, Manaus, Brazil.

Aguilera PA, Caputi AA (2003) Electroreception in *G. carapo:* detection of changes in waveform of the electrosensory signals. J Exp Biol 206:989–998.

Aguilera PA, Castello ME, Caputi AA (2001) Electroreception in *Gymnotus carapo*: differences between self-generated and conspecific-generated signal carriers. J Exp Biol 204:185–198.

Albert JS (2001) Species diversity and phylogenetic systematics of American knifefishes (Gymnotiformes, Teleostei). University of Michigan Misc Publ Mus Zool 190:1–127.

Albert JS (2002) Eternal vigilance on an Amazon floodplain. Curr Biol 12:R442–R443.

Albert JS (2003a) Family Apteronotidae. In: Reis RE, Kullander SO, Ferraris CJ Jr (eds), Checklist of the Freshwater Fishes of South and Central America. Porto Alegre: Edipucrs, pp. 503–508.

Albert JS (2003b) Family Sternopygidae. In: Reis RE, Kullander SO, Ferraris CJ Jr (eds), Checklist of the Freshwater Fishes of South and Central America. Porto Alegre: Edipucrs, pp. 493–497.

Albert JS, Campos-da-Paz R (1998) Phylogenetic systematics of American knifefishes: a review of the available data. In: Malabarba L, Reis RE, Vari RP, de Lucena CAS, de Lucena ZMS (eds), Phylogeny and Classification of Neotropical Fishes. Porto Alegre: Museu de Ciências e Tecnologia, pp. 409–438.

Albert JS, Crampton WGR (2001) Five new species of *Gymnotus* (Teleostei: Gymnoti-

formes) from an Upper Amazonian floodplain, with descriptions of electric organ discharges and ecology. Ichthyol. Explor Freshwaters 12:241–226.
Albert JS, Crampton WGR (2003a) Seven new species of the Neotropical electric fish *Gymnotus* (Teleostei, Gymnotiformes) with a redescription of *G. carapo* (Linnaeus). Zootaxa 287:1–54.
Albert JS, Crampton WGR (2003b) Family Hypopomidae. In: Reis RE, Kullander SO, Ferraris CJ Jr (eds), Checklist of the Freshwater Fishes of South and Central America. Porto Alegre: Edipucrs, pp. 500–502.
Albert JS, Fink WL (1996) *Sternopygus xingu,* a new species of electric fish (Gymnotoidei, Teleostei), from South America, with comments on the phylogenetic position of *Sternopygus*. Copeia 1996:85–102.
Albert JS, Fink WL (in review) Phylogenetic relationships of fossil Neotropical electric fishes (Osteichthyes: Gymnotiformes) from the Upper Miocene of Bolivia. J Vert Paleontol (in review).
Albert JS, Miller RR (1995) *Gymnotus maculosus*: a new species of electric fish from Middle America (Teleostei: Gymnotoidei), with a key to the species of *Gymnotus*. Proc Biol Soc Wash 108:662–678.
Albert JS, Lannoo MJ, Yuri T (1998) Testing hypotheses of neural evolution in gymnotiform electric fishes using phylogenetic character data. Evolution 52:1760–1780.
Albert JS, Fernandes-Matioli FM, de Almeida-Toledo LF (1999) A new species of *Gymnotus* (Gymnotiformes, Teleostei) from Southeastern Brazil: towards the deconstruction of *Gymnotus carapo*. Copeia 1999:410–421.
Albert JS, Froese R, Paulay D (2000) The brains table. In: Froese R, Paulay D (eds), FishBase 2000, Concepts, Design and Data Sources. Manila: ICLARM, pp. 234–237.
Albert JS, Lovejoy NR, Crampton WGR (2005a) Miocene tectonism and the separation of cis- and trans-Andean basins: evidence from Neotropical fishes. J. South Am Earth Sci (in press).
Albert JS, Crampton WGR, Thorsen DH, Lovejoy NR (2005b) Phylogenetic systematics and historical biogeography of the Neotropical electric fish *Gymnotus* (Teleostei: Gymnotiformes). Systemat Biodivers (in press).
Alves-Gomes JA (1997) Informações preliminares sobre a bio-ecologia de peixes eléctricos (Ordem Gymnotiformes) em Roraima. In: Barbosa RI, Ferreira EJG, Castellón EG (eds), Homen, Ambiente e Ecologia no Estada de Roraima. Manaus: INPA, pp. 509–555.
Alves-Gomes JA (1999) Systematic biology of gymnotiform and mormyriform electric fishes: phylogenetic relationships, molecular clocks and rates of evolution in the mitochondrial rRNA genes. J Exp Biol 202:1167–1183.
Alves-Gomes JA (2001) The evolution of electroreception and bioelectrogenesis in teleost fish: a phylogenetic perspective. J Fish Biol 58:1489–1511.
Alves-Gomes JA, Orti G, Haygood M, Heiligenberg W, Meyer A (1995) Phylogenetic analysis of the South American electric fishes (order Gymnotiformes) and the evolution of their electrogenic system—a synthesis based on morphology, electrophysiology, and mitochondrial sequence data. Mol Biol Evol 12:298–318.
Anderson MJ (1987) Molecular differentiation of neurons from ependyma-derived cells in tissue cultures of regenerating teleost spinal cord. Brain Res 388:131–136.
Anderson MJ, Waxman SG (1981) Morphology of regenerated spinal cord in *Sternarchus albifrons*. Cell Tissue Res 219:1–8.
Assad C, Rasnow B, Stoddard PK, Bower JM (1998) The electric organ discharges of the gymnotiform fishes: II. *Eigenmannia*. J Comp Physiol A 183:419–432.

Assunção MI, Schwassman HO (1995) Reproduction and larval development of *Electrophorus electricus* on Marajó Island (Pará, Brazil). Ichthyol Explor Freshwaters 6: 175–184.

Barthem RB (1985) Ocorrencia, distribuicao e biologia de peixes da Baia de Marajo, Estuario Amazonico. Bol Mus Paraense Emilio Goeldi Zool 2:49–69.

Bass AH (1986) Electric organs revisited. In: Bullock TH, Heiligenberg W (eds), Electroreception. New York: John Wiley & Sons, pp. 13–70

Bastian J (1986) Electrolocation: behavior, anatomy, physiology. In: Bullock TH, Heiligenberg W (eds), Electroreception. New York: John Wiley & Sons, pp. 577–612.

Bell C, Bodznick D, Montgomery J, Bastian J (1997) The generation and subtraction of sensory expectations within cerebellum-like structures. Brain Behav Evol 50:17–31.

Bennett MVL (1971) Electric organs. In: Hoar WS, Randal DJ (eds), Fish Physiology. New York: Academic Press, p. 346–491.

Benton MJ (1993) The Fossil Record 2. London: Chapman and Hall, 845 pp.

Bermingham E, Martin AP (1998) Comparative mtDNA phylogeography of neotropical freshwater fishes: testing shared history to infer the evolutionary landscape of lower Central America. Mol.Ecol 7:499–517.

Bermingham E, McCafferty SS, Martin AP (1997) Fish biogeography and molecular clocks: perspectives from the Panamanian Isthmus. In: Kocher TD, Stepien CA (eds), Molecular Systematic of Fishes. San Diego: Academic Press, pp. 113–128.

Black-Cleworth P (1970) The role of electrical discharges in the non-reproductive social behavior of *Gymnotus carapo* (Gymnotidae, Pisces). Animal Behav Monogr 3:1–77.

Boulenger GA (1898) On a collection of fishes from the Rio Juruá. Brazil. Trans Zool Soc Lond 14:421–428.

Campos-da-Paz R (2003) Family Gymnotidae. In: Reis RE, Kullander SO, Ferraris CJ (eds), Checklist of the Freshwater Fishes of South and Central America. Porto Alegre: Edipucrs, pp. 483–486.

Campos-da-Paz R, Albert JS (1998) The gymnotiform "eels" of tropical America: a history of classification and phylogenetic studies of South American electric knifefishes (Teleostei: Ostariophysi). In: Malabarba L, Reis RE, Vari RP, de Lucena CAS, de Lucena ZMS (eds), Phylogeny and Classification of Neotropical Fishes. Porto Alegre: Museu de Ciências e Tecnologia, p. 401–417.

Campos-da-Paz R, Costa WJEM (1996) *Gymnotus bahianus* p. nov., a new gymnotid fish from Eastern Brazil (Teleostei: Ostariophysi: Gymnotiforms), with evidence for the monophyly of the genus. Copeia 1996:937–944.

Campos-da-Paz R, Paepke HJ (1994) On *Sternarchorhamphus hahni*, a member of the rhamphichthyid genus *Rhamphichthys* (Ostariophysi: Gymnotiformes). Ichthyol Explor Freshwaters 5:155–159.

Caputi AA (1999) The electric organ discharge of pulse gymnotiforms: the transformation of a simple impulse into a complex spatiotemporal electromotor pattern. J Exp Biol 202:1229–1241.

Caputi A, Aguilera P (1996) A field potential analysis of the electromotor system in *Gymnotus carapo*. J Comp Physiol a 179:827–835.

Caputi AA, Aguilera PA (2003) Probability and amplitude of novelty responses as a function of the change in contrast of the reafferent image in *G. carapo*. J Exp Biol 206:999–1010.

Caputi A, Macadar O, Trujillo-Cenóz O (1989) Waveform generation of the electric organ discharge in *Gymnotus carapo*. 3. Analysis of the fish body as an electric source. J Comp Physiol A 165:361–370.

Caputi A, Macadar O, Trujillo-Cenóz O (1994) Wave-form generation in *Rhamphichthys rostratus* (L) (Teleostei, Gymnotiformes). J Comp Physiol A 174:633–642.

Caputi AA, Silva AC, Macadar O (1998) The electric organ discharge of *Brachyhypopomus pinnicaudatus*. Brain Behav Evol 52:148–158.

Carr CE, Maler L (1986) Electroreception in gymnotiform fish: central anatomy and physiology. In: Bullock TH, Heiligenberg W (eds), Electroreception. New York: John Wiley & Sons, pp. 319–374.

Carter GS, Beadle C (1931) The fauna of the swamps of the Paraguayan Chaco in relation to its environment. II. Respiratory adaptations in the fishes. Zool J Linn Soc Lond 37:327–366.

Castelnau F (1855) Poissons nouveaux ou rares récueillem pendant l'Expedition dans les parties centrales de l'Amérique du Sud, de Rio de Janeiro a Lima, et de Lima au Pará. Paris P. Bertrand, Libraire-Editeur, 112 pp.

Castro RMC, Casatti L (1997) The fish fauna from a small forest stream of the upper Paraná River basin, southeastern Brazil. Ichthyol Explor Freshwaters 7:337–352.

Chardon M, de la Hoz E (1974) Towards an improved classification of the gymnotiform fishes by the use of the splanchnocranium characters. Acta Biol Jugoslav, Beograd 6: 15–25.

Cox-Fernandes C (1998a) Sex related morphological variation in two species of apteronotid fishes (Gymnotiformes) from the Amazon River Basin. Copeia 1998:730–735.

Cox-Fernandes C (1998b) Detrended canonical correspondence analysis (DCCA) of electric fish assemblages in the Amazon. In: Val AL, Almeida-Val VMF (eds), Biology of Tropical Fishes, Manaus: INPA, pp. 21–39.

Cox-Fernandes C, Lundberg JGL, Riginos C (2002) Largest of all electric-fish snouts: hypermorphic facial growth in male *Apteronotus hasemani* and the identity of *Apteronotus anas* (Gymnotiformes: Apteronotidae). Copeia 2002:52–61.

Crampton WGR (1996a) Gymnotiform fish: an important component of Amazonian flood plain communities. J Fish Biol 48:298–301.

Crampton WGR (1996b) The electric fish of the Upper Amazon: ecology and signal diversity. Oxford: Department of Zoology, The University of Oxford.

Crampton WGR (1998a) Electric signal design and habitat preferences in a species rich assemblage of gymnotiform fishes from the Upper Amazon Basin. Anais Acad Brasil Ciên 70:805–847.

Crampton WGR (1998b) Effects of anoxia on the distribution, respiratory strategies and electric signal diversity of gymnotiform fishes. J Fish Biol 53:502–520.

Crampton WGR (1999) Os peixes da Reserva Mamiraua: diversidade e historia natural na planicie alagavel da Amazonia. In: Queiroz HL, Crampton WGR (eds), Estrategias para Manejo de Recursos Pesqueiros em Mamiraua. Sociedade Civil Mamiraua/CNPq, Brasilia, pp. 10–36.

Crampton WGR, Albert JS (2003) A redescription of *Gymnotus coropinae*, an often misidentified species of gymnotid fish, with notes on ecology and electric organ discharge. Zootaxa 348:1–20.

Crampton WGR, Albert JS (2005) Evolution of electric signal diversity in the gymnotiform fishes. In: Collin SP, Kapoor BG, Ladich F, Moller P (eds), Fish Communication. New York: Science Publishers (in press).

Crampton WGR, Hopkins CD (2005) Nesting and paternal care in the weakly electric fish *Gymnotus* (Gymnotiformes: Gymnotidae) with descriptions of larval and adult electric organ discharges of two species. Copeia (in press).

Crampton WGR, Lovejoy NR, Albert JS (2003) *Gymnotus ucamara*: a new species of Neotropical electric fish from the Peruvian Amazon (Ostariophysi: Gymnotidae). Zootaxa 277:1–18.

Crampton WGR, Thorsen DH, Albert JS (2004a) *Steatogenys ocellatus*, a new species of Neotropical electric fish (Gymnotiformes: Hypopomidae) from the lowland Amazon Basin. Copeia 2004:78–91.

Crampton WGR, Hulen K, Albert JS (2004b) *Sternopygus branco*, a new species of Neotropical electric fish (Gymnotiformes: Sternopygidae) from the lowland Amazon basin, with descriptions of ecology and electric organ discharges. Copeia 2004:244–258.

Crampton WGR, Thorsen DH, Albert JS (2005) Three new species from a diverse assemblage of *Gymnotus* (Gymnotidae) in the Upper Amazon. Copeia (in press).

de Almeida-Toledo LF (1998) Cytogenetic markers in Neotropical freshwater fishes. In: Malabarba L, Reis RE, Vari RP, de Lucena CAS, de Lucena ZMS (eds), Phylogeny and Classification of Neotropical Fishes. Porto Alegre: Museu de Ciências e Tecnologia, pp. 583–588.

de Almeida-Toledo LF, Daniel-Silva MFZ, F Foresti, Toledo SA (1998) Sex related heterochromatin in two neotropical fish species, order Gymnotiformes, presenting X1X2Y sex chromosome system. Cytogenet Cell Genet 81:68.

de Almeida-Toledo LF, Daniel-Silva MFZ, Lopes CE, Toledo SD (2000a) Sex chromosome evolution in fish. II. Second occurrence of an X1X2Y sex chromosome system in Gymnotiformes. Chromosome Res 8:335–340.

de Almeida-Toledo LF, Foresti F, Daniel MFZ, Toledo SA (2000b) Sex chromosome evolution in fish: the formation of the neo-Y chromosome in *Eigenmannia* (Gymnotiformes). Chromosoma 109:197–200.

de Almeida-Toledo LF, Daniel-Silva MFZ, Moysés CB, Fonteles SBA, Lopes CE, Akama A, Foresti F (2002) Chromosome evolution in fish: sex chromosome variability in *Eigenmannia virescens* (Gymnotiformes: Sternopygidae). Cytogenet Genome Res 99: 164–169.

de Pinna MCC (1998) Phylogenetic relationships of Neotropical Siluriformes. In Malabarba L, Reis RE, Vari RP, de Lucena CAS, de Lucena ZMS (eds), Phylogeny and Classification of Neotropical Fishes. Porto Alegre: Museu de Ciências e Tecnologia, pp. 279–330.

Diaz de Gamero ML (1996) The changing course of the Orinoco River during the Neogene: a review. Palaeogeogr Palaeoclimatol Palaeoecol 123:385–402.

Duque-Caro H (1990) Major Neogene events in Panamaic South America. In: Tsuchi R (ed), Pacific Neogene Events, Their Timing, Nature and Interrelationships. Tokyo: Tokyo University Press, pp. 101–114.

Dye JC, Meyer JH (1986) Control over the electric organ discharge. In: Bullock TH, Heiligenberg W (eds), Electroreception. New York: John Wiley & Sons, pp. 71–102

Eernisse DJ, Kluge AG (1993) Taxonomic congruence versus total evidence, and amniote phylogeny inferred from fossils, molecules, and morphology. Mol Biol Evol 10:1170–1195.

Eigenmann CH (1905) *Sternarchorhamphus* gen. nov. In: Eigenmann CH, Ward DP (eds), The Gymnotidae. Proc Wash Acad Sci 7:157–186.

Eigenmann CH, Allen WR (1942) Fishes of Western South America. Lexington: University of Kentucky, 494 pp.

Eigenmann CH, Ward DP (1905) The Gymnotidae. Proc Wash Acad Sci 7:157–186.

Ellis MM (1912) Order Glanencheli, Family X. Gymnotidae. In: Eigenmann CH (ed), The Freshwater Fishes of British Guiana. Mem Carnegie Mus 5:422–442.

Ellis MM (1913) The Gymnotid Eels of Tropical America. Mem Carnegie Mus 6:109–195.

Eschmeyer WN (2003) Catalog of Fishes. San Francisco: California Academy of Sciences.

Evans M (1929) Some notes on the anatomy of the electric eel, *Gymnotus electrophorus*, with special reference to a mouth breathing organ and the swim-bladder. Proc Zool Soc Lond 2:17–22.

Fernandes-Matioli FMC, de Almeida-Toledo LF (2001) A molecular phylogenetic analysis in *Gymnotus* species (Pisces: Gymnotiformes) with inferences on chromosome evolution. Caryologia 54:23–30.

Fernandes-Matioli FMC, de Almeida-Toledo LF, Toledo SA (1998a) Natural triploidy in the Neotropical species *Gymnotus carapo* (Pisces: Gymnotiformes). Caryologia 51: 319–322.

Fernandes-Matioli FMC, Marchetto MCN, de Almeida-Toledo LF, Toledo, SA (1998b) High intraspecific karyological conservation in four species of *Gymnotus* (Pisces: Gymnotiformes) from southeastern Brazilian basins. Caryologia 51:221–234.

Fernandes-Matioli FMC, Matioli SR, de Almeida-Toledo LF (2000) Species diversity and geographic distribution of *Gymnotus* (Pisces: Gymnotiformes) by nuclear (GGAC)(n) microsatellite analysis. Genet Mol Biol 23:803–807.

Ferraris CJ (2003) Family Rhamphichthyidae. In: Reis RE, Kullander SO, Ferraris CJ (eds), Checklist of the Freshwater Fishes of South and Central America. Porto Alegre: Edipucrs, pp. 492–493.

Finger TE (1986) Electroreception in catfish: behavior, anatomy, and physiology. In: Bullock TH, Heiligenberg W (eds), Electroreception. New York: John Wiley & Sons, pp. 287–318.

Finger TE, Karten HJ (1978) The accessory optic system of teleosts. Brain Res 135: 144–149.

Fink SV, Fink WL (1981) Interrelationships of the ostariophysan fishes (Teleostei). Zool J Linn Soc 72:297–353.

Fink SV, Fink WL (1996) Interrelationships of ostariophysan fishes (Teleostei). In: Stiassney M, Parenti LR, Johnson GD (eds), Interrelationships of Fishes. New York: Academic Press, pp. 209–249.

Fonteles-Santos SBA, Fernandes-Matioli FMC, Lopes CE, Akama A, de Almeida-Toledo LF (2002) *Electrophorus electricus* (Gymnotiformes: Electrophoridae): características citogenéticas, genéticas e moleculares. 48 Congresso Nacional de Genética (GA110), realizado no período de 17 a 20 de setembro de 2002, Águas de Lindóia-SP.

Franchina CR, Hopkins CD (1996) The dorsal filament of the weakly electric Apteronotidae (Gymnotiformes, Teleostei) is specialized for electroreception. Brain Behav Evol 47:165–178.

Franchina CR, Stoddard PK (1998) Plasticity of the electric organ discharge waveform on the electric fish *Brachyhypopomus pinnicaudatus* I. Quantification of day-night changes. J Comp Physiol A 183:759–768.

Garcia M (1995) Aspectos ecológicos dos peixes das águas abertas de um lago no arquipélago das Anavilhanas, Rio Negro, AM, INPA/UFAM, Unpublished MSc. Thesis, 99 pp.

Gayet M, Meunier F (2000) Rectification of the nomenclature of the genus name *Ellisella*

Gayet and Meunier, 1991 (Teleostei, Ostariophysi, Gymnotiformes) in *Humboldtichthys Nom. Nov.* Cybium 24:104.
Gayet M, Meunier FJ, Kirschbaum F (1994) *Ellisella kirschbaumi* Gayet and Meunier, 1991, gymnotiform e fossile de Bolivie et ses relations phylogénétiques au sein des formes actuelles. Cybium 18:273–306.
Gill TN (1864) Ichthyological notes. Proc Acad Nat Sci Phil 16:151–154.
Giora J (2004) Biologia reprodutiva e hábito alimentar de *Eigenmannia trilineata* López and Castello, 1966 (Teleostei, Sternopygidae) do Parque Estadual de Itapuã, Rio Grande do Sul. Universidade Federal do Rio Grande do Sul, Brasil, 90 pp.
Gotter AL, Kaetzel MA, Dedman JR (1998) *Electrophorus electricus* as a model system for the study of membrane excitability. Comp Biochem Physiol 119:225–241.
Goulding M, Carvalho ML, Ferreira EG (1988) Rio Negro, Rich Life in Poor Waters. The Hague, Netherlands: SPB Academic Publishing, 200 pp.
Goulding MJ, Bartham R, Ferreira E (2003) The Smithsonian Atlas of the Amazon. Washington, DC: Smithsonian Institution Press, 253 pp.
Greenwood PH (1984) What is a species flock? In: Echelle AA Kornfield I (eds), Evolution of Fish Species Flocks. Orono, ME: University of Maine at Orono Press, pp. 13–25.
Gregory-Wodzicki KM (2000) Uplift history of the Central and Northern Andes: a review. Geol Soc Am Bull 112:1091–1105.
Guerrero J (1997) Stratigraphy, sedimentary environments, and the Miocene uplift of the Colombian Andes. In: Kay RF, Hadden RH, Cifelli RL, Flynn JJ (eds), Vertebrate Paleonotology in the Neotropics: The Miocene Fauna of La Venta, Colombia. Washington, DC, Smithsonian Institution Press, pp. 15–43.
Hagedorn M (1985) Ecology and behaviour of a pulse-type electric fish, *Hypopomus occidentalis* (Gymnotiformes, Hypopomidae), in a fresh-water stream in Panama. Copeia 1985:324–335.
Hagedorn M, Keller C (1996) Species diversity of gymnotiform fishes in Manu Bioreserve, Pakitza, Peru. In: Wilson DE, Sandoval A (eds), Manu: The Biodiversity of Southeastern Peru. Washington, DC: Smithsonian Institution Press, pp. 483–502.
Heiligenberg WF (1980) Species specificity of electric organ discharges in sympatric gymnotoid fish of the Rio Negro. Acta Biol Venezol 10:187–203.
Heiligenberg WF, Bastian A (1986) Jamming avoidance responses. Model systems for neuroethology. In: Bullock TH, Heiligenberg WF (eds), Electroreception. New York: John Wiley & Sons, pp. 613–649.
Henderson PA, Walker I (1986) On the leaf litter community of the Amazonian blackwater stream Tarumã-Mirim. J Trop Ecol 2:1–17.
Hoedeman JJ (1962) Notes on the ichthyology of Surinam and other Guianas, 11. New gymnotiform fishes from Surinam and French Guiana, with additional records and a key to the groups and species from Guiana. Bull Aquat Biol Amst 3:97–107.
Hoorn C (1996) Miocene deposits in the Amazonian foreland basin. Science 273:122–123.
Hoorn C, Guerrero J, Sarmiento GA, Lorente MA (1995) Andean Tectonics as a Cause for Changing Drainage Patterns in Miocene Northern South-America. Geology 23: 237–240.
Hopkins CD (1972) Sex differences in electric signaling in an electric fish. Science 176: 1035–1037.
Hopkins CD (1983) Functions and mechanisms in electroreception. In: Northcutt RG,

Davis RE (eds), Fish Neurobiology, Vol. 1. Ann Arbor, MI: University of Michigan Press, pp. 215–259.

Hopkins CD (1999) Design features for electric communication. J Exp Biol 202:1217–1228.

Hopkins CD, Heiligenberg WF (1978) Evolutionary designs for electric signals and electroreceptors in gymnotiform fishes of Surinam. Behav Ecol Sociobiol 3:113–134.

Hopkins CD, Comfort NC, Bastian J, Bass A (1990) Functional analysis of sexual dimorphism in an electric fish, *Hypopomus pinnicaudatus*, order Gymnotiformes. Brain Behav Res 35:350–367.

Hulen K, Crampton WGR, Albert JS (2005) Phylogenetic systematics and historical biogeography of the Neotropical electric fish *Sternopygus* (Gymnotiformes, Teleostei). Systemat Biodivers (in press).

Ituralde-Vincent MA, MacPhee RDE (1999) Paleogeography of the Caribbean region implications for Cenozoic biogeography. Bull Am Mus Nat Hist 238:1–95.

Jordan DS, Evermann BW (1896) The fishes of North and Middle America. Bull US Nat Mus 47:i–lx, 1–954.

Julian D, Crampton WGR, Wolgemuth S, Albert JS (2003) Oxygen consumption by weakly electric Neotropical fishes. Oecologia 137:502–511.

Junk WJ, Robertson BA (1997) Aquatic invertebrates. In: Junk W (ed), The Central Amazon Floodplain: Ecology of a Pulsing System. New York: Springer-Verlag, pp. 279–298

Kaup JJ (1856) Family Gymnotidae. In: Catalogue of Apodal Fishes. Br Mus Nat Hist, pp. 124–142.

Kirschbaum F (1977) Electric organ ontogeny: distinct larval organ precedes the adult organ in weakly electric fish. Naturwissenchaften 64:387–388.

Kirschbaum F (1995) Taxonomy, zoogeography, and general ecology of South American knifefishes (Gymnotiformes). In: Moller P (ed), Electric Fishes. London: Chapman and Hall, pp. 446–464

Kirschbaum F, Meunier FJ (1981) Experimental regeneration of the caudal skeleton of the glass knifefish *Eigenmannia virescens* (Rhamphichthyidae, Gymnotoidei). J Morphol 168:121–135.

Kirschbaum F, Wieczorek L (2002) Entdeckung einer neuen Fortpflanzungs-strategie bei südamerikainischen Messerfischen (Teleostei: Gymnotiformes: Gymnotidae): Maulbrüten bei *Gymnotus carapo*. Verhalt Aquarienfisch 2:99–107.

Kornfield I, Smith PF (2000) African cichlid fishes: model systems for evolutionary biology. Ann Rev Ecol Systemat 31:163–196.

Korringa M (1970) A new gymnotiform fish from the Rio Tocantins, Brazil. Proc Cal Acad Sci 38:265–271.

Lacépède BGE (1798–1803) Histoire Naturelle des Poissons, 5 volumes, Paris.

Lasso C, Rial AB, Lasso-Alcalá O (1997) Bioecological aspects of the taxocenosis of electric fishes (Ostariophysi: Gymnotiformes), in the Apure Llanos of Venezuela. Acta Biol Venezuél 17:7–29.

Lazar GP, Toth P, Szabo T (1987) Retinal projections in gymnotiform fishes. J Hirnforsch 28:13–26.

Liem KF, Eclancher B, Fink WL (1984) Aerial respiration in the banded knife fish *Gymnotus carapo* (Teleostei: Gymnotoidei). Physiol Zool 57:185–195.

Linnaeus C (1758) Systema Naturae, Ed. X, Vol. 1, 246. Holmiae: Laurentii Salvii, 824 pp.

Lorenzo D, Velluti JC, Macadar O (1988) Electrophysiological properties of abdominal electrocytes in the weakly electric fish *Gymnotus carapo*. J Comp Physiol A 162: 141–144.

Lovejoy NR, deAraújo MLG (2000) Molecular systematics, biogeography, and population structure of Neotropical freshwater needlefishes of the genus *Potamorrhaphis*. Mol. Ecol 9:259–268.

Lovejoy NR, Bermingham RE, Martin AP (1998) Marine incursions into South America. Nature 396.421–422.

Lundberg JG (1993) African-South American freshwater fish clades and continental drift: problems with a paradigm. In: Goldblatt P (ed), Biological Relationships Between Africa and South America. New Haven: Yale University Press, pp. 156–199.

Lundberg JG (1997) Freshwater fishes and their paleobiotic implications. In: Kay RF, Hadden RH, Cifelli RL, Flynn JJ (eds), Vertebrate Paleonotology in the Neotropics: The Miocene Fauna of La Venta, Colombia. Washington, DC: Smithsonian Institution Press, pp. 67–92.

Lundberg JG (1998) The temporal context for the diversification of Neotropical fishes. In: Malabarba L, Reis RE, Vari RP, de Lucena CAS, de Lucena ZMS (eds), Phylogeny and Classification of Neotropical Fishes. Porto Alegre: Museu de Ciências e Tecnologia, pp. 49–68.

Lundberg JG, Mago-Leccia F (1986) A review of *Rhabdolichops* (Gymnotiformes, Sternopygidae), a genus of South American freshwater fishes, with descriptions of four new species. Proc Acad Nat Sci Phil 138:53–85.

Lundberg JG, Lewis WM, Saunders JF, Mago-Leccia F (1987) A major food web component in the Orinoco River channel: evidence from planktivorous electric fishes. Science 237:81–83.

Lundberg JG, Cox-Fernandes C, Albert JS, Garcia M (1996) *Magosternarchus*, a new genus with two new species of electric fish (Gymnotiformes: Apteronotidae) from the Amazon River basin, South America. Copeia 1996:657–670.

Lundberg JG, Marshall LC, Guerrero J, Horton B, Malabarba MCSL, Wesselingh F (1998) The stage for Neotropical fish diversification: a history of tropical South American rivers. In: Malabarba L, Reis RE, Vari RP, de Lucena CAS, de Lucena ZMS (eds), Phylogeny and Classification of Neotropical Fishes. Porto Alegre: Museu de Ciências e Tecnologia, pp. 13–48.

Mago-Leccia F (1994) Electric Fishes of the Continental Waters of America. Biblioteca de la Academia de Ciencias Fisicas, Matematicas, y Naturales, Caracas, Venezuela 29: 1–206.

Mago-Leccia F, Lundberg JG, Baskin JN (1985) Systematics of the South American freshwater genus *Adontosternarchus* (Gymnotiformes, Apteronotidae). Contrib Sci LA County Mus Nat Hist. 358:1–19.

Maldonado-Ocampo JA, Albert JS (2003) Species diversity of gymnotiform fishes in Colombia. Biota Colomb 4:145–163.

Marrero C (1987) Notas preliminares acerca de la historia natural de los peces del Bajo Llano. I. Comparacion de los habitos alimentarios de tres especies de peces gymnotiformes en el Rio Apure (Edo Apure, Venezuela). Rev Hydrobiol Trop 20:57–63.

Martin AP, Bermingham E (2000) Regional endemism and cryptic species revealed by molecular and morphological analysis of a widespread species of Neotropical catfish. Proc R Soc Lond B 267:1135–1141.

McAnelly L, Silva A, Zakon HH (2003) Cyclic AMP modulates electrical signaling in a weakly electric fish. J Comp Physiol A 189:273–282.

Mills A, Zakon HH, Marchaterre MA, Bass AH (1992) Electric organ morphology of

Sternopygus macrurus, a tone-type weakly electric fish with a sexually dimorphic EOD. J Neurobiol 23:920–932.

Moller P (1995a) Basic elements of electrocommunication systems. In Moller P (ed), Electric Fishes: History and Behavior. London: Chapman and Hall, pp. 167–198.

Moller P (1995b) Electric Organs. In Electric Fishes: History and Behavior. London: Chapman and Hall, pp. 385–402.

Motta-Bührnheim C, Cox-Fernandes C (2001) Low seasonal variation of fish assemblage in Amazonian rain forest streams. Ichthyol Explor Freshwaters 12:65–78.

Müller J, Troschel FH (1848) Horae Ichthyologicae. Beschreibung und Abbildung neuer Fische. Berlin: Drittes Heft, 28 pp.

Nijssen H, Isbrücker IJH (1972) On *Hypopygus lepturus*, a little known dwarf gymnotid fish from South America (Pisces, Cypriniformes, Gymnotoidei). Zool Meded 47:160–176.

Paxton CGM, Crampton WGR, Burgess P (1996) Miocene deposits in the Amazonian foreland basin. Science 273:122–123.

Rasnow B, JM Bower (1996) The electric organ discharges of the gymnotiform fishes I. *Apteronotus leptorhynchus*. J Comp Physiol A 178:383–396.

Reid S (1983) La biologia de los bagres rayados *Pseudoplatystoma fasciatum* y *P. tigrinum* en la cuenca del rio Apure, Venezuela. Rev. UNELLEZ Cienc Tecnol 1:13–41.

Reis RE (1998) Systematics, biogeography, and the fossil record of the Callichthyidae: a review of available le data. In: Malabarba L, Reis RE, Vari RP, de Lucena CAS, de Lucena ZMS (eds), Phylogeny and Classification of Neotropical Fishes. Porto Alegre: Museu de Ciências e Tecnologia, pp. 351–362.

Silva A, Quintana L, Galeano M, Errandonea P (2003) Biogeography and breeding in Gymnotiformes from Uruguay. Environ Biol Fishes 66:329–338.

Stewart DJ, Ibarra M, Barriga-Salazar RB (2002) Comparison of deep-river and adjacent sandy-beach fish assemblages in the Napo River Basin, Eastern Ecuador. Copeia 2002: 333–343.

Stoddard PK (1999) Predation enhances complexity in the evolution of electric fish signals. Nature 400:254–256.

Stoddard PK, Rasnow B, Assad C (1999) Electric organ discharges of the gymnotiform fishes: III. *Brachyhypopomus*. J Comp Physiol A 184:609–630.

Sullivan JP (1997) A phylogenetic study of the Neotropical hypopomid electric fishes (Gymnotiformes: Rhamphichthyoidea). Unpublished Ph.D. Thesis, Duke University.

Triques ML (1993) Filogenia dos gêneros de Gymnotiformes (Actinopterygii, Ostariophysi), com base em caracteres esqueléticos. Comun Mus Cienec PURCS, Zool Porto Alegre 6:85–130.

Triques ML (1996a) *Eigenmannia vicentespelae*, a new species of cave dwelling electrogenic Neotropical fish (Ostariophysi: Gymnotiformes: Sternopygidae). Rev Aquariol 23:1–4.

Triques ML (1996b) *Iracema caiana*, new genus and species of electrogenic Neotropical freshwater fish (Rhamphichthyidae: Gymnotiformes: Ostariophysi: Actinopterygii). Rev Aquariol 23:91–92.

Triques ML (1997) *Stegostenopos cryptogenys*, new genus and species of Hypopomidae electrogenic neotropical fish from the Rio Negro system, Brazil (Actinopterygii: Ostariophysi: Gymnotiformes). Rev Aquariol 24:31–36.

Triques ML (1999) Three new species of *Rhamphichthys* Muller et Troschel, 1846 (Ostariophysi: Gymnotiformes: Rhamphichthyidae). Rev Aquariol 26:1–6.

Trujillo-Cenóz O, Echagüe JA, Macadar O (1984) Innervation pattern and electric organ discharge waveform in *Gymnotus carapo*. J Neurobiol 15:273–281.

Unguez GA, Zakon HH (1998) Phenotypic conversion of distinct muscle fiber populations to electrocytes in a weakly electric fish. J Comp Neurol 399:20–34.

Vari RP (1988) The Curimatidae: a Lowland Neotropical fish family (Pisces: Characiformes); Distribution, Endemism, and Phylogenetic Biogeography. In: Vanzolini P, Heyer WR (eds), Proceedings of a Workshop on Neotropical Distribution Patterns. Rio de Janeiro: Academia Brasileira de Ciencias, pp. 343–377.

Vari RP (1995) The Neotropical fish family Ctenoluciidae (Teleostei: Ostariophysi: Characiformes): Supra and interfamilial phylogenetic relationships, with a revisionary study. Smithsonian Contrib Zool 564:1–97.

Vari RP (1998) Higher level phylogenetic concepts within Characiformes (Ostariophysi): a historical review. In: Malabarba L, Reis RE, Vari RP, de Lucena CAS, de Lucena ZMS (eds), Phylogeny and Classification of Neotropical Fishes. Porto Alegre: Museu de Ciências e Tecnologia, pp. 111–122.

Vari RP, Weitzman SH (1990) A review of the phylogenetic biogeography of the freshwater fishes of South America. In: Peters G, Hutterer R (eds), Vertebrates in the Tropics. Bonn: Museum Alexander Koenig, pp. 381–393.

Watson D, Bastian J (1979) Frequency response characteristics of electroreceptors in the weakly electric fish *Gymnotus carapo*. J Comp Physiol A 134:191–202.

Westby GWM (1988) The ecology, discharge diversity and predatory behaviour of gymnotiform electric fish in the coastal streams of French Guiana. Behav Ecol Sociobiol 22:341–354.

Winemiller KO, Adite A (1997) Convergent evolution of weakly electric fishes from floodplain habitats in Africa and South America. Environ Biol Fish 49:175–186.

Zakon HH (1986) The electroreceptive periphery. In: Bullock TH, Heiligenberg W (eds), Electroreception. New York: John Wiley & Sons, pp. 103–156.

14

Electric Organs and Their Control

Angel Ariel Caputi, Bruce A. Carlson, and Omar Macadar

1. Introduction: Why Aquatic Animals Evolve Electrogenic Systems

It has been known for more than 2000 years that some fish can induce painful, numbing sensations when handled. In the late 1600s, it was discovered that this sensation resulted from the production of strong electrical currents produced by the fish (Kellaway 1946). In *On the Origin of Species by Means of Natural Selection* (1859), Darwin paid special attention to difficulties with his theory, devoting an entire chapter to the subject. The evolution of electrogenesis was one such area of concern. He noted the difference in structure, innervation, and electrical power of different electric organs (EOs), and the absence of geological evidence for a possible common ancestor between the different taxa exhibiting EOs. He concluded that rather than being homologous structures, the EOs of different fishes should be considered analogous structures that evolved independently. He also raised questions about potential functions for weak EOs. Although electric emissions from weakly electric fish had not yet been discovered, Darwin believed that a functional, weak EO was a necessary step in the evolution of strong EOs. About a century later, the rigorous study of a fish that was found to produce weak electric fields provided answers to such questions (Lissman 1951). Behavioral experiments clearly demonstrated that weak EOs function in electrosensory imaging of the environment (active electrolocation) and communication (Lissman 1958; Lissman and Machin 1958). Since these initial discoveries, there has been a wealth of research into the biology of active electrosensory and electromotor systems.

1.1 Evolution of Electroreception and Electrogenesis

All aquatic animals generate weak electric fields as a result of the uneven distribution of ions between the interior of the animal and its environment (Kalmijn 1974). In teleost fish, DC potentials may reach amplitudes in the hundreds of microvolts to millivolt range near the mouth, and ventilatory movements of the head lead to a characteristic AC modulation of this field that is detectable up to a distance of a few centimeters (Barham et al. 1969; Kalmijn 1974). Such biological sources of electrical potential serve as important cues for predatory fish with electroreceptors and were likely a powerful selective force in electroreceptor evolution (Kalmijn 1987).

More recently, AC potentials resulting from respiratory movements in the round stingray (*Urolophus halleri*) have been shown to serve as an important cue for males to locate females buried under the sand during the breeding season (Tricas et al. 1995). It is likely that similar kinds of electrical cues served as precursors to the evolution of genuine electric signals in the ancestors of extant electrogenic species. Once electrical information could be used for identifying and locating conspecifics, the stage was set for signal evolution and ritualization of electric cues into stereotyped weak electric organ discharges (EODs) for use in communication (Otte 1974). In the sea lamprey (*Petromyzon marinus*), respiration results from highly synchronized muscle activity, which leads to a triphasic potential reaching amplitudes of 200 to 300 μV at a distance of 1.5 to 2 cm (Kleerokoper and Sibakin 1956). Such synchronized muscle activity is a plausible evolutionary precursor to specialized EOs, which are thought to have evolved independently at least six times (Bass 1986). They are composed of highly modified muscle or nerve cells termed electrocytes, whose summated activity results in an EOD.

1.2 Weakly Electric Fish: Electric Energy as a Carrier of Information

Measured close to the fish's body, field amplitudes of weakly electric fish are generally on the order of a few hundred millivolts per centimeter. Weak EOs are known to exist in four different groups of fish: the African mormyriforms (*Gymnarchus niloticus* and the Mormyridae, or "elephant nose fish"), the South American gymnotiforms (knifefish), several species of siluriforms (catfish), and the rajids (skates).

The EODs of mormyriforms and gymnotiforms are produced nearly continuously throughout the life of the animal and are highly constant within individuals, showing stereotypic variation that may be used to determine species (Hopkins and Bass 1981), sex (Bass and Hopkins 1983), relative dominance (Carlson et al. 2000), and possibly even individual identity (McGregor and Westby 1992). Variation in the patterning or frequency of the EOD plays an important role in electrical communication during behaviors such as aggression, courtship, and mating (Hagedorn 1986; Hopkins 1986; Carlson 2002a). In addition, active electrolocation provides information on the size, shape, distance, and electrical properties of objects (Bastian 1986).

The EODs of skates and catfish are generally less stable within individuals and in some cases are produced only rarely throughout life (Bratton and Ayers 1987; Hagedorn et al. 1990). The available evidence suggests that they play a role in communication (Bratton and Ayers 1987; Baron et al. 1994), although a role in active electrolocation cannot be ruled out. It is likely that these relatively variable EODs are similar to those of a primitive stage in the evolution of electric organs in mormyriforms and gymnotiforms.

1.3 Strongly Electric Fish: Electric Energy as a Weapon

The evolution of strong electric organs represented the addition of a new function to EODs: that of offensive and defensive weaponry. The electric fields produced by the electric eel (*Electrophorus electricus*), electric rays (torpedinids), and African electric catfish (*Malapterurus electricus*) can reach amplitudes of hundreds of volts per centimeter, and are extremely effective in stunning both predator and prey (Bauer 1979; Belbenoit et al. 1979; Belbenoit 1986; Lowe et al. 1994). Certain species of stargazers (genus *Astroscopus*) produce somewhat weaker fields (up to tens of volts per centimeter), and although evidence suggests they play a role in predation, their effectiveness is unclear (Pickens and McFarland 1964). In some species, strong EODs may also play a role in electrical communication (Rankin and Moller 1986).

2. What Is the Output of the Electromotor System?

One common method for measuring the output of the electrogenic system is to record in water the temporal course of the head-to-tail associated field of the EOD in the absence of surrounding objects (Coates et al. 1954). In most weakly electric fish these fields consist of species-specific waveforms emitted repetitively. These waveforms can be broken down into segments that are referred to as EOD wave components. The head-to-tail recording of the EOD is of great help in taxonomy and gender identification, but it oversimplifies the description of the electric field. Potentials recorded in the water along the fish body show site-specific, multiphasic waveforms that reveal the true spatiotemporal complexity of the EOD (Bennett 1971; Caputi 1999). The head-to-tail EOD is therefore insufficient to describe the stimulus for the electrosensory system when the interest is to find changes in reafferent waveform in the presence of objects or to analyze the communication signals of conspecifics.

This introduces the general problem of defining the output of an effector organ. The simplest and most general parameter for assessing this output is the external power, defined as the amount of energy transferred from the individual to the surrounding environment per unit time. However, external power is a synthetic magnitude in several senses.

First, it is the synthesis of three quantities: the mass or charge moved, the speed, and the acceleration at which the mass or charges are moved. In the case of an electromotor system, external power is the product of the drop in voltage across an external load (the work necessary to move a single charge) times the generated current (the number of charges moved per unit of time).

Second, the effector organs are connected to the external loads through passive tissues having their own filtering properties. Aponeurotic sheaths, tendons, skeleton, and tegument make it possible to transmit muscle action to the external loads. Owing to the impedance of these elements part of the generated mechanical work is dissipated internally. Similarly, part of the electrical work generated by the electrogenic tissue is dissipated internally because of current flow through internal tissues and skin.

Third, in real systems energy transfer does not occur at a single point. At the soles of the feet and at the pads of the fingers (and even the whole surface of the body in contact with the air or water when the speed causes enough resistance), mechanical energy is transferred according to a spatiotemporal pattern. Thus, the power transferred to the environment in every instant is the spatial synthesis (or integration) on all relevant surfaces. In the case of the electromotor system, the power results from the integration of current density and electric potential throughout the skin surface of the animal.

2.1 The Nervous System Controls Electromotive Force Generation

Electric fish inspired the famous Voltaic pile, which was described by its designer as an "artificial electric organ." At that time it was recognized that the complete description of such a system requires at least two parameters describing the "force" required to expel a certain "quantity" of electricity through the skin (for an historical review see Moller 1995). In fact, the fish body may be represented as a battery having an electromotive force (EMF) and an internal resistance (R_i) connected in series (Cox and Coates 1938; reviewed by Caputi 1999; Fig. 14.1) When a fish is isolated in air, and its rostral and caudal extremes are connected by a resistive load (R_{load}), the generated current is a linear and decreasing function of voltage across the load. When the connection is open, current is null and voltage is equivalent to the EMF. It is important to note that the EMF is not the battery output but its ability for generating a given output when connected to a load. When the external load resistor decreases in value the current increases up to a maximum at short circuit. The equivalent circuit for the fish consists of a time variant EMF with both poles connected to fixed internal and external resistors. The amount of energy transferred to the environment is at a maximum when both resistors are equal.

The larger the resistance of the internal tissues, the smaller the maximum current (and therefore the power) that the fish body is able to yield (Fig. 14.1B). To illustrate this feature Figure 1B compares the voltage versus current plots obtained from two fish of different size. Small fish (filled symbols) and large fish (open symbols) generate the same electromotive force but the larger fish can deliver a larger amount of current, because internal resistance decreases with body size (Caputi et al. 1989; Caputi and Budelli 1995).

It is clear that the electromotor system (i.e., the neural network and the electric organ under their control) is only a part of the electrogenic system, which also

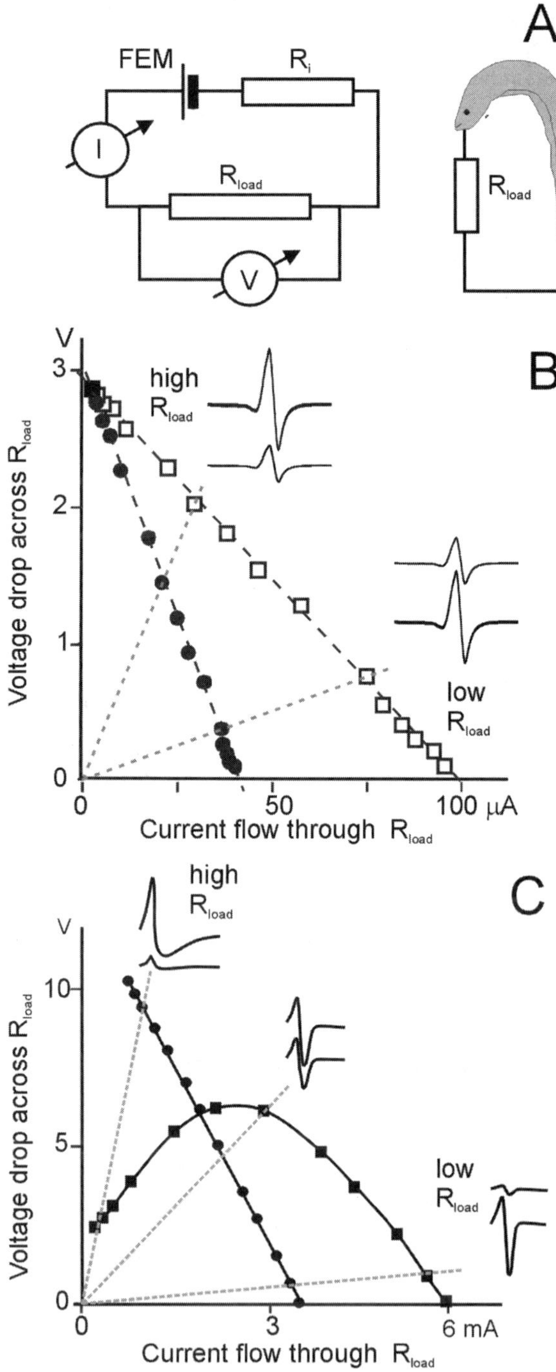
A
FEM
R_i
I
R_{load}
V
R_{load}
B
V
high
R_{load}
low
R_{load}
Voltage drop across R_{load}
Current flow through R_{load}
0
50
100 μA
0
1
2
3
C
V
high
R_{load}
low
R_{load}
Voltage drop across R_{load}
Current flow through R_{load}
0
3
6 mA
0
5
10

includes nonelectrogenic tissues that set the internal resistance. The nervous system directly controls the EO and determines its stereotyped time course of activation. When the load is exclusively internal (fish tissues), the electrical output resulting from the EOD is the equivalent EMF of the fish body. However, the electrical output is contingent on the external impedance and thus the nervous system does not control the actual output of the electrogenic system but only its EMF (Caputi 1999).

Furthermore, minor but behaviorally significant changes in EMF may depend on the external load. As shown in the example in Figure 14.1C, late waves of the EOD do not result from the neural activation of the EO but are a consequence of the initial, externally dependent EO output. In these cases the function relating current and voltage is not linear, indicating that either electromotive force, internal resistance, or both, are dependent on external load and the analogy with a simple electric battery is unsatisfactory (Bell et al. 1976; Caputi et al. 1998; see Section 4.2). On a different time scale the impedance of the fish's tissues is controlled by the complex homeostatic system under environmental and neuroendocrine control (see Section 4.2). Interestingly, fish growth is a harmonic process that keeps the generated EMF by adult animals as a constant species-specific feature independent of body length.

2.2 Geometry of the EO and Nonelectrogenic Tissue Determines the Field Shape

If the EO were relatively short and immersed in a medium of homogeneous impedance, the external field could be approximated by a dipole. However, the bodies of freshwater fish have an internal resistance much lower than the external resistance, while the reverse is true for marine fish (Kalmijn 1974). This discordance between the impedance of internal tissues and the external medium means that one of the main determinants of the external field is the species-specific body geometry.

FIGURE 14.1. Equivalent electric source parameters. (**A**) The fish body can be considered as a simple battery in which V is the voltage drop across the load resistance R_L, and I is the current through the circuit (*left*). The air-gap recording system that allows one to evaluate the equivalent electromotive force (EMF) and internal resistance (R_I) in electric fish is shown in the schematic diagram at the right. (**B**) The plot shows the positive peak voltage across R_L as a function of the corresponding peak current, measured in two differently sized fish [*Gymnotus carapo* of 8 cm (*filled circles*); 20 cm (*open squares*)]. Note that, in spite of the difference in length, the ordinate intersection (the electromotive force, EMF) is approximately the same while the slope of the line (R_i) decreases with animal length. V = EMF − $R_{i\,I}$. (Modified from Caputi 1999.) (**C**) In *Gnathonemus petersii,* while the positive wave (*circles*) the negative phase exhibits a maximum (*squares*). This is because loading the EO causes a change in the emitted waveform: the negative phase is a function of the current generated during the positive phase. (Modified from Bell et al. 1976.)

Pulse mormyrids have a relatively short EO localized at the caudal peduncle, which makes it easier to understand this posteffector conditioning of EO output. The specific conductance of the fish's body is more than 20 times larger than that of the surrounding water, whereas the transverse surface area and the length of the main body are much larger than the EO (Budelli et al. 2002; Caputi et al. 2002). For these reasons the resistance that the body opposes to current flow in the rostral direction is less than the resistance that the water opposes; thus most of the current is funnelled through the fish's body (Castello et al. 2000). This means that the passive fish body acts equivalently to an additional electric source located rostrally.

The situation is more complex when the EO is not localized but is instead distributed along an extended portion of the body. This is the case in most gymnotiforms and the mormyriform *Gymnarchus niloticus*. In these cases each longitudinal portion of the EO can be considered an independent source, having its own internal resistance and being loaded by different portions of the fish either in the rostral or caudal directions.

Therefore, geometry plays an even more important role in fish having extended EOs than in pulse mormyrids. Gymnotiforms and *Gymnarchus* have a fusiform shape with an asymmetric tapering angle to both poles. In the caudal two thirds of the body, the cross-sectional area and the cross-sectional area/perimeter ratio decrease from head to tail. This causes a parallel increase of the generator series resistance and the body attenuation coefficient. As a consequence of the increase in the ratio between series and load resistance, for an identical distance, the body attenuation coefficient causes a larger voltage decay in the caudal than in the rostral direction. In this way, fish geometry orients currents longitudinally and polarizes current flow in a caudal to rostral direction (Caputi et al. 2002).

Because of the difference in internal resistance, caudal regions of the EO behave as "current" sources (high internal resistance), while rostral regions behave as "voltage" sources (low internal resistance). As a consequence, geometry and impedance are weighting factors for the contribution of every fish portion to the whole EOD-associated field (Caputi et al. 2002). Moreover, in these fishes the EO is not synchronously activated and in some species different EO portions do not generate identical waveforms. Therefore, the external field shows a complex, site dependent temporal waveform (see Section 4.2).

In conclusion, there is not a single measure of electromotor output and the most comprehensive way to assess the output of the electromotor system is to measure the EMF and internal resistance distribution along the fish to construct a model of the fish body as an electric source. This kind of model was recently implemented and applied for calculating the EOD-associated field in different circumstances and also to calculate electric images of nearby objects and neighboring conspecifics (Budelli et al. 2002; Caputi 2004).

3. Two-Step Strategy for Electrogeneration

The temporal course of the EMF is a stereotyped spatiotemporal function organized at the spinal and peripheral levels and repetitively triggered by the discharge of a command nucleus located in the brain stem. The discharge of the command nucleus activates a relay structure that projects to the spinal structures. Thus, the electromotor system is organized as two networks hierarchically linked at the command-relay level. The superior network receives multiple influences including those from sensory modalities and internal motor collaterals, and has a single output: the synchronized discharge of the command nucleus. This event decides the activation time of the lower network that in turn transforms the series of command impulses into a series of stereotyped EODs.

3.1 Classification of Electric Fish According to Their EODs

The relationship between the duration of the EOD and its timing have permitted the differentiation of two main groups of weakly electric fish. In *wave-type fish*, the discharge interval is regular and similar to the duration of the EOD. The electrogenic output therefore has a continuous sinewavelike shape. In these fish the discharge frequencies range from about 100 to 1000 Hz depending on the species. The inter-EOD interval shows a very small coefficient of variation (less than 0.0001 in the resting fish, Bullock 1986). Wave-type fish are represented in both main groups of weakly electric fish. In mormyriforms they are represented by a single species (*Gymnarchus niloticus*) and in gymnotiforms by two families (*Stermopygidae* and *Apteronotidae*). In these fish, the command neurons function as coupled oscillators (Moortgat et al. 2000). The relay neurons project on spinal neurons innervating the electrocytes (the electromotorneurons, EMNs), driving them at the command frequency.

The rest of the weakly electric fish are referred to as *pulse-type fish* because the interval between command discharges is much larger than the duration of the EOD and the electrogenic output consists of a series of discrete beats of similar waveform. Pulse-type fish are represented in the two main groups of weakly electric fish. Their electrogenic strategy is based on completely different mechanisms at the three levels of the system: (1) the command nucleus of pulse-type mormyrids show an irregular pattern of discharge, in contrast to the pacemaker of pulse-type gymnotids, which exhibits a very regular discharge rate (coefficient of variation less than 0.05); (2) pulse-type mormyrids have a short and localized EO that generates a synchronous and uniform discharge while gymnotiforms have extended organs producing different regional waveforms in a rostrocaudal sequence; and (3) their different body geometry causes a different distribution of internal resistance and posteffector spatial filtering of the generated EOD.

The less studied weakly electric fishes (skates and catfishes), and the strong discharging fishes, may also be considered pulse-type fish. In the case of skates the command drives an EO located at the tail. The location of the EO poses a similar electrogenic problem as in pulse-type mormyrids. However, the peculiar shape of the fish body and the relatively low resistance of the habitat may cause important although still unexplored posteffector conditioning.

In conclusion, electric fish show a two-step strategy for electrogenesis based on two networks hierarchically linked at a command nucleus. According to their electrogenic strategy, they can be classified into three main types: wave-type fish (some gymnotiforms and *Gymnarchus niloticus*), pulse-type fish with localized organs and irregular discharge timing (pulse mormyrids, rajids, and some siluriforms), and pulse-type fish with extended organs and regular discharge timing (pulse gymnotiforms).

3.2 Variations in the EOD Waveform Within a Species

The EO and its innervation pattern and the neural command wiring seem to be organized to assure a constant species-specific waveform that does not vary across individuals of the same species and along the life span of each individual. In general it could be stated that every discharge of the command nucleus is followed after a fixed time by a stereotyped discharge of the EO. Within a large confidence interval around the average discharge rate of the command the spinal network is able to control the EO to generate the species-specific waveform.

Nevertheless, simpler waveforms have been described for larval electric fish. In addition, the electric power of the EOD increases with fish length and weight, both of which are dependent on age. In some aggressive species (i.e., *Gymnotus carapo*), it is common to find individuals with regenerated tails. Some small individual variation in waveform has been observed in relation to regeneration.

Sex differences in waveform have been reported in many different species (Hopkins 1972; Hagedorn and Carr 1985). These changes are generally associated with reproductive cycles, which are about annual in most cases, and are regulated by sex steroid hormones (Zakon 1993). Amplitude and duration changes in the EOD have been observed in *Brachypopomus,* where action potential propagation that provokes the late head-negative wave is under endocrine and thermal control (Silva et al. 2002). In the wave fish *Sternopygus*, the change in waveform (slower frequency in the maturing male) results from an EOD with a longer duration combined with a slower repetition rate. As described below, this implies a dual action of androgens both at central and peripheral structures (Zakon 1993).

Water temperature can also modulate the EOD waveform. During winter undifferentiated *Brachyhypopomus pinnicaudatus* ("pez ratón") show high waveform sensitivity to temperature but sensitivity decays in spring to a minimum in the reproductive summer months. In the temperate zone, water temperature exhibits circadian and annual cycles covering a large range (6 to 30°C). Silva et al. (2002) hypothesized that sustained warm temperature is an environmental cue for sexual maturation and that low waveform sensitivity to temperature might be a requirement for sexual communication, recognition, and mating. Present evidence supports this hypothesis: lowering waveform sensitivity to temperature could be induced both by testosterone treatment and by acclimation to maintained high water temperature (28°C; see review by Silva et al. 2002).

Finally, a tradeoff between active electrolocation and communication occurs during social interaction. In these circumstances command rate may increase in the form of transient bursts (so-called "chirps") or decrease in the form of long pauses. These modulations of the EOD waveform constitute communication signals during sexual and aggressive interactions. Chirps are accompanied by changes in the waveform and amplitude of the EOD that probably jeopardize the ability to electrolocate. Pauses not only block active electrolocation but also are followed by changes in the peripherally organized late waves (Schuster 2000).

4. Organization of the EOD Waveform

The electromotor system is organized in parallel to the skeletomotor system. Most fish have an EO composed of series of electrocytes derived from muscle tissue that are controlled by electromotor neurons located in the spinal cord. The Apteronotidae evolved a different electrogeneration strategy. In adults the EO is completely of neural origin. It consists of an array of electromotor neuron axons running in parallel along the fish's body. Its electrogenic mechanism is discussed in previous reviews of this topic and is not treated in detail here (Bennett 1968, 1971; Bass 1986).

In fish having myogenic EOs the EMF is the sum of postsynaptic currents and action currents generated in hundreds of electrocytes. As in the skeletomotor system, each electromotor neuron innervates a set of synergistic effector units (a set of similarly oriented electrocytes in series). In most cases, however, each electrocyte receives innervation from several electromotor neurons. Because of this convergence on the same effector structures (a set of electrocyte faces), the sum of activities evoked by a set of individual electromotor neurons is smaller than the activity evoked by the synchronized activation of the same set. An additional difference from the skeletomotor system of vertebrates derives from the fact that the effector target of synergistic electromotor neurons is a subcellular structure; in some species a single electrocyte implements antagonistic actions depending on the activated region of its membrane. In pulse gymnotiforms, different electromotor neuron types innervating opposite faces of the same electrocyte elicit coordinated antagonist effects (see Section 4.3). Therefore, to organize the EOD waveform, the nervous system has to coordinate the activity of different electromotor neuron types to activate similarly oriented electrocyte faces synchronously and sequentially activate opposite oriented electrocyte faces.

4.1 Contribution of Single Electrocytes to Waveform Generation: Roles of Shape, Ion Channel Repertoire, and Innervation Pattern

Three elements are important to determine the contribution of a single electrocyte to the discharge of the whole EO: (1) electrocyte morphology, (2) repertoire and distribution of ion channels, and (3) innervation pattern.

As currently believed for most excitable cells the electrical activity of electrocytes depends on the change in conductance of different populations of "membrane channels" having differential conductance for specific ion types (Hille 2001). As observed in other cells, channel density is heterogeneously distributed over the surface of the electrocytes, and therefore different electrocyte regions may have different current components. These components summate in the extracellular space to generate the contribution of each individual electrocyte to the EOD.

The basic rule for understanding current flow resulting from electrocyte activation was stated long before currents were measured in individual electrocytes. Paccini observed that "the innervated faces of the electroplaques become negative during the discharge whatever the orientation of the organ" (cited by Keynes et al. 1961). In the resting situation there is no external current flow because of the prominence of potassium conductance. During and shortly after synaptic activation, localized changes in various conductances produce an uneven distribution of the membrane parameters that results in a "macroscopic" external current. Because of the decrease in net EMF (or even inversions) and conductance near the synaptic contact the negative interior of the electrocyte becomes "electrically closer" to the innervated region.

These changes in potential may cause regenerative action potentials that propagate to other regions of the electrocyte membrane. In different species, electrocytes adopt different shapes, and within species, electrocytes may have different shapes in different regions of the EO. As a consequence, local attenuation (generally constant in a fine cylinder such as an axon, and estimated by the length constant) governing action potential propagation depends on the shape and membrane region of the electrocyte. Combinations of all these elements, associated with the localization of the synaptic contacts and their activation timing determine the propagation of electrical activity on the cell surface and the waveform of the electrocyte output characteristic of the region of the EO in each species.

Electrocyte activity results in different types of locally generated waveforms. There is a broad repertoire of waveforms displayed by the various types of electrocytes. They are described extensively elsewhere (Bullock and Heiligenberg 1986; Kramer 1990; Moller 1995; Hopkins 1999). Figure 14.2 shows some examples illustrating the principal electrogenic mechanisms at the electrocyte level.

Monophasic waveforms depend on the activation of the innervated face. In some fish, such as strong marine electric fish, rajids and the electric eel *Elec-*

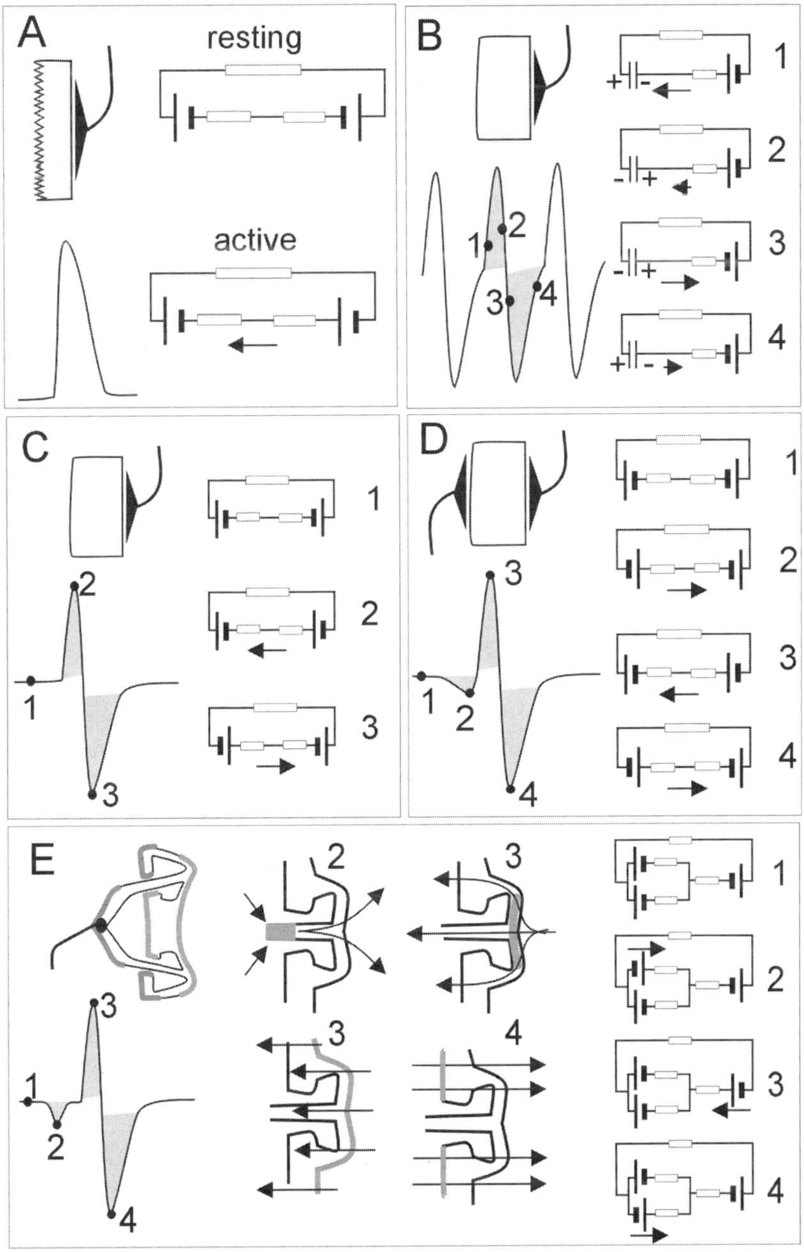

FIGURE 14.2. Different strategies for generating the elementary components of the EOD waveform at the electrocyte level (as suggested by Bennett 1961, 1971). (**A**) Monophasic waveforms. (**B**) Biphasic waveforms: electrocyte resonance. (**C**) Biphasic waveforms: the propagation of action potentials. (**D**) Bi- and triphasic wavefoms generated by double innervated electrocytes. (**E**) Triphasic waveforms generated by stalked electrocytes.

trophorus electricus (Fig. 14.2A), the electrocytes receive innervation on a single face and generate postsynaptic potentials or action potentials restricted to the innervated face. The opposite face has very low resistance. As explained above, the innervated face becomes transiently negative and the resulting EOD is monophasic.

A derived wave with reverse polarity can be generated in the face opposite to the innervated one. This could be implemented by different mechanisms. In *Gymnarchus niloticus* and other wave-type fish, electrocytes receiving innervation on a single face generate postsynaptic potentials or action potentials that do not propagate to the opposite face. However, because of its large resistance the admittance of the opposite membrane is mainly capacitive. As a consequence negativity occurs transiently at the innervated face and the current flow accumulating at the opposite face changes polarity. Subsequently, while the innervated face returns to the resting situation the noninnervated face discharges, resulting in a biphasic EOD. The charging-discharging period is similar to the pacemaker interval, resulting in a continuous sinewavelike discharge (Fig. 14.2B). Alternatively, in most fish exhibiting biphasic waveforms, electrocytes receiving innervation on a single face generate postsynaptic potentials or action potentials that propagate to the opposite face. As a consequence negativity occurs transiently at the innervated face and subsequently at the opposite face, again leading to a biphasic EOD. The asymmetry in area and channel repertoire between both faces generates asymmetry between both phases of the discharge (Fig. 14.2C).

Local multiphasic waveforms could be generated by two different mechanisms that evolved separately. The electrocytes of pulse mormyrids exhibit a singular topological organization. They are thin fenestrated discs that give rise to many stalks that fuse at the site of innervation (Bass 1986). In the example shown in Figure 14.2E, the stalks arise from the caudal face, then reverse course and penetrate the disc to receive innervation on the anterior side. Synaptic activity on the stalks generates a small head negative current. This is followed by a strong head-positive phase resulting from action potential propagation to the caudal face and subsequently by a head-negative phase due to action potential propagation to the rostral face of the main disc, leading to a triphasic EOD (Fig. 14.2E).

Pulse gymnotiforms (*Gymnotus* or "morenita" and *Rhamphychthys* or "pez bombilla") evolved doubly innervated electrocytes, which not only allows them to generate multiphasic waveforms, but also to coordinate a temporally complex discharge. Head-negative waves result from postsynaptic potentials or action potentials elicited in the rostral faces by terminal branches of electromotor nerves and head-positive waves result from action potentials evoked at the caudal faces by a delayed activation volley. A final head-negative wave resulting from the propagation of the action potential to the rostral face may also be present in some electrocytes (Fig. 14.2D).

4.2 Peripheral Mechanisms for Waveform Plasticity

Although neural activation of the EO seems to be a very robust mechanism for producing a constant EOD waveform, changes in the properties of the effector cells are the main mechanisms involved in waveform plasticity. Water conductivity and temperature naturally affect the EOD waveform. The former changes the electric load and thus the current flow associated with the EOD. In some species, such as the elephant nose fish, *Gnathonemus petersii,* the flow of current underlying the early phase of the EOD is involved in stimulating the noninnervated face of the electrocytes. As seen in Figure 14.1C, in the initial phase, the voltage and current are linearly related, indicating that the number of active electrocytes that are recruited by the neural volley is constant and independent of the external load. However, recruitment of the opposite electrocyte faces and the equivalent EMF for the late wave depend on the external current, which is highly affected by water conductivity. When the load increases the current generated by the innervated faces decreases dramatically. Below a certain water conductivity, the currents generating the first phase are subthreshold for stimulating the opposite face. This explains why at the late phase the voltage increases as a function of current as electrocytes are being recruited, reaches a maximum about complete EO activation and finally when conductivity is high it decays as a function of current approximating a straight line (Fig. 14.1C). Thus, any factor that regulates the excitability of the noninnervated faces and the shape of the action potential will modify the EOD waveform.

In pulse mormyrids and pulse gymnotids, action potential propagation from the caudal to the rostral face is under endocrine control (Bass 1986; Moller 1995).

Although the EOD waveforms of immature *Brachyhypopomus pinnicaudatus* are not sensitive to load, the late waves of both the male and female EODs are relatively enhanced when the EO is loaded by low resistances. This indicates that the electrogenic activity is enhanced during sexual differentiation (Caputi et al. 1998). Temperature appears to diminish the responsiveness of the noninnervated face, causing a reduction in the late wave in nondifferentiated animals. However, this reduction is compensated in differentiated fish by an opposing hormone-dependent effect as evidenced by the increase in the late wave in low conductivity. Androgen-induced changes in electrocyte membrane excitability and action potential shape have been related to sex-specific features of the EOD waveform (Hagedorn and Carr 1985; Bass and Volman 1987).

A more complex phenomenon occurs in wave gymnotiforms, in which androgens cause a coherent change in both EOD waveform and frequency (Zakon 1993). The best-studied example is *Sternopygus*, which emits a sexually dimorphic wave-type EOD. In these fish the frequency is lowered by systemically applied androgens, which also leads to a corresponding increase in the duration of single EODs, while estrogens have the exact opposite effect. A recent study suggests that steroid-mediated changes in EOD duration and frequency result from independent peripheral and central actions, respectively (Few and Zakon 2001).

Androgen-induced changes in the EOD may also be paralleled by morphological changes in the electrocytes (Bass 1986). In pulse-type mormyrids, the degree of surface invaginations is generally greater in species with longer duration EODs. Among species with sexually dimorphic EODs, natural males and females treated with gonadal steroid hormones have longer-duration EODs and thicker electrocytes with more surface invaginations along the anterior face. Presumably, these morphological changes affect the capacitance of the electrocyte membrane, thereby increasing its time constant of activation.

4.3 The EO as a Population of Electrocytes

The electromotor output results from the sum of activities of subcellular units, electrocyte faces similarly oriented and orderly recruited by the electromotoneuron pools. Membrane properties and the innervation-activation pattern determine each electrocyte contribution to the regional waveform. Electrocyte size and density determine amplitude. Owing to the impedance of nonelectrogenic tissues, electrocyte location determines both the amplitude of the discharge and the ratio between internal and external work.

There are multiple types of EOs according to the structural organization of their electrocyte population types. In one extreme some fish exhibit a uniform population of electrocytes densely packed in a short localized EOs, in the other multiple populations differing in type, size, and density are distributed along an extended EO. The following examples describe the main types of EO following a criterion of increasing complexity in organizational features.

4.3.1 Case 1. The Uniform EO of Pulse-Type Mormyrids

In pulse-type mormyrids the EO is located within the caudal peduncle and consists of four columns of electrocytes. Almost the entire cross section of the caudal peduncle is made up of EO. Each column of electrocytes occupies one quadrant and the only other structures are the spinal column, the tendons to the fin, and the skin. The electrocytes are a homogeneous population of thin, multinucleated discs, each containing a series of stalks that arise from the posterior or anterior face and fuse to form a single large stalk that receives electromotor neurons innervation (Bass 1986). The stalks actively propagate action potentials to the electrocytes, where both the anterior and posterior faces are also active (Bennett 1971). There is wide diversity in the morphology of this stalk system across species, and electrocytes may be classified according to the site of innervation (anterior versus posterior) and as penetrating, doubly penetrating or nonpenetrating, based on whether the stalks reverse course and penetrate the electrocyte face once or twice before fusing (Hopkins 1999). These differences in morphology are directly linked to species differences in EOD polarity and number of phases. Because the EO is formed from a homogeneous population of electrocytes, the EOD can be understood as a simple amplification of the electrical activity of a single electrocyte.

4.3.2 Case 2. The Extended EO of Wave-Type Gymnotiforms

Wave gymnotiforms such as *E. virescens* have a single type of electrocyte innervated only on their caudal face. However, their extended EO is not homogeneous: electrocyte size and separation diminish exponentially from head to tail. Although the EOD waveform is similar among different regions, the amplitude is not. Consistently, the amplitude of the local EOD shows an exponential increase along the EO, indicating that the amplitude of the local EOD is proportional to the number of generating units.

4.3.3 Case 3. The Complex EO of Pulse-Type Gymnotiforms

Pulse-type gymnotiforms show up to four different degrees of complexity: type, innervation, size, and density are not constant along the EO, causing different local features in the EOD (see reviews by Macadar 1993 and Caputi 1999).

Three paradigmatic examples are shown in Figure 14.3 to illustrate different

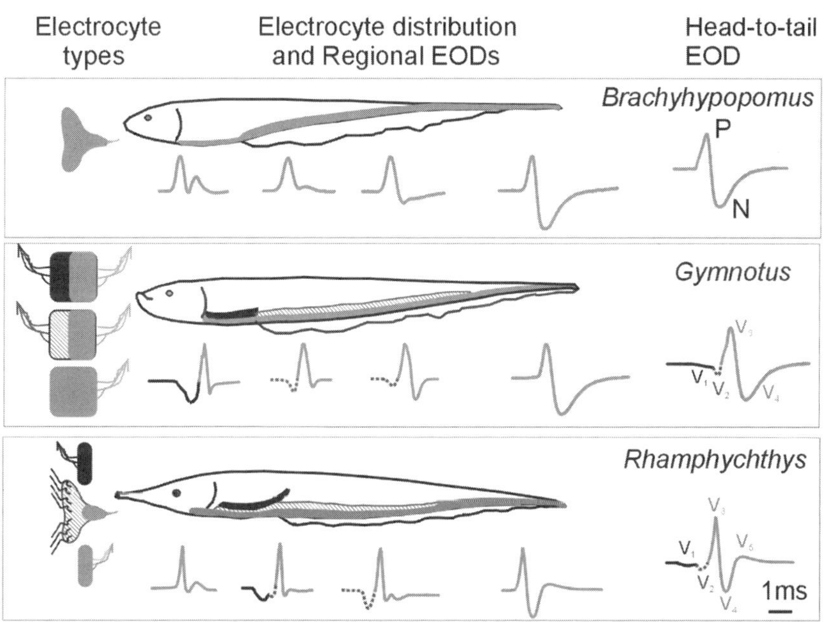

FIGURE 14.3. Comparing different anatomical and functional complexity of the EO in pulse gymnotids. This semischematic drawing (modified from Caputi 1999) correlates different electric organ (EO) portions with samples of the regional electric organ discharges (EODs) recorded in air-gap conditions in three species. It also illustrates the distribution of the different electrocyte types along the EO.

degrees of complexity. The simplest example is *Brachyhypoppomus pinnicaudatus*, which exhibits singly innervated electrocytes differing in size, density, and membrane responses along the EO. Two pairs of tubes extend from the tip of the tail to the anal papilla while the other rows do not extend beyond the rostral edge of the anal fin. All the electrocytes are caudally innervated (Trujillo-Cenoz et al. 1984). This simple innervation plan is consistent with the observed biphasic head-to-tail EOD waveform and with the spatiotemporal pattern of the EMF. Nevertheless, owing to size and density the amplitude of the local discharge increases exponentially from head to tail. In addition, the ratio between the amplitude of the late wave-components and the amplitude of their preceding waves is not constant along the fish body. This feature depends on the membrane properties of electrocytes. Whereas large electrocytes located at the rostral regions of the EO produce a monophasic local waveform, small and caudally located electrocytes generate a biphasic local waveform (Fig. 14.3). As explained in Section 3, late wave components are due to the propagation of the neurally elicited action potentials from the innervated to the noninnervated face. This indicates that the intinsic properties determining the efficiency of action potential propagation along the electrocyte membrane are correlated with electrocyte location along the EO.

A remarkable feature that appears in pulse-type gymnotiforms is the presence of doubly innervated electrocytes. This strategy allows the nervous system to control the timing of activation of different electrogenic structures and construct complex electromotor outputs. The best-studied example is the EO of *Gymnotus carapo*, which consists of eight longitudinal tubes running parallel, on parasagittal planes, four on each side of the body. The dorsal tube on each side (tube 1) has doubly innervated electrocytes in its abdominal and central portions but not in its caudal portion. The other tubes contain only caudally, singly innervated electrocytes. Tube 1 (containing the doubly innervated electrocytes) and tube 2 extend within the abdominal wall to the anal papilla while the others do not. On the other hand, all electrocytes in the caudal portion of the EO are innervated exclusively on the caudal face. Thus, the EO of *Gymnotus carapo* has three different zones according to their innervation pattern (Fig. 14.3). Four wave components (V_1 to V_4; Fig. 14.3) are present in *G. carapo*. V_1 results from the postsynaptic potentials at the rostral faces of abdominal doubly innervated electrocytes; V_2 results from the action potentials at the rostral faces of doubly innervated electrocytes at the central region; V_3 results from the action potential elicited at the caudal faces of most electrocytes; and V_4 results from propagation of this action potential to the rostral faces.

An even more complex example is the EO of *Rhamphychthys rostratus* (“pez bombilla”). This fish shows three types of electrocytes with dissimilar innervation pattern: caudally innervated, rostrally innervated, and marginally caudally innervated. These doubly innervated electrocytes show three distinctive features: (1) the nerve fibers arriving at the rostral side form a regular palisade all over the electrocyte margin; (2) synaptic junctions occur in the marginal membrane; and (3) the caudal faces bear large ridges receiving most of the caudal innervation (Fig. 14.3C). In the subopercular region, the EO of *Rhamphychthys rostratus* consists of a pair of tubes containing only caudally innervated electrocytes. In the abdominal region the EO consists of three pairs of tubes. Each pair contains one of the described electrocyte types. In the central region, most tubes contain doubly innervated electrocytes and in the caudal 25% exclusively caudally innervated electrocytes occur. This electrocyte variety supports a more complex EOD pattern that exhibits five wave components. V_1 results from the activity of the rostral faces of rostrally innervated electrocytes; V_2 results from the activities of rostral faces of marginally caudally innervated electrocytes; and V_3 results from the activities of caudal faces of most electrocytes. V_4 and V_5 are not directly elicited by neural activity and probably result from repetitive activity of single innervated electrocytes resulting from action potential propagation from one face to the other.

4.4 Coordination of Electrocyte Activity: The Neural Network

To organize the stereotyped pattern of EO activation, the nervous system must determine the timing of activation of the electrocyte faces. As explained above, EOs show great variability in their organization. While some taxonomic groups (i.e., pulse mormyrids) use the simple strategy of summing the synchronized activities of a large population of electrocytes, each one generating a similar response to their synaptic drive, in other groups (i.e., pulse gymnotiforms) the EOD is the sum of subcellular responses to a highly organized neural activation pattern. In these last cases (1) the same subset of electrocytes faces must be activated in a very fast rostrocaudal sequence along the EO and (2) different subsets of electrocyte faces located at different sites along an extended organ have to be recruited in the temporally correct order. From the simplest to the most complex case, performance of these tasks depends on interactions among three types of excitable cells: the relay neurons (RN), the electromotor neurons (EMNs), and the electrocytes (Fig. 14.4A). RNs receive the EOD command and project onto electromotor neurons, which in turn project onto electrocytes. Each subset of excitable cells may be considered a layer in a network arranged with a lattice hierarchy. At both projection levels there is convergence and divergence, ensuring by a distributed control strategy, the timing of activation of electromotor neurons and electrocyte faces. In addition to the neural mechanisms of synchrony, the electric interaction between neighboring electrocytes facilitates synchrony (Albe-Fessard and Buser 1950).

The synchronous activation of equally oriented faces of the same kinds of electrocytes requires equalization of the time taken by the neural volley from the pacemaker to the electrocyte face response. Several mechanisms act synergistically to equalize this time: whereas in pulse-type mormyrids the most important is the similarity in length of spinal and peripheral pathways resulting from the occurrence of a short EO localized at the tail, in gymnotiforms the presence of a long and in some cases heterogeneous EO requires time equali-

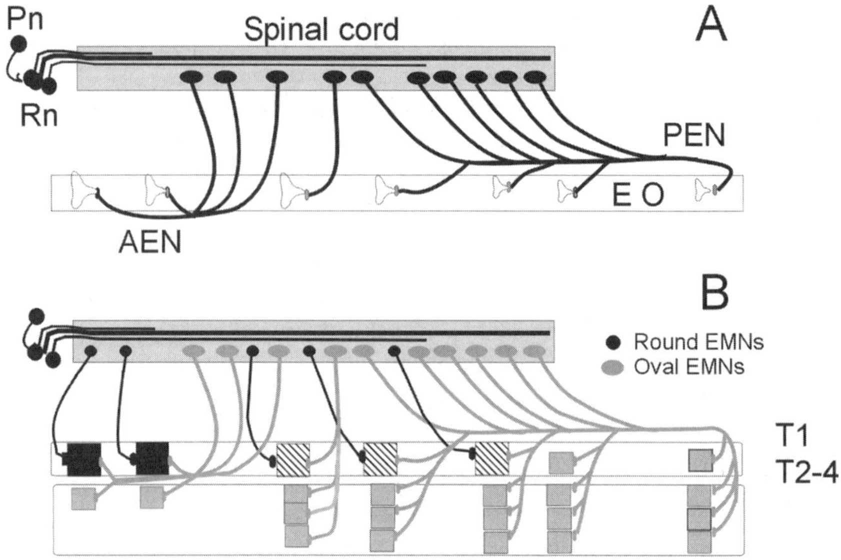

FIGURE 14.4. The neural network organizing the EOD waveform in pulse gymnotids. (**A**) *Brachyhypopomus* is the simplest example to show the main elements of the network. The descending tract contains fibers of different length, the longest fiber being the thickest. There is a single pool of electromotor neurons whose domain along the spinal cord is shorter in length than the EO. This difference in length is accounted rostrally by a recurrent nerve (anterior electromotor nerve, AEN) and caudally by an antecurrent nerve (posterior electromotor nerve, PEN). All electrocytes are innervated in their caudal faces. (**B**) While synchronism is similar to that in *Brachyhypopomus*, in *Gymnotus* wave sequencing depends on the orderly recruitment of two different electromotoneuron pools, one of small round neurons (*black*) and a second of large oval neurons (*grey*) innervating the rostral and caudal faces of the electrocytes, respectively. (Modified from Caputi 1999.) Note the different distributions of both electromotor neuron types and the distribution of different types of electrocytes along the different tubes composing the EO (T1 to T4; see also Fig. 14.3).

zation at the different stages of the neural network as those suggested by Bennett (1971) and experimentally described by Albe-Fessard (1954), Trujillo-Cenoz et al. (1986), Lorenzo et al. (1990), and Caputi and Trujillo-Cenóz (1994). The relative importance of these mechanisms depends on the species and can be characterized by field potential analysis in the intact fish (Caputi and Aguilera, 1996). It is important to note that rather than synchronizing the activity of similar electrocyte faces the nervous system provides the EO with an activation sweep running about 1 km/s along the EO and yielding a head to tail delay of 0.2 to 0.4 ms which is in the neighborhood of a cycle in a wave fish discharging at 300 Hz.

The organized recruitment of different sets of homologous faces generating the different waves composing the EOD appears to be based on the presence of different electromotor neuron pools corresponding to each subset of electrocyte faces. In *Gymnotus carapo,* wave sequencing appears to be based on two mechanisms related to electromotor neuron size and location. First, Hennemann's size principle (Hennemann 1957) appears applicable to the electromotor system. As mentioned, some pulse gymnotiforms, such as *Gymnotus carapo,* have doubly innervated electrocytes and exhibit two morphological types of electromotor neurons (Fig. 14.4B): ones with small-round somata and others with large-oval somata (Caputi and Trujillo-Cenoz 1994). The rostral regions of the spinal cord contain exclusively small-round electromotor neurons; in the intermediate portions, large and small neurons coexist; and the caudal region contains only large neurons. Furthermore, electromotor neuron location along the spinal cord is correlated with their projection site along the EO (Caputi and Trujillo-Cenoz 1994). Thus, large electromotor neurons innervating a large population of caudally innervated electrocytes would be recruited later and with lower safety factor than small electromotor neurons projecting to the rostral face of rostrally innervated electrocytes. Therefore head negative waves will precede head-positive waves (Caputi and Trujillo-Cenóz 1994).

A second mechanism for sequencing the activation of opposite faces is that the rostral faces of the abdominal electrocytes are innervated by a pool of small electromotor neurons located rostral to the electromotor neurons innervating the caudal faces of the same electrocytes (Trujillo-Cenoz et al. 1984; Caputi and Trujillo-Cenoz 1994). As a consequence, the neural path from the pacemaker to the rostral faces of abdominal electrocytes is much shorter than the path serving all other electrocytes.

The process of EO activation in *Gymnotus carapo* is one of the best known examples of motor pattern organization (see review by Caputi 1999). This example illustrates how a simple network synchronizes and sequences about a thousand independent effector units transforming a single impulse into a complex stereotyped electromotor pattern.

5. Timing Control of the EOD: Roles in Communication and Active Electrolocation

Wave-type fish produce very regular firing patterns, with interval durations approximately equal to pulse duration, thus resulting in a continuous quasi-sinusoidal discharge of 150 to 1200 Hz, depending on species, sex, and behavioral state. Pulse gymnotiforms generate pulses of 1 to 5 ms, separated by intervals of ten (or more) times longer duration. EODs are repeated at quite regular frequencies (coefficient of variation less than 0.05). While gymnotiforms and *Gymnarchus* modulate a regular pulse train (frequency control), pulse mormyrids utilize pulse-by-pulse control (interval control). Correspondingly, the former have a pacemaker command with an endogenous rhythm that may be modulated by descending prepacemaker centers, while the latter have a com-

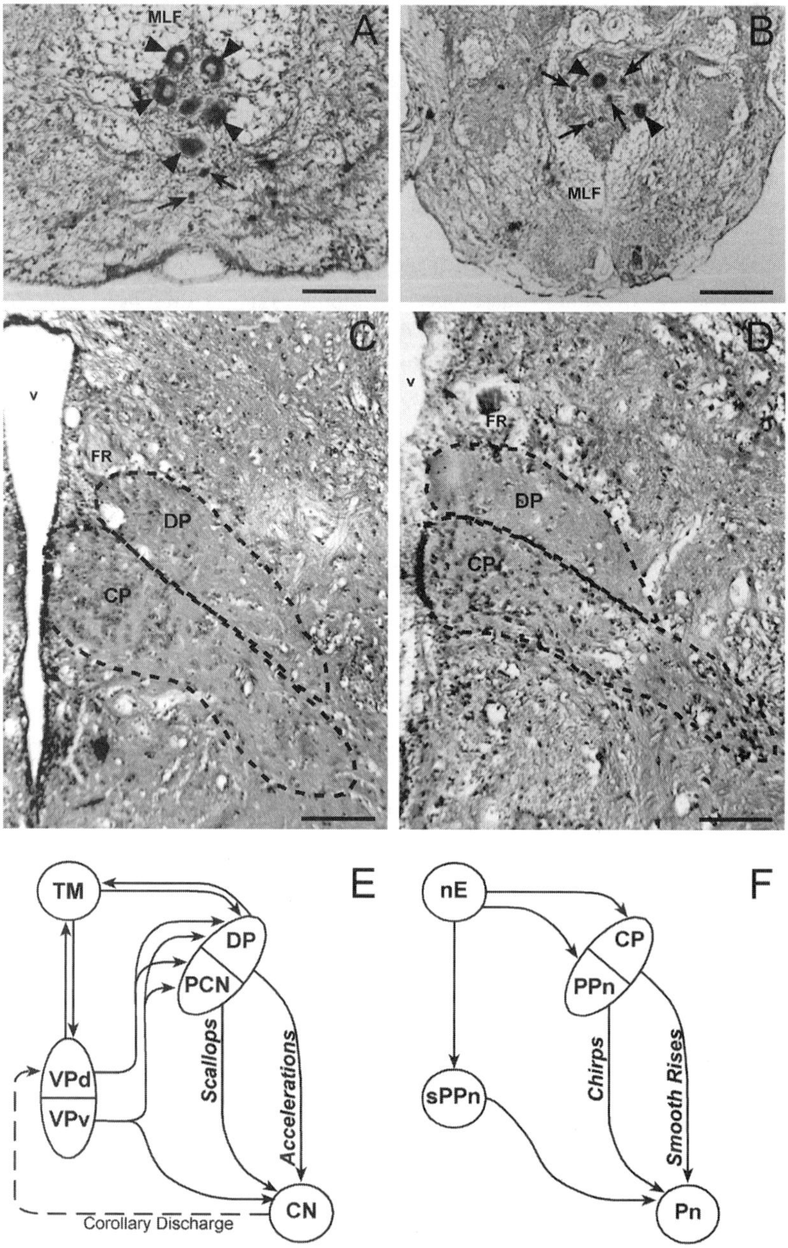

FIGURE 14.5. Comparison of gymnotiform and mormyrid electromotor networks. Transverse sections through the hindbrains of (**A**) the pulse-type mormyrid *Brienomyrus niger* and (**B**) the wave-type gymntiform *Eigenmannia* virescens. Cells and fibers are labeled

mand nucleus whose discharge results from the integration of incoming descending influences.

Long-lasting frequency variations have been described in relation to seasonal changes in water temperature, and they are probably also influenced by the reproductive cycle (Dunlap et al. 2000; Silva et al. 2002). A circadian cycle of EOD frequency has also been observed. EOD amplitude and frequency increase during the night, which has been attributed to the nocturnal increase in motor activity (Moller 1995). Similar changes in the EOD can be induced by the administration of serotonin (Stoddard et al. 2003).

Short-term frequency modulations occur in different behavioral contexts related to the preservation or improvement of electrolocation ability: environmental exploration (electrolocation), foraging, identification and entering of hiding places, identification of novel stimuli, and detection and escape from predators. In all of these behavioral contexts, an increase in the discharge rate increases the information per unit time gathered by the active electrolocating system.

The EOD of a nearby conspecific distorts the electric image produced by the self-generated discharge, resulting in a deleterious effect to electrolocation abilities. Avoidance of such jamming is solved with different strategies: frequency increases or decreases in wave-type fish, phase locking in pulse mormyrids, and persistent frequency change and transient frequency increases in pulse-type gymnotiforms.

Modulations of the temporal patterning of emitted discharges are also the main communication signals in electric fish. During agonistic displays or courtship, sudden transient increases and decreases in frequency, as well as discharge interruptions may occur. In some species, transient increases in frequency are associated with a decrease in EOD amplitude and distortion of its waveform.

5.1 Common Solutions for Timing Control in Gymnotiforms and Mormyriforms

In both gymnotiforms and mormyriforms, the medullary neurons that drive EOD production are located in unpaired midline nuclei (Fig. 14.5A, B; Bennett et al. 1967a,b; Bell et al. 1983). In addition, there is extensive electrotonic coupling among neurons within each level of the EOD command chain (horizontal

---

FIGURE 14.5. *Continued*
with Bodian staining. The location of command/pacemaker neurons and relay neurons are indicated by *arrows* and *arrowheads*, respectively. Scale bar = 100 μ MLF, Medial longitudinal fasciculus. Transverse sections through the dorsal thalamus of (**C**) *B. niger* and (**D**) *E. virescens*. Cells and fibers are labeled with Bodian staining. The dorsal posterior nucleus (DP) is located just dorsal to the central posterior nucleus (CP). Scale bar = 65 μm. (**E**) Summary diagram of the mormyrid electromotor system, indicating the pathways for scallop and acceleration production. (**F**) Summary diagram of the gymnotiform electromotor system, indicating the pathways for chirp and smooth rise production.

coupling), as well as between levels (Bennett et al. 1967a,b; Dye and Meyer 1986).

In both groups of fish, there are two different types of midline medullary neurons in the electromotor pathway: command/pacemaker neurons that control EOD production and relay neurons that receive input from the command/pacemaker neurons and project down the spinal cord to the electromotor neurons (Fig. 14.5A, B). Compared to the command/pacemaker neurons, the relay neurons are extremely large, with correspondingly thick axons (Szabo and Enger 1964; Ellis and Szabo 1980; Bell et al. 1983). This large size of the axons presumably maximizes the effectiveness of electromotor neuron driving. However, as soma size increases, input resistance decreases, and it therefore takes large amounts of current to exact the significant change in membrane potential required for precise control of EOD timing. Overcoming this problem, both fish groups have the decision center (command or pacemaker) one level upstream in smaller neurons, where descending inputs do not need to deal with low input resistance. Synchronous input from these command/pacemaker neurons easily provides sufficient current to drive the large relay neurons.

The primary input to the command/pacemaker nucleus is from a column of cells that starts rostrally within the dorsal thalamus, and then heads caudally and ventrolaterally toward the mesencephalon. In gymnotiforms, it is the centroposterior nucleus of the thalamus that has been co-opted by the electromotor system (Fig. 14.5D; Zupanc and Maler 1997), while in pulse mormyrids, it is the dorsal posterior nucleus (Fig. 14.5C; Carlson 2002b). In both cases, the rostral cells within the dorsal thalamus are relatively small with thin intrinsic dendrites, while the caudal cells are relatively large with thick extrinsic dendrites. Furthermore, these morphological and anatomical differences are likely linked directly to their divergent roles in modifying behavioral output: in both cases, the rostral cells are responsible for driving smooth accelerations, while the caudal cells are responsible for driving transient, intense bursts (Fig. 14.5E, F).

In both groups of fish, the command/pacemaker nucleus receives an additional input from a relatively small group of cells within the mesencephalon, the sublemniscal prepacemaker nucleus (sPPn) in gymnotiforms and the ventral subdivision of the ventroposterior nucleus (VPv) in pulse mormyrids (Fig. 14.5E, F; Bell et al. 1983; Metzner 1999; Carlson 2002b). However, these two nuclei are quite distinct anatomically, as VPv is a small group of large cells located within the torus semicircularis, and sPPn is a group of small cells located just ventral to the lateral lemniscus. Whereas the sPPn is typically involved in driving frequency decreases or interruptions, the role of VPv in pulse mormyrids is unclear.

The descending inputs to the command/pacemaker nucleus appear to be largely regulated by a single sensorimotor interface in both groups. In pulse mormyrids, it is the tectum mesencephali, whereas in gymnotiforms, it is the nucleus electrosensorius (Fig. 14.5E, F; Metzner 1999; Carlson 2002b). These nuclei receive input from multiple sensory regions, and thereby may provide a pathway for electrosensory, auditory, lateral line, and visual information to influence electromotor output.

5.2 Control of Timing in Gymnotiformes: Behavioral Patterns and Neural Circuit

The timing patterns most frequently displayed by Gymnotiforms may be grouped into four behavioral contexts (Fig. 14.6):

1. Novelty responses: Transient, smooth frequency increases (Fig. 14.6A) are observed as a response to novel stimuli in different species of pulse gymnotiforms, particularly in *Gymnotus carapo* (Bullock 1969). The response increases frequency by 5 to 10 Hz and decays smoothly in about 3, and is dependent on the magnitude of the stimulus and the previous history of stimulus presentation (Caputi et al. 2003).

2. Pacemaker acceleration associated with the escape response: Escape reactions (particularly valuable in avoiding predators) in teleost fish depend on a set of giant reticulospinal neurons among which the Mauthner cell is the most conspicuous. During escape reactions, large accelerations occur (Fig. 14.6B). This frequency increase can be interpreted as an enhanced sampling rate aimed at improving electrolocating ability when the animal is performing a fast movement that requires exploration of a novel environment (Falconi et al. 1995).

3. Jamming avoidance responses: The detrimental effect of EODs generated by a nearby conspecific is avoided in wave-type fish by shifting the EOD frequency away from an interfering signal of similar frequency (less than 4 Hz difference). Heiligenberg and co-workers have studied this behavior described as the *jamming avoidance response* (JAR) and identified the neural circuit subserving this function in wave-type fish (see Heiligenberg 1991).

In pulse-type fish, the interference is limited to the occurrence of coincident EODs. Coincidences will occur more frequently if both signals have similar frequencies. *Gymnotus carapo* increases its frequency to avoid jamming by a lower frequency conspecific (Capurro et al. 1998), but it is unable to lower its discharge rate if the interfering signal has a higher frequency. On the other hand, *Gymnotus carapo* also shows a novel strategy: the fish with a faster firing rate is able to transiently increase its frequency just before the occurrence of a coincident discharge (Fig. 14.6C). These transient frequency increases lower the probability of repeated coincidences that are most deleterious to electrolocation ability (Capurro et al. 1998), although single coincidences are not lowered significantly.

4. Communication displays: During agonistic displays or courtship, sudden transient increases in frequency, transient decreases and discharge interruptions occur. In some species, special transient increases in frequency are associated with a decrease in EOD amplitude and a distortion of its waveform. In *Brachyhypopomus pinnicaudatus*, these signals (called "chirps") are emitted mostly by courting males and have been shown to have communication value. A variation from "weak" to "strong" chirps has been described, implying different

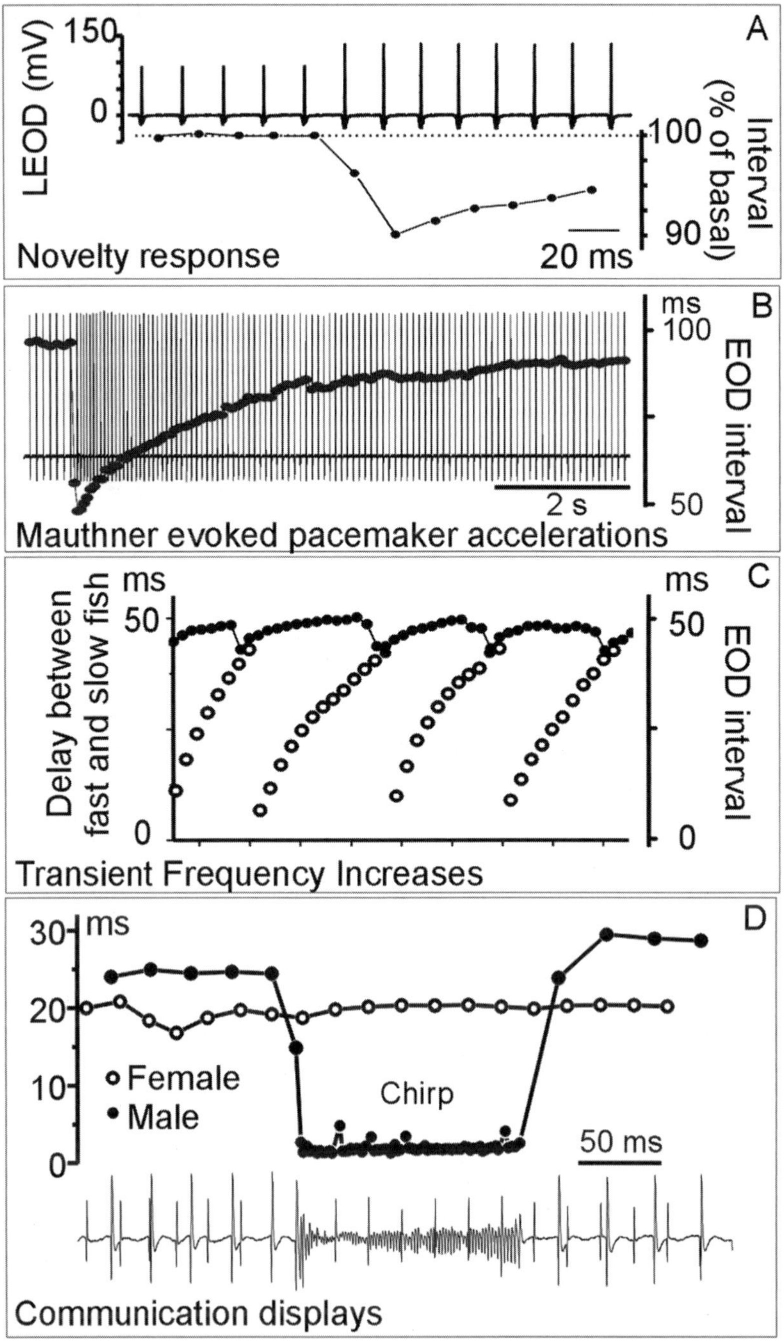
A
150
LEOD (mV)
0
Interval (% of basal)
100
90
20 ms
Novelty response
B
ms
100
EOD interval
2 s
50
Mauthner evoked pacemaker accelerations
C
ms
Delay between fast and slow fish
50
0
ms
50
EOD interval
0
Transient Frequency Increases
D
30
ms
20
10
0
Female
Male
Chirp
50 ms
Communication displays

degrees of frequency increase associated with corresponding amplitude attenuations and waveform distortions (Kawasaki and Heiligenberg 1989). In similar behavioral and physiological conditions, females of this species are observed to interrupt their discharge from 0.5 to several seconds (Fig. 14.6D).

5.2.1 Structural and Functional Bases of EOD Frequency Modulation in Gymnotiforms

The command signal for EOD generation is initiated in an unpaired medullary nucleus with the functional properties of a cellular pacemaker (Bennett et al. 1968; Dye and Meyer 1986). This nucleus is composed of two cell types: pacemaker neurons (PMn), which are small cells with short axons and a slow depolarizing membrane potential that is initiated after each action potential and reaches threshold to generate the next spike; and relay neurons (Rn), which are larger neurons with large axons that descend the spinal cord and terminate on the spinal electromotor neurons; they have a flat interspike membrane potential and generate an action potential only when pacemaker neurons excite them (Bennett 1968; Smith 1999). Both cell types are clearly separated into two parts of the nucleus in pulse-type fish, with pacemaker neurons being dorsal, while they are intermingled in wave-type species. Another difference between wave and pulse Gymnotiforms in the cellular organization of the pacemaker nucleus is that the ratio in numbers of PMn to Rn is large (close to 2) in wave-type fish, while it is small (close to 0.5) in pulse fish. PMn axons make mixed (electric and chemical) excitatory synaptic connections with Rn. Electrical coupling among PMn is relatively large compared to that among Rn (Dye and Meyer 1986). A third type of cell, called parvocells, was described in wave fish (Smith et al. 2000). These parvocells have been labeled with an antibody against the neuronal calcium-binding protein, parvalbumin, and were dye coupled to the other two main cell types. They receive electrotonic inputs from and make chemical synapses back onto pacemaker and relay cells. The function of these neurons in regulating the EOD frequency is unknown.

FIGURE 14.6. Control of EOD timing in pulse gymnotids, behavioral patterns. (**A**) Novelty response. Left axis and continuous trace: local EOD; right axis and open circles: inter-EOD interval (species *G. carapo,* modified from Caputi et al., 2003). (**B**) Mauthner evoked transient acceleration. *Left axis* and *continuous trace*: local EOD; *right axis* and *open circles*: inter-EOD interval (species *G. carapo*, figure provided by M. Borde). (**C**) The jamming avoidance strategy in *Gymnotus carapo* consists of brief frequency increases that the faster fish of a pair generates when the slower fish's EOD is about to coincide with the former. The figure shows the inter-EOD intervals of the faster fish (*open circles*) and the phase of the slower relative to the faster (*filled circles*). These transient frequency increases (TFIs) lower the probability of repeated coincidence (figure provided by A. Capurro). Mating *Brachyhypopomes pinnicaudatus* generate sudden changes of discharge train; female (*open circles*), male (*filled circles*). (**D**) Males generate sharp increases in frequency that usually decrease EOD amplitude and distort waveform. These signals are produced in bursts of up to 50 per minute and can last 200 ms.

The pharmacological profiles of the ionic currents in the pacemaker nucleus of gymnotiforms are similar to those of ionic currents involved in other cellular pacemakers (Smith 1999). The oscillatory interaction of these ionic currents dramatically influences the pacemaker and the EO firing frequency is probably a target of steroid hormone action in producing sexual dimorphisms in EOD frequency.

The rate modulations described above depend on the descending influences of excitatory and inhibitory connections to the Pn. Tract tracing has demonstrated the existence of a bilateral prepacemaker nucleus (PPn) in the diencephalon, at each side of the midline, which extends ventrolaterally from the centroposterior nucleus of the thalamus (CP). CP/PPn can be divided into a dorsomedial subnucleus composed of small ovoidal cells and a ventrolateral part composed of large multipolar neurons. Another input to the Pn comes from the sublemniscal prepacemaker nucleus (sPPn), a small, bilateral group of cells located ventral to the lateral lemniscus. Most projections from the sPPn are glutamatergic and terminate on the relay cells of the Pn (Kawasaki et al. 1988; Kawasaki and Heiligenberg 1989).

Gradual accelerations in EOD rate can be induced in *Brachyhypopomus* by stimulation of a portion of the dorsomedial region of CP/PPn, while stimulation of a different (smaller) portion gives rise to decelerations and even interruptions. Both discharge modulations occur with maintenance of the normal activation sequence (pacemaker-relay), as demonstrated by intracellular recordings (Kawasaki and Heiligenberg 1989), and they depend on glutamatergic excitation and γ-aminobutyric acid-ergic (GABAergic) inhibition of PMns, respectively.

Stimulation of the ventrolateral region of CP/PPn induces the generation of chirps by acting directly on relay cells while PMns do not alter their discharge rhythm. Sudden interruptions of the EOD are provoked by activation of sPPn by acting directly on relay cells that become strongly depolarized, while PMns keep their normal rhythm. When the relay cell repolarizes the resting activity is resumed (Kawasaki and Heiligenberg 1989).

In the wave-type fish *Eigenmannia*, EOD accelerations associated with the JAR are induced by glutamatergic excitatory inputs to Pn from the dorsomedial region of CP/PPn (Metzner 1999). sPPn provides tonic glutamatergic excitatory input to Pn: EOD decelerations associated with the JAR are induced by GABAergic inhibition of sPPn and a consequent removal of tonic excitation.

The sudden increase in discharge rate induced by Mauthner cell activation depends on glutamatergic excitation of pacemaker cells involving both *N*-methyl-*D*-aspartate metabotropic receptors and non-NMDA metabotropic receptors (Curti et al. 1999).

In addition to sexual hormone effects to the pacemaker there are other hormone-dependent structural differences between male and female brains related to differences in behavior. Nonspawning males consistently chirp, whereas females do not. This behavioral dimorphism is paralleled by sex differences in CP/PPn that regulates chirping behavior. Neurogenesis and changes in the morphology of dendritic trees in PPn was proposed by Zupanc (1999) as the bases of seasonal changes associated with courtship and the associated production of chirps.

5.3 Control of Timing in Pulse Mormyrids: Behavioral Patterns and Neural Circuit

Unlike gymnotiforms, which produce EODs at highly regular rates, mormyrid electrical behavior is characterized by variable timing (Fig. 14.7A. At rest, there is a general tendency toward EOD intervals of around 100 to 300 ms (Teyssedre and Boudinot 1987), and because they are nocturnal, EOD intervals are generally lower and less variable at night (Carlson 2002a). However, there can be extensive variation around this basic pattern, as the sequence of pulse intervals may fluctuate from less than 10 to several hundred milliseconds (Fig. 14.7A). As in gymnotiforms, pulse mormyrids increase their EOD rates while actively probing their environment (von der Emde 1992). They also produce transient rate increases as a novelty response (Post and von der Emde 1999) and use phase-locking (echo response) as a jamming avoidance strategy (Russell et al. 1974).

The electrical behavior of mormyrids is analogous to action potential activity

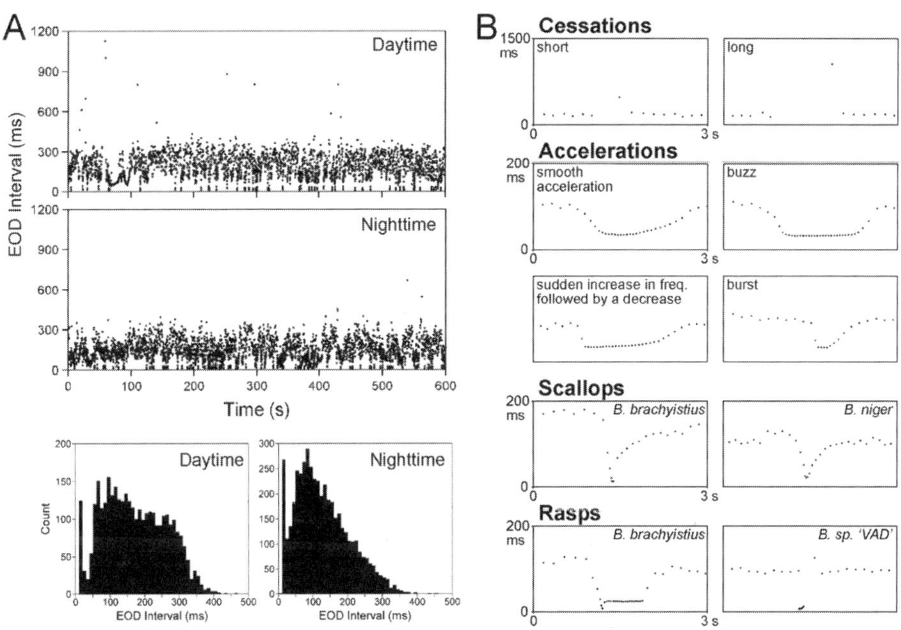

FIGURE 14.7. The sequence of pulse intervals (*SPI*) in mormyrid electromotor output. (**A**) Long-term SPI plots from a single *Brienomyrus brachyistius* during the day and night. Below the SPI plots are histograms of EOD interval distributions. (**B**) Phasic modulations of the SPI that have been implicated in a variety of behavioral contexts (see Carlson 2002a for details).

in single neurons: there is a stereotyped pulse of electricity produced with variable timing. The temporal patterning of action potential activity can encode a wide range of complex information in the nervous system, and the sequence of pulse intervals similarly provides pulse mormyrids with a rich repertoire of social signals (Carlson 2002a). Several studies have described different phasic patterns that are correlated with a variety of social contexts such as aggression, submission, courtship, and spawning (Hopkins 1986; Carlson 2002a). These patterns include cessations in electromotor output, as well as a wide variety of bursts (Fig. 14.7B). In Brienomyrus brachyistius, three basic burst types have been described: accelerations, scallops, and rasps (Carlson and Hopkins 2004a). Accelerations are modest, sustained decreases in interval to values of around 25 to 40 ms that are quite variable in both duration and in the minimum EOD intervals reached during the burst. Scallops are highly stereotyped, phasic bursts that begin with a sudden drop in interval to values of 10 to 20 ms, followed by an immediate increase back to resting intervals. Rasps have an initial scalloplike phase followed by an accelerationlike tail, supporting the notion that novel signal types may arise in evolution by combining preexisting signal types, while also suggesting a combinatorial mechanism for their production.

These different displays are associated with different functions in social behavior: scallops may function as an advertisement signal, accelerations as an aggressive display, rasps as a male courtship display, and long cessations as a submissive display (Carlson 2002a; Carlson and Hopkins 2004a). Thus, much as the nervous system encodes information into temporal patterns of action potentials, pulse mormyrids appear to encode social information into temporal patterns of EOD production.

5.3.1 Structural and Functional Bases of Inter-EOD Interval Control in Pulse Mormyrids

Production of each EOD is initiated in a command nucleus (CN) located on the ventral midline of the medulla (Fig. 14.8A) (Bell et al. 1983; Grant et al. 1986, 1999). Immediately dorsal to CN is the medullary relay nucleus (MRN), which receives a direct projection from CN, as well as an indirect projection through the bilateral bulbar command-associated nucleus (BCA). The axons of MRN neurons project down the spinal cord to terminate on the electromotor neurons (EMNs) that drive the electric organ. Unlike gymnotiforms and *Gymnarchus*, the pulse-type mormyrids electromotor network also gives rise to a corollary discharge pathway that originates in BCA and relays the timing of EOD production to higher brain centers. Details on this pathway and its role in electrosensory processing are discussed elsewhere in this volume (Bell and Maler, Chapter 4 and Kawasaki, Chapter 7).

Although the electric organ is silenced by drugs that immobilize fish for physiological recording, electromotor output can be monitored by placing an electrode next to the tail and recording the field potential that results from the synchronous activation of electromotor neurons, which occurs about 5 ms before the EOD in freely behaving fish. Neurons in both CN and MRN fire in a strict 1:1, phase-locked relationship with electromotor output (Fig. 14.8B) (Grant et al. 1986). Unlike gymnotiforms, there is no evidence that CN functions as a pacemaker. Instead, neurons within CN integrate descending inputs, and because they are tightly electrotonically coupled (Elekes and Szabo 1985), the nucleus essentially functions as one large integrate-and-fire unit (Carlson 2002a).

The mesencephalic precommand nucleus (PCN) and the thalamic dorsal posterior nucleus (DP) provide the primary excitatory input to CN (Carlson 2002b, 2003; Carlson and Hopkins 2004b). These two nuclei form a bilateral column of cells that start rostrally in the dorsal thalamus and extend caudally and ventrolaterally into the rostral mesencephalon. CN also receives bilateral input from 10 to 15 large multipolar neurons along the ventral edge of the toral ventroposterior nucleus (VPv). The dorsal region of the same nucleus (VPd) contains relatively small, ovoid neurons that provide strong inhibitory input to DP/PCN (Carlson 2002b, 2003; Carlson and Hopkins 2004b). This region, in turn, re-

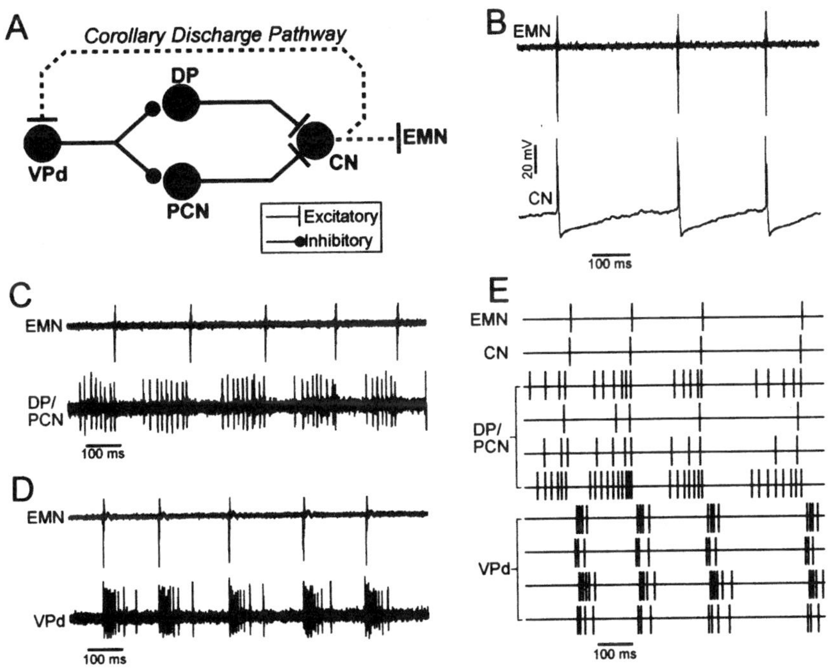

FIGURE 14.8. Mormyrid electromotor network. (**A**) Simplified schematic of electromotor connectivity (for details on the anatomy see Bell et al. 1983 and Carlson 2002b). (**B**) Intracellular recording from a CN neuron in relation to the electromotor neuron volley in *B. brachyistius*. (**C**) Extracellular activity of a single DP/PCN neuron in relation to the electromotor neuron volley in *B. brachyistius*. (**D**) Extracellular activity of a single VPd neuron in relation to the electromotor neuron volley in *B. brachyistius*. (**E**) Simulated simultaneous activity patterns in electromotor neurons, CN, DP/PCN, and VPd.

ceives a projection from the medullary command associated nucleus (MCA), an important relay of the corollary discharge pathway. The pulse mormyrids electromotor pathway can therefore be broken down into a relatively simple functional circuit: CN integrates descending excitatory input from DP/PCN, and an internal collateral associated with the corollary discharge pathway provides negative feedback to DP/PCN via VPd (Fig. 14.8A).

Neurons in DP/PCN and VPd exhibit characteristic patterns of activity in relation to the electromotor neuron cycle (von der Emde et al. 2000; Carlson 2003). DP/PCN neurons vary widely in their average firing rates, but each unit has an increased probability of firing about 4 to 10 ms before the electromotor neuron volley, and then goes silent for about 50 to 150 ms before resuming activity prior to the next volley (Fig. 14.8C). VPd neurons are silent throughout the entire cycle, except for firing a stereotyped burst of action potentials starting right around the time of electromotor neuron activation, which exactly coincides with the silent period in DP/PCN units (Fig. 14.8D). Thus, our working model of the underlying mechanism for regulating the sequence of pulse intervals is that this network functions as a relatively simple central pattern generator: DP/PCN fires tonically to depolarize CN and drive EOD production, which activates a recurrent inhibitory pathway that silences DP/PCNand imparts a baseline rhythm to electromotor output (Fig. 14.8E).

If DP and PCN both provide excitatory input to CN and contain neurons with similar basic properties, this raises the question of why there are two separate populations. In *B. brachyistius*, stimulating DP with the excitatory neurotransmitter L-glutamate results in smooth increases in EOD rate and regularity similar to accelerations, while stimulation of PCN induces transient, intense bursts similar to scallops (Carlson and Hopkins 2004b). This suggests that these two nuclei are involved in generating distinct communication displays (Fig. 14.9). Thus, much as many sensory systems use "labeled lines" to code for specific stimulus features, so too can motor systems employ "labeled lines" that are dedicated to generating specific motor acts. The anatomical separation of these pathways provides a relatively simple mechanism for generating different displays. Because rasps appear to result from a combination of scallop and acceleration production, their production likely relies on activation of both groups of neurons (Fig. 14.9).

The question that arises at this point is what differs between the neurons in DP and PCN and how does this translate into different effects on electromotor output for *B. brachyistius*? There is extensive variation in the activity patterns of individual DP/PCN neurons, ranging from units that are almost completely silent to those that fire tonically at high rates (Carlson 2003). Relatively silent neurons get recruited during accelerations, which increases the number of descending units driving CN, leading to increases in the rate and regularity of EOD output. During scallops, however, units that are already firing at high rates experience a further increase in activity, which may saturate the membrane potential of CN neurons, leading to an intense, stereotyped burst of activity (Carlson 2003). Thus, DP neurons appear to have a lower baseline activity level

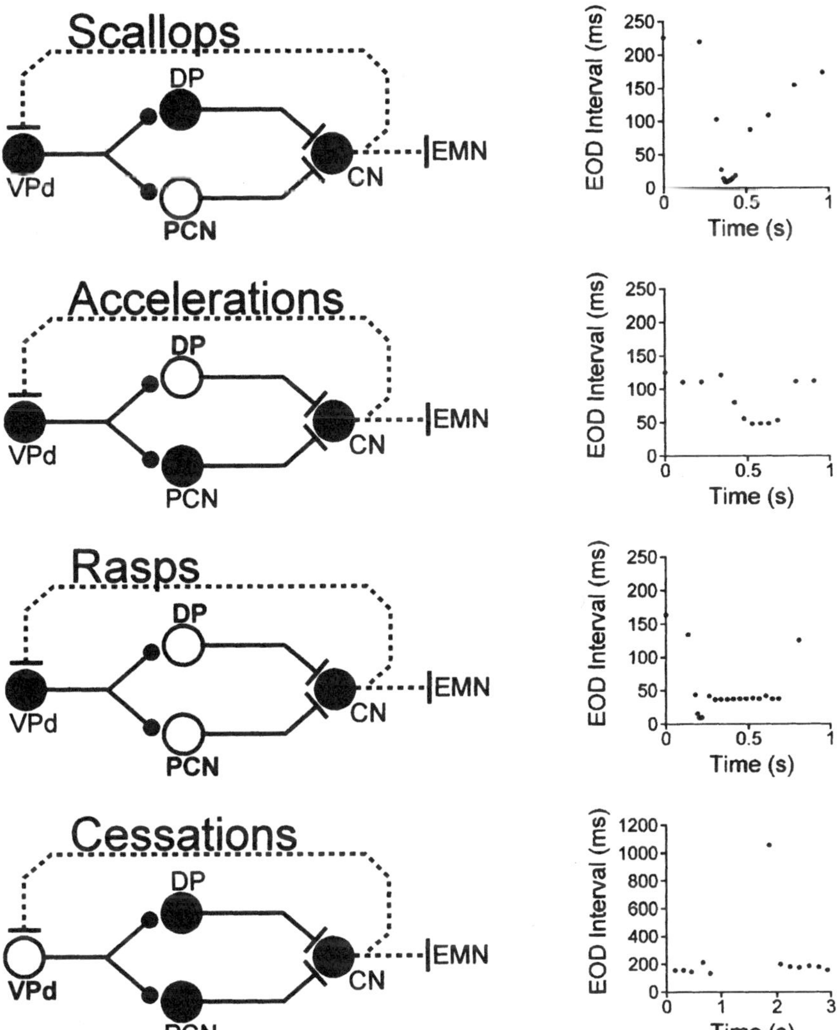

FIGURE 14.9. Central control of electric communication displays in mormyrids. In the left column, simplified schematics of the electromotor network, taken after Fig.14.8A are shown. For each row, the nucleus/nuclei thought to be responsible for generating a given electrical display are highlighted, with an example of the respective display shown as an SPI plot in the right column.

compared to PCN neurons, which leads to different patterns of output when activating each population.

Stimulation of VPd using L-glutamate typically results in an immediate cessation of electromotor output (Carlson and Hopkins 2004b), suggesting that cessations may be driven by increases in the inhibitory feedback provided by VPd (Fig. 14.9). This inhibition appears to be GABAergic, as preliminary evidence indicates that PCN neurons are surrounded by terminals containing GABA (Niso et al. 1989), and iontophoresis of GABA into DP or PCN causes a reduction in EOD rates (Carlson and Hopkins 2004b).

The bursts produced by VPd neurons show little variation from cycle to cycle, and there is no correlation between this variation and the resting sequence of pulse intervals (von der Emde et al. 2000; Carlson 2003). During both scallops and accelerations, however, the activity of these units decreases markedly (Carlson 2003), suggesting that their production results at least in part via disinhibition of DP/PCN. In support of this hypothesis, application of the $GABA_A$ receptor blocker bicuculline methiodide to either DP or PCN results in tonic bursting behavior (Carlson and Hopkins 2004b). Thus, it appears that recurrent inhibition regulates the resting cadence of the sequence of pulse intervals, while disinhibition frees DP and PCN from their normal rate-limiting factor, providing a permissive mechanism for driving bursts. Furthermore, after blocking inhibitory feedback, electromotor output patterns resulting from stimulating DP and PCN with L-glutamate are indistinguishable (Carlson and Hopkins 2004b). This suggests that differences in the strength of recurrent inhibition to DP and PCN are responsible for differences in their baseline activity levels and therefore different effects on electromotor output when stimulated.

6. Common Principles in the Organization of Skeletomotor and Electromotor Systems

The phylogenetic evidence strongly indicates that mormyriforms and gymnotiforms share no common electrogenic or electrosensory ancestors (Bullock et al. 1983). Thus, their electromotor systems are thought to have evolved independently (Darwin 1859, Bullock et al. 1983). It is therefore quite remarkable to find multiple, striking similarities in the electromotor systems of these two groups (Hopkins 1995). Such convergent evolution supports the argument that there are certain basic design features underlying the evolution of vertebrate neural circuits that may be typified by the electromotor systems of these fish.

Electric fish have proven to be an excellent model system for studying the central mechanisms for generating motor output. Many vertebrate behaviors involve several complex interacting elements, making it difficult to form a direct link between the production of a behavior and its neural control. Electric fish are unique in this respect, since their electrical behavior may be characterized by just two parameters: the waveform of each EOD and its frequency or temporal pattern of production. The relative simplicity of timing control in electric fish makes them an excellent model system for understanding how vertebrate neural circuits generate different temporal patterns. Thus, research on the central control of electric signaling behavior has proven successful in addressing general issues in vertebrate motor control, and several basic principles have emerged that likely apply to more complex systems.

6.1 Central Pattern Generators: Neurons and Networks for Rhythmic Behaviors

The electromotor output of both mormyriforms and gymnotiforms is characterized by inherent rhythmicity. Despite the many similarities between the electromotor systems of these two groups of fish, rhythmic motor output is generated in distinctly different ways. In gymnotiforms and *Gymnarchus*, regular EOD rates are maintained by a set of pacemaker neurons (Kawasaki 1994; Metzner 1999). Thus, rhythmic motor output relies on endogenous neuronal characteristics. By contrast, the command neurons of pulse mormyrids do not have any pacemaker properties. Instead, they integrate tonically active descending inputs that drive EOD production (Grant et al. 1999, Carlson 2002a). These descending units receive input from a recurrent inhibitory pathway that silences them after each EOD, and this negative feedback imparts a baseline rhythm to EOD production. Thus, rhythmic motor output in this case appears to rely predominantly on network features. These two different mechanisms for generating rhythmic motor output exemplify common strategies that have been described for a number of invertebrate central pattern generators (Marder and Calabrese 1996; Marder and Bucher 2001).

6.2 Central Pattern Generators: Coordination of Agonist and Antagonist Effectors

The presence of spinally coordinated motor programs is a control strategy present in all vertebrates. The reflex control of posture and movement was known before Descartes (1637) formalized this idea in "Les passions de l'âme." The open loop control of motor patterns was first put in evidence by Lashley (1917), who described a spinal cord injured patient that was able to move his hindlimbs despite complete anesthesia below the injury level of the spinal cord. This observation led to the identification of spinally organized units in motor programs. The coordinated activation of these units by descending commands appears to be a general strategy for skeletomotor control. The control systems of the EOD waveform goes from the extreme simplicity of the effector response to a nerve volley in pulse mormyrids to a complex electromotor act that involves the central coordination of different electromotor pools by a fixed motor program that is genetically determined and stored in the circuitry of the spinal cord in pulse gymnotiforms (Caputi 1999).

The electromotor system illustrates various general principles for the organization of the neural circuits. First, the system is redundant; several mechanisms converge synergistically for the implementation of the delay lines subserving the synchrony of homologous faces of the electrocytes generating each wave. Second, the orderly recruitment of electromotor neurons appears to be dependent on their size and represents another example of Hennemann's size principle (Hennemann 1957) occurring in the skeletomotor system of vertebrates and invertebrates. Smaller neurons having larger input resistance fire earlier than large neurons in the same stimulus conditions. Third, the different electromotor pools along the spinal cord of some pulse gymnotiforms are collectively controlled by a common descending drive also involved in skeletomotor control (De Luca and Erim 1994). The specific architecture of the system and the different excitability of electromotor neuron types allows that a single, relatively synchronous descending volley provokes a timed recruitment of different effector units. This type of organization relieves the nervous system from the necessity to monitor and control every unit in a separated way and therefore enhances efficacy (Lemon 1988). Finally, as observed in skeletomotor systems the electromotor output depends on the transduction properties of the effector cells and the filtering properties of the passive tissues linking the effector units with the exterior. Posteffector filtering of muscle fiber output is an important factor for controlling muscle performance and enhancing mechanical efficacy (Hoffer et al. 1992) These structures are also the target for sexual hormones, as found for muscles, tendons, and bones in the vertebrate skeletomotor system.

7. In Conclusion

Electric fish have evolved a peculiar effector system that serves as a weapon and as a carrier of information for the electrosensory system. These fish create an electric field that allows them to communicate with each other and to detect objects that have a different impedance than the surrounding water. Therefore, understanding electrogenesis is a necessary step for understanding the action-perception electrosensory cycle. The electromotor output of the two main groups of electric fish, mormyriforms and gymnotiforms, is characterized by species-specific EODs of stereotyped waveform, generated according to specific temporal patterns. Organizing the electromotor units and regulating their emission patterns are the two main functions of the electromotor system. The emission pattern results from the effect of a small number of brain stem and thalamic nuclei on a single electrotonically coupled bulbar nucleus where EOD timing is determined. This single command activates a robust, hard wired, spinal and peripheral system responsible for the constancy of EOD waveform. While EOD emission patterns may be highly variable, EOD waveform is mainly modulated by slow changes depending on hormonal or acclimation effects. The comparative study of the neural organization of these two signal features reveals a number of different strategies for motor organization. Most regular EOD emission patterns are maintained by a set of pacemaker neurons while the most flexible and centrally modulated patterns are organized by a command nucleus that integrates tonically active descending inputs that drive EOD production. The coordinated activation of spinally organized units by descending commands appears to be a general strategy for skeletomotor control that illustrates three general principles for the organization of spinal motor units: (1) a common descending drive for the timed recruitment of spinal units, (2) the role of Hennemann's size principle in motorneuron recruitment, and (3) the presence of redundant, synergistic mechanisms for securing the final output.

Abbreviations

AC	Alternating currents
AEN	Anterior electromotor nerve
BCA	Bulbar command-associated nucleus
CN	Command nucleus
CP	Centroposterior nucleus of the thalamus
DC	Constant current
DP	Thalamic dorsal posterior nucleus
EMF	Electromotive force
EMN	Electromotor neurons
EO	Electric organ
EOD	Electric organ discharge
JAR	Jamming avoidance response
MCA	Medullary command associated nucleus
MLF	Medial longitudinal fasciculus
MRN	Medullary relay nucleus
PCN	Mesencephalic precommand nucleus
PEN	Posterior electromotor nerve
PMn	Pacemaker neurons
PPn	Pre-pacemaker nucleus
R_{load}	Load resistance
Ri	Internal resistance
Rn	Relay neurons
sPPn	Sublemniscal prepacemaker nucleus
VP	Ventroposterior nucleus
VPd	Dorsal region of the toral ventroposterior nucleus
VPv	Ventral region of the toral ventroposterior nucleus

References

Albe-Fessard D (1954) Novelle étude des latences spinales dans le dispositif de commande des organs électriques chez *Electrophorus electricus*. Anais Acad Brasil Cienc 26:187–192.

Albe-Fessard D, Buser P (1950) Etude de l'interaction électrique entre deux fragments d'organe de Torpille (*Torpedo marmorata*). J Physiol (Paris) 42:528–529.

Barham E, Huckabay W, Gowdy R, Burns B (1969) Microvolt electric signals from fishes and the environment. Science 164:965–968.

Baron VD, Morshnev KS, Olshansky VM, Orlov AA (1994) Electric organ discharges of two species of African catfish (*Synodontis*) during social behaviour. Anim Behav 48:1472–1475.

Bass AH (1986) Electric organs revisited: Evolution of a vertebrate communication and orientation organ. In: Bullock TH, Heiligenberg W (eds), Electroreception. New York: John Wiley & Sons, pp. 13–70.

Bass AH, Hopkins CD (1983) Hormonal control of sexual differentiation: changes in electric organ discharge waveform. Science 220:971–974.

Bass AH, Volman SF (1987) From behavior to membranes: testosterone-induced changes in action potential duration in electric organs. Proc Nat Acad Sci USA 84:9295–9298.

Bastian J (1986) Electrolocation Behavior, Anatomy and Physiology. In: Bullock TH, Heiligenberg W (eds), Electroreception. New York: John Wiley & Sons, pp. 577–612.

Bauer R (1979) Electric organ discharge (EOD) and prey capture behavior in the electric eel, *Electrophorus electricus*. Behav Ecol Sociobiol 3:311–319.

Belbenoit P (1986) Fine analysis of predatory and defensive motor events in *Torpedo marmorata* (Pisces). J Exp Biol 121:197–226.

Belbenoit P, Moller P, Serrier J, Push S (1979) Ethological observations on the electric organ discharge behavior of the electric catfish, *Malapterurus electricus* (Pisces). Behav Ecol Sociobiol 4:321–330.

Bell CC, Bradbury J, Russell CJ (1976) Electric organ of a mormyrid as a current and voltage source. J Comp Physiol 110:65–88.

Bell CC, Libouban S, Szabo T (1983) Pathways of the electric organ discharge command and its corollary discharges in mormyrid fish. J Comp Neurol 216:327–338.

Bennett MVL (1961) Modes of operation of electric organs. Ann NY Acad Sci 94:458–509.

Bennett MVL (1968) Neural control of electric organs. In: Ingle D (ed), The Central Nervous System and Fish Behavior. Chicago: University of Chicago Press, pp. 147–169.

Bennett MVL (1971) Electric organs. In: Hoar WS, Randall DJ (eds), Fish Physiology. London: Academic Press, pp. 347–491.

Bennett MVL, Pappas G, Gimenez M, Nakajima Y (1967a) Physiology and ultrastructure of electrotonic junctions. IV. Medullary electromotor nuclei in gymnotiform fish. J Neurophysiol 30:236–300.

Bennett MVL, Pappas G, Aljure E, Nakajima Y (1967b) Physiology and ultrastructure of electrotonic junctions. II. Spinal and medullary electromotor nuclei in mormyrid fish. J Neurophysiol 30:180–208.

Bratton BO, Ayers LJ (1987) Observations on the electric organ discharge of two skate species (Chondrichthyes: Rajidae) and its relationship to behaviour. Environ Biol Fish 20:241–254.

Budelli R, Caputi AA, Gomez L, Rother D, Grant K (2002) The electric image in *Gnathonemus petersii*. J Physiol Paris 96:421–429.

Bullock TH (1969) Species differences in effect of electroreceptor input on electric organ pacemakers and other aspects of behavior in electric fish. Brain Behav Evol 2:85–118.

Bullock TH, Heiligenberg W, eds (1986) Electroreception. New York: John Wiley & Sons.

Bullock TH, Bodznick DA, Northcutt RG (1983) The phylogenetic distribution of electroreception: evidence for convergent evolution of a primitive vertebrate sense modality. Brain Res Rev 6:25–46.

Capurro A, Macadar O, Perrone R, Pakdaman K (1998) Computational model of the jamming avoidance response in the electric fish *Gymnotus carapo*. Biosystems 48: 21–27.

Caputi AA (1999) The electric organ discharge of pulse gymnotiforms: the transformation of a simple impulse into a complex spatiotemporal electromotor pattern. J Exp Biol 202:1229–1241.

Caputi AA (2004) Contributions of electric fish to the understanding sensory processing by reafferent systems J Physiol (Paris) 98:81–97.

Caputi A, Aguilera P (1996) A field potential analysis of the electromotor system in *Gymnotus carapo*. J Comp Physiol A 179:827–835.

Caputi A, Budelli R (1995) The electric image in weakly electric fish: I. A data-based model of waveform generation in *Gymnotus carapo*. J Comput Neurosci 2:131–147.

Caputi A, Trujillo-Cenoz O (1994) The spinal cord of *Gymnotus carapo*: the electromotoneurons and their projection patterns. Brain Behav Evol 44:166–174.

Caputi A, Macadar O, Trujillo-Cenoz O (1989) Waveform generation in *Gymnotus carapo*. III. Analysis of the fish body as an electric source. J Comp Physiol A 165:361–370.

Caputi A, Silva A, Macadar O (1998) The effect of environmental variables on waveform generation in *Brachyhypopomus pinnicaudatus*. Brain Behav Evol 52:148–158.

Caputi AA, Castelló ME, Aguilera P, Trujillo-Cenóz O (2002) Electrolocation and electrocommunication in pulse gymnotiforms: signal carriers, pre-receptor mechanisms and the electrosensory mosaic. J Physiol Paris 96:493–505.

Caputi AA, Aguilera P Castelló ME (2003) Probability and amplitude of novelty responses as a function of the change in contrast of the reafferent image in *G. carapo* J Exp Biol 206:999–1010.

Carlson BA (2002a) Electric signaling behavior and the mechanisms of electric organ discharge production in mormyrid fish. J Physiol (Paris) 96:405–419.

Carlson BA (2002b) Neuroanatomy of the mormyrid electromotor control system. J Comp Neurol 454:440–455.

Carlson BA (2003) Single-unit activity patterns in nuclei that control the electromotor command nucleus during spontaneous electric signal production in the mormyrid *Brienomyrus brachyistius*. J Neurosci 23:10128–10136.

Carlson BA, Hopkins CD (2004a) Stereotyped temporal patterns in electrical communication. Anim Behav 68:867–878.

Carlson BA, Hopkins CD (2004b) Central control of electric signaling behavior in the mormyrid *Brienomyrus brachyistius*: segregation of behavior-specific inputs and the role of modifiable recurrent inhibition. J Exp Biol 207:1073–1084.

Carlson BA, Hopkins CD, Thomas P (2000) Androgen correlates of socially induced changes in the electric organ discharge waveform of a mormyrid fish. Horm Behav 38:177–186.

Castello ME, Aguilera PA, Trujillo-Cenoz O, Caputi AA (2000) Electroreception in *Gymnotus carapo*: pre-receptor processing and the distribution of electroreceptor types. J Exp Biol 203:3279–3287.

Coates CW, Altamirano M, Grundfest H (1954) Activity in electrogenic organs of knifefishes. Science 120:845–846.

Cox R, Coates C (1938) Electrical characteristics of the electric tissue of the electric eel *Electrophorus electricus* (Linnaeus). Zoologica 23:203–212.

Curti S, Falconi A, Morales FR, Borde M (1999) Mauthner cell-initiated electromotor behavior is mediated via NMDA and metabotropic glutamatergic receptors on medullary pacemaker neurons in a gymnotiform fish. J Neurosci 19:9133–9140.

Darwin C (1859) On the Origin of Species by Means of Natural Selection, or the Preservation of Favoured Races in the Struggle for Life. London: John Murray.

DeLuca CJ, Erim Z (1994) Common drive of motor units in regulation of muscle force. Trends Neurosci 17:299–305.

Descartes R (1637) Les passions de l'âme.

Dunlap KD, Smith GT, Yekta A (2000) Temperature dependence of electrocommunication signals and their underlying neural rhythms in the weakly electric fish, *Apteronotus leptorhynchus*. Brain Behav Evol 55:152–162.

Dye JC, Meyer JH (1986) Central control of the electric organ discharge in weakly electric fish. In: Bullock TH, Heiligenberg W (eds), Electroreception. New York: John Wiley & Sons, pp. 71–102.

Elekes K, Szabo T (1985) The mormyriforms brainstem. III. Ultrastructure and synaptic organization of the medullary "pacemaker" nucleus. Neuroscience 15:431–443.

Ellis DB, Szabo T (1980) Identification of different cell types in the command (pacemaker) nucleus of several gymnotiform species by retrograde transport of horseradish-peroxidase. Neuroscience 5:1917–1929.

Falconi A, Borde M, Hernandez-Cruz A, Morales FR (1995) Mauthner cell–initiated abrupt increase of the electric organ discharge in the weakly electric fish *Gymnotus carapo*. J Comp Physiol A 176:679–689.

Few WP, Zakon HH (2001) Androgens alter electric organ discharge pulse duration despite stability in electric organ discharge frequency. Horm Behav 40:434–442.

Grant K, Bell CC, Clausse S, Ravaille M (1986) Morphology and physiology of the brainstem nuclei controlling the electric organ discharge in mormyriforms fish. J Comp Neurol 245:514–530.

Grant K, von der Emde G, Gomez-Sena L, Mohr C (1999) Neural command of electromotor output in mormyrids. J Exp Biol 202:1399–1407.

Hagedorn M (1986) The ecology, courtship, and mating of gymnotiform electric fish. In: Bullock TH, Heiligenberg W (eds), Electroreception. New York: John Wiley & Sons, pp. 497–525.

Hagedorn M, Carr C (1985) Single electrocytes produce a sexually dimorphic signal in the South American electric fish *Hypopomus occidentalis* (Gymnotiformes Hypopomidae). J Comp Physiol A 156:511–524.

Hagedorn M, Womble M, Finger TE (1990) Synodontid catfish: a new group of weakly electric fish. Brain Behav Evol 35:268–277.

Heiligenberg WF (1991) Neural Nets in Electric Fish. Cambridge: MIT Press.

Hennemann E (1957) Relation between size of neurons and their susceptibility to discharge. Science 126:1345–1346.

Hille B (2001) Ion Channels of Excitable Membranes, 3rd ed. Sunderland, MA: Sinauer Associates.

Hoffer JA, Caputi AA, Pose IE. (1992) Activity of muscle proprioceptors in cat posture and locomotion:relation to EMG, tendon force and the movement of fibers and aponeurotic segments. In: Jami L, Pierrot-Deseilligny, E, Zytnicki D. (eds), Muscle Afferents and Spinal Control of Movement. Oxford: Pergamon Press.

Hopkins CD (1972) Sex differences in electric signaling in an electric fish. Science 176: 1035–1037.

Hopkins CD (1986) Behavior of Mormyridae. In: Bullock TH, Heiligenberg W (eds), Electroreception. New York: John Wiley & Sons, pp. 527–576.

Hopkins CD (1995) Convergent designs for electrogenesis and electroreception. Curr Opin Neurobiol 5:769–777.

Hopkins CD (1999) Design features for electric communication. J Exp Biol 202:1217–1228.

Hopkins CD, Bass AH (1981) Temporal coding of species recognition signals in an electric fish. Science 212:85–87.

Kalmijn AJ (1974) The detection of electric fields from inanimate and animate sources other than electric organs. In: Fessard A (ed), Handbook of Sensory Physiology. Heidelberg: Springer-Verlag, pp. 147–200.

Kalmijn AJ (1987) Detection of weak electric fields. In: Atema J, Fay R, Popper A, Tavolga W (eds), Sensory Biology of Aquatic Animals. New York: Springer-Verlag, pp. 151–186.

Kawasaki M (1994) The African wave-type electric fish, *Gymnarchus niloticus*, lacks corollary discharge mechanisms for electrosensory gating. J Comp Physiol A 174: 133–144.

Kawasaki M, Heiligenberg W (1989) Distinct mechanisms of modulation in a neuronal oscillator generate different social signals in the electric fish *Hypopomus*. J Comp Physiol A 165:731–741.

Kawasaki M, Maler L, Rose GJ, Heiligenberg W (1988) Anatomical and functional organization of the prepacemaker nucleus in gymnotiform electric fish: the accommodation of two behaviors in one nucleus. J Comp Neurol 276:113–131.

Kellaway P (1946) The part played by electric fish in the early history of bioelectricity and electrotherapy. Bull Hist Med 20:112–137.

Keynes RD, Bennett MVL, Grundfest H (1961) Studies on the morphology and electrophysiology of electric organs. II. Electrophysiology of the electric organ of *Melapterurus electricus*. In: Chagas C, Paes de Carvalho A (eds), Bioelectrogenesis. Amsterdam: Elsevier, pp. 102–112.

Kleerokoper H, Sibakin K (1956) An investigation of the electrical 'spike' potentials produced by the sea lamprey (*Petromyzon marinus*) in the water surrounding the head region. J Fish Res Bd Can 13:175–383.

Kramer B (1990) Electrocommunication in Teleost Fishes: Behavior and Experiments. New York: Springer-Verlag.

Lashley KS (1917) The accuracy of movement in abscence of excitation from the moving organ. Am J Physiol 43:169–194.

Lemon R (1988) The output map of the motor cortex. Trends Neurosci 11:501–506.

Lissman HW (1951) Continuous electrical signals from the tail of a fish, *Gymnarchus niloticus*. Nature 167:201–202.

Lissman HW (1958) On the function and evolution of electric organs in fish. J Exp Biol 35:156–191.

Lissman HW, Machin KE (1958) The mechanism of object location in *Gymnarchus niloticus* and similar fish. J Exp Biol 35:451–486.

Lorenzo D, Sierra F, Silva A, Macadar O (1990) Spinal mechanisms of electric organ discharge synchronization in *Gymnotus carapo*. J Comp Physiol A 167:447–452.

Lowe CG, Bray RN, Nelson DR (1994) Feeding and associated electrical behavior of the Pacific electric ray *Torpedo californica* in the field. Mar Biol 120:161–169.

Macadar O (1993) Motor control of waveform generation in *Gymnotus carapo*. J Comp Physiol A 173:728–729.

Marder E, Bucher D (2001) Central pattern generators and the control of rhythmic movements. Curr Biol 11:R986–R996.

Marder E, Calabrese R (1996) Principles of rhythmic motor pattern generation. Physiol Rev 76:687–717.
McGregor PK, Westby GWM (1992) Discrimination of individually characteristic electric organ discharges by a weakly electric fish. Anim Behav 43:977–986.
Metzner W (1999) Neural circuitry for communication and jamming avoidance in gymnotiform electric fish. J Exp Biol 202:1365–1375.
Moller P (1995) Electric Fishes: History and Behavior. New York: Chapman and Hall.
Moortgat KT, Bullock TH, Sejnowski TJ (2000) Precision of the pacemaker nucleus in a weakly electric fish: network versus cellular influences. J Neurophysiol 83:971–983.
Niso R, Serrier J, Grant K. (1989) Mesencephalic control of the bulbar electromotor network in the mormyriforms *Gnathonemus petersii*. Eur J Neurosci (Suppl) 2:176.
Otte D (1974) Effects and functions in the evolution of signaling systems. Annu Rev Ecol Syst 5:385–417.
Pickens PE, McFarland WN (1964) Electric discharge and associated behaviour in the stargazer. Anim Behav 12:363–367.
Post N, von der Emde G (1999) The "novelty response" in an electric fish: response properties and habituation. Physiol Behav 68:115–128.
Rankin C, Moller P (1986) Social behavior of the African electric catfish, *Melapterurus electricus*, during intra- and interspecific encounters. Ethology 73:177–190.
Russell CJ, Myers JP, Bell CC (1974) The echo response in *Gnathonemus petersii* (Mormyridae). J Comp Physiol 92:181–200.
Schuster S (2000) Changes in electric organ discharge after pausing the electromotor system of *Gymnotus carapo*. J Exp Biol 203 (Pt 9):1433–1446.
Silva A, Quintana L, Ardanaz JL, Macadar O (2002) Environmental and hormonal influences upon EOD waveform in gymnotiform pulse fish. J Physiol Paris 96:473–484.
Smith GT (1999) Ionic currents that contribute to a sexually dimorphic communication signal in weakly electric fish. J Comp Physiol A 185:379–387.
Smith GT, Lu Y, Zakon HH (2000) Parvocells: a novel interneuron type in the pacemaker nucleus of a weakly electric fish. J Comp Neurol 423:427–439.
Stoddard PK, Markham MR, Salazar VL (2003) Serotonin modulates the electric waveform of the gymnotiform electric fish *Brachyhypopomus pinnicaudatus*. J Exp Biol 206:1353–1362.
Szabo T, Enger P (1964) Pacemaker activity of the medullary nucleus controlling electric organs in high frequency gymnotiform fish. Z Vgl Physiol 49:285–300.
Teyssedre C, Boudinot M (1987) Rhythmicity as an intrinsic property of the mormyrids electromotor command system. Physiol Behav 41:201–207.
Tricas TC, Michael SW, Sisneros JA (1995) Electrosensory optimization to conspecific phasic signals for mating. Neurosci Lett 202:129–132.
Trujillo-Cenoz O, Echague JA, Macadar O (1984) Innervation pattern and electric organ discharge waveform in *Gymnotus carapo* (Teleostei Gymnotiformes). J. Neurobiol 15: 273–282.
Trujillo-Cenoz O, Echague JA, Bertolotto C, Lorenzo D (1986) Some aspects of the structural organization of the spinal cord of *Gymnotus carapo* (Teleostei Gymnotiformes) I. The electromotor neurons. J Ultrastruct Mol Struct Res 97:130–143.
von der Emde G (1992) Electrolocation of capacitive objects in four species of pulse-type weakly electric fish. 2. Electric signaling behavior. Ethology 92:177–192.
von der Emde G, Sena LG, Niso R, Grant K (2000) The midbrain precommand nucleus of the mormyriforms electromotor network. J Neurosci 20:5483–5495.

Zakon HH (1993) Weakly electric fish as model systems for studying long-term steroid action on neural circuits. Brain Behav Evol 42:242–251.
Zupanc GK (1999) Neurogenesis, cell death and regeneration in the adult gymnotiform brain. J Exp Biol 202 (Pt 10):1435–1446.
Zupanc GKH, Maler L (1997) Neuronal control of behavioral plasticity: the prepacemaker nucleus of weakly electric gymnotiform fish. J Comp Physiol A 180:99–111.

Index

SPRINGER HANDBOOK OF AUDITORY RESEARCH *(continued from page ii)*

Volume 22: Evolution of the Vertebrate Auditory System
Edited by Geoffrey A. Manley, Arthur N. Popper and Richard R. Fay

Volume 23: Plasticity of the Auditory System
Edited by Thomas N. Parks, Edwin W Rubel, Richard N. Popper, and Richard R. Fay

For more information about the series, please visit www.springer-ny.com/shar.